D0212246

Biomes and Ecosystems

Biomes and Ecosystems

EDITOR
Robert Warren Howarth
Cornell University

ASSOCIATE EDITOR
Jacqueline E. Mohan
University of Georgia, Athens

Volume 2
Articles

Acacia-Commiphora Bushlands and Thickets – Guyanan Savanna

SALEM PRESS
A Division of EBSCO Publishing
Ipswich, Massachusetts

GREY HOUSE PUBLISHING

Cover photo: Rock Islands, Koror, Palau, Pacific. © Casey Mahaney

Biomes and Ecosystems, 2013, published by Grey House Publishing, Inc., Amenia, NY, under exclusive license from EBSCO Publishing, Inc.

The paper used in these volumes conforms to the American National Standard for Permanence of Paper for Printed Library Materials, X39.48-1992 (R1997).

LIBRARY OF CONGRESS CATALOGING-IN-PUBLICATION DATA

Biomes and ecosystems / Robert Warren Howarth, general editor ; Jacqueline E. Mohan, associate editor.
volumes cm.
Includes bibliographical references and index.
ISBN 978-1-4298-3813-9 (set) -- ISBN 978-1-4298-3814-6 (volume 1) -- ISBN 978-1-4298-3815-3 (volume 2) -- ISBN 978-1-4298-3816-0 (volume 3) -- ISBN 978-1-4298-3817-7 (volume 4) 1. Biotic communities. 2. Ecology. 3. Ecosystem health. I. Howarth, Robert Warren. II. Mohan, Jacqueline Eugenia
QH541.15.B56B64 2013
577.8'2--dc23

2013002800

ebook ISBN: 978-1-4298-3818-4

First Printing

PRINTED IN THE UNITED STATES OF AMERICA

Produced by Golson Media

Contents

T

U

V

Part 2: Articles

Acacia-Commiphora Bushlands and Thickets

Category: Grassland, Tundra, and Human Biomes.
Geographic Location: Africa.
Summary: The acacia-commiphora landscape is dominated by thorny vegetation that transforms after rains; it is increasingly affected by drought and human changes.

The tangled, spiny, often leafless vegetation found in the dry, low-lying areas of east Africa is typical of the warm, semi-arid habitat known as the acacia-commiphora woodland and bush, which extends over much of northern and northeastern Kenya, parts of eastern Ethiopia, and through Somalia. This habitat is a transitional zone between two of east Africa's major ecological communities: the east African savannas and grasslands, and the dry lands and semi-deserts of the Horn of Africa.

A scarcity of water underpins the entire ecology of the acacia-commiphora woodland and bushland habitats. Therefore, extreme measures of survival are a feature of plant and animal communities in this area. The dominant acacia and commiphora trees, which give this woodland its name, form thickets with many other deciduous shrubs and trees that remain leafless for as much as nine months of the year, alongside succulents like *Euphorbia* spp. Only by shutting down in this way can the trees conserve enough energy to withstand the area's prolonged dry spells. As a result, these woodlands seem to be in a suspended state for long periods.

Transformation After Rainfall

A transformation occurs in the acacia-commiphora bushland after rains. The gray and seemingly lifeless trees suddenly burst into leaf, putting out buds that develop quickly into flowers so that pollination, seed production, and fruiting can all be accomplished rapidly.

Following rains, the acacia-commiphora bushland soon fills with flowers, creepers, and insect life. New grasses spring up from the red dust and from sandy patches in stonier, more open ground. From rhizomes and tubers, or from seeds that lay buried in the soil, herbs and other plants emerge. Wildflowers abound in this zone, quickly carpeting the ground and bush. The common species include *Thunbergia*, *Ipomoea* spp., *Aloe* spp., *Boophone disticha*, *Craterostigma* spp., and the labiates *Ocimum* and *Orthosiphon.*

This time is crucial for wildlife and livestock in this region, as food and water are briefly abundant. The typically dry streams and pans fill rapidly and turn into frantic sites of mating and breeding for insects including damselflies and mosquitoes, as well as large numbers of vocal frogs that are not typically associated with these dry zones. Some of these pans hold water for several months and become important reservoirs that keep thirsty animals watered well into the dry months.

At times of rain, leopard tortoises move about, making the most of the opportunity to lay their eggs in the rain-softened soil. Many other reptiles and invertebrates are also active and breeding at this time in the acacia-commiphora bushland.

Soon after the brief rains, the temporary pools, streams, and other water sources dry up. As another dry, dormant period begins, the elephants—a keystone species in the acacia-commiphora bushland—move to other areas where they know that they can find water. The commiphora trees lose their leaves and conduct photosynthesis through their peeling, papery bark, which has a layer of chlorophyll-bearing cells beneath it.

These bare commiphora myrrh trees, like other vegetation in the acacia-commiphora bushland, may remain leafless for months at a time in dry seasons. (Thinkstock)

Specialist tree-browsers, such as the reticulated giraffe and gerenuk, use their adaptive advantage of height to feed on vegetation.

Euphorbias and other succulent plants such as *Sansevieria,* which appear in low, spiky clumps, are among the few evergreen plants on view after the shutdown of acacia-commiphora leafy vegetation is complete. Dryland-adapted species of antelope like the lesser kudu and the tiny dik-dik are able to survive on these fleshy plants and on the occasional Boscia tree. Otherwise, the woodlands revert to a monotonous gray tangle of bare limbs for the rigors of another long dry season.

Effects of Livestock and Wildlife

In this land of extremes, overgrazing by livestock animals is a constant threat to the equilibrium of the acacia-commiphora bushland ecology. Overstocking and mismanagement can easily tilt the balance from a productive relationship to one that is damaging to both the environment and to the pastoralists, who, with their livestock herds, depend on the fragile resources the habitat provides. Overgrazing quickly results in capped ground—hard, impermeable soil that has little chance of reestablishing cover. Imbalances can become most obvious during long droughts, when livestock (mainly cattle) perish in large numbers.

Elephants also have an increasing effect on acacia-commiphora bushland vegetation. They leave many acacia trees stripped of their bark and push over commiphora trees. Because many of the areas previously open to elephants are unavailable today, their impact on the acacia-commiphora bushland is significant, creating large zones with dead or dying and damaged trees.

The acacia-commiphora woodlands depend for their stability on the extreme conditions that they embody at different times of the year. Abundant browsing and grazing after rain mitigates the effect of large herbivores by encouraging them to disperse widely through the landscape, allowing the acacia-commiphora bushland plenty of scope to propagate itself. When the vegetation shuts down in the absence of water, elephants and other large browsing herbivores leave the woodlands, sparing the trees and shrubs.

Bee Habitats

For one insect group, the acacia-commiphora bushland is one of the richest habitats in Africa. The diversity of native bees is very high, with several hundred species present. More than 50 species of bees can be found foraging simultaneously on a single flowering acacia. The acacia-commiphora woodlands are renowned the world over for yielding exceptionally high volumes of the finest-quality honey from managed honeybees, especially in years when *Acacia mellifera* is flowering profusely.

Acacia-commiphora bushland provides habitat for hundreds of species of bees. (Thinkstock)

The acacia-commiphora bushland is an interesting habitat that is undergoing rapid changes due to the impact of climate change (extensive droughts), population growth, and increasing settlement, as well as large-scale conversion of areas for charcoal production.

DINO J. MARTINS

Further Reading

Beentje, Henk Jaap. *Kenya Trees, Shrubs, and Lianas.* Nairobi: National Museums of Kenya, 1994.

Bennun, L. and P. Njoroge. *Important Bird Areas in Kenya.* Nairobi, Kenya: East Africa Natural History Society, 1999.

Martins, Dino J. "Foraging Patterns of Managed Honeybees and Wild Bee Species in an Arid African Environment: Ecology, Biodiversity, and Competition." *International Journal of Tropical Insect Science* 24, no. 1 (2004).

Shorrocks, Bryan. *The Biology of African Savannahs.* Oxford, UK: Oxford University Press, 2007.

Acadia Intertidal Zones

Category: Marine and Oceanic Biomes.
Geographic Location: North America.
Summary: This classic rocky intertidal ecosystem on the northeastern coast of the United States, though well protected by the National Park Service, still faces threats from climate change and invasive species.

Acadia National Park, founded in 1916, is a predominantly coastal reserve protected by the U.S. National Park Service, located in central coastal Maine. The park protects nearly 40 miles (65 kilometers) of rocky coastline, and its relatively high tidal range of greater than 10 feet (3 meters), combined with the geology of the bedrock, helps create an extensive intertidal ecosystem along the shores of the park.

This system is home to a wide range of marine plants, including rockweed, kelp, and algae (green, brown, and red); as well as a high diversity of marine invertebrates including snails, mussels, crabs, and urchins; and seabird species that typically function as top predators in the system. Acadia National Park is generally very well protected from human exploitation and isolated from large cities, so the health of this biome is generally considered to be fairly high, although it does face increasing threats from invasive species, increased tourist traffic, and climate change.

Acadia National Park constitutes roughly 73 square miles (189 square kilometers) of mostly rocky coastline. The majority of the park—about 47 square miles (122 square kilometers)—is located on Mount Desert Island. The island is also home to the largest nearby center of population: the town of Bar Harbor, with a permanent population of only about 5,500—although the national park plays host to more than 2 million visitors per year.

The remainder of the park consists of numerous smaller islands including Isle au Haut, Baker Island, and Bar Island, as well as approximately 3.5 square miles (9 square kilometers) on the nearby Schoodic Peninsula. Approximately three-quarters of the land managed by the park

A National Park Service ranger shows a group of tourists a rocky shoreline at Monument Cove in Acadia National Park. The park and its almost 40 miles of coastline provide a habitat for as many as 338 species of birds, many of which feed off intertidal life. (National Park Service/Marc Neidig)

is owned by the National Park Service, including numerous private land donations, most notably about 15 square miles (39 square kilometers) donated by John D. Rockefeller. The remaining one-quarter is private land managed under conservation easements.

The rocky coastline can be attributed to several factors. Mount Desert Island is predominantly comprised of geologically young, hard, and chemically resistant granite that has been heavily scoured by recent glacial activity. Ice more than a half-mile (0.8 kilometer) thick covered the island during the last glaciation, roughly 20,000 years ago, creating a system of ridges and U-shaped valleys. This young hard rock also collides with the older stratified rocks of the Bar Harbor formation, creating shatter zones. The high rugosity of this coastline and a large tidal range of greater than 10 feet (3 meters) make for an extensive system of rocky habitats throughout the supralittoral fringe, the intertidal zone, and the shallow subtidal, all of which play host to a diverse assemblage of animal and plant life.

The rocky intertidal zone is an important natural resource and is somewhat biologically similar in temperate regions around the world, with congener species often filling the same ecological niches in different regions. Acadia is perhaps one of the best-known examples of this ecosystem.

The intertidal zone is almost universally defined as the zone between the highest high tide (Mean high high water or MHHW) and the lowest low tide (mean low low water or MLLW) Although the majority of this biome is regularly covered and exposed by the advance and retreat of the tides, it encompasses a broad range of extremes: very hot and very cold temperatures; broad ranges of salinity; and varying exposure to air, water, wind, waves, and predation.

Intertidal Plant and Animal Life

Rocky intertidal zones provide critical habitat and forage for many terrestrial and marine plant and animal species. Rocky crevasses, undersides of rocks, and tide pools in the intertidal zone support an array of hardy species that have adapted to withstand periodic exposure to air, subzero winter temperatures, and the force of pounding waves.

A broad range of organisms thrives in the dynamic area between the high and low tide marks, where the land meets the sea. Plants and animals growing in the rocky intertidal zone must cope with both aquatic and terrestrial environments to survive, and most species are adapted to a specific niche environment within the intertidal.

A combination of physical and biological factors determines where each species is found. Important physical factors that control distribution include light level, temperature, salinity, desiccation, and wave action. Biological factors that control distribution include competition for space, susceptibility to predation, and availability of prey. Often, the upper or lower range of a plant or animal species is determined by the temperature, salinity, or exposure tolerance of its principal predator or prey species.

Birds generally are the top predators in the intertidal ecosystem, preying on a wide variety of creatures that are exposed during low tide. In Acadia, a wide range of bird species are present, including gulls, terns, common eiders, cormorants, oystercatchers, and a variety of small shorebirds. Inside the park, 338 bird species have been recorded.

These predators rely on food sources such as small fish trapped in tide pools; grazers like sea urchins, snails, and limpets that feed on algae; filter feeders like barnacles and mussels, which typically stick to the rocks and feed on plankton washed in with the tide; and predators like crabs and starfish, which also prey on filter feeders and grazers.

The rocky coast of Acadia National Park also contains many kinds of marine algae and plants. In the splash zone and upper intertidal zone, spray from high tide moistens the algal mats of green and blue-green algae. The most common group throughout the intertidal is often rockweed, though more exposed areas or areas where rockweed cannot establish a foothold have a wide range of green, brown, and red algae. The lower intertidal zone and lower intertidal pools, which are rarely completely desiccated, contain many kinds of seaweed, including sea lettuce, kelp, and Irish moss.

Ecosystem Threats

Though human effects on this ecosystem are relatively low, due in part to strong protection and in part to relative isolation from large centers of human population, the Acadia intertidal system is not without threats from human influence. One major threat to this system comes from the impact of the roughly 2 million visitors to the national park each year. Visitors may affect bird behavior by disrupting nesting sites or feeding activity, and persistent foot traffic on rocky shorelines can affect barnacle and mussel colonization rates.

Another major threat comes from invasive species. In the past 30 years alone, two invasive crab species (green and Asian shore crabs), three species of tunicate (a type of sea squirt), and several plant and algae species have colonized the shoreline, competing with native species for space and food. Often, these invaders have few or no natural predators, and they may disturb the delicately balanced interspecies interactions that determine each organism's niche.

Temperate intertidal zones also face pressure from climate change. Although sea-level rise is generally considered to be slow enough that intertidal organisms can adapt over time, warming water and changing weather patterns expand the range of more temperate and subtropical species (such as the blue crab) and contract the range of boreal species (like the American lobster). Acadia National Park is on the border between a temperate and a boreal climate, and as such, it may be more susceptible to this type of shift.

With continued vigilance from federal, state, and local governments, research agencies, various nongovernmental organizations, and park visitors, researchers can continue to study and protect this beautiful and critically important transitional habitat and its inhabitants for generations to come.

Jason Krumholz

Further Reading

Acadia National Park. http://www.nps.gov/acad/index.htm.

Cammen, Leon M. and Peter F. Larsen. *An Ecological Characterization of Intertidal Resources of Acadia National Park: Macrofauna.* Boston: National Park Service, North Atlantic Region, Office of Scientific Studies, 1992.

Kaiser, James. *Acadia: The Complete Guide: Mount Desert Island & Acadia National Park.* Ringgold, GA: Destination Press, 2010.

U.S. Geological Survey (USGS). "Nonindigenous Aquatic Species." 2009. http://nas.er.usgs.gov/queries/SpeciesList.aspx?Group=&Sortby=1&state=ME.

Ad Dahna Desert

Category: Desert Biomes.
Geographic Location: Middle East.
Summary: The Ad Dahna Desert comprises the central corridor of the greater Arabian Desert; this hot desert ecosystem has vast expanses of sand and sand dunes, oases ringed with date palms, and dromedary camels.

The Ad Dahna Desert is a bow-shaped corridor of land that comprises the central component of the larger Arabian Desert located within the Arabian Peninsula in southwestern Asia. It is the dominant feature of the central portion of Saudi Arabia. The Ad Dahna Desert is over 620 miles (1,000 kilometers) long and less than 50 miles (80 kilometers) wide. The desert falls under the biome of deserts and xeric shrublands. The Arabian Desert has been prized for centuries due to the presence of valuable mineral resources including petroleum, natural gas, phosphates, and sulfur. Herding, hunting, and other human encroachments have resulted in species extinction and threaten the desert's future environmental health.

Saudi Arabia is the largest country within the Arabian Peninsula, with more than half of its land comprised of desert. The approximately 900,000-square-mile (2.3-million-square-kilometer) Arabian Desert covers the majority of the Arabian Peninsula, mainly within Saudi Arabia. The Ad Dahna Desert serves as the central connection between two other larger components of the Arabian Desert, the northern portion known as the An Nafud (Great Nafud) Desert, and the southeastern portion known as the Rub' al-Khali (Empty Quarter) Desert.

The Ad Dahna Desert has a hot, dry climate and is scientifically characterized as a hot desert. Like all deserts, it is ecologically marked by its aridity, with humidity rates generally averaging between 10 percent to 30 percent. During the day, the land absorbs the heat of the relentless sun, but the desert experiences significant temperature drops in the night as heat loss rapidly accelerates after sunset due to the general lack of cloud cover. Droughts can be years long in duration. Brief periods of rain allow desert plants and animals to survive, but may also create flash floods.

The Ad Dahna Desert contains plateaus of rock, boulders, gravel, and vast stretches of golden yellow sands as well as towering sand dunes. Desert soil is generally coarse and rich in minerals. The sands consist primarily of silicates such as quartz and feldspar. One of the desert's most distinctive features are the horizontal bands of tall sand dunes, also known as veins, which constitute approximately 20 percent of the entire Arabian Desert. The sand dunes are formed and shifted into a variety of changing shapes as the result of the work of water and, most notably, of winds. They are constantly migrating. Iron oxides coating the sand grains sometimes give the dunes red, orange, and purple hues.

Underneath the desert sands lie plains of either gravel or gypsum. Beneath these plains stretches a vast network of underground chambers, passageways, and caves, slowly formed over the centuries by the work of rainwater permeating the rocks. The water carved the rocks and pooled in underground aquifers. Water that remains close to the surface support oases, islands of life in the arid region. Local residents refer to these underground caves as *dahls.* The process also forms towering stalactites and stalagmites. Although Bedouin nomads have utilized the underground caves for their water resources, a prized commodity for its scarce nature, they were not fully explored or geologically surveyed until the late 20th century.

Flora and Fauna

Although the desert's plant and animal species must adapt to its harsh environment, a surprising diversity of life flourishes in limited numbers. Fossils lend support to scientific theories that the Arabian Peninsula once contained wetlands where desert now dominates. Brief periods of rainfall provide for short growing seasons. Various types of algae, fungi, and lichens grow in the region while *Calligonum crinitum, Cyperus conglomeratus, Dipterygium glaucum, Limeum,* and *Arabicum* are common plant species. Among the most common perennials are mimosas, acacias, and aloe. There are few trees; those that are present, such as tamarisks and date palms, grow on the desert's outskirts or oases.

The most symbolic of the Ad Dahna Desert's animals is the dromedary, a one-humped Arabian camel, known for its ability to survive for long periods without water. Other animal species include sand gazelles, white or Arabian oryx, and sand cats. Hyenas, jackals, foxes, civets, hares, golden sand rabbits, ratels, jerboas, mice, rats, porcupines, and hedgehogs can also be present. Among the reptiles, monitors, spiny-tailed lizards, dabbs, geckos, collared lizards, agamids, dammsa, and skinks are common lizards, while horned vipers and cobras are relatively common snakes. Desert oases can support small fish, newts, salamanders, toads, frogs, and turtles.

Both local and migratory birds are present. Native bird species include larks, sand grouse, Arabian coursers, chats, wrens, bustards, owls, falcons, ravens, kestrels, eagles, and vultures. Migratory birds from Europe, Africa, and India pass through on their seasonal journeys between warmer and cooler climates. Eagles can also occasionally be found within the Ad Dahna. Desert insects include flies, mosquitoes, fleas, lice, roaches, ants, termites, beetles, mantids, butterflies, moths, caterpillars, locusts, scorpions, ticks, and spiders.

Environmental Stresses

The Ad Dahna Desert faces numerous environmental threats from human habitation. Historically, Bedouin nomads have been the most well known of the desert's human occupants, successfully adapting to its harsh environment. The Bedouins are primarily herders, raising sheep, goats, donkeys, and horses. They rely heavily on the largely domesticated camel for travel and survival. Overgrazing by livestock herds, primarily camel and goats, has led to significant habitat destruction.

Residents have undertaken efforts to control the region's insects, including the anopheles mosquito, which carries malaria, and the locust, which has caused devastation to crops and vegetation. Hunters' use of motor vehicles in the second half of the 20th century has greatly reduced the presence of targeted mammals such as the gazelle in the region. The striped hyena, jackal, honey badger, and ostrich have become extinct within the desert. Other animals with greatly reduced populations, such as the white oryx and ibex, have slowly begun to recover their numbers in the 21st century.

Centuries of mining and drilling to extract the Arabian Desert's valuable natural resources of oil, natural gas, phosphates, and sulfur have led to pollution and habitat destruction. Instability and conflict in the region encompassing the Ad Dahna Desert have also threatened its ecological health. Most notable were the deliberate and accidental oil spills in nearby Iraq and Kuwait that resulted from the 1991 Gulf War. Desert air, soil, and groundwater all suffered contamination. Ongoing climate change, desertification, and population pressures on the region's scarce water resources endanger both the environment and its human occupants.

There are no large areas of the Ad Dahna or larger Arabian Deserts under legal environmental protection, although that is likely to change in the future. Many environmentalists and scientists feel that the desert ecosystem is at a critical or endangered state due to habitat destruction and species extinction. National governmental hunting regulations, the establishment of wildlife preserves, and the breeding in captivity and reintroduction of endangered animals such as the white oryx and sand gazelle have helped the desert's animal populations to recover. The oryx has been protected since 1961.

Marcella Bush Trevino

Further Reading

Allan, J. A. and Andrew Warren. *Deserts: The Encroaching Wilderness: A World Conservation Atlas*. New York: Oxford University Press, 1993.

Edgell, H. Stewart. *Arabian Deserts: Nature, Origin, and Evolution*. New York: Springer, 2006

Jones, Toby Craig. *Desert Kingdom: How Oil and Water Forged Modern Saudi Arabia*. Cambridge, MA: Harvard University Press, 2010

Oldfield, Sara. *Deserts: The Living Drylands*. Cambridge, MA: MIT Press, 2004.

Aden, Gulf of

Category: Marine and Oceanic Biomes.
Geographic Location: Middle East.
Summary: The Gulf of Aden is part of a major shipping route for international trade, including oil transportation. Consequently, it faces both rapid economic development and increasing environmental threats.

Tectonically, the funnel-shaped Gulf of Aden is located between the African and Arabian plates, extending the Great Rift Valley. This tropical zone between the southern coast of the Arabian Peninsula and the Horn of Africa, surrounded by the arid coastlines of Yemen, Somalia, and Djibouti, possesses extraordinary biotic richness. In the northwest, it connects with the Red Sea through the Bab-el-Mandeb Strait. In the east, it opens into the Arabian Sea, a northern part of the Indian Ocean. At its widest point, the Gulf of Aden is 217 miles (349 kilometers) from north to south and stretches about 783 miles (1,260 kilometers) from east to west, covering a total surface area of 158,300 square miles (409,995 square kilometers). In the northwest, it is limited by the Red Sea, more precisely by a line joining Husn Murad and Ras Siyan.

The depth of the Gulf of Aden increases from 2,867 feet (874 meters) in the west to 16,500 feet (5,029 meters) in the east, where the gulf opens into the Indian Ocean. This markedly deep system so close to land is due to the area's conjunction with the regional geographically dominant feature of the Great Rift Valley, of which the Gulf of Aden can be considered an extension. A small part of its water flows underground, moving westward into the saline Lake Assal in Djibouti. A predominantly diurnal tide occurs, with an extreme range of about 10 feet (3 meters) at Aden and Djibouti.

The mountains surrounding the Gulf of Aden reach elevations of more than 8,000 feet (2,438 meters): Shimbiris, in Somalia, is 7,927 feet (2,416 meters) above sea level, and Jabal Thamar, in Yemen, is 8,248 feet (2,514 meters) above sea level. Coastal environments here range from rocky outcrops and sand beaches to muddy inlets and drowned river mouths. Intertidal features include seagrass beds, salt marshes, and mangroves. Hard shell and leatherback turtles are present in some abundance, with nesting grounds ranging from the Bab-el-Mandeb Strait in the west to the island chain of Socotra at the eastern end of the gulf, where it joins with the Arabian Sea. The beaches of Socotra are of particular importance as an egg-laying area for the loggerhead sea turtle (*Caretta caretta*) population in the northwestern Indian Ocean and Arabian Sea.

The photo shows the steep elevations found along the arid coastline leading from Yemen to Oman on the Gulf of Aden. In Yemen, Jabal Thamar Mountain reaches 8,248 feet (2,514 meters) above sea level. (Thinkstock)

Fisheries and Coral

Temperatures vary seasonally between 59 degrees F (15 degrees C) and 82 degrees F (28 degrees C). Additionally, monsoons influence wind patterns, which affect the water exchange between the Red Sea and the Gulf of Aden. During the winter (October to May), the northeast monsoon winds blow into the Gulf of Aden and lead surface water into the Red Sea from the south-southeast. The water inflow into the Red Sea is triggered by high evaporation rates and low precipitation, causing high salinity. During the summer (June to September), the southwest monsoon winds cause intense upwelling of deep, cold, nutrient-rich ocean water along the southern Arabian coastline. Abundant available nutrients inhibit coral growth but allow kelp beds to thrive; hence, coastal fisheries are rich.

Despite upwelling water and sandy shorelines, the gulf's coral reefs are surprisingly diverse, especially around the Socotra archipelago, albeit not as splendid as the reef systems in the Red Sea. Altogether, 194 species of corals have been recorded along the Saudi Arabian coast. Other habitat types include seagrass beds, limited mangrove communities, salt marshes, and salt flats. Seagrass diversity is relatively high compared with that of surrounding areas, constituting productive ecosystems that provide feeding grounds for green turtles and many species of sea cucumbers. Turtles are caught and eaten opportunistically throughout the region, so all the marine turtles of the region are considered to be threatened, and both the green and hawksbill turtles are declared endangered species. Among the mollusks, cuttlefish and dried sea cucumbers are the main exports. Commercially important invertebrates include rock and deep-sea lobster species. Due to temperature conditions and high salinity levels, the diversity of mangroves is low.

Environmental Stress and Response

All shipping routes from the Mediterranean Sea to the Indian Ocean use the Suez Canal and, thus, the Gulf of Aden. Currently, about 21,000 vessels pass through the gulf each year. About 11 percent of the world's petroleum—3 million barrels of oil—passes through the Gulf of Aden each day. Main port cities are Aden, Yemen; Djibouti City, Djibouti; and Zeila, Berbera, and Boosaaso, Somalia.

Most of the coastal areas and the waters of the region are considered to be biologically intact, but increasing anthropogenic activities introduce new threats. Power and desalination plants, refineries, and chemical plants along the coastline leak their effluents—including oil, organic pollutants, heavy metals, heated brine, and cooling water—into the gulf, often in uncontrolled fashion. Pressures on the marine ecosystems include increased salinity and particularly chlorine and thermal pollution.

Spilled hydrocarbons smother corals. Drilling mud from oil exploitation can suffocate reefs and disrupt their growth. In 1997–98, reefs were affected by the worldwide coral bleaching event. Furthermore, corals have been mined for the manufacture of cement and lime, as well as for use as jewelry. Chlorinated and organophosphorus pesticides, insecticides, and herbicides, introduced into the marine environment by agricultural runoff, cause eutrophication. Further pressure on corals comes from a natural vector: the population of crown of thorns starfish (*Acanthaster planci*), for which hard corals are the main sustenance. The crown of thorns is known to thrive when its predators are overfished or displaced by habitat disruption. This has occurred periodically in the Gulf of Aden region.

The problems of pollution and sewage call for sustainable management measures in the industrial and agricultural sectors, but no regulations exist in the fishery sector to ensure sustainable reproduction. Currently, blast fishing with homemade explosives composed of fertilizer, fuel, and fuse caps inserted into empty beer bottles causes massive damage to corals. When the reef structure has been weakened, it is much more vulnerable to wave action, and the reef becomes unable to protect the coastline. Another destructive fishing method involves the use of cyanide to stun fish, allowing fishers to collect them and sell them to the live-fish trade. Monitoring could prevent overfishing of shark and ray, as well as numerous by-catch varieties. Declines in catches have been reported for diverse species, including Indian mackerel, shark, cuttlefish, shrimp, and Trochus.

To organize effective regional cooperation, the Programme for the Environment of the Red Sea and Gulf of Aden (PERSGA) was initiated in 1974 by the Arab League Educational, Cultural and Scientific Organization (ALECSO) and the United Nations Environment Programme (UNEP). In 1982, the Regional Convention for the Conservation of the Red Sea and Gulf of Aden Environment—the Jeddah Convention—was adopted by the neighboring nations of Egypt, Djibouti, Jordan, Eritrea, Palestine, Saudi Arabia, Somalia, Sudan, and Yemen. It was complemented by relevant provisions of various internationally recognized Law of the Seas bodies and covenants.

The Jeddah Convention and its associated protocols through 2006 include a range of features aimed at preventing and mitigating such threats as oil spill pollution and other features that promote conservation of biological diversity. This latter goal is supported by the ongoing development of protected areas in the Red Sea and Gulf of Aden. An overall Strategic Action Plan was implemented in 1999 as a coordinated effort by PERSGA, UNEP, and related regional and global bodies. This plan strives to strengthen administrative structures and regional cooperation; to reduce navigation risks and maritime pollution; and to support sustainable use of living marine resources, habitat and biodiversity conservation, development of a network of marine protected areas, and management of an integrated coastal zone.

MANJA LEYK

Further Reading

Feidi, Izzat H. "Mechanism for Fisheries Management in the Red Sea and Gulf of Aden." *Arab Agricultural Statistics Yearbook 2012*, no. 1 (2012).

Groombridge, B. and M. D. Jenkins, eds. *The Diversity of the Seas: A Regional Approach.* WCMC Biodiversity Series No. 4. Cambridge, UK: World Conservation Press, 1996.

Vogt, H. "Coral Reefs and Coral Bleaching in the Region, Reefs in the Red Sea and Gulf of Aden." *Al Sanbouk* 12, no. 1 (2000).

Adriatic Sea

Category: Coastal Seas Biome.
Geographic Location: Northeastern arm of the Mediterranean Sea.
Summary: A historically important sea, the Adriatic, with its complex biodiversity, circulation, and geology, plays an important role in the greater Mediterranean Sea region.

An arm of the Mediterranean Sea, the Adriatic Sea separates the Italian Peninsula from the Balkan Peninsula. The Adriatic Sea is bordered by Italy to the west and north; and by Slovenia, Croatia, Bosnia and Herzegovina, Montenegro, and Albania to the east. The Adriatic stretches some 500 miles (800 kilometers) from the Gulf of Venice—the northernmost extent of the Mediterranean—to the Strait of Otranto, a 45-mile (72-kilometer) channel separating Italy and Albania while connecting the Adriatic to the Ionian Sea area of the Mediterranean.

The Adriatic Sea is relatively narrow throughout, never spanning more than 120 miles (200 kilometers) east to west. Its surface area is about 53,500 square miles (138,600 square kilometers); the sea is fed by freshwater flows from a catchment area nearly twice that size, or 90,750 square miles (235,000 square kilometers). Indeed, the Adriatic supplies up to one-third of the freshwater flow received by the entire Mediterranean, mitigating the salinity of the greater body of water. From the shallows near the Venice lagoon and Bay of Trieste in the north, depths grow deeper moving southward, with a range from about 80 feet (25 meters) to 4,035 feet (1,240 meters).

The western Adriatic coast tends to be alluvial or terraced,while the eastern coast is highly indented with pronounced karstification, the geologic term for chemical and mechanical erosion by water on soluble bodies of rock such as the limestone found here. This type of erosion results in sinkholes, towers, caves, and a complex subsurface drainage system, and facilitates significant groundwater contributions to the Adriatic Sea. More than 1,000 islands dot the eastern coast. Dominant circulation in the Adriatic is clockwise.

It is estimated that the Adriatic's entire volume is exchanged into the Mediterranean Sea through the Strait of Otranto every three to four years, a very short period and likely due to the combined contribution of rivers and submarine groundwater discharge. The Po River flowing west to east across Italy's north is the greatest single contributor of freshwater flows to the Adriatic, at 28 percent, and the largest deposition of sediments as well. Submarine springs along the Balkan, or Dalmatian, coast together contribute another 29 percent of freshwater flows.

The climate of the Adriatic Sea is Mediterranean, with hot dry summers, and mild wet winters. The predominant winter winds are the *bora* and *scirocco*. The *bora* brings cold, dry continental air from the mountains east of the Adriatic Sea. The *scirocco* brings humid and warm air from northern Africa.

Biota

The Adriatic Sea contains more than 7,000 species including many unique, rare, and endangered ones. The diversity and degree of specialization found here is due in part to the Dalmatian coastal islands, karst complexes, and submarine water flows, including some geothermal springs. Several marine protected areas have been established along this coast to protect the karst structures and their habitats.

The Kornati Islands National Park, for example, was established by Croatia in 1980 to protect 89 islands (since expanded to 109) and their marine environs. Among the signature protected species in the park is the noble pen shell (*Pinna nobilis*), a quite large, elongated clam. Kornati Park also hosts about half of the 682 known deepwater flora families of the Adriatic, such as red, brown, and green algae (*Rhodophyta, Phaeophyta,* and *Chlorophyta*). The park is home to an estimated 2,500 or more types of deepwater and tidewater species, including at least 177 species of mollusk; 127 species of bristle worms (*Polychaeta*); 64 species of sea stars, urchins, and cucumbers (*Echinodermata*); and 160 species of fish.

There are also more than 20 species of corals. Sponges are also found in relative diversity, although diminished from past human exploitation. Terrestrial species in the park include Dubrovnik knapweed (*Centaurea ragusina*), an endemic (evolved specifically and uniquely to a biome) plant that forms a centerpiece of the coastal scrub vegetation here; olive trees are by far the leading cultivar. Representative fauna include seagulls, cormorants, and owls; their prey in the form of snails and lizards; and some 60 or more species of moth and butterfly (*Lepidoptera*).

The Adriatic Sea features numerous species rated from declining to critically endangered, with representatives across the spectrum: various rockweeds and seagrasses including *Cystoseira zosteroides, Cystoseira spinosa, Zostera noltii,* and *Posidonia oceanica*; bivalves such as *Gibbula nivosa* and the ribbed Mediterranean limpet (*Patella ferruginea*); the European eel (*Anguilla anguilla*); Adriatic sturgeon (*Acipenser sturio*); the beluga or great sturgeon (*Huso huso*); the great white shark (*Carcharodon carcharias*); sea turtles such as the loggerhead (*Caretta caretta*) and leatherback (*Dermochelys coriacea*); and marine mammals including various cetacean species and the Mediterranean monk seal (*Monachus monachus* or *Monachus albiventer*).

Avian species under stress range from raptors like the white-tailed sea eagle (*Haliaeetus albicilla*), greater spotted eagle, (*Aquila clanga*), and golden eagle (*Aquila chrysaetos*); to typical seabirds such as the great white pelican (*Pelecanus onocrotaius*), Dalmatian pelican (*Pelecanus crispus*), little tern (*Sterna albifrons*), and Balearic shearwater (*Puffinus mauretanicus*); and wetlands-oriented species including purple heron (*Ardea purpurea*), European flamingo (*Phoenicopterus ruber roseus*), common crane (*Grus grus*), glossy ibis (*Plegadis falcinellus*), and northern pintail (*Anas acuta*).

More than two-thirds of the major commercially important fish species in the Adriatic are considered overfished. Those whose populations are thought to be in the "safe" range—even though they have in several cases seen marked declines in recent-year catch volume—include bogue (*Boops boops*), sprat (*Sprattus sprattus*), pilchard and sardinella (*Sardina pilchardus, Sardinella aurita*), jack mackerel and horse mackerel (*Trachurus picturatus, Trachurus trachurus*).

Human Interaction

There is a long history of human settlement and use in the Adriatic Sea and adjacent coasts. Human activities have influenced marine ecosystems since Roman times, with this influence increasing in the 19th and 20th centuries. Today, almost all of the original marine resources have been reduced to less than half of their former abundance, with large and midsized fish species being most affected. These changes may also make the overall ecosystem less resilient and more vulnerable in the future, as the regional effects from global warming gain in intensity, for example.

Fisheries, marine transport, and tourism are important throughout the Adriatic Sea. Technological improvements in the fishing fleet and increased activity driven in part by higher prices have resulted in a blanket decline in the fish catch rates per boat. Coastal pollution and eutrophication—excess nutrient inflow, typically from agricultural and municipal runoff—have been additional factors in declining fisheries yields. In many cases in the more urbanized northern Adriatic, fish kills have occurred due to algal blooms and low-oxygen conditions.

Shipping trade goods is another Adriatic Sea industry that dates from Roman times and before. Significant cargo port facilities ring the northern Adriatic; today, tourism is an added generator of mechanized sea traffic and threats to coastal ecosystems. Seeking to encourage "green" practices, the Denmark-based Foundation for Environmental Education has found acceptance by the operating authorities of several hundred Adriatic Sea marinas and beaches of its Blue Flag certification program. These local businesses and governments seem to realize that the extra short-term expense to contain fuel spills, prevent debris dumping, and restrict coastal development will in the long run protect their investment in the healthy ecosystem qualities that attract customers to their facilities.

Venice illustrates a worst-case scenario for polluted coastal waters in the Adriatic, where shipping, transportation, farming, manufacturing, and wastewater combine to heavily pollute the sea. Venice also contains oil refineries, which fortunately have avoided a significant spill that would have extreme consequences on its marshes, species, commercial fisheries, and tourism. There are hydrocarbon resources underlying other sections of the Adriatic Sea, and various controversial proposals for developing them. Just beyond the Strait of Otranto, Northern Petroleum, a United Kingdom concern, was poised in 2012 to commence seafloor drilling off the coast of the Puglia region, after receiving approval by the Italian Ministry of Environment—even though the 2,550-square-mile (6,600-square-kilometer) area of the sea in question adjoins a host of variously conserved zones including a Marine Protected Area; a Specially Protected Area of Mediterranean Importance; a National Natural Reserve; and, under a European Union policy centerpiece known as Natura 2000, nine Sites of Community Importance.

Magdalena Ariadne Kim Muir

Further Reading

Coll, Marta, et al. "Biodiversity of the Mediterranean Sea: Estimates, Patterns, and Threats." *PLoS ONE* 5, no. 8 (2010).

Lotze, Heike K., Marta Coll, and Jennifer A. Dunne. "Historical Changes in Marine Resources, Food-Web Structure and Ecosystem Functioning in the Adriatic Sea, Mediterranean." *Ecosystems* 14, no. 2 (2011).

Sala, Enric. "The Past and Present Topology and Structure of Mediterranean Subtidal Rocky-Shore Food Webs." *Ecosystems* 7, no. 4 (2004).

Aegean Sea

Category: Marine and Oceanic Biomes.
Geographic Location: Mediterranean Sea.
Summary: The Aegean Sea, in the northeastern corner of the Mediterranean Sea, is a marine province that shares many climate and biome characteristics with temperate North Atlantic realms.

Within the Mediterranean Sea basin, the Aegean Sea forms an embayment surrounded on the west and north by Greece, on the east by Turkey, and on the south by the island of Crete. The area of the Aegean is 82,625 square miles (214,000 square kilometers). It reaches a maximum depth, near Crete, of 11,624 feet (3,543 meters). The area has a complex topography marked by deep trenches and elevated structures; the eastern Mediterranean is a remnant of an ancient body of water known as the Tethys Ocean. The waters are generally nutrient poor, sustaining an oligotrophic environment (an environment that is characterized by relatively slow growth and metabolic activity).

To the northeast, the Aegean connects through Turkey to the Black Sea via the Sea of Marmara and the straits of the Dardanelles and the Bosphorus. The Aegean is dotted with thousands of mountain peaks peeping up from the seafloor. These island chains and the numerous cragged gulfs and bays surrounding them are unsurpassed in utility to seafarers. The islands include Crete and Euboea (the Mediterranean's fifth-largest and sixth-largest, respectively), the Sporades, the Cyclades, the Saronic, the Dodecanese, and the Northern and Eastern groups.

Salinity and Temperature

Like the Mediterranean as a whole, the Aegean Sea flows counterclockwise, its salty waters rising along its eastern boundary and then falling along its northern boundary, carrying the waters of the Black Sea down along its western Greek shoreline and out into the larger basin. The Aegean is influenced by water exchange with such regional sources as the Adriatic and Ionian Seas, as well as flow from the Nile River delta and the Red Sea via the Suez Canal. Changes in saltwater composition over time have periodically altered the circulation patterns and the fundamental ecology of the Aegean basin area. Rising global temperatures now contribute to fluctuating carbon dioxide (CO_2) levels, in concert with CO_2 escaping from volcanic sources.

One result is acidified marine zones that trigger losses in biodiversity and alter food-web functions. This situation is complicated by the Aegean basin being a somewhat enclosed province; pollutants such as heavy metals cannot be readily discharged and over time their levels are concentrated. These factors contribute to worrisome changes in ecology including the erosion of coral ecosystems and the extinction of species such as the sea turtle, the monk seal, and the Atlantic tuna, all of which spawn only in the Mediterranean.

The effects of climate change on pelagic (deep sea) marine ecologies have only begun to be documented; some effects are unexpectedly beneficial to various life forms. Changes of interest include variations in the metabolic rates of marine organisms, and environmental conditions including changes in currents, column stratification, and nutrient production, all variables that affect food-web structure, population distribution, and community dynamics within an ecosystem.

Researchers marked a "transient event" in the area occurring in two phases during the 1990s. As a result of increased salinity by climate forcing, a mass of dense, warm water formed in the southern Aegean. This process was followed by a drop in temperature of 0.7 degrees F (0.4 degrees C), creating another deeper layer of dense water. These changes triggered an upwelling of deep bottom water, creating an intermediate stratum

A girl tosses a stone into the water off the Greek island of Paros in the central Aegean Sea. The waters of the Aegean Sea are very clear because of low levels of phytoplankton, which also contribute to the 46 percent lower total biomass of the eastern versus the western Mediterranean. (Thinkstock)

of nutrient-rich water now known as the Transitional Mediterranean Water. This change, in turn, set into motion a threefold increase in biological activity along the continental shelf and upper slope.

Plankton and Human Needs

The total biomass of the Eastern Mediterranean is 46 percent lower than densities found at the western edge. Since the construction of the Aswan High Dam in Egypt and a major dam on the Ebro River in Spain, freshwater inputs into the Mediterranean Sea were significantly reduced. Continued flows of saltwater from the Atlantic Ocean and Red Sea contributed fractional increases in salinity. This, in turn, affects the quality of marine life throughout the region. The Atlantic Ocean waters that spill into the Mediterranean are poorer in nutrients, having served to feed the ocean's population of plankton. As the waters move eastward toward the Aegean, bacteria and other organisms feed on the remaining nutrients, leaving little left for plankton and other marine life. Bacteria are thought to account for about 56 percent of all organisms found on the shallow shores of the Aegean Sea.

The Suez Canal is effectively a passageway for the migration of over 300 species of Erythraen marine biota from the Red Sea into the Mediterranean, an immigration that adds to environmental changes in the Aegean. This is known as the Lessepsian Migration, named after Ferdinand de Lesseps, the French mastermind of the 19th-century Suez Canal project. Exotic macrophytes, invertebrates, and fish include the bivalve *Brachiodontes pharaonis*, the gastropods *Cerithium scabridum* and *Strombu persicus*, the rock oyster *Spondylus spinosus*, the jellyfish *Rhopilema nomadica*, penaeid prawns, the American blue crab, mullids, the barracuda *Spyraena chrysotaenia*, thriving populations of clupeids, and pearl mollusks.

The competitive displacement of autochthonous species is also the result of coastal mariculture farming. Farm stocks include the Pacific oyster *Crassostrea gigas*, the soft-shell clam *Mya arenariea*, and the pearl oyster *Pinctada radiata*. Changes in bathymetric populations are a direct result of migration and displacement. Human population densities and subsequent coastal development have further decimated marine biota and altered fragile littoral (shoreside) ecosystems.

The general absence of phytoplankton contributes to the amazing clarity of the Aegean's waters. The most notable of the sea fauna include the blennies (Blenniidae), Cardinal Fish (Apogonidae), eels, picarels, electric rays, groupers (Serranidae), lizard fish, red mullets, and scorpion fish. The larger sea mammals include dolphins and seals. Divers and snorkelers enjoy touring the Aegean to study and to admire the habitats of sponges and sea urchins, octopus squids, and cuttlefish. Algae, kelps, and seagrasses are also common to the area.

Coastal Species

The Aegean Sea, its archipelago, Greece, and Turkey together form a distinct ecological region classified by the World Wildlife Fund (WWF) as a zone of sclerophyllous and mixed forests (designated PA1201). The eastern border of this ecoregion extends to plains and lowlands in Turkey, while Greece is replete with sparse, mountainous landscapes. Callabrian pine (*Pinus brutia*) and maquis are the dominant vegetation; north to south, the maquis are varied and include stands of *Arbutus andrachne, A. unedo, Spartium junceum,* and *Laurus nobilis*. Stands of sweetgum and Datça palms, endangered native species, are also found. Varied species of oak grow in areas where the pine stands have been removed.

Each of these types of forests provide habitat for European woodland birds; pine forests echo with the calls of wrens, blackbirds, tits, jays, and Krüper's nuthatch, a bird only found in southeast Turkey. Warblers, buntings, and partridges make their habitat in maquis scrublands. Other native wildlife include loggerhead turtles, wolves, foxes, and boars. The entire basin is noted for its olive trees and viticulture.

Victoria M. Breting-Garcia

Further Reading

Colasimone, Luisa, ed. *MedWaves: Focus on Biodiversity.* Athens: United Nations Environment Programme, Mediterranean Action Plan, 2010.

Danovaro, Roberto, Antonio Dell'Anno, Mauro Fabiano, Antonio Pusceddu, and Anastasios Tselepides. "Deep-Sea Ecosystem Response to Climate Changes: The Eastern Mediterranean Case Study." *Trends in Ecology & Evolution* 16, no. 9 (2001).

Evert, Sarah. "The Mediterranean: Beneath the Surface." *Chemical & Engineering News* 90, no. 15 (2012).

Gritti, E.S., B. Smith, and M. T. Sykes. "Vulnerability of Mediterranean Basin Ecosystems to Climate Change and Invasion by Exotic Plant Species." *Journal of Biogeography* 33, no. 1 (2006).

Pichon, X. Le, J. Angelier, M. F. Osmaston, and L. Stegena. "The Aegean Sea." *Philosophical Transactions of the Royal Society of London, Mathematical and Physical Sciences* 300, no. 1454 (1981).

Şekercioğlu, Çağan Hakki, Sean Anderson, Erol Akçay, and Raşit Bilgin. "Turkey's Rich Natural Heritage Under Assault." *Science* 23, no. 12 (2011).

Spalding, Mark D., et al. "Marine Ecoregions of the World: A Bioregionalization of Coastal and Shelf Areas." *Bioscience* 57, no. 7 (2007).

Alaska Peninsula Montane Taiga

Category: Forest Biomes.
Geographic Location: North America.
Summary: The Alaska Peninsula montane taiga is a rugged terrain hosting elevation-variegated plants and fauna ranging from vast migratory bird rookeries to the huge Kodiak bear.

The Alaska Peninsula montane taiga covers some 18,500 square miles (48,000 square kilometers) of rugged vertical terrain along the southern slopes of the Alaska Peninsula, the Kodiak Archipelago, and as far west as Unimak Island in the eastern Aleutian Islands chain. Situated below and around volcanic peaks of up to 8,000 feet (2,500 meters), floral and faunal features of this biome thrive in elevations of 4,000 feet (1,200 meters) and higher.

Taiga is a characteristic northern coniferous forest type, with great stretches found across North America, Europe, and Asia. The Alaska Peninsula montane taiga exists in a climate of cold, snowy winters and rainy, humid, warm summers; due to its coastal position, the high-elevation harshness here is somewhat reduced by the moderating influence of Pacific Ocean temperature effects.

Taiga forests typically lack the great biological diversity of many tropical and temperate forest systems. The Alaska Peninsula montane taiga biome, however, does host a tremendous summer insect population which in turn attracts vast bird migrations during nesting season.

Vegetation Communities

Supporting its flora types, the Alaska Peninsula montane taiga enjoys a moderate scale of seasonal temperatures as well as a high precipitation regime of from 24 to 130 inches (60 to 330 centimeters) along the coasts to more than 160 inches (400 centimeters) in the highest elevations. The temperatures in winter range between 12 to 34 degrees F (minus 11 to 1 degree C), with summer ranges from 42 to 59 degrees F (6 to 15 degrees C). There is little in the way of permafrost terrain, but glaciers are found on the high peaks.

Upper slopes of the biome feature dwarf scrub vegetation dominated by black crowberry (*Empetrum nigrum*), along with such species as wandering daisy (*Erigeron peregrinus*), mountain white radish or cathaleaf avens (*Geum calthifolium*), yellow-flowered sedge (*Carex anthoxanthea),* Nootka reedgrass (*Calamagrostis nutkaensis*), and the moss *Pleurozium schreberi*. Toward the lower elevations, various willows (*Salix* spp.) become common; lower still are found stands of green alder (*Alnus viridis)* and, especially in the floodplains, balsam poplar (*Populus balsamifera*).

Animal Populations

The Alaska Peninsula montane taiga supports considerable herds of caribou, as well as moose, Arctic ground squirrel *(Spermophilus parryii)*, and Alaskan hare (*Lepus othus).* The Kodiak bear (*Ursus arctos middendorffi*) is an outstanding endemic—

unique to this biome—species, considered among the largest bears in the world. The Kodiak and other brown bears gather periodically in alluvial plains here for salmon runs in summer and fall. At the Katmai National Park and Preserve, this activity is a focal point for tourists and scientists alike, who annually congregate at Hallo Bay, Geographic Harbor, Swikshak Lagoon, Moraine Creek, and related areas. Adjacent to Katmai is the McNeil River State Game Sanctuary, containing the full, 35-mile length of this river, which is also a prime bear-watching hot spot.

Migratory waterfowl are another key component of the Alaska Peninsula montane taiga, with estimated colony sizes of 500,000 tufted puffins (*Fratercula cirrhata*) on Unimak Island, and 650,000 common murres (*Uria aalge*) and 500,000 northern fulmars (*Fulmarus glacialis*)in the Semidi Islands.

Pressures and Threats

Leading human activities in the area include commercial fishing and processing—by far the most extensive industry—along with mining and subsistence activities such as hunting and fishing. There are many small communities along the water which rely mostly on fishing combined with hunting marine mammals. There is a fairly limited amount of coal and petroleum extraction, as well as scattered mining of gold, lead, silver, and copper. There are also several active or former military installations that house toxic waste storage facilities.

Taiga systems are generally susceptible to wildfires; trees in these areas adapt by growing thick bark. Therefore, the fires will burn away the upper canopy of the trees and let sunlight reach the ground. Through this cycle, new plants will grow and provide food for animals that would have otherwise been hard-pressed to find sufficient nutrition from a conifer-dominant landscape. The Alaska Peninsula montane taiga, doused regularly with ocean spray and high humidity, is generally not as susceptible to wildfires as other taiga regions in Europe and Asia.

Violent winter storms are frequent, however, driving hillside erosion. Infrequent but sometimes extensive volcanic activity in this Ring of Fire geographic area produces quantities of ash that can blanket an area and stunt plant growth.

The World Wildlife Fund, in its in-depth study of the area, concluded that the Alaska Peninsula montane taiga biome is almost entirely intact, with minor habitat loss and degradation being confined to the localized effects of development around the small communities and villages along the coast. Much of the area is protected, including such zones as the Alaska Peninsula National Wildlife Refuge, Izembek National Wildlife Refuge, Kodiak National Wildlife Refuge, Aniakchak National Monument and Preserve, Becharof National Wildlife Refuge, as well as the Katmai and McNeil reserves noted above.

Still there is the threat of gradual habitat degradation from existing ranching activity, and the subsequent release of feral cattle on some islands, as well as predation effects from feral foxes. Brown bears in the biome are considered in jeopardy from overhunting; stabilizing the brown bear population along the McNeil River is one of the prioritized biodiversity conservation measures in the region. Another goal is to consolidate Kodiak National Wildlife Refuge holdings using *Exxon Valdez* settlement funds to purchase properties.

Further threats of climate change and variable weather events also hold potential threats to this biome. The taiga ecosystem is maintained by its precipitation volume and its dependable, moderated temperatures for each season. With the looming threat of climate change, this balance may be altered and could sharply impact this ecosystem in unprecedented ways.

Christina Bonanni

Further Reading

Hulten, Eric. *Flora of Alaska and Neighboring Territories: A Manual of the Vascular Plants.* Palo Alto, CA: Stanford University Press, 1968.

Scheffer, Victor B. and Olaus Johan Murie. *Fauna of the Aleutian Islands and Alaska Peninsula.* Charleston, SC: Bibliobazaar Nabu Press, 2011.

Vitt, Dale H., Janet E. Marsh, and Robin B. Bovey. *Mosses, Lichens, and Ferns of Northwest North America.* Auburn, WA: Lone Pine Publishing, 1988.

Albemarle-Pamlico Sound

Category: Marine and Oceanic Biomes.
Geographic Location: Eastern region, United States.
Summary: The Albemarle-Pamlico Sound is the second-largest estuary system in the United States, providing habitat for fish and shellfish. Land-use activities pose challenges to water quality and aquatic resources here.

The Albemarle-Pamlico Sound is located in the mid-Atlantic coastal region of North Carolina and is protected from the Atlantic Ocean by a sliver of sandy barrier islands known as the Outer Banks (and their easternmost extension, Cape Hatteras). It is the second-largest estuary system in the United States, after the Chesapeake Bay. The sound supports diverse, abundant fish and shellfish populations; the Albemarle and Pamlico regions historically developed around these natural resources. Commercial and recreational fishing are important components of the social and economic vitality of the coastal area. The water quality of the Albemarle-Pamlico Sound, however, has been influenced by land use in the watershed.

The watershed contributing to the sound is approximately 31,000 square miles (80,290 square kilometers), composed of more than 9,000 miles (14,484 kilometers) of freshwater rivers and streams, as well as 1.5 million acres (607,028 hectares) of estuarine habitat. The average depth in the sound is 13 feet (4 meters). The watershed drains 36 counties in North Carolina and 16 counties in Virginia. Six major river basins flow into the sound: The Pasquotank, Chowan, Roanoke, Tar-Pamlico, Neuse, and White Oak Rivers provide freshwater flow to the sound, which is connected to the Atlantic Ocean by three inlets.

This mix of freshwater and saltwater creates diverse habitats that include bald cypress wetlands; brackish, freshwater, spartina, and salt marshes; abundant submerged aquatic vegetation; forested wetlands; and an interesting, unique habitat referred to as pocosin or southeastern shrub bog. *Pocosin* is a Native American term that means "elevated or high swamp." Pocosins accumulate layers of peat, an infertile layer of organic material on the bottom of the wetland. This layer of peat results in plant growth that is less robust than typical growth in more fertile conditions. The Albemarle-Pamlico Sound watershed supports more pocosin habitat than any other place in the world.

Fish Habitat and Nursery

The Albemarle-Pamlico Sound provides habitat and nursery waters for more than 75 species of fish and shellfish. Salinity is one of the major influences on the distribution of fish species in the sound; it varies constantly, ranging from

A U.S. Fish and Wildlife officer stands by an eastern North Carolina pond pine pocosin habitat that was retreating in 2011 because of saltwater intrusion caused by sea-level rise from climate change. The coast from Cape Hatteras to north of Boston is experiencing sea-level rise at three to four times the global average. (U.S. Fish and Wildlife Service/Steve Hillebrand)

nearly 0 percent in the lower portions of the rivers to 25 percent near the inlets on the Outer Banks. Salinity is affected by freshwater inflow, season, tides, and wind direction and intensity.

Fish species that are important commercially and recreationally in the sound include striped bass, flounder, speckled trout, redfish, menhaden, and croaker. Up to 90 percent of the commercially important finfish species in the greater mid-Atlantic region rely on the estuarine habitats of the Albemarle-Pamlico Sound for spawning, development, and forage. Shrimp, blue crabs, oysters, and scallops are additional commercially and recreationally valued species in the estuary.

The sound is habitat for a fish community that has evolved reproductive strategies using the rivers that enter the sound and the tidal currents that originate in the Atlantic Ocean. Catadromous fish species such as the American eel live in the freshwater rivers and creeks in the Albemarle-Pamlico Sound watershed as adults. The eels migrate from freshwater through the sound to the Atlantic Ocean and on to the Sargasso Sea to reproduce. The immature eels, or *elvers*, use ocean currents to return to the sound to develop; as they mature, they use the sound for foraging and refuge

Anadromous fish species in the sound include striped bass, American shad, and several herring species. Anadromous fish live in the saltwater habitat of the Atlantic Ocean and migrate through the sound to spawn in freshwater rivers and creeks. The reproductive strategy of these species is to spawn in the water column so that the fertilized eggs drift with the currents downstream to the sound. The larval fish develop and grow in the sound before migrating back to the Atlantic Ocean. Other species—such as blue crab, gray trout, red drum, spot, flounder, and shrimp—spawn in the ocean. Tidal currents carry the eggs and larval life stages into the sound to find food and protection from predators.

Water-Quality Challenges

The Albemarle-Pamlico Sound faces several complex water-quality issues. The North Carolina Environmental Management Commission identified water-quality problems that negatively affect aquatic resources in the watershed; it determined that nearly 33 percent of the freshwater streams are affected by land-use activities that impair water quality. The primary causes of water-quality impairment in the basin are excessive sediment; fecal coliform bacteria; and excessive nutrient loading, especially nitrogen that derives largely from runoff of the region's extensive pig farming operations. These conditions result in harmful algal blooms, anoxic conditions, and fish kills.

Nutrients are substances that help plants and animals grow. Two nutrients that are critical for development are nitrogen and phosphorous; these nutrients are also present in plant fertilizer and animal and human waste. Nutrient loading in the watershed has been widely linked to agricultural and urban runoff, atmospheric deposition in the form of acid rain, and point and nonpoint discharges that include sewage-treatment-plant effluent.

The effects of nutrient loading often promote abundant algal blooms, which deplete the water column of dissolved oxygen. Low dissolved oxygen results in stressed environmental conditions that can kill fish directly or that allow bacteria and viruses to more readily infect finfish and shellfish populations. Excessive nutrient loading has resulted in periodic fish-consumption advisories in the sound to protect human health.

Sea-Level Rise

Another looming threat to the Albemarle-Pamlico Sound is a combination of sea-level rise due to global warming, and land subsidence that is notable along the mid-Atlantic coast. Recent studies have confirmed that the coastal area from Cape Hatteras to north of Boston is logging sea-level rise at three to four times the global average. With the cape and its associated barrier islands of the Outer Banks under such pressure—and the concomitant higher storm surges and heavier hurricane threat—conservationists must plan for responses to the potential of increased salinity in the delicately balanced biome of the sound.

Thomas A. Shervinskie

Further Reading

Kennish, Michael J. and Hans W. Paerl, eds. *Coastal Lagoons: Critical Habitats of Environmental Change.* Boca Raton, FL: CRC Press, 2010.

McCarthy, Annette Marie. "Fate and Distribution of Current-Use Pesticides in the Albemarle-Pamlico Estuarine System of North Carolina." North Carolina State University, 2003. http://repository.lib.ncsu.edu/ir/bitstream/1840.16/5908/1/etd.pdf.

Sallenger, Asbury H. Jr., Kara S. Doran, and Peter A. Howd. "Hotspot of Accelerated Sea-Level Rise on the Atlantic Coast of North America." *Nature Climate Change* 2, no. 6 (2012).

United States Environmental Protection Agency (EPA). *Ecosystem Services Research Program (ESRP) Albemarle-Pamlico Watershed and Estuary Study (APWES) Research Plan.* Washington, DC: EPA Office of Research and Development, 2010.

Albert, Lake (Africa)

Category: Inland Aquatic Biomes.
Geographic Location: Africa.
Summary: Lake Albert, a component of the African Great Lakes ecosystem, is home to high biological diversity and a source of key socioeconomic benefits to the region.

Lake Albert, a part of the Great Rift Valley of east Africa, was formerly called Lake Mobutu Sese-Seko. It is located on the border between Uganda and the Democratic Republic of Congo (DRC), with a length of 100 miles (161 kilometers) and a width of 20 miles (32 kilometers); the surface area is 2,046 square miles (5,300 square kilometers). The depth of the water in Lake Albert reaches 167 feet (51 meters) in some places, while its surface elevation is 2,024 feet (617 meters) above mean sea level.

Nearly all of the inflow from the lake drainage basin plus direct precipitation onto the lake itself are lost by evaporation from the lake surface, but the inflow of the Semliki River, which enters the lake from the southwest, accounts for the net gain. This river connects the smaller Lake Edward to Lake Albert after flowing a distance of about 155 miles (250 kilometers) down the Great Rift Valley to the west of Ruwenzori Mountain. Lake Albert is also fed seasonally by the upper reaches of the White Nile River, which loops in and out of its northern tip, and is denoted as the Albert Nile.

The mean annual temperature of the lake region is about 79 degrees F (26 degrees C), and average annual rainfall in the area is 34 to 40 inches (864 to 1,016 millimeters). Because of the high rates of sedimentation and evaporation, the lake water is quite saline and contains free phosphate.

Uganda lies on the eastern, northern, and southern banks of Lake Albert, while the western lakeshore belongs to the Democratic Republic of Congo. Uganda and the DRC share 54 percent and 46 percent of the lake, respectively, and the lake is part of chronic border disputes. Nevertheless, the lake contributes both to the greater region's ecosystem and provides many socioeconomic benefits for the people.

The landscape is shaped by escarpments, forests, savanna, grasslands, wetlands, and deltas; this combination underlies the lake's exceptional biodiversity. Elephants, buffaloes, hippopotami, crocodiles, and antelopes are among the larger animals living in the area, especially on the Semliki Plains south of the lake, and along the northern shores of Lake Albert. The protected areas of Albert Lake in Uganda include one the nation's leading tourist attractions and game reserves, Murchison Falls National Park. Here are found buffaloes, lions, leopards, and elephants in relative abundance. In the DRC, the closest protected area to Lake Albert is Virunga National Park, a World Heritage Site that ranges along the Semliki River and in the Lake Edward environs. Populations of antelopes, crocodiles, elephants, and mountain gorillas are found there.

Human Activity and Environmental Threats

Although the water of Lake Albert is saline, it supports the local human and livestock populations not only by supplying drinking water, but also by income sources and food security to the communities that border it. Indeed, the abundance of salt has itself spawned a minor, if vital, local industry. Lake Albert contains a great variety of fish. Commercial catches, however, are largely comprised of three major species: *Alestes baremose, Hydrocynus forskahli,* and *Lates niloticus.*

The region's population is dependent on natural resources or subsistence agriculture, cattle herding, and logging, but those resources are increasingly stressed by overuse and degradation. Rudimentary methods of natural resource exploitation further increase the pressure on the environment. Security concerns and lack of government structures, particularly on the DRC side, hinder adequate management of protected areas. As a result, national parks face deforestation due to illegal logging and insecurity due to poaching activities.

Planned oil extraction and processing in the area also poses great risk to the environment. Test drilling has begun in Murchison Falls National Park and the Kabwoya Wildlife Reserve in Uganda, but no method of handling the resulting toxic wastewater and mud has yet been found. Furthermore, animals are substantially disturbed by the drilling activities. The operation of a landing site with several flights per day in the Kabwoya Wildlife Reserve, the construction of roads to the sites and drilling stations, and the increased population density in the immediate area greatly disturb the animals' breeding grounds. Human in-migration from Uganda and the DRC, the residences and other infrastructure of drilling company employees, and the presence of large military and police forces in this chronically strife-torn area are added pressures. Outside the protected areas, especially on the DRC side of the lake, the situation is worse; many plants and animals are threatened or have gone extinct.

Political Pressures

The loss of fertile land from overgrazing in Uganda has emboldened some of the local population to cross borders; some Congolese in turn let their cattle graze in Uganda, in part to avoid political violence on their side of the frontier. According to a study by the Ugandan Nile Discourse Forum, Ugandan farmers must pay high taxes to cross borders and to receive access to fertile land in the DRC, whereas Congolese migrants are not obliged to pay taxes in Uganda for cattle grazing. This situation causes additional inequalities and suspicion in both population groups.

Furthermore, there has been an ongoing trend of land acquisition by migrants for cattle keeping and other agricultural activities in the areas around Lake Albert. This trend has been a source of conflict and tension. Unconfirmed reports indicate that the powers behind the land acquisitions are investors who want to acquire land with underground oil resources for possible sale or compensation later on, which has made the indigenous people hostile even to genuine migrants.

From a geographic and political perspective, the region reflects major transboundary concerns related to natural resources and bilateral relations. A shared history of conflict, poor livelihoods, border disputes, and recent oil findings in the region jeopardize management of common resources and normal political relations. The occupation of the DRC's eastern district, Ituri, by Ugandan forces between 1998 and 2003 is recalled as a major disruption in the region. Environmental degradation linked to oil extraction, as well as increasing pressure on natural resources such as water and land, directly affects the fragile ecosystem and local livelihoods. With the promise and challenge of oil exploration, the Ugandan and Congolese governments have engaged in first steps to normalize their relationship.

Suman Singh

Further Reading

Odada, Eric O. and Daniel O. Olago, eds. *The East African Great Lakes: Limnology, Palaeolimnology, and Biodiversity.* New York: Springer, 2011.

Serruya, Colette and Utsa Pollingher. *Lakes of the Warm Belt.* Cambridge: Cambridge University Press, 1983.

Von Sarnowski, Andrea. *The Artisanal Fisheries of Lake Albert and the Problem of Overfishing.* Mainz, Germany: Johannes Gutenberg Universitat, 2004.

Westerkamp, Meike and Annabelle Houdret. *Peacebuilding Across Lake Albert—Reinforcing Environmental Cooperation Between Uganda and the Democratic Republic of Congo.* Brussels, Belgium: Initiative for Peacebuilding, 2010.

Albertine Rift Montane Forests

Category: Forest Biomes.
Geographic Location: Africa.
Summary: Remote and rugged, the Albertine Rift spans a wide range of habitats, from rainforest to glaciers, active volcanoes to deep lakes; the montane forests here are very rich in biodiversity.

The Albertine Rift covers more than 115,831 square miles (300,000 square kilometers) in central east Africa. This area is one of the most species-rich regions in Africa and also one of the most scenic, including the Ruwenzori Mountains, the Virunga Mountain volcanoes, and the Semliki Valley. The Albertine Rift is one of Africa's most biodiversity-rich—and threatened—ecosystems. There is no clear-cut definition of the Albertine Rift, but the area it covers begins north of Lake Albert between Arua and Pakwach (West Nile, Uganda) and extends southward, following the course of the Rift Valley to the Lendu Plateau and the lower reaches of the Kibali and Ituri Rivers. It includes the forests of western Uganda and Kigezi (Uganda), north and south Kivu in the Democratic Republic of Congo, parts of western Rwanda and Burundi, Itombwe to Marungu in western Katanga in the Democratic Republic of Congo, areas of western Tanzania, and part of northwestern Zambia.

The Albertine Rift has been marked by political turmoil that has included massive displacement of peoples. Habitat destruction and other issues, such as the bushmeat trade, are very real and direct threats to species in the region. The Albertine Rift also has one of the highest population-growth trends in eastern and central Africa. Most of the population of this area is largely engaged in rural subsistence agriculture. One of the most threatened and famous animals on the planet, the mountain gorilla, lives within the Albertine Rift on the slopes of the Virunga volcanoes, which are among the steepest in Africa.

The Virunga volcanic range, like the Albertine as a whole, is in the westernmost part of the immense East Africa Rift Valley system that cuts down across the continent from the Horn of Africa at the juncture of the Red Sea and the Gulf of Aden. Due to volcanic activity and the sinking and rising of parts of the continental plate, diverse and dramatic landscapes have formed. The flanks of the mountains and volcanoes of the Albertine Rift are covered with distinctive bands of vegetation that change higher up the slopes. At lower altitudes is dense forest, with montane forest typically occurring above this zone, where the trees are covered with moss and ferns. Above the montane forest is a zone of giant bamboo, followed by moorland and rock and/or ice if the mountains are tall enough.

A gorilla mother holds her infant at the National Volcano Park in Rwanda. While the mountain gorilla's numbers are growing again in the Albertine Rift, they have been seriously threatened by habitat destruction and hunting. (Thinkstock)

Plant and Animal Life

The levels of endemism (evolving uniquely to fit an isolated biome) among species is very high here for all groups: plants, invertebrates, birds, and mammals. A total of 13 species of dragonflies and damselflies are unique to the Albertine Rift, for example, and some 117 butterfly species have been described as being endemic to the area. The forests of the Albertine Rift are mostly unexplored when it comes to invertebrates. Bwindi Impenetrable Forest in Uganda alone has five species of stingless bees. Many more species of invertebrates await discovery and formal description by scientists.

Bird life is abundant in the region, and the forests are home to a large number of endemic and near-endemic species. A total of 41 species of birds are endemic to the Albertine Rift, including the Rwenzori Turaco, Albertine Owlet, Congo Bay Owl, Red-throated Alethe, and Rockefeller's Sunbird.

Some 34 species of endemic mammals live here; most are small mammals such as shrews, rats, and bats. A few of the larger ones include the Rwenzori duiker, which lives at high altitudes in the Rwenzori massif, and the striking golden monkey, which is confined to the Virunga volcanoes. The well-known mountain gorilla (*Gorilla gorilla beringei*) is an endemic subspecies. The gorillas were seriously threatened with extinction due to habitat destruction and hunting. Thanks to energetic conservation efforts, mountain gorilla numbers are currently increasing, and the threats now come from managing the needs of a growing human population around the gorilla habitat, as well as emerging threats from diseases and climate change. A few larger mammals are near-endemic species, such as Grauer's gorilla (*Gorilla beringei graueri*) and a species of monkey called L'Hoest's guenon.

The Albertine Rift montane forests are centers of geologic activity, with particularly active volcanoes, as well as innumerable endemic species that must face the effects of a growing population recovering from many decades of conflict. These factors combine to make this region one of the most important for conservation and sustainable development today.

DINO J. MARTINS

Further Reading

Byaruhanga, A., P. Kasoma, and D. Pomeroy. *Important Bird Areas in Uganda.* Kampala, Uganda: East Africa Natural History Society, 2001.

Eilu, Gerald, David L. N. Hafashimana, and John M. Kasenene. "Density and Species Diversity of Trees in Four Tropical Forests of the Albertine Rift, Western Uganda." *Diversity & Distributions* 10, no. 4 (2004).

Eilu, Gerald, David L. N. Hafashimana, and John M. Kasenene. "Tree Species Distribution in Forests of the Albertine Rift, Western Uganda." *African Journal of Ecology* 42, no. 2 (2004).

Vedder, A., L. Naughton-Treves, A. J. Plumptre, L. Mubalama, E. Rutagarama, and W. Weber. "Conflict and Conservation in the African Rain Forest." In W. Weber, L. J. T. White, A. Vedder, and L. Naughton-Treves, eds. *African Rain Forest Ecology and Conservation.* New Haven, CT: Yale University Press, 2001.

Aldabra Atoll

Category: Marine and Oceanic Biomes.
Geographic Location: Africa.
Summary: Aldabra is a pristine atoll with a high degree of endemism that makes it a United Nations Educational, Scientific and Cultural Organization (UNESCO) World Heritage Site and a prize for scientists and naturalists.

Aldabra is among the very largest coral atolls in the world, with an area of approximately 174 square miles (450 square kilometers). It is part of the Seychelles island group in the western Indian Ocean, and is located 684 miles (1,100 kilometers) southwest of Mahé, the main island, and 261 miles (420 kilometers) north of Madagascar. The atoll is actually four large coral islands that enclose a shallow lagoon. These islands formed in the late Quaternary and now are surrounded by coral reefs. Climate can vary throughout the year, with maximum mean temperatures of 88 degrees F (31 degrees C) in December and minimum temperatures of 72 degrees F (22 degrees C) in August. An average 43 inches (1,100 millimeters) of rainfall occurs each year, with the northwest monsoon winds bringing the heaviest rainfall from November to March. The terrain is rocky, with little soil, and freshwater is virtually absent, so native organisms have adapted to the abundant ocean water. The tides have major influences on all life, especially within the lagoon, which drains regularly.

The Aldabra atoll is one of the most isolated and pristine biomes on Earth. Its location and environment have kept it from much human disturbance. These factors have also contributed to the extreme degree of endemism (evolution of unique species that fit an isolated biome) across many taxonomic levels on Aldabra.

Flora and Fauna

The atoll is home to large seabird colonies; breeding sea turtles, including the threatened green sea turtle (*Chelonia mydas*) and hawksbill sea turtle (*Eretmochelys imbricate*); the last Western Indian Ocean flightless-bird species, the white-throated rail (*Dryolimnas cuvieri*); and the largest population of giant tortoises on Earth. The atoll is a refuge for 13 endemic bird taxa, including two species—the Aldabra brush warbler (*Nessilas aldabrana*) and the Aldabra drongo (*Dicrurus aldabranus*)—and 11 subspecies. An endemic lizard race (*Phelsuma abbotti abbotti*) and an endemic fruit bat race, the Aldabra Flying Fox (*Pteropus seychellensis aldabrensis*), also inhabit the atoll. The most widely known of the endemic species is the Aldabra giant tortoise (*Dipsochelys dussumieri*), which has a healthy population of more than 100,000 individuals.

Offshore reefs house many species of corals, fish, and marine invertebrates. Mangroves and seagrass beds grow along the shoreline. The land supports around 180 species of flowering plants, with 20 percent being endemic.

Conservation and Management

Overall, Aldabra has been spared the fate of many other island environments and is relatively intact. The atoll is not conducive to human presence, which has deterred settlement and exploitation. Development was briefly considered in the 1960s, however, when the British government considered constructing an air-staging outpost until international outcry from scientists caused the proposed development to be abandoned.

The Royal Society of London began a full inventory of the atoll's features that lasted until 1979. At that point, management was handed over to the Seychelles government and, specifically, to the Seychelles Island Foundation (SIF), which has been successful in conserving the biome through consistent monitoring and management. Tourism is strictly regulated, and the logistics of visiting the atoll have deterred most prospective visitors. The largest threats to this biome are invasive species (including cats, rats, and goats), global climate change, and oil spills from increasing energy development in the Indian Ocean region.

Aldabra's conservation was further ensured by its designation as a UNESCO World Heritage Site in 1982 and as an endemic bird area by BirdLife International. International support has also come from the Global Environment Facility of the World Bank. Funding to renovate the research station and to provide the resources for a complete management, science, and conservation plan has been influential in establishing Aldabra as a location of immense scientific activity.

Perhaps the gravest threat on the horizon is global warming; with much of the atoll no more than 6 or 7 feet (about 2 meters) above sea level—the high-

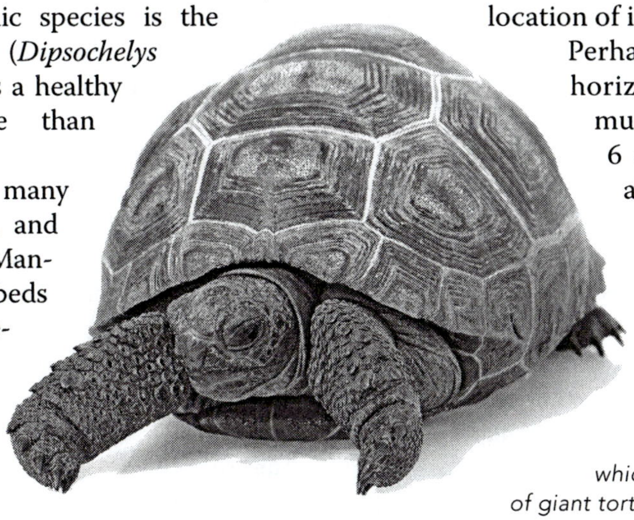

There are 100,000 Aldabra giant tortoises like this one currently living on the atoll, which is home to the largest number of giant tortoises anywhere. (Thinkstock)

est ground is just 26 feet (8 meters)—this danger is very real in light of sea-level rise as well as the increase in cyclone intensity due to higher ambient water temperatures. Indeed, sea-level rise is cited as a major threat in various studies of endangered species on Aldabra, in particular for those biota that inhabit either the intertidal realm or even the more inland niches that in the case of Aldabra are extremely close to the sea's direct influence. Among such fauna are the snail species *Kaliella aldabra* and *Cyathopoma picardense.*

Scientific Studies

The biome is well suited to scientific investigations, especially of evolution. In fact, Aldabra was described by the U.S. Academy of Sciences as "an ideal location for the scientific study of evolutionary processes in a relatively closed environment." Indeed, a large amount of research has been done on the evolution of the endemic species and populations of this atoll. Most research has focused on the terrestrial environment—specifically the birds, giant tortoises, and invertebrate communities.

The unique marine environment has been comparatively understudied, with only about 25 percent of the published scientific work involving these ecosystems. This shortcoming is minor, however, compared with the conservation concerns of many other imperiled biomes worldwide. No doubt the Aldabra atoll will persist into the future, due to its conservation designation and its importance to scientific inquiry.

Daren C. Card

Further Reading

BirdLife International. "Important Bird Areas Factsheet: Aldabra Atoll." 2012. http://www.birdlife.org/datazone/sitefactsheet.php?id=6799.

Newberry, D. McC. *Numerical Classification of Mixed-Scrub Vegetation on Aldabra Atoll.* Washington, DC: Smithsonian Institution, 1981.

Rainbolt, Raymond E., et al. "Greater Flamingos Breed on Aldabra Atoll, Republic of Seychelles." *Wilson Bulletin* 109, no. 2 (1997).

Seychelles Islands Foundation. "Aldabra." 2012. http://www.sif.sc/index.php?langue=eng&rub=4.

United Nations Educational, Scientific and Cultural Organization (UNESCO). "Aldabra Atoll." http://whc.unesco.org/en/list/185.

Verheyen, Roda. *Climate Change Damage and International Law: Prevention Duties and State Responsibility.* Leiden, Netherlands: Brill Academic Publishers, 2005.

World Wildlife Fund. "Afrotropics–Afrotropics (AT1301)." 2012. http://www.worldwildlife.org/science/wildfinder/profiles//at1301.html.

Allegheny Highlands Forests

Category: Forest Biomes.
Geographic Location: North America.
Summary: The Allegheny highlands are home to a series of unique forest flora and fauna, dominated by Eastern hemlock and American beech, found throughout the Allegheny Plateau of Pennsylvania and New York State.

The Allegheny highlands forests are a temperate to boreal hardwood biome composed largely of two dominant tree species: the Eastern hemlock (*Tsuga canadensis*) and American beech (*Fagus grandifolia*), and this forest type is well represented in the Allegheny National Forest. Much of this forest was cleared in the early 20th century, leaving relict pockets throughout northeastern Pennsylvania. Furthermore, logging left behind large piles of slash and debris, which piled up and led to catastrophic fires that tended to promote early-successional species such as aspen rather than late-successional species like Eastern hemlock. Because this forest is now restricted to such a small area—less than 1 percent of its original range—careful attention should be focused on conservation strategies.

The topography of this biome is characterized by both glaciated and unglaciated sections, which in turn influence the distribution and composition of the vegetation. The glaciated section's bedrock is composed of sandstone, siltstone, and shales; this topography features rounded hillsides, ridges, and valleys. The unglaciated section contains sharper topographic relief; the bedrock is similar but also includes limestone parent material.

The Allegheny highlands forest community is found in the Allegheny Plateau in New York and Pennsylvania, specifically in the mountain ranges of the Catskills, Poconos, Finger Lakes, Allegheny, and Appalachians. Also found here are some unusual habitats such as high-elevation wetlands and shale barrens. The climate, characterized by warm summers and cool winters, is considered to be a humid continental climate. The soils are largely acidic, which is preferential for hemlock and beech tree species, and have thick organic layers that decompose slowly.

Vegetation Community

The two dominant tree species—Eastern hemlock and American beech—compose roughly 60 percent of all forest overstory tree species within the Allegheny highlands forests. Eastern hemlock is the most common tree species in approximately 2,471,054 acres (1 million hectares) of forest from the southern Appalachian Mountains north into southern Canada and west to the Great Lakes region. Strikingly, this abundant species is disappearing rapidly (within decades across its entire range) due to infestation by an exotic pest: the hemlock woolly adelgid (*Adelges tsugae*). Hemlock is a late-successional tree and an irreplaceable foundation species (a type of species that is the dominant primary producer in an ecosystem).

American beech is a shade-tolerant species that is slow-growing and prefers moist, well-drained acidic soils, making it a complementary species to the acidic environment commonly associated with the Eastern hemlock. American beech provides an important food source—beechnuts—for a variety of wildlife species. American beech is currently threatened by the beech bark disease complex, which is leading to a decline in its abundance and importance within the Allegheny highlands forests.

Associated vegetation forest types include sugar maples (*Acer saccharum*), which in some parts will replace Eastern hemlock as the dominant overstory species, leading to beech–maple stands, which are more commonly found in drier soils. Associated tree species within the Allegheny highlands include red maple (*Acer rubrum*), yellow birch (*Betula allegheniensis*), black birch (*Betula lenta*), white ash (*Fraxinus americana*), and black cherry (*Prunus serotina*). These associated species are found both in hemlock–beech stands and beech–maple stands, and generally are indicative of nutrient-rich soils.

Today, the Allegheny forests where this relict forest community still exists are more often composed of black cherry and maple trees than of the hemlocks and beeches that dominate the Allegheny highlands. The understory community within these forests largely consists of shade-tolerant herbaceous species such as Canada mayflower (*Maianethemum canadense*), partridgeberry (*Mitchella repens*), wintergreen (*Gaultheria procumbens*), and lady's slipper (*Cypripedium acaule*), accompanied by hardy ericaceous shrubs such as mountain laurel (*Kalmia latifolia*) and highbush blueberry (*Vaccinium corymbosum*).

Important Biota

The Allegheny highlands support a wide variety of terrestrial and aquatic organisms. Important plant species include Fraser's marsh, St. John's wort (*Triadenum fraseri*), bunchberry (*Cornus canadensis*), Catawba rhododendron (*Rhododendron catawbiense*), and the flame azalea (*Rhododendron calendulaceum*).

The forest is home to a variety of mammals, including the black bear (*Ursus americanus*), snowshoe hare (*Lepus americanus*), red and gray fox (*Vulpes vulpes, Urocyon cinereoargenteus*), beaver (*Castor canadensis*), mink (*Neovison vison*), and muskrat (*Ondatra zibethicus*). The white-tailed deer has played an important role in the development of the Allegheny highlands forest after logging and has strongly influenced the trajectory of succession and community structure

Tree Pests and Pathogens

The Allegheny highlands forest is severely threatened by invasive pests and pathogens. The two dominant tree species are both being widely affected by invasive organisms.

The Eastern hemlock is experiencing a widespread chronic infestation of the hemlock woolly adelgid and its co-occurring invasive, the elongate hemlock scale. Hemlock woolly adelgid is a sap-sucking insect that can infest a tree quickly, produce numerous offspring in a single generation, and lead to rapid hemlock mortality, or it can create chronic infestation over a decade as hemlock health declines.

The American beech is currently facing serious threats from beech bark disease, which is both fungal and insect-based. Trees present with cankers that girdle the tree and split open the bark, increasing the likelihood of infection and infestation by other pathogens and pests. Beech bark disease occurs in three temporal phases: the dispersal of the scale insect, which affects beech bark; the colonization of the bark by *Neonectria*; and the fungal decomposition phase.

by browsing preferentially. The timber rattlesnake (*Crotalus horridus*) is among the larger of many reptilian species supported by the Allegheny highlands forest. A suite of avian species depends on this forest for vital habitat and food resources, including migratory songbirds such as thrushes (*Catharus* spp.), warblers (*Dendroica* spp.), and raptors (*Accipitor* spp.). This forest also provides important habitat for insects including dragonflies, damselflies, moths, and butterflies.

Special Features

The Allegheny highlands are home to several unique places worth trekking out to, including several nationally designated areas including the Hickory Creek Wilderness, the Allegheny Islands Wilderness, the Allegheny and Clarion Wild and Scenic Rivers, the Allegheny National Recreation Area, the North Country Scenic Trail, and Buzzard Swamp. The Allegheny National Forest encompasses a variety of topographic features, and its protection ensures the protection of a vital watershed within the Allegheny region, thus promoting the protection of animal and plant communities that live within that region.

Human Threats

Humans can be a threat to the health and maintenance of the Allegheny highlands forests. Over the past century, the landscape was transformed, initially due to logging, then by targeted conservation efforts such as the Civilian Conservation Corps, and more recently by development. People have increased the habitat fragmentation within this forest due to road construction, development for recreational activities, and forest management for wood products. Furthermore, pollution from urban centers such as Chicago and Cleveland travels eastward toward the Allegheny highlands, negatively influencing plant and animal health and community resilience to environmental disturbances. The most recent harsh stress comes from the rush to drill and apply hydrofracture techniques to extract natural gas from the Marcellus Shale formation that underlies much of this biome.

Global climate change, induced by anthropogenic pollution over large scales, will also have devastating effects on this ecosystem as the vegetation and animal communities are exposed to unprecedented climate regimes, which will likely lead to species displacement over the long term; a warmer climate also leads to longer windows of seasonal opportunity for invasive insects and their linked bacterial and fungal disease threats.

Relena R. Ribbons

Further Reading

Braun, E. Lucy. *Deciduous Forests of Eastern North America.* Caldwell, NJ: Blackburn Press, 1950.

Eyre, F. H., ed. *Forest Cover Types of the United States and Canada.* Bethesda, MD: Society of American Foresters, 1980.

Kudish, Michael. *The Catskill Forest: A History.* Fleischmanns, NY: Purple Mountain Press, 2000.

Alps Conifer and Mixed Forests

Category: Forest Biomes.
Geographic Location: Western and west-central Europe.
Summary: An extensive European mountain forest environment with high biodiversity and a rich, long-lived cultural history, this biome is threatened by climate change and other human impacts.

The Alps conifer and mixed forests are temperate forests also known as European-Mediterranean montane mixed forests. This type of ecosystem contains temperate coniferous and mixed forests, with a geography that can include deep valleys, Mediterranean characteristics, or alpine characteristics depending on specific location within the Alps. The watersheds of three rivers, the Po, the Danube, and the Rhine, are prominent in this conifer and mixed forest ecosystem.

The geology of the range is relatively young when compared to other mountains. For example, the Great Smoky Mountains in the United States are between 200 and 300 million years old. The Alps were formed by Pleistocene glaciation within the period from about 2.6 million to 11,700 years ago. As a result of the glaciation, the Alps have a unique architecture, one that is responsible for much of the biological and ecological variation through the mountain range. The Alps mountain range runs from west to east through many of the countries in both western and eastern Europe, including Switzerland, Germany, Austria, Slovenia, France, Italy, Liechtenstein, and the Principality of Monaco. The total length of the range is about 750 miles.

The existence of conifer and mixed forests in this alpine region is clearly demarcated by what is known as the tree line, or timberline. The timber line is the elevation above which trees do not grow. In the Alps the timberline is between 5,700 and 7,200 feet (1,737 to 2,195 meters), though the exact elevation differs widely throughout the range. Tree line elevation is known to shift by slope, aspect, latitude, rain shadow, and other additional factors; thus there is no single, commonly accepted figure. Those trees that grow nearest the timberline are often subject to the harshest of the elements, and become dense and twisted by the extreme variations in climate. Generally the trees at timberline are of coniferous nature. These types of trees are known as *krummholz*, meaning "crooked wood" or "twisted wood" in German. In the European Alps, such krummholz trees are not only formed by the elements, but are also genetically predisposed to this type of stunted physiology.

Treasured Flora

Below the timberline, vast amounts of mountainous land are densely blanketed by conifer and mixed forests. Alps forests are renowned for having a high biodiversity of plants and animals. For example, 39 percent of the plant species in all of Europe can be found in the Alps. All together there are about 13,000 different species of plants distributed throughout the European Alps. These consist of many vascular plants, as well as fungi, mosses, and lichens. Of the vascular plants, 400 native species have been identified. This array of plant biodiversity has earned the region notoriety as a plant biodiversity hot spot. About 15 percent of the total forested land of the Alps is protected by legislation in order to mitigate damage to the forest ecosystem. The Global 200 Initiative of the World Wildlife Fund (WWF), the International Union for Conservation of Nature, and the Ramsar Convention have all recognized the Alps conifer and mixed forests as a wilderness area of great importance.

The primary deciduous, broadleafed tree species present in the Alps forests include beech (*Fagus* spp.), ash (*Fraxinus* spp.), sycamore maple (*Acer pseudoplatanus*), and oak (*Quercus* spp.). As angiosperms, these trees are called hardwoods; note that this designation is not necessarily related to the actual hardness of the wood. Alps deciduous forests are sparsely distributed; centuries of logging by inhabitants have taken a toll on these trees, and it is rare to find woodlands composed exclusively of deciduous trees. Coniferous tree species distributed throughout the Alps include larch (*Larix* spp.)—technically a deciduous conifer—pine (*Pinus* spp.), spruce, and fir. As gymnosperms, these trees are known as softwoods. Coniferous softwoods are the primary type of harvested forest tree here. Fortunately, much of Europe's timber demand is supplied by managed conifer plantations.

Characteristic Fauna

Some of the best-known animals in the Alps are the ungulates, or hoofed mammals, such as the chamois (*Rupicapra rupicapra*) and the alpine ibex (*Capra ibex*). The total number of Alps mammals, including these charismatic megafauna, is about 80. There are also 200 bird species, 21 amphibian species, only one of which, Lanza's Alpine salamander (*Salamandra lanzai*), is endemic (specifically adapted and unique to a particular biome). The Alps are home to 15 reptile species as well.

Recent reestablishment of wolves (*Canis lupus*) in the Italian and Western Alps has brought with it a number of ecological implications, as well as concerns about the integration of these animals into a landscape with human inhabitants. At the same time, the diverse geography found in the Alps conifer and mixed forests places physical limits on growing populations of wolves. In the Eastern Alps, for example, the presence of large lakes inhibits the movement of wolf populations, effectively placing a check on unmanaged growth.

Human Interface

Alps forests contribute numerous ecosystem services to the Alps region as a whole. Ecosystem services are those aspects of ecosystems that provide an intrinsic benefit to human health and quality of life. For forest ecosystems, these may be air filtration, water purification, erosion protection, biodiversity, carbon sequestration, and opportunities for recreation. Alps conifer and mixed forests also act as formidable protection against avalanche damage to alpine settlements. Portions of the Alps have prohibited or set limits on forest cutting activities and even access to the forest to reflect this important benefit. According to researcher Roland Olschewski, an example of this can be found in the Swiss municipality of Andermatt, Canton Uri, where residents have been banned from forest harvesting practices since 1397 C.E.

Forest ecologists have traditionally recognized a number of factors responsible for forest damage, with wildfire, pests, invasive species, and wind among them. The high elevation locations of Alps conifer and mixed forests means that avalanches are an additional and inevitable risk for both forests and humans. However, just as wildfire can serve a crucial function in forest regeneration and ecological succession, so can avalanche events play a similar role. As forest trees are uprooted by the events, space is opened in the forest canopy for new organisms to establish themselves. In addition, the disrupted landscape topography associated with these fallen trees dynamically affects forest soils and microbial ecology. Ecological succession is often instigated in this way as some individuals of an abundant species are removed or damaged, allowing for new growth to occur.

Much of the forested environment in the Alps is no longer pristine or untrammeled by humans; indeed much of it has been inhabited for thousands of years. While there are still portions of pristine and preserved wilderness, the Alps conifer and mixed forest landscape has undergone extensive change since the advent of human settlement in the region. With settlement came a pastoral way of life in which inhabitants survived through agriculture, herding, and grazing. This well-established cultural pattern appears to have begun shifting in the late 20th century and continues to this day. Current landscape change in many parts of the Alps seems to be pointed to a slow but steady reforestation of the region, rather than the steady deforestation that can be found in many other forests across the globe.

In the Swiss Alps, researcher P. P. Germann notes that between 1985 and 1995 there has been annual growth of between 0.5 and 1 percent in forested lands, amounting to roughly 3,860 to 4,050 square miles (1 million to 1.05 million hectares). Land abandonment, which has become increasingly common as small-scale farm operations become unprofitable, is largely responsible for this phenomenon. Land abandonment has also promoted a shift in the tree line as forest regenerates at higher elevations. Recent academic research has shown that some of this tree line shift has been found to be related to climate change effects in the Alps as well. If climate change effects are exacerbated over time, this physical shifting of the tree line will likely continue upward, potentially at increasing rates.

Forestry in Alps conifer and mixed forests has long been a way of life for the region's inhabitants. Traditional forest harvesting practices have all but been replaced by mechanization in many European forests. This response to recent globalization has forced many forest managers to produce at higher speeds and lower costs in order to stay in business. The economic advantages of modern agriculture and forestry have long been known: costs of labor are lower, and the speed with which trees can be removed and sold as timber is much more rapid.

While the use of modern forestry machinery has for the most part become standard practice, modern sustainable forest management has become increasingly prominent throughout the Alps as well. Austrian Federal Forests (AFF), which manages much of Austria's natural environment including the conifer and mixed forests, has a sustainable forest management plan in effect. In 2010, AFF also spearheaded a five-year biodiversity program to promote public awareness of biodiversity threats and implement protection measures.

Tourism and recreation are other important sources of income for Alps inhabitants. One side effect of tourism, however, is that the continued use of certain locations damages the ecosystem. Added together, the effect of tourism in many different locales puts an added burden on the environment. Climate change is predicted to negatively affect tourism in the Alps in the future, since alpine and winter sports depend on a very limited range of environmental conditions to operate. Ski hills need certain amounts of snow at regular intervals to attract visitors, and they need visitors to make a profit. Without the first factor in place, owners and operators will be hard pressed to maintain their facilities.

Physical and political boundaries often intersect, making it difficult for lawmakers and natural resource professionals to adequately manage all resources across all boundaries. This is particularly true in the case of Alps conifer and mixed forests spread across many countries. Often countries are able only to implement domestic policies regarding forest use and protection. In 2007, regional national parks legislation was established in Switzerland to help manage the Swiss landscape, and also incorporate economic considerations and sustainable management. This initiative has attempted to synthesize land use within a single, coherent policy that is more easily implemented.

Pressures on the biological and ecological hallmarks of this region, which threaten to also disrupt its unique European culture, are showing the need for increased environmental protection. Perhaps the greatest danger posed to the Alps conifer and mixed forests is climate change. Climate change impacts, most of which are unknown or at least uncertain in the long-term future, may greatly affect the ecology of the conifer and mixed forests. Considerations of climate change will likely place a heavy burden on lawmakers and natural resource managers in their efforts to manage its effects on some of Europe's most extensive forests.

CHRISTOPHER PETERS

Further Reading

Gerber, J. D. and P. P. Knoepfel. "Towards Integrated Governance of Landscape Development: The Swiss Model of Regional Nature Parks." *Mountain Research and Development* 28, no. 2 (2008).

Germann, P. P. and P. P. Holland. "Fragmented Ecosystems: People and Forests in the Mountains of Switzerland and New Zealand." *Mountain Research and Development* 21, no. 4 (2001).

Marucco, F. F. and E. B. McIntire. "Predicting Spatio–temporal Recolonization of Large Carnivore Populations and Livestock Depredation Risk: Wolves in the Italian Alps." *Journal of Applied Ecology* 47, no. 4 (2010).

Olschewski, R., P. Bebi, M. Teich, U. Wissen Hayek, and A. Grêt-Regamey. "Avalanche Protection by Forests—A Choice Experiment in the Swiss Alps." *Forest Policy & Economics* 17, no. 19–24 (2012).

Spinelli, R. and N. Magagnotti. "The Effects of Introducing Modern Technology on the Financial, Labour and Energy Performance of Forest Operations in the Italian Alps." *Forest Policy & Economics* 13, no. 7 (2011).

Amazon River

Category: Inland Aquatic Biomes.
Geographic Location: South America.
Summary: The world's largest river is a diverse ecosystem providing a home to numerous species found nowhere else.

The Amazon is the largest river by flow volume in the world, accounting for one-fifth of the world's total river flow. Accordingly, it has the largest drainage basin—about 2.7 million square miles (6,992,968 square kilometers) in area. It is so large that if accounted independently, two of its 1,000 tributaries—the Rio Negro and the Madeira River—would be among the 10 largest rivers in the world. Nearly one-sixth of all the freshwater that drains into an ocean passes through the Amazon. Its width varies from one to six miles (two to 10 kilometers) when the river is low, expanding beyond 30 miles (48 kilometers) in the wet season.

The Amazon begins in the Andes Mountains and flows for about 4,000 miles (6,437 kilometers) through South America before entering the Atlantic Ocean in an estuary some 150 miles (241 kilometers) wide. Interestingly, at no point is the Amazon crossed by a bridge—not because of its width, which has been traversable by modern engineering for a century, but because so much of the Amazon passes through rainforest that there is little demand for a crossing. The river is about 11 million years old and has had its current shape for about 2.4 million years, according to a 2009 study of sediment columns.

The Amazon's size leads to habitats found nowhere else. The Piramutaba catfish, for example, one of the larger catfishes in the Amazon, migrates more than 2,000 miles (3,219 kilometers) from its nursery in the Guianan–Amazon mangroves to its upper-Amazon spawning grounds. The Amazon Basin is home to as many as 25 percent of the world's terrestrial species, and its flora accounts for about 15 percent of the world's land-based photosynthesis activity. Massive numbers of species remain unidentified. For a long time, the idea persisted that the white-water rivers of the Amazon were plentiful with fish, whereas the darker waters like the Rio Negro were "hunger rivers," void of most life. It has become clear that this is an oversimplification and that the black rivers are home to significant turtle populations, in addition to supporting fisheries.

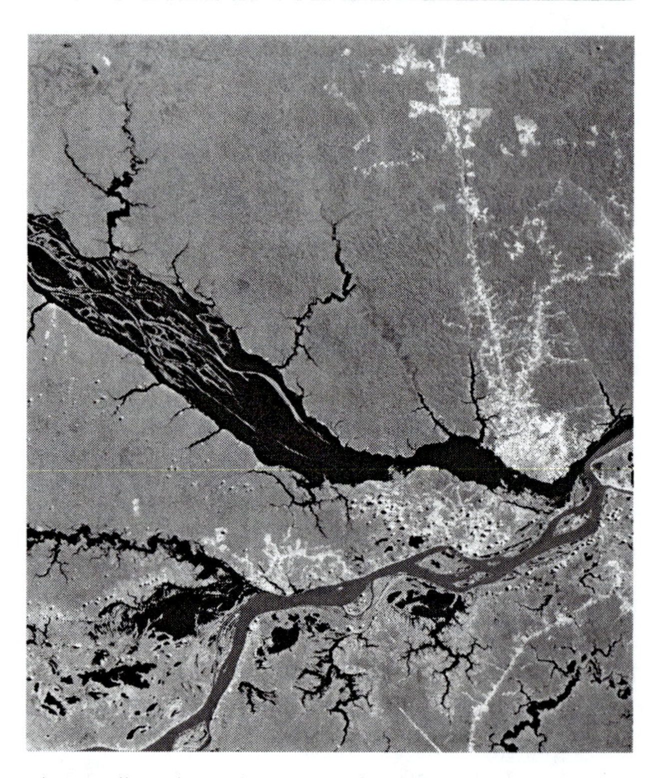

This satellite photo shows two of the Amazon's 1,000 tributaries, the Rio Negro and the Rio Solimões, meeting near Manaus, Brazil, to form the Amazon River. (Thinkstock)

Temperature, rainfall, and climate vary throughout the large region, but it is generally warm and humid. Because the wave and tidal energy of the massive river is sufficient to carry most of its sediments to sea, it doesn't form a true delta. Instead, it empties directly into the turbulent Atlantic Ocean, which rapidly carries the silt away.

The Amazon is joined to the Orinoco River basin by the Casiquiare canal. The Casiquiare is a distributary of the Orinoco and flows into the Rio Negro, one of the Amazon's tributaries.

Surrounding Forests

Terrestrial ecosystems in the region include the moist broadleaf forests of the Rio Negro Campinarana, Iquitos Varzea, Gurupa Varzea, Marajo Varzea, Purus Varzea, and Monte Alegre Varzea, all of which are classified as threatened or endangered. Dramatic topographical variations have led to a rich diversity of Amazonian life, and for a long time, public discussion of biodiversity and conservation implicitly or explicitly centered on the Amazon. In the south, a dearth of human settlement and roads have kept the habitat intact. In the north, logging, mining, and increased settlement have posed hazards. The south has the greatest endemic richness, with 11 endemic mammal species (of about 250) and 17 endemic bird species (of nearly 800).

The forests are full of plants that have significant commercial value, including mahogany (*Swietenia macrophylla*), balsam (*Myroxylon balsamum*), rubber (*Hevea brasiliensis*), strychnine (*Strychnos asperula*), and tagua nut (*Phytelephas microcarpa*). The south has the greatest number of palm species.

Widespread Amazonian fauna include tapirs (*Tapirus terrestris*), jaguars (*Panthera onca*), capybaras (*Hydrochoeris hydrochaeris*), and kinkajous (*Potos flavus*), as well the endangered woolly monkey (*Lagothrix lagotricha*), giant otter (*Pteronura brasiliensis*), giant anteater (*Myrmecophaga tridactyla*), and ocelot (*Leopardus pardalis*). The giant otter was once common throughout most of South America, but its population has declined to as little as 2,000, primarily in the protected areas of the Amazon Basin, as a result of being hunted for its fur.

Amazonian deforestation has been a concern for centuries. The Amazonian rubber boom contributed substantially to deforestation in the late 19th century and throughout the 20th century, due to the introduction of the pneumatic tire and the many commercial uses for rubber. Today, cattle ranchers are responsible for much of the deforestation, as demand for beef began a swift rise in the mid-20th century (principally due to the growing worldwide middle class—an important customer base—and to improvements in refrigeration and transport that expanded beef markets) and has stayed steady ever since.

The increasing demand for biofuel and other soy products has also increased deforestation by soybean farmers; in some areas, other livestock farmers and miners are the primary threats to the forest. Highway construction and land speculation, especially in the Brazilian stretch of the Amazon, are certainly factors, but outside Brazil, the problem of deforestation is thornier in its social and economic implications, because the laborers responsible for it are so often poor and deeply dependent on whatever work has displaced the forest. Each of these deforestation vectors has seen acceleration in recent decades—yielding an increasing stress level on the essential biology of the Amazon rainforest and other riverine ecosystems.

Even apart from the destruction of forest habitats and the various pollutants introduced by the purpose to which the deforested land is put, deforestation affects the greater ecosystem by impacts on the water cycle. The conversion of forest to pasture or farmland leads to water running off into the river without being recycled through the trees; deforestation contributes to drought and wildfires, and reduces the rainfall that is critical to the water supply of the heavily populated parts of Brazil and Argentina. These direct effects of deforestation are seen in turn to contribute to global warming—both by the increased emanation of greenhouse gases from the industrial activities and more fires, and by depleting the carbon-absorbing power of the forest base. The feedback loop is completed when higher global air temperature and dislocation of normal precipitation patterns contribute to drought and drives up the spread of wildfires.

Upper Amazon Rivers and Streams

The major aquatic habitats of the Amazon are located in the Upper Amazon Rivers and Streams ecoregion, consisting of the muddy white-water tributaries in the west of the basin and the more-nutrient-poor black rivers of the Guiana Highlands. Hundreds of endemic (uniquely evolved to fit a locally biome niche) fish species can be found in this zone of the nearly 1,500 total fish species found throughout the Amazon. They include many ostariophysan fishes, including species from the siluroid catfish (*Siluridae*), minnow (*Cyprinidae*), and charcin (*Characidae*) families. Ostariophysan

This fisherman was hunting piranha on Lake Iranduba in the Amazon region near Manaus, Brazil. Of the 20 species of piranha in the Amazon River, only five are actually a threat to humans. (World Bank)

fishes are noteworthy for the way their gas-filled swim bladders amplify sound waves in water and transmit them to interear vertebrae, enhancing their sense of hearing.

Other fish species in the region include piranhas, electric eels, and Loricariid catfish. The Napo River alone is home to more than 500 species of fish. The piranha is perhaps the most famous Amazonian fish, and the razor-sharp teeth responsible for that fame have been used as cutting tools by Amazonian tribes for millennia. Less famous is their importance to the ecosystem, as they eat both dead fish and dead animals whose decomposition would otherwise pollute the river. Only five of the river's 20 piranha species are dangerous to humans. The shallow waters are home to the anaconda, one of the largest snake species, which remains low in the water, with just its nostrils above the surface.

The Upper Amazon is one of the most jeopardized regions of the Amazon. The introduction of hydroelectric dams, while providing a relatively green source of power, has modified river flow and altered the movements of migratory fish species. Deforestation during construction has increased erosion of riverbanks, and when the cleared land is converted to pasture or farmland, it may contribute contaminants to the water in the form of fertilizer, pesticide, animal feces, detergents, and other pollutants.

A more recent threat is the Andes gold rush. Once the base of the Andes, where the Amazon begins, was a pristine environment even as other parts of the rainforest were threatened by deforestation. Now humans have dug thousands of mines there in search of gold. Though ranching and logging account for a greater amount of deforestation, gold miners—who have destroyed perhaps 100,000 acres (40,469 hectares) of rainforest in Peru so far—burn the forest in which they work, including trees more than a millennium old, and strip away the Earth's surface to a depth of 50 feet (15 meters). This damage is much harder to recover from than that of slash-and-burn agriculture. The mercury used in recovering gold from silt leaches into the watershed and is eventually taken up by Amazonian fish.

Most of the gold mines in Peru (as many as 98 percent) are illegal, making regulation a difficult remedy to apply. On a typical day, let alone a lucky one, a barely skilled laborer–turned–gold miner can earn twice as much as he or she would make in a month at his or her old job. The skyrocketing cost of gold has led many people to leave their old jobs and join into informal bands of miners who divide their findings among themselves. Other mine workers are teenagers, sold into servitude by their impoverished parents. Legal, commercial operations are larger in scale, with a typical mine removing 16 dump-truck loads of rock and soil every hour, 18 hours a day. Though legal operations are required to pay environmental remediation fees, the fees do nothing to stop the changes to, and destruction of, the local Amazonian ecosystem.

Amazonian River Dolphins

The river dolphins are perhaps the best known of the Amazon's endemic species. In Portuguese, they are called *botos*, a name that has come to refer to the paraphyletic clade of freshwater dolphins found throughout South America. The pink river dolphin (*Inia geoffrensis*) can be found in the Amazon, Orinoco, and Araguaia river systems, all in South America. It has a hump instead of a fin and pink skin due to the proximity of capillaries to the surface of the skin. It feeds on fish, small turtles, shrimp, and crustaceans. The pink river dolphin is abundant in the flooded forests and is well adapted to using its long flexible neck and long snout to hunt through hollow logs and thick patches of submerged vegetation to find fish. When the floodwaters recede, the dolphins locate themselves in the freshwater lakes of the forest or return to the main river channels.

Easily spotted because of their pale pink skin and the relative proximity of the flooded forests to Amazonian population centers, the dolphins have been the subjects of numerous legends in Amazonian cultures, including the folk belief in many cultures that river dolphins are shape-shifters who become human men at night and seduce human women.

River-dolphin organs, especially their eyeballs and genitals, are dried and used as powerful fetishes. Researchers have discovered that although most of the fetishes bought and sold in inland communities actually come from livestock, those sold closer to the delta are in fact taken from the gray river dolphin. Humans have easy access to corpses of gray river dolphins because of the frequency with which they are caught along with fish. Though these myths and magical uses may seem harmful to the river dolphin's health, it is thought that this belief in the dolphin's power—and the implication that it may be related, via those seductions, to the local human populations—has discouraged native Amazonians from hunting it and helped to keep populations stable. Many tribes consider it taboo to kill or eat a dolphin, even in times of famine. Other dolphin myths resemble the stories of fairies in Europe, including the myth of an underwater enchanted land to which a human swimming alone could be whisked away if he or she was not careful.

The gray river dolphin (*Sotalia fluviatilis*) or costero is found only in the Amazon Basin and is classed in the same family (*Delphinidae*) as oceanic dolphins. It most closely resembles the bottlenose dolphin and possesses an electroreceptive sense. It is usually found close to estuaries and inlets, in small groups of 10 to 30 that swim in tightly knit groups and show signs of a highly developed social structure. Breaching, hopping, somersaulting, and tail-splashing are common behaviors, but these dolphins are generally more skittish around humans than marine dolphins are.

As with bottlenose dolphins, infanticide has been witnessed in costero populations, in which a calf is separated from its mother and killed by a group of males. The behavior has rarely been observed—only twice with bottlenose dolphins and once with costeros—and the leading theory for now is that the calf is killed to make the female receptive to mating. Costeros are commonly caught in coastal fishing nets and sometimes killed for use as shark bait. Their predators are the bull shark and the orca.

Amazonian Flooded Forests

The Amazon's cycle of flooding leads to the most extensive system of floodplain habitats in the world, as the river rises more than 30 feet (nine meters) during the wet season. The flooded forested areas of north-central South America encompass an area about twice the size of California. These areas are home to a large number of freshwater fish, reptiles, and other aquatic fauna that migrate into the newly flooded area, including two species of the freshwater dolphins characteristic of the Amazon. Floodplain lakes and floating meadows are replenished, and freshwater fish feed on fruit dropping from the trees of the flooded forests. Indeed, the reproduction cycle of many floodplain trees depends on their fruit being eaten, and seeds dispersed, by these fish.

Floodplain fish include the tambaqui (a fruit eater), arawana, pirarucu, arapaima, dourada catfish, tucunare, and *Lepidosiren paradoxa,* one of the few lungfish to evade extinction. Four species of the threatened uakari monkey (genus *Cacajao*) are among the floodplain mammals, as well as the pink and gray river dolphins and the manatee. The black caiman (*Melanosuchus niger*) is a local member of the alligator family; once nearly extinct due to hunting, the species has partially recovered its population, and it is classified as Conservation Dependent. The caiman depends primarily on fish for its diet, but larger adults will feed on mammals like tapirs and deer that come to the riverbank to drink. Jaguars prey on the juveniles.

The endangered Arrau turtle (*Podocnemis expansa*), the largest side-neck turtle, also migrates into the flooded forests, feeding on plants. In the dry season, the turtles travel in large numbers to find nesting areas, laying clutches of eggs in sandbanks. Most of the young are consumed almost immediately by predators; the survivors migrate to the floodplains once more.

The annual action of waters flowing into and then out of the floodplains contributes some of the Amazon River's trace mineral content. A 2004 study of the mineral content of the river, lakes, and leaves of the Ilha de Marchantaria forest affirmed that concentrations of manganese and copper could not be explained by tributary mixing or in-stream processes, and were likely the result of sediment–water and plant–water interfaces. Though the results were less clear, the same could be true for concentrations of iron, aluminum, and rubidium. It may be that some of the plant species endemic to the region in fact depend on or benefit from these trace minerals, with which they would not otherwise be provided.

Lateral exchange with the floodplain also has significant effects on the carbon abundance and oxidation of the mainstream. The rate of respiration in the river is high, supported by organic material. Carbon-processing exchanges are most active at the places where floodplain waters interact with the mainstream water, which have a carbon and water mass balance anomaly. Most respiratory carbon dioxide (CO_2) is lost to the atmosphere, as is common in riverine ecosystems.

The flooded forests are jeopardized by logging of the virola (also known as epena, patricia, and cumala), used by tribes in the Amazon and Orinoco Basins for its resin, which includes hallucinogenic alkaloids such as the most potent forms of DMT. Another threat is logging of the kapok (*Ceiba pentandra*), the seed pods of which produce large amounts of lignin- and cellulose-rich fiber with numerous commercial applications.

Bill Kte'pi

Further Reading

Bayley, Peter B. "Understanding Large River-Floodplain Ecosystems." *BioScience* 45, no. 3 (1995).

Eisenberg, J. F. and K. H. Redford, eds. *Mammals of the Neotropics: The Central Neotropics.* Chicago: University of Chicago Press, 1999.

Foley, Jonathan A., et al. "Amazonia Revealed: Forest Degradation and Loss of Ecosystem Goods and Services in the Amazon Basin." *Frontiers in Ecology and the Environment* 5, no. 1 (2007).

Godar, Javier, Emilio Jorge Tizado, and Benno Pokorny. "Who is Responsible for Deforestation in the Amazon? A Spatially Explicit Analysis Along the Transamazon Highway in Brazil." *Forest Ecology and Management* 267, no. 58 (2012).

Henderson, A. *The Palms of the Amazon.* New York: Oxford University Press, 1995.

Richey, Jeffrey E., et al. "Biogeochemistry of Carbon in the Amazon River." *Limnology and Oceanography* 35, no. 2 (1990).

Smith, Nigel J. J. *Amazon Sweet Sea: Land, Life, and Water at the River's Mouth.* Austin: University of Texas Press, 2002.

Thorp, James H. "The Riverine Productivity Model: An Heuristic View of Carbon Sources and Organic Processing in Large River Ecosystems." *Oikos* 70, no. 2 (1994).

Viers, Jerome, et al. "The Influence of the Amazonian Floodplain Ecosystems on the Trace Element Dynamics of the Amazon River Mainstem." *Science of the Total Environment* 339, no. 1–3 (2005).

Amu Darya River

Category: Inland Aquatic Biomes.
Geographic Location: Asia.
Summary: The Amu Darya is the longest river in central Asia. Its lower reaches have been so negatively affected by human engineering that it now can be regarded as an ecological tragedy.

The Amu Darya is one of the most iconic rivers of the Old World, shimmering like the Nile, Euphrates, and Tigris Rivers. Known to the Greeks as the Oxus and to the Arabs as the Gihon or Jeihoon (one of the four rivers of paradise), the Amu Darya rushes down from the high peaks of the Pamir, Hindu Kush, and Alayskiy mountains—all part of the western portion of the Himalayas—and gently flows in a northwesterly direction through the central Asian lowlands toward the Aral Sea, which was the world's fourth-largest inland sea before it was strangled in recent decades by irrigation dams, canals, and reservoirs. Its sister river, Syr Darya, flows on a course roughly parallel to the Amu but drains a much smaller volume of water into the northern section of the sea.

Human beings have lived along both rivers for more than 3,000 years. Alexander the Great, Zoroaster, Genghis Khan, and the Russian czars spent considerable time exploring these environs at a time when they were least transformed by culture. The ecoregions within the desert biome that set the conditions that give life to the Amu Darya have changed dramatically, and so have the long-term ecological characteristics defining this once-celebrated river.

The Darya (the Persian word for *river*) emerges within a dry and desert biome, which is subdivided into eight large-scale ecoregions. The 1,600-mile-long (2,575-kilometer-long) Amu, named after the medieval city of Amul, has a drainage basin of 193,051 square miles (500,000 square kilometers) that joins three of these regions: Tropical/Subtropical Desert Altitudinal Zone, Tropical/Subtropical Steppe Altitudinal Zone, and Temperate Desert Province Mid-Latitude. The climate and landforms of each region enable certain kinds of plants and animals to thrive, including those in rivers, seas, and other water bodies. It is this interrelationship that defines the biogeography of the Amu Darya, yet human cultures' participation in this relationship should be emphasized to understand the present ecological condition of the river.

Although the location of the source of the Amu Darya is debatable, most geographers agree that the river begins at the confluence of the Pandj and Vakhsh Rivers, on the southwestern border of Tajikistan. These tributaries, along with others joining the main channel, flow from the highest reaches of the western portion of the Himalayas in what is called a Tropical/Subtropical Desert Altitudinal Zone, where melting alpine glaciers and snow provide up to 80 percent of the Amu's water. This cold, arid, high-altitude region has little vegetation, mainly lichens and mosses that grow on boulders and rocks. Shrubs, herbaceous plants, and grasses can flourish within the region, but at lower altitudes and often in small patches. Meadows form; trees such as juniper and poplar take root at still lower altitudes. In the vicinity of the confluence of the Amu's major tributaries, where the average height of the mountains has decreased significantly and the general location of the Amu and its tributaries trends slightly southward, the river becomes part of the Tropical/Subtropical Steppe Altitudinal Zone. The transition is almost imperceptible.

Dam Construction

The rivers of the altitudinal regions move swiftly, from high peaks to low, through Afghanistan, Kyrgyzstan, Tajikistan, and Uzbekistan, cutting V-shaped valleys into the landscape. Such natural architecture offers an ideal setting for the construction of dams for hydroelectricity, agricultural irrigation, and flood prevention. Because the economies of the altitudinal regions depend deeply on local agriculture, the water of the Amu is a precious resource and, therefore, a cause of contention. The building of dams anywhere in the mountain drainage basin—a surface area that constitutes about half the total area of the basin—is the most contentious issue.

By the end of the 20th century, a network of dams had been built across the Amu, but mostly beyond the mountains. During the era when most of central Asia was under the control of the Soviet Union, however, the tallest of these dams—in fact, the tallest in the world at 984 feet (300 meters)—was erected on the Vakhsh River at Norak (Nurek), Tajikistan. Another hydroelectric dam, the Rogan project, was in the works on the same river until the fall of the Soviet Union. Now an independent Tajikistan intends to complete the project, and if successful, Rogan would be the tallest in the world. Yet as global warming reduces Himalayan glaciers, populations in host countries are increasing, and so is the demand for water.

Eventual peace in war-torn Afghanistan, a country that contributes more water to the Amu than any other, will also place heavy demands on the river as restarted economic development will carry the need for hydropower, irrigation, and drinking water. For these reasons, governments including that of Turkmenistan are attempting to negotiate an arrangement that promotes fair and sustainable use of the river. If unsuccessful, these countries will lose more than the riverine political boundaries that separate them.

Desert Region and Evaporation

If the altitudinal ecoregions are the source of water nourishment, the temperate desert region is the grounds for depletion not only of water, but also of the river's biodiversity. When it reaches the temperate desert, where differences in seasonal temperatures can be extreme and water bodies turn to ice during the winter, the Amu becomes one of the world's most turbid rivers, suspending a high amount of sediment scraped from high-mountain-valley walls by the intense force of the river system's currents. Gradually, as the drainage basin levels, the Amu's current slows, and the channel not only becomes wider, but also divides and braids as deposition of sand and silt creates numerous intermittent islands, ponds, and lakes. The constant shifting of alluvium deposition makes the river hardly navigable. Endemic fish species, however, such as the Amu Darya sturgeon and its much smaller cousin, the shovelnose sturgeon, both of the caviar kind, still ply these muddy waters. Their presence, however, like that of other fish species (namely pike, trout, and roach), is becoming less evident.

Since the mid-20th century, evaporation, irrigation, and infiltration have been the primary influences on the life of the river. As the Amu flows toward and through the Kyzyl Kum and Kara Kum deserts, an area that receives about 2,500 to 3,000 hours of sunshine and less than four inches (10

A tractor waits beside the Amu Darya River in Uzbekistan in 2007. The river is threatened by growing demands for irrigation of cultivated land. Few forests remain along the river, with two percent of the forest being protected by reserves such as the Badai-Tugai in Uzbekistan. (Flickr/AudreyH)

centimeters) of rainfall annually, much of its water evaporates into the atmosphere. The barrage of dams, canals, and reservoirs diverts the channels mainly for irrigation of cultivated land. Earlier, the Soviets realized their goal of making central Asia one of the largest cotton-producing regions in the world, but they did so with unsustainable technology and practices. Canals seeped water, and the resulting flooding of large swaths of desert not only increased evaporation—thus turning floodplain and desert into infertile salt flats—but also layered the surface with agrochemicals that contaminated and grossly changed the Amu's biota.

Tugai Forests

The influences of flooding and agrochemical contamination together have had a devastating effect on one of the most unique riparian ecosystems in the world: the Tugai forests. In the temperate desert region of central Asia, these deciduous forests are like oases lining the banks of rivers and narrowly expanding across floodplains. Trees such as willows, buckthorn, poplars, oleasters, and tamarix are stabilizers around which grow lianas, as well shrubs such as briar roses and honeysuckles. Interspersed throughout are herbaceous meadows, tussocks of reeds, and meadow-swamp wetlands.

These verdant sanctuaries are habitats for mammals including the dormouse, badger, wild boar, jackal, Tolai hare, and tamarix gerbil. Less common are the otter, Bukhara deer, hyena, and goitered gazelle. The Caspian tiger, once positioned at the apex of the food chain, has not been seen for more than 50 years. Among common amphibians and reptiles are the Anatolian frog, central Asian cobra, Caspian turtle, and Turkestan gecko. Common birds of the Tugai are shikra, pheasant, scops owl, kestrel, penduline tit, and cuckoo. In wetland ecosystems, common species are the pelican, cormorant, heron (with the rarer aquatic species being gulls), white-headed duck, marbled teal, and flamingo. The wetlands offer temporary accommodations for more than 150 species of birds, most flying in from Kazakhstan and western Siberia to winter, nest, or rest and recoup before migrating onward.

At the beginning of the 20th century, the gallery-like forests along the Amu Darya once covered an area of around 200 square miles (518 square kilometers). At the beginning of the 21st century, all that remains is 20 square miles (52 square kilometers), and in that area, only a few forests are intact and flourishing.

Water pollution and depletion have dramatically changed the habitat and the river's ecosystem to the point where both are now nearly devoid of their most common denizens: birds and fish, respectively. At present, just two percent of the forest along the Amu is protected in several small preserves, the largest of these being the Badai-Tugai and Kyzyl-Kum Reserves, both in Uzbekistan.

Tugai forests, reed jungles, lush lakes and bogs, and swamps enclosed between sand dunes were once characteristic of the distributaries that formed the Amu Delta. Salt- and drought-resistant plants now dominate the dry and gritty landscape, which is deflating for lack of water. Of the 178 animal species that have evolved with this ecosystem, only 38 have survived the recent rapid changes imposed on this ecoregion.

The mid-latitude westerly winds still blow over the delta and Aral Sea, but they offer little moisture for the thirsty winds that have always carried water across the deserts and into the majestic Himalayas, where it falls as rain and snow, streams into the upper Amu and its tributaries, and then begins its journey back to the sea. The winds are drier now, carrying salt and pesticides. The shimmering Oxus no longer reaches the sea.

Ken Whalen

Further Reading

Colegrave, Bill. *Halfway House to Heaven: Unravelling the Mystery of the Majestic Oxus.* London: Bene Factum Publishing, 2011.

Shahgedanova, Maria. *The Physical Geography of Northern Eurasia.* Oxford, UK: Oxford University Press, 2003.

Veregin, Howard. *Rand McNally. Goode's World Atlas.* 22nd edition. Upper Saddle River, NJ: Prentice Hall, 2009.

Amur River

Category: Inland Aquatic Biomes.
Geographic Location: Asia.
Summary: The Amur River and its tributaries are a biologically diverse ecosystem, but economics and international clashes have threatened the environment in a variety of ways.

Known as the "Russian Mississippi" because of its size and importance within northeastern Asia, the Amur River Asia is 1,800 miles (2,897 kilometers) long. The tenth-longest river in the world, the Amur is one of the largest free-flowing (undammed) rivers in the Eastern Hemisphere. The inclusion of the Amur's headwaters stretches its length to 2,700 miles (4,345 kilometers). Generally flowing toward the southeast, the Amur River is formed by the confluence of the Shilka and Argun Rivers along Russia's Siberian border with China. Known as the Heilongjiang or Black Dragon River on the Chinese side of the river, the river has historically served China as a conduit for cultural diffusion.

From the region east of Lake Baikal in Russia, the Amur flows eastward toward the Pacific. After reaching the Chinese border, the Amur heads northeastward, where it traverses 650 miles (1,046 kilometers) to the Tartar Strait in the Pacific Ocean, connecting the Sea of Japan with the Sea of Ohotsk. The Amur has several branches that have formed into numerous floodplains, creeks, and oxbow lakes. The lands surrounding the Amur are made up of desert, steppe, tundra, and taiga.

The major tributaries in the northern portion of the Amur are the Zoya, Bureya, and Amgun Rivers. In the south, the main tributaries are the Sungari and Usssuri, which also flow along the Russia–China border. The major ports of Khabarovsk, Komsomolsk, and Niboayevsk are all located within Russia. The Heilongjian (Amur) and Kaluga sturgeon that are used in caviar live in the waters of the Amur, making the river a valuable commercial fishery asset. For much of the year, the frozen water is nonnavigable, but from May to November, the entire river is open to travel. The Amur River is being threatened by overfishing and population depletion, yet much of the Russian area remains basically undeveloped in the 21st century.

A six-week-old Amur leopard, one of the most critically endangered animals in the Amur River region. There may be as few as 30 left in the wild. (Thinkstock)

Rich Base of Wildlife

Millions of years ago, molten lava was pushed up into the Amur River, forming large boulders throughout its waters. Those boulders in some places served as a canvas for prehistoric rock carvings; today's human population has made its own mark on the river, not always a positive one. The waters of the Amur River are filled with a mixture of temperate and northern flora and fauna, but the ecosystems have been threatened by human greed and a lack of agreement over allotted resources. International nongovernmental organizations (NGOs) have been instrumental in protecting the fragile environment of the Amur.

Some experts suggest that only the Mississippi River has a greater wealth of biodiversity than the Amur River. Some 95 percent of the entire world population of the Oriental white stork live there, along with 65 percent of red-crowned cranes and half of all white-napped cranes. At least 108 fish species live in the waters of the Amur. Permanent residents include the grass carp, black carp, sky gazer, snakehead, tamien, mirror carp, Amur sturgeon, and Great Siberian sturgeon. Salmon spawn by the river and its tributaries.

Although the lands around the Amur are considered to be a horticultural paradise, much is still unknown about the plant life that is native to the area, which has a climate similar to that of the American Midwest. Known flora include the purpleblow maple, Manchurian stripebark maple, Korean maple, Amur cork tree, willow, paper birch, Mongolian oak, and conifers such as the Korean pine. Flowers are abundant, including blue cranesbill, campion, white bog orchid, Siberian iris, ladies' bedstraw, cotton grass, bell flower, day lilies, and gooseneck loose strife.

Human Pressures

Northward and eastward of the Amur River and its tributaries is the Russian Far East, one of the largest wilderness areas in the world and home to a large variety of ecosystems. With few major human settlements to upset the balance of nature, the area is home to 15 species of cranes, almost half of the world population of wild Pacific salmon, and some 400 Siberian tigers. About a fourth of the world's forest cover is found here.

To the south, the story is far different. Some 75 million people live in northeast China's Amur basin. Because of rapid deforestation since the end of the 20th century, small border towns have grown into industrial areas that prosper from a largely illegal timber trade. That timber is sold to retailers such as Walmart that buy everything from baby cribs to candleholders from the Chinese manufacturers, whose efforts have denuded the forests along this stretch of the Amur. China now ranks as the world's top exporter of wood products, but is paying a stiff price for such practices. Soil erosion and deforestation are considered to be major challenges now. On the other hand, when China banned commercial logging in 17 provinces, Russia launched its own timber boom. By 2009, China was importing 5 billion pounds (2.3 billion kilograms) of wood per year from the Primorski Krai region of Russia. At least half of that wood may be produced by illegal logging practices.

Historically in the region, the 17th-century Amur was a magnet for adventurers, vagabonds, and river pirates intent on benefiting from the wealth of beaver, Arctic foxes, and sables that were there for the taking along the shores. Furs at one point accounted for one-third of the treasury assets of the Russian Empire. Peasants also were attracted to the Amur River because they believed that its rich resources could bring an end to their poverty.

By 1689, Russia had been forced to cede the area around the Amur to the Chinese by the Treaty of Nerchinsk. On the Russian side of the river, the land settled into obscurity, and the Amur lay virtually forgotten, with its ecosystems virtually untouched by humans. It was "rediscovered" two centuries later and was seized on as a vital link to the Pacific Ocean despite its harsh climate, with long, bitter winters and short, mosquito-laden summers. In 1858, Russia recovered the left bank under the Treaty of Aaigun. By the 1970s, Russia and China had become entrenched political rivals, and border conflicts frequently broke out. A modern agreement over the Amur was finally reached in 2005.

Elizabeth Rholetter Purdy

Further Reading

"The Amur's Siren Song." *The Economist* 393, no. 8662 (2009).

Bassin, Mark. *Imperial Visions: Nationalist Imagination and Geographical Expansion in the Russian Far East, 1840–1865.* Cambridge, UK: Cambridge University Press, 1999.

Federman, Adam. "Border Lands." *Earth Island Journal* 23, no. 4 (2009).

Ivanov, Grigoriĭ Ivanovich. *Classifications of Soils of Primorye and Amur River Region.* Washington, DC: United States Department of Agriculture, 1973.

Maxwell, Neville. "How the Sino-Russian Boundary Conflict Was Finally Settled." *Critical Asian Studies* 39, no. 2 (2007).

Anatolian Conifer and Deciduous Mixed Forests

Category: Forest Biomes.
Geographic Location: Asia Minor.
Summary: Anatolian conifer and deciduous mixed forests are differentiated across elevations;

much of this once-diverse habitat has been heavily affected by human use.

The Anatolian conifer and deciduous mixed forests that once comprised a wealth of temperate zone oak- and pine-dominated mountain landscapes have been stressed and broken up by centuries of human activity. Anatolia, or the Asian portion of Turkey, is bounded by the Black Sea to the north, the Mediterranean to the south, and the Aegean to the west. Its land borders touch the nations of Georgia, Armenia, Iran, Iraq, and Syria. The northern Anatolian Mountains are a key formation affecting local climate and habitat; another is the presence of isolated volcanic cones greater than 9,800 feet (more than 3,000 meters) in elevation in central and eastern Anatolia.

Anatolia is often broken into three climatic regions with associated forest types: northern Anatolia; southern Anatolia; and inner, eastern, and southeastern Anatolia. The north has a humid and cold–humid climate, with precipitation abundant throughout the year but at a minimum in early summer. The western and southwestern tracts have a Mediterranean climate with little rain in spring and winters tending to be wet. The inner, eastern, and southeastern sections share a semi-arid climate with most of the limited precipitation coming in late spring.

Due to the variable landscapes of Anatolia, various soil types occur, with those in the mountainous areas most distinctly favoring the growth of both deciduous and coniferous forests. Across much of the region, vegetation has been severely affected by human activity; the extent of forests today is less than in the past due to destruction of trees for fuel, construction, conversion to agriculture, and a gradual resultant conversion to steppe ecosystems in much of the region. Perhaps due to this trend, less research has been produced on Anatolian conifer and deciduous mixed forests than on the better-sustained temperate forests in other, climate-similar bands of Europe and Asia. The World Wildlife Fund lists 25 animal species of conservation concern dwelling in the Anatolian conifer and deciduous mixed forest ecoregion.

Northern Anatolia

The region to the northwest along the Black Sea is highly populated by people now, but the scattered occurrence of trees such as the oaks *Quercus ilex* (holly oak) and *Quercus suber* (cork oak) suggest that this zone may once have been an evergreen mixed forest. Today, forests of beech trees featuring *Fagus orientalis* (Oriental beech) are common. Also, *Alnus glutinosa* (common alder), *Prunus* (cherry) species, and *Carpinus betulus* (hornbeam) form forested canopies, with *Rhododendron* often occupying the understory. The temperature along the coast varies from 50 to 57 degrees F (10 to 14 degrees C). Inland, at elevations of 328 to 6,562 feet (100 to 2,000 meters), temperatures are colder, varying from 43 to 50 degrees F (six to 10 degrees C).

The forests in northeastern Anatolia, bordered by Georgia and the Black Sea, are profoundly influenced by the high mountains. These temperate coniferous forests and mixed deciduous forests consist of common alder, Oriental beech, *Castanea sativa* (sweet chestnut), *Abies nordmanniana* (Nordmann fir, one of the tallest trees in Europe at 256 feet or 78 meters), *Pinus sylvestris* (Scotch pine), and *Picea orientalis* (Oriental spruce). The composition of these forests varies with elevation and the associated temperatures and precipitation. The broadleaf species *Quercus petrea* subspecies *iberica* (sessile oak), sweet chestnut, maples, and Oriental beech occur from 1,640 to 3,940 feet (500 to 1,200 meters). At elevations of 3,940 to 4,920 feet (1,200 to 1,500 meters), Oriental spruce becomes dominant. These forests are defined by markedly high rainfall, averaging 59 to 98 inches (1,500 to 2,500 millimeters) per year.

Southern Anatolia

South Anatolia contains the Mediterranean phytogeographical region. Here, mean annual temperatures are warmer, averaging 57 to 66 degrees F (14 to 19 degrees C) north to south. The mean January temperature is 41 to 50 degrees F (5 to 10 degrees C), and the mean July temperature exceeds 68 degrees F (20 degrees C). Annually, 16 to 98 inches (400 to 2,500 millimeters) of precipitation occurs here, mostly during the winter months. With a gain in elevation, precipitation increases, and temperature decreases.

Pinus brutia (Turkish pine) dominates the forests from sea level up to 1,312 feet (400 meters) in the north and 765 to 875 feet (700 to 800 meters) in the south, due to its tolerance of drought and fire. These trees are extremely tall, potentially reaching more than 328 feet (100 meters) on the south-facing slopes of the Taurus Mountains. Mixed forests of *Pinus nigra* (black pine), *Cedrus libani* (cedar of Lebanon), and *Juniperus excels* (Greek juniper) occur from 2,620 to 6,560 feet (800 to 2,000 meters). Oak forests of *Quercus libani* (Lebanon oak), *Quercus frainetto* (Hungarian or Italian oak), *Quercus cerris* (Turkey oak), and *Quercus pubescens* (pubescent oak) occur at 2,620 to 3,940 feet (800 to 1,200 meters) in the eastern parts of the Taurus Mountains.

Inner, Eastern, and Southeastern Anatolia

Inner and eastern Anatolia are semi-arid, with cold, somewhat snowy winters and hot, dry summers. Lower elevations and valleys of the inner and eastern Anatolian region receive too little precipitation to support forest ecosystems. Higher elevations receive a bit more precipitation, at 20 to 24 inches (500 to 600 millimeters), which is enough to support trees that are tolerant of dry conditions. By comparison, most temperate forests globally receive 31 to 47 inches (800 to 1,200 millimeters) of precipitation per year. The dry forests of inner and eastern Anatolia are characterized by oak-dominated forests on dry to xeric sites—largely *Quercus pubescens* (pubescent oak) and *Quercus infectoria* (gall oak).

As precipitation increases at higher elevations of more than 3,940 feet (1,200 meters), oak trees co-occur with *Pinus nigra* subspecies *pallasiana* (black pine) and juniper species. Higher-elevation sites with more precipitation are dominated by black pine. As soils become more degraded or drier, or where black pine has been removed, mixed forests of pine, oak, and juniper grow. These forest types have been heavily affected by human land use, including deforestation, and have largely been destroyed at higher elevations and on north-facing slopes. Such areas tend to transition from forest to steppe ecosystems dominated by grasses and shrubs.

Less-dry areas in the middle and western mountains of eastern Anatolia are some of the most productive oak forests in present-day Turkey. Where these oak forests still occur, they are dominated by oak species adapted to moist soil conditions. These species include *Q. brantii* (Brant's oak), *Q. libani* (Lebanon oak), *Q. robur* subspecies *pendunculiflora* (Pedunculate oak), and sessile oak. Many, if not most, of these forest stands have been consumed by humans for wood fuel and oak-leaf feed for goats.

The southeastern Anatolian region contains the hottest and driest regions in Turkey. Trees in these places are similar to those of the dry forests of inner and eastern Anatolia, dominated or codominated by dry-adapted oak and pine species. The annual temperature is 63 to 64 degrees F (17 to 18 degrees C), with rainfall slightly over 20 inches (500 millimeters). This region has two dominant forest types: Turkey oak and *Quercus* (oak) species. The pine forest generally occurs at drier, lower altitudes, and the oaks appear in the upper subhumid elevations.

Paul T. Frankson
Jacqueline E. Mohan

Further Reading

Archibold, O. W. *Ecology of World Vegetation*. London: Chapman & Hall, 1995.

Atalay, I. and R. Efe. "Structural and Distributional Evaluation of Forest Ecosystems in Turkey." *Journal of Environmental Biology* 31, no. 61 (2010).

Mitchell, W. A. and A. Irmak. "Turkish Forest Soils." *Journal of Soil Science* 8, no. 184 (1957).

Anatolian Steppe, Central

Category: Grassland, Tundra, and Human Biomes.

Geographic Location: Middle East.
Summary: The central Anatolian steppe is a varied ecosystem that is home to a variety of unique halophytic flora.

Anatolia is a geo-historical term denoting the westernmost region of Asia on the Mediterranean coast, which has historically been home to Hittite, Persian, Greek, Assyrian, Armenian, Seljuk Turk, and Ottoman civilizations, and is currently composed primarily of the Republic of Turkey. The region is topographically diverse, including conifer and deciduous forests, temperate broadleaf mixed forests, sclerophyllous and mixed forests, and the montane conifer and deciduous forests around the Taurus mountains in the south.

The central Anatolian steppe is an ecosystem that includes marshes, rivers, freshwater and saline bodies, salt steppe, and halophytic vegetation (vegetation adapted to saline conditions). It is one of the major centers of endemic (adapted to specific local biome niches) plant species in Turkey and includes 10 separate Important Bird Areas designated by BirdLife International as being important to a threatened-bird species. There are no highlands or mountains; the altitude averages 3,281 feet (1,000 meters), while plains and river basins are the dominant land formations. The prevailing climate is characterized by cold winters and long, hot, dry summers.

The photo shows the scale-like leaves and many-branched stems of Salicornia europaea *(common glasswort). This type of halophyte, which is vegetation that has adapted to saline conditions, covers 95 percent of the ground in parts of the Anatolian steppe. (Wikimedia/Kristian Peters)*

Three separate areas of steppe vegetation make up the ecoregion: Tuz Lake, the Karapinar Plain, and the combined Porsuk and Sakarya River basins. Tuz Lake, at the center of the region, is the second-largest in central Anatolia. The Kizilirmak River, the longest river fully contained within Turkey's boundaries, provides the northern and eastern borders of the Tuz Lake area as it flows along a very roundabout route into the Black Sea. The Karapinar Plain lies south of Tuz Lake, beyond the Obruk Plateau. The Haymana and Cihanbeyli plateaus separate Tuz Lake from the Porsuk and Sakarya Rivers. The entire steppe region is in turn surrounded by deciduous forest ecoregions. Very little woody vegetation grows in the central Anatolian steppe; the soil and water are too saline here, and the local flora is well adapted to the salt levels.

Tuz Lake is home to 12 endemic vascular plant taxa. In the summer, much of the lake dries up, exposing a thick layer of salt more than one foot (three meters) deep. Salt concentrations decrease the further one is from the lake, so different ranges of species have developed in concentric circles around the lake, in bands of decreasing salinity. Nearest the lake, where the water has most recently receded, is *Salicornia europaea,* which covers 95 percent of the ground in some parts of the steppe. This species is better known as common glasswort, and like other species of *Salicornia* (sometimes known as samphire or sea bean), it is a succulent halophyte with small, scale-like leaves on jointed stems with many small branches. Glasswort is used as a food plant by both animals and humans, and is noticeably salty and crisp.

Moving farther from the water, the next plant species is *Halocnemum strobilaceum,* which covers much of the ground in its band, followed by patches of *Atropis distans, Limonium gmelinii, L. globuliferum, Juncus maritimus,* and *Plantago crassifolia,* mixed in with more patches of glasswort and *Halocnemum.* The next vegetation band, in muddy, sandy soils, is a mix of *Petrosimonia birandii* and other *Salicornia* species. Beyond that, on the slopes around the lake, are *Frankenia hirsuta, Kochia prostrata, Petrosimonia brachiata, Atriplex tatarica,* and *Salsola vermiculata.* Finally, at the least-saline fringes, are more than a dozen additional salt-tolerant species, such as *Salsola inermis, Aelropus lagopoides, Camphorosma monspeliaca,* and *Limonium iconium.*

In the Karapinar plains, the soil is less saline, and *Limonium anatolicum* is dominant. Marsh meadows are formed with *Juncus maritimus, Limonium globuliferum,* and *Tamarix gracilis,* and reeds and cyperus dominate in the freshwater areas. Many of the cyperus communities have a large number of endemic species. Where the water table is high, *Juncus heldreichanus, Aeolropous litoralis,* and *Pucinella convulata* thrive.

Threatened Species

Threatened avian species in this biome include both the great and little bustard. Larks are common, represented in at least six species, of which the Asian short-toed lark (*Calandrella cheleensis*) is restricted to the fringes of the salt lakes in the central plateau. Birds living in one of the region's 10 Important Bird Areas and key to conservation include the pygmy cormorant, white pelican, Dalmatian pelican, little bittern, squacco heron, purple heron, glossy ibis, greater flamingo, spoonbill, marbled teal, red-crested pochard, ferruginous duck, white-headed duck, pallid harrier, lesser kestrel, crane, collared pratincole, kentish plover, greater sand plover, spur-winged plover, gull-billed tern, and little tern.

Threatened nonavian fauna species include the marbled polecat. Other mammals that range here include the golden hamster, badger, and gray wolf. Dominant reptiles are *Agama stellio, Mabuya aurata, Typlops vermicularis, Coluber najadum,* and *C. numnifer. Phoxinellus crassus* is an endemic fish of the Tuz Lake basin, living in the streams that feed into the lake.

The major threat to the ecological sustainability of many of these species and the region as a whole is the overconsumption of freshwater resources. Hydroelectric dam construction and diversion of water for agricultural irrigation and civil engineering use has made an increasing impact across the region.

Bill Kte'pi

Further Reading

Finncioglu, Huseyn K., Steven S. Seefeldt, and Bilal Sahin. "The Effects of Long-Term Grazing Exclosures on Range Plants in the Anatolian Region of Turkey." *Environmental Management* 39, no. 3 (2007).

Isik, Kani. "Seasonal Course of Height and Needle Growth in Pinus Nigra Grown in Summer-Dry Central Anatolia." *Forest Ecology and Management* 35, no. 3–4 (July 1990).

Kirwan, Guy, Barbaros Demirci, Hilary Welch, Kerem Boyla, Metehan Ozen, Peter Castell, and Tim Marlow. *The Birds of Turkey.* New York: A&C Black, 2009.

Ocel, Isil, Ender Yurdakulol, Yuksel Keles, Latif Jurt, and Atilla Yildiz. "Role of Antioxidant Defense System and Biochemical Adaptation on Stress Tolerance of High Mountain and Steppe Plants." *Acta Oecologica* 26, no. 3 (December 2004).

Andaman Sea

Category: Marine and Oceanic Biomes.
Geographic Location: Southeast Asia.
Summary: The Andaman Sea is a reef- and coastal seagrass-rich basin that connects the South China Sea with the Indian Ocean via the Straits of Malacca.

The Andaman or Burma Sea is a marginal sea adjoining Sumatra, Thailand, and Burma; it is situated in the eastern Indian Ocean and transitions southward into the Straits of Malacca (between Sumatra and the Malay Peninsula). The

Andaman Sea borders the western margin of the Sunda continental shelf (6 to 16 degrees north, 93 to 98 degrees east). It has more or less an elliptical shape, covering the equivalent of approximately 308,882 square miles (approximately 800,000 square kilometers), with its longest axis oriented approximately north–south. Further adjacencies include the deltaic plain of the Irrawaddy River to the north, and the Andaman and Nicobar Islands to the west.

The Andaman combines scarcely known oceanographic and ecological patterns with an unfortunately all-too-well-known, highly active tectonic location—it was heavily affected by the 2004 Indian Ocean seaquake and tsunami. The Andaman also boasts a strategic position for shipping trade routes. The name *Andaman* probably originated from *Handuman,* the Malay form of *Hanumān,* a Hindu apelike deity. This name possibly refers to the aboriginal people of the Andaman Islands, a unique and ancient human lineage that migrated from Africa more than 50,000 years ago and is now almost completely extinct.

Geology and Recent Seismic Activity

The Andaman Sea is a back-arc marginal basin that originated along the oblique Sunda subduction zone in the Miocene. Several transform faults oriented north–south and an active oceanic ridge separate the Burma microplate to the west from the Sunda plate to the east. Barren Island in the Andaman is an active volcano.

This tectonic scenario was the stage for the 2004 event which featured a catastrophic 9.2 magnitude quake with the epicenter in the southern portion of the Burma microplate, about 62 miles (100 kilometers) off the northwest Sumatran coast. This massive tsunami killed some 200,000 people from Thailand and Indonesia to Sri Lanka and India. Large slip earthquakes of this type, although relatively rare and usually not associated with tsunamis, continue to occur in this active seismic area.

Oceanography and Meteorology

The average depth of the Andaman Sea is about 3,609 feet (1,100 meters). A depression 6,562 to 13,123 feet (2,000 to 4,000 meters) deep, 62 to 186 miles (100 to 300 kilometers) wide, and 466 miles (750 kilometers) long is oriented north–south along the oceanic ridge.

The seasonal patterns of the rainforest climate that is prevalent along its shores is dominated by the South Asian monsoon, with a dry winter or northeastern monsoon in December through February and a rainy summer or southwestern monsoon season during May through September. The northeastern monsoon is much weaker and less persistent. During transitional periods, in spring and fall, weather is more unpredictable, with lower average wind speeds and occasional extreme winds. Along the mainland coast, average annual rainfall is the equivalent of 138 inches (3,500 millimeters).

The distribution and fluctuations of sea surface temperatures directly influence the effects of global climate cycles such as the El Niño Southern Oscillation and the Indian Ocean Dipole on neighboring continental areas. Such patterns are influenced by river discharge, strong tidal mixing and internal waves, and exchange of surface and deepwater from the Bay of Bengal.

Coastal Systems and Continental Shelf

To the north, the Irrawaddy, Sittang, and Thanlwin (or Salween) Rivers discharge on average more than 144 cubic miles (600 cubic kilometers) of freshwater and more than 386 million tons (350 million metric tons) of sediments per year, resulting in the progradation, or outgrowth, of the eastern portion of this delta, and deeply affecting the seasonal hydrological patterns and habitat suitability of the northern basin. South of the Gulf of Martaban, the Burmese Tanintharyi coast includes the Mergui Archipelago. Due to the influence of freshwater inputs along the coast, nearshore systems here are often characterized by mangrove forests and seagrass beds, while fringing and patch coral reefs colonize offshore islands.

Moving southward along the Thai Andaman Sea coast, the shoreline is more regular, with limited river inputs and abundant coral reefs fringing both near-shore and offshore islands. On the west side of the basin, the Andaman and Nicobar Islands—territories of India—are also fringed by mangrove formations and seagrass beds, being characterized by 2,510 square miles (6,500 kilometers) of coral reefs. This body of reefs constitute 88 percent of all India's coral reefs.

Anthropogenic Impact

The Andaman Sea fisheries are exploited by four countries whose Exclusive Economic Zones (EEZs) extend over this basin: Thailand; Burma; India, through its Andaman and Nicobar Island populations; and Indonesia, through the island of Sumatra. All these countries are known to fish to some degree in others' EEZs, generating harsh economic and legal disputes.

On both the west and east sides of the basin, coral reefs and seagrass beds are increasingly affected by illegal and destructive fishing practices, including blast fishing, poisoning, trawling, push-netting, and long-line fishing. The Thai coastal ecosystems are also affected by habitat destruction and silting-up from land erosion, coastal and inland deforestation, damming, and rapid development due to the infrastructure needs of the growing tourism industry.

Gianluca Polgar

Further Reading

Hall, Robert and Jeremy D. Holloway, eds. *Biogeography and Geological Evolution of SE Asia.* Leiden, Netherlands: Backhuys Publishers, 1998.

Ramasamy, V., P. S. Rao, K. H. Rao, N. S. Thwin Swe Rao, and V. Raiker. "Tidal Influence on Suspended Sediment Distribution and Dispersal in the Northern Andaman Sea and Gulf of Martaban." *Marine Geology* 208, no. 1 (2004).

Shankar, D., P. N. Vinayachandran, and A. S. Unnikrishnan. "The Monsoon Currents in the North Indian Ocean." *Progress in Oceanography* 52, no. 1 (2002).

Temple, R. C. *Imperial Gazetteer of India Provincial Series: Andaman and Nicobar Islands.* Calcutta, India: Superintendent of Government Printing, 1909.

Varkey, M. J., S. N. Murty, and A. Suryanarayana. "Physical Oceanography of the Bay of Bengal and Andaman Sea." *Oceanography and Marine Biology* 34, no. 1 (1996).

Andean Montane Forests, Northwestern

Category: Forest Biomes.
Geographic Location: South America.
Summary: The northwestern Andean montane forests are among the most diversely populated forest regions in the world.

Among the most biodiverse forest regions in the world is the northwestern Andean montane forest. Consisting of tropical and subtropical moist broadleaf forests in a large area of western South America, from northwestern Colombia to Ecuador, the forest includes numerous animal and plant populations that were forced to readapt after glacial periods isolated them from their earlier ecoregions. Rainfall in this region is among the most plentiful in the world. The complex topography of the area and recurring altitudinal migration

A spectacled bear taking a bath in a stream. The northwestern Andean montane forest where these bears live has one of the highest levels of both biodiversity and incidence of endemic species in the world. (Thinkstock)

of vegetation as a result of shifts in climate have set up ideal conditions for speciation.

The result has been today's present diversity and an unusually high level of endemism (species evolved specifically to fit an isolated biome). The ecoregion is not only one of the most biodiverse on Earth, but also home to one the highest percentages of endemic species: Half of the plant species found here are found only here. Nearby ecoregions include the Magdalena Valley montane forests, Venezuelan Andes montane forests, Cauca Valley montane forests, Santa Marta montane forests, and Eastern Cordillera Real montane forests. The encroachment of human activity has interrupted the expanse of forest, as some of it has been cleared for farmland or settlements.

Iconic Bear Species

One of the important species of the region is the spectacled bear (*Tremarctos ornatus*), also known as the Andean bear, ukuo, or ucumari. The last surviving member of the short-faced bear family, which was common in the Middle and Late Pleistocene epoch, it is small for a bear species but the largest nontapir mammal in South America. Its short face, small ears (set far back), and facial markings make it faintly resemble a raccoon, though not all spectacled bears have spectacled facial coloring. Males are nearly twice as big as females.

The bears are nonterritorial but tend to live in isolation, and with the exception of mothers protecting their young, they are not known to attack humans; no human death by spectacled bear has been recorded. Spectacled bears build platforms in trees, as places to rest and to store food. The behavior may have developed during the Pleistocene epoch, when adult bears had more predators to conceal themselves from.

The bears rely more heavily on plants than most bear species do and are consumers of many plant food sources that other animals have difficulty opening or digesting, including cacti, palm nuts, orchid bulbs, forest fruit, and unopened palm leaves. They may travel to higher elevations in search of berries. Although meat normally makes up only about 5 percent of the bear's diet, its prey may include deer, cattle, and horses, and local

Threatened Bird Species

The forests are also home to the sword-billed hummingbird (*Ensifera ensifera*), the only member of its genus and the only bird with a bill longer than the rest of its body. The hummingbird has so adapted to feed on Passiflora flowers, which are long and heavy enough to hang straight down; the hummingbird approaches them from below to feed on their nectar. Proportionally, the hummingbird's keel and sternum are very large, and its flight muscles are powerful to keep its long bill from unbalancing it in flight.

The ecoregion has 37 Important Bird Areas designated by BirdLife International. Several of the restricted-range birds in the ecoregion are classified as threatened or near-threatened. Notable endemic species include the near-threatened species chestnut wood quail (*Odontophorus hyperythrus*), black-thighed puffleg (*Eriocnemis derbyi*), and crescent-faced antpitta (*Grallaricula lineifrons*); the vulnerable species rusty-faced parrot (*Hapalopsittaca amazonina*), giant antpitta (*Grallaria gigantea*), and bicolored antpitta (*Grallaria rufocinerea*); the endangered brown-banded antpitta (*Grallaria milleri*); and the critically endangered indigo-winged parrot (*Hapalopsittaca fuertesi*).

The white-rimmed brush finch (*Atlapetes leucopis*) falls into the Least Concern classification. Though the bird is somewhat rare, its population is stable and is believed to be over the threshold population to be considered Vulnerable (10,000 mature adults).

BirdLife International considers the region as a whole to be extremely threatened, particularly by deforestation and large-scale agricultural expansion (not only pasture land, but also commercial coffee, banana, and sugarcane plantations). There is also evidence of hunting pressure on bird populations. The ecoregion's long development of highly localized, highly specific habitats, although leading to the extreme speciation of the area, also means that when land is cleared, no suitable substitute may be available for the habitats destroyed in the process.

farmers have been known to shoot bears on sight to protect their cattle.

One of the palms on which the spectacled bear feeds is the wax palm or white palm (*Copernicia alba*), found throughout South American forests, usually in dense, naturally occurring monocultures. The lightweight, semi-hard wood is used for utility poles, and the waxy coating of the leaves has long been used by native South Americans to make candles. Its globular fruit is a dark berry containing a long ovoid seed that is eaten by the bears and other mammals. A growing danger to existing wax palm forests is the growing interest in the tree as a biodiesel crop. This interest would not jeopardize the species itself, of course, but monoculture plantation forests such as those in which the oil palm is grown are less hospitable to the habitats of other species that develop around the naturally occurring monocultures of the Andean wax palm forests.

Tapir Species

Four species of tapirs live in South America, three of which are large and native to rainforests. The fourth, the woolly or mountain tapir (*Tapirus pinchaque*), is the smallest and the only one to live in the Andean montane forests. About 6 feet (2 meters) long and 3 feet (1 meter) tall at the shoulder, tapirs are built stoutly and weigh as much as 500 pounds (227 kilograms). They have small tails and the long, flexible proboscises characteristic of their genus. Herbivores, they feed on ferns, umbrella plants, grasses, and pineapples.

The wax palm depends on them to disperse its seeds. Though mountain tapirs are not the only animals to feed on wax palm seeds, their digestive system is fairly inefficient, and they tend to defecate near bodies of water—a combination that makes them very adept at germinating in their dung the seeds of plants they have eaten. The wax palm and the highland lupine both decline dramatically when mountain tapir populations are reduced.

Bill Kte'pi

Further Reading

Keating, Philip. "Fire Ecology and Conservation in the High Tropical Andes: Observations From Northern Ecuador." *Journal of Latin American Geography* 61 (2007).

Malcolm, Jay. "Global Warming and Extinctions of Endemic Species from Biodiversity Hotspots." *Conservation Biology* 20, no. 2 (2006).

Mark, Bryan. "Tracing Increasing Tropical Andean Glacier Melt with Stable Isotopes in Water." *Environmental Science and Technology* 41, no. 20 (2007).

Norris, Ken and N. Harper. "Extinction Processes in Hot Spots of Avian Biodiversity and the Targeting of Pre-Emptive Conservation Action." *The Royal Society Proceedings, Biological Sciences* 271, no. 1 (2004).

O'Dea, Niall. "How Resilient are Andean Montane Forest Bird Communities to Habitat Degradation?" *Biodiversity and Conservation* 16, no. 4 (2007).

Andean Páramo Grasslands

Category: Grassland, Tundra, and Human Biomes.
Geographic Location: Central and South America.
Summary: The Andean páramo grasslands is a highly biodiverse tropical Andean ecosystem and cultural landscape that provides essential, especially water-related, services to large rural and urban populations.

Typical páramos are tropical wet grasslands approximately 10,499 to 15,420 feet (3,200 to 4,700 meters) above sea level, although several natural factors influence the altitude of the lower limit, such as proximity to wetter or drier lowland areas, rain-shadow effects, and soil type, as well as human use of the land. The combination of high altitude and tropical location creates a unique environment with daily seasonality ("winter every night, summer every day"), a generally humid climate and cold temperatures, and a surprisingly high level of biodiversity. Páramos provide important ecosystem services, including water provision and carbon sequestration, and are also important cultural landscapes. As recognition of their value has increased, so have efforts to improve their management and conservation.

Páramos cover 13,514 square miles (35,000 square kilometers) in South America; Colombia has the largest extension, and Ecuador has the greatest extension as a percentage of the country's total area (6 percent). Although páramos are ecological units, much diversity exists between the northern limit in Colombia and Venezuela and the southern limit in Peru. Even within one country, variations are remarkable, including extremely wet and very dry páramos. Human use is also an important factor that influences this diversity; páramos form a complex mosaic from pristine to disturbed areas.

Espeletia pycnophylla *plants in a grassland area in El Angel, Ecuador, a country whose total area is 6 percent páramo.* Espeletia, or frailejones, *plants can grow as tall as 33 feet (10 meters). (Wikimedia/Thomas van Hengstum)*

Biodiversity

The distribution of páramos is similar to a series of islands in the landscape, promoting the development of high levels of biodiversity and endemism. Plants and animals have adapted to extreme conditions such as high ultraviolet radiation, cold, wind, limited water uptake due to low temperatures, and wide daily temperature variations. Tussock grasses are the dominant life form in grasslands; trees of the genera *Polylepis, Buddleja,* and *Weinmannia* are generally scarce but can form extensive forests within the páramo. The vegetation also includes giant rosettes, especially of the genus *Espeletia,* which are unusual plants commonly known as *frailejones* that reach up to 33 feet (10 meters), as well as dwarf shrubs and cushion plants. Animal life includes several iconic species such as Andean condors and spectacled bears; a wide variety of birds, such as hummingbirds and buzzard eagles; and mammals such as rabbits, foxes, wildcats, and (occasionally) Andean tapirs.

Soil Characteristics

One of the most important characteristics of páramos are their soils, which function like sponges that can hold more water than their own dry weight. The combination of geological materials with organic matter that decomposes slowly in the cold, wet conditions is the basis for the principal ecosystem function of páramos. Water retained and released by this soil is used both by local communities and downstream for irrigation, drinking water, and hydropower. Large Andean cities such as Mérida, Bogotá, Quito, and Cajamarca, as well as many others, depend substantially on páramos for their water. These soils also have high carbon contents and thus play a role as a carbon sink that can help mitigate climate change. Other ecosystem services of páramos have to do with their at-times breathtaking landscapes and biodiversity, which are the basis for ecotourism activities.

Human Use of Páramos

People have used páramos since pre-Columbian times, and a "páramo culture" exists that includes cultivation of crops such as potatoes and other Andean tubers. Incas developed the use of llamas and alpacas, and established irrigation systems with páramo water. The European conquest decimated and subjugated the indigenous population, and began the extensive use of páramos for livestock, especially sheep and cattle, with large herds transforming certain páramos dramatically.

Some human activities—including burning for grazing, deforestation (clearing for pastures and firewood extraction), exotic-tree plantations (especially pine trees for timber), and the advance of agriculture into páramos—can affect páramo ecosystem functions, especially in less humid and more accessible areas. Other activities that have an effect are mining, uncontrolled tourism, hunting, and infrastructure related to water and roads. All these activities affect the soil, water, biodiversity, and landscape, and thus affect one or more ecosystem services. The effect of climate change, perceptible through the reduction of glaciers, is not yet well understood but could have severe consequences for páramo soil and water.

Conservation Efforts

The contrast between conserved and disturbed páramos has prompted a reaction from communities, government agencies, and civil society. Several ongoing efforts are seeking to manage the cultural landscape with practices including intensification of agricultural activities in lowland zones, use of alpacas and llamas instead of cattle and sheep, eradication of exotic-tree plantations, ecotourism initiatives, as well as conflict management among actors with competing interests and political processes from local to regional scales. Some projects include the concept of compensation for ecosystem services, in which lowland users pay or otherwise compensate upland people's conservation and land management efforts that enhance production of water and other ecosystem services.

Not long ago, páramos were considered to be almost-useless places, but they have acquired strategic status in the past two decades. Much more is yet to be known, discussed, and implemented to guarantee their long-term sustainable management and conservation.

Patricio Mena-Vásconez
Kathleen A. Farley

Further Reading

Balslev, Henrik and James L. Luteyn, eds. *Páramo: An Andean Ecosystem Under Human Influence.* London: Academic Press, 1992.

Hofstede, Robert, Pool Segarra, and Patricio Mena-Vásconez, eds. *Los Páramos del Mundo.* Quito, Ecuador: Global Peatland Initiative/NC-IUCN/EcoCiencia, 2003.

Mena-Vásconez, Patricio, Galo Medina, and Robert Hofstede, eds. *Los Páramos del Ecuador: Particularidades, Problemas, y Perspectivas.* Quito, Ecuador: Proyecto Páramo/Abya Yala, 2001.

Andean Puna Grasslands, Central

Category: Grassland, Tundra, and Human Biomes.
Geographic Location: South America.
Summary: The Puna grassland is a mountain grasslands and shrublands biome. The dominant vegetation consists of shrubs, often also including grasses, herbs, and geophytes (plants with nutrition-storing abilities often manifesting as underground bulbs).

Grasslands occupy the central Andes Mountains of Peru and extend south as far as northern Argentina and Chile. This type of ecosystem is considered to be one of the eight natural regions in Peru. The slopes and ravines of greater altitude and slope, including forest residues (the Polylepis species of queñual) harbor a large part of the diverse flora and fauna found in the area.

The overall territory of high-Andean grassland, located at an altitude of 9,843 to 15,748 feet (3,000 to 4,800 meters), runs throughout the Andes, starting at the border with Ecuador, crossing through all of Peru, and reaching Bolivia. At a lower altitude in the area of Piura (10,171 feet or 3,100 meters) and Cajamarca (10,499 feet or 3,200 meters), the biome is called a jalca or páramo, and extends to a latitude of approximately nine degrees south.

The central Andean puna itself is a mountain grasslands and shrubland ecoregion in the Andes of southern Peru, Bolivia, and northern Argentina. The landscape in this ecoregion consists of high mountains with permanent snow and ice, meadows, lakes, plateaus, and valleys. It transitions to the central Andean wet puna to the north and the central Andean dry puna to the south. Elevations range from 10,499 to 21,654 feet (3,200 to 6,600 meters). The landscape is characteristically mountainous, with snowy peaks, U-shaped glacial valleys or high headwater valleys, plains, and lakes. The plateaus are dominated by a landscape that combines typical prairie grassland, dominant in the landscape, with patches of forest, scrubs, and wetlands, limited by permanent snow line or the jalca region.

Peruvian Differences

Peru has different climate regimes and two types of puna: the wet (center) and dry (southwest). Annual rainfall varies widely, ranging from 6 to 31 inches (150 to 800 millimeters) from north to south. Other features of the puna landscape and vegetation are groups of Puya Raimondi remnants, high-altitude wetlands, and bush and shrub vegetation consisting of Baccharis and Gynoxis species; denser formations of the Oreoboluys and Calamagrostis species, reminiscent of highlands artificial grass; hundreds of high-altitude lakes; and a high-Andean alpine vegetation zone formed by very slow growing plants that have adapted to cold weather and the harsh highlands environment.

The Peruvian Andes, because of its sheer size, orientation, altitude, and topography, is the main physical system that structures the distribution of rainfall. Due to the location of deserts and plateaus as wet basin headwaters, its role in the continental water regulation system is essential. Temperatures in the Andes vary daily between extreme highs and lows, with a marked dry season that is more pronounced south of Peru. In the daily cycles and within seasons, temperatures fall below 32 degrees F (0 degree C), with conditions of frost and hail making agriculture and even livestock breeding high-risk ventures. Overall rainfall ranges from 8 to 39 inches (200 to 1,000 millimeters), affecting water availability in both the microtopography and climate processes that are global in nature, which unexpectedly generate droughts and floods.

Puna, Páramo, and Jalca

The differentiation among puna, páramo, and jalca is still under discussion, but among the general characteristics that differentiate them is the degree of dryness. The puna grasslands are seasonal and dry, because they do not receive sufficient rainfall to maintain a vegetation coverage, whereas páramos are more like sponges that act like reservoirs of water in the form of permafrost.

These ecosystems occur in the tropical Andes, in the northern part of Peru, and in southern Ecuador. The landscape is quite rugged in the upper parts of the western Cordillera Mountains, with the presence of gallery forests, queñuales, quishuares, and alders forming part of the headwaters of the basin. The weather is cold, characterized by high soil moisture and permanent cloud cover, with frequent precipitation often exceeding 59 inches (1,500 millimeters) per year.

Puna Grasslands

The puna grasslands are also located in the Andean highlands, but only from the Cordillera Blanca south, where the wet punas begin forming a transition zone between the jalca or moor and the dry puna of Central and South America. It is difficult to establish a defined altitude at which the highlands begin, but they are generally considered to start at around 12,467 feet (3,800 meters) and can reach an altitude of 15,748 feet (4,800 meters). The puna ecoregions extend into slots in the highest parts of the Andean mountain system, and their locations are critical in terms of their impact on lower areas. Because of the height of these ecoregions, the youth of their soils, and their complex topography, which is characterized by steep slopes prone to erosion, they are extremely fragile both biophysically and socially.

The strategic importance of this ecoregion lies in its position along the Andes as the headwater of innumerable peaceful rivers that form part of the Amazon Basin. As a result, strategic management is affecting all conservation areas located downstream. In addition, these high grassland ecoregions can be seen as an ecological corridor connecting valleys that would otherwise be segmented. In recent years, concern has grown over glacial pullback that many see as a symptom of global climate change, another front in the challenge to establish a sustainable control regime to mitigate human effects on the ecological infrastructure.

Alexandra M. Avila

Further Reading

Kricher, John. *Tropical Ecology.* Princeton, NJ: Princeton University Press, 2011.

Mountain Institute and American Alpine Club. *Alpine Conservation Association: Protection and Restoration of Ecosystems in the World for Future Generations.* Elkins, WV: The Mountain Institute, 2006.

Recharte, Jorge, Luis Albán, Roberto Arévalo, Enrique Flores, Luis Huerta, Miguel Orellana, Luis Oscanoa, and Pablo Sánchez. *The Paramo, Jalcas and Punas Group: Institutions and Actions That Benefit the Highland Andean Communities and Ecosystems.* Lima, Peru: The Mountain Institute in Peru, 2002.

Rodriguez, C. T. and J. L. Quispe. *Domesticated Camelids, the Main Animal Genetic Resource of Pastoral Systems in the Region of Turco, Bolivia.* Rome, IT: Food and Agriculture Organization, Inter-Departmental Working Group on Biological Diversity for Food and Agriculture, 2007.

Andros Barrier Reef

Category: Marine and Oceanic Biomes.
Geographic Location: North Atlantic.

Summary: The Andros Barrier Reef, one of the world's largest coral reef complexes, is a valuable habitat for many species in the western North Atlantic Ocean.

Andros, the largest island of the Bahamas, possesses a reef complex lying along its entire eastern coastline. The reef is a somewhat discontinuous structure that extends for approximately 135 miles (217 kilometers) from Joulter Cays southward to South Cay. This reef complex is the second-largest in the western Atlantic and considered to be the third-largest reef in the world.

Although the Andros reef is sometimes referred to as a barrier reef, its structure and origins indicate that it is not truly a barrier reef in the strictest sense. Barrier reefs, as described by Charles Darwin, are formed over long periods of time by the simultaneous growth of shallow-water corals along a tropical coastline and the slow subsidence of that coastline and the adjacent land mass. The resulting reef lies many miles offshore, with a deep lagoon separating the reef from the adjacent land.

By contrast, the Andros reef complex parallels the island coastline but is separated from it by a relatively narrow and shallow lagoon. The reef likely originated from corals growing along the seaward edge of a newly submerged rocky platform that was created during a time of lower sea levels. Such reefs are common in the Caribbean and are considered to be hybrid structures that are sometimes referred to as bank-barrier reefs.

Reef Habitats

The predominant habitats along the reef are the shallow reef crest and the deeper fore reef that extends toward the open ocean. Occasional patch reefs also occur behind the reef crest in the lagoon. The reef crest is dominated by colonies of Elkhorn coral (*Acropora palmata*). The thick branches of Elkhorn coral are ideally suited to withstand the breaking waves common on the reef crest. Star corals (*Montastrea* spp.) predominate in the fore reef zone. These colonies form boulder-like structures or may be somewhat flattened at increasing depths. The region known as the Tongue of the Ocean lies a short distance seaward from the fore reef. Here, the ocean floor drops precipitously to depths of approximately 6,000 feet (1,829 meters).

Fish Species

More than 160 species of fish inhabit the Andros reef complex. Generally, the reef crest possesses higher densities of fish than the fore reef. The fish community is dominated by species of grunts, parrotfish, snappers, tangs, and surgeonfish, as well as large schools of smaller fishes such as silversides, herrings, and anchovies.

Groupers are also found along the length of the reef, but in lower numbers than other species of fish. Groupers are major predators on the reef and are currently under stress throughout the area due to commercial fishing in the Caribbean. The Andros reef is especially noted for the large spawning aggregations of Nassau grouper that occur in the central and southern portions of the complex. Although solitary for most of their lives, groupers may travel each year from as far as 100 miles (161 kilometers) away to arrive at the reef for their mass spawning events, which occur during the full moon in December and January. Spawning aggregations vary in size but can include thousands of fish. As the sun sets, mates gather in smaller subgroups within the aggregation, and the release of gametes occurs. Offspring are transported across the reef and into the protected waters of the lagoon. Young groupers use the quiet seagrass beds within the lagoon to feed and grow before returning to the reefs and rocky bottom areas inhabited by their parents.

A hawksbill turtle swims among hundreds of small fish above a coral reef in the Bahamas. The Andros reef complex is home to over 160 different species of fish. (Thinkstock)

Human Activities

Fishing pressure along the reef is considered to be light to moderate compared with other regions of the western North Atlantic and the Caribbean. Although groupers and snappers are the primary fishes targeted, both commercial and subsistence fisherman will also take grunts, hogfish, triggerfish, and barracuda. Groupers are particularly vulnerable to capture during spawning aggregations, and their protection during that time of year is important to the sustainability of the fishery. Andros Island is sparsely populated, and local land uses do not appear to greatly affect the reef.

Ecosystem Stresses

Regional and global stresses to the reef have been documented. A lack of herbivorous fish and invertebrates, for example, has likely allowed the overgrowth of algal turfs on some parts of the reef. These turfs cover rocks and rubble and can reduce sites where new coral colonies can become established. This is particularly true of the fore reef, which does not exhibit as many herbivorous fish species as the reef crest.

Populations of another important herbivore, the long black-spined sea urchin (*Diadema antillarum*), diminished severely during a Caribbean-wide disease outbreak in the mid-1980s. This decline has resulted in a predominance of algal turfs over many reefs in the region. Urchin populations are slowly beginning to recover on the Andros reef but have not yet reached sufficient sizes to act as effective controls for the algae.

Some coral diseases also occur along the Andros reef complex. Evidence of disease is most prevalent in star corals and brain corals on the fore reef. Black-band disease and white plague have both been documented. Black-band disease is a consortium of sulfur-metabolizing bacteria and blue-green algae that invades the coral tissue and leaves bare skeleton in its wake. The pathogen involved in white plague has not been identified, but the results are similar to that of black-band disease, as coral tissue death exposes the colony's skeleton. These conditions can kill or cause partial death of coral colonies. Even if a coral is not completely killed, these diseases can reduce the colony's reproductive ability and open its skeletal structure to bioeroding organisms such as sponges and marine worms.

Finally, coral bleaching, a stress on coral reefs worldwide, has been observed on the Andros reef. Bleaching, which can lead to coral mortality, is the loss of symbiotic algae that are normally found within the coral's tissues. The resident algae impart a color to the colony that varies from yellow to golden-brown to shades of green. Following bleaching, the colony appears to be white, as the white skeleton can be seen through the transparent tissues of the coral. Bleaching is a generalized response to many types of environmental stress but is most often observed on reefs during periods of elevated water temperature. Although no large-scale bleaching events have been documented in Andros, areas of the reef crest are sometimes affected. Despite comparative field research, it is not understood exactly how the Andros has escaped the large-scale bleaching events that have impacted other reefs around the world. Coral colonies can recover from partial bleaching, but even in recovery, the growth and reproductive ability of corals are negatively affected.

Despite the environmental stresses just discussed, the Andros reef is generally considered to be in good condition relative to other reefs in the western North Atlantic and the adjoining Caribbean Sea areas. In 2002, the government of the Bahamas set aside more than 64,000 acres (25,900 hectares) of this reef complex to create two national parks. The establishment of these marine protected areas is an important first step in conserving a beautiful and fragile tropical marine resource.

M. Drew Ferrier

Further Reading

Andros Conservancy and Trust. "The Andros Barrier Reef." 2010. http://www.ancat.net/index.php?option=com_content&view=article&id=64&Itemid=76.

Kaplan, Eugene. *A Field Guide to the Coral Reefs: Caribbean and Florida.* New York: Houghton Mifflin, 1982.

Kramer, Philip, et al. "Assessment of Andros Island Reef System, Bahamas, Part I: Stony Corals and Algae." *Atoll Research Bulletin* 496, no. 1 (2003).

Kramer, Philip, et al. "Assessment of Andros Island Reef System, Bahamas, Part II: Fishes." *Atoll Research Bulletin* 496, no. 1 (2003).

Society for the Conservation of Fish Aggregations. "The Nassau Grouper." http://www.scrfa.org/index.php/about-fish-spawning-aggregations/aggregating-species/the-nassau-grouper.html.

An Nafud Desert

Category: Desert Biomes.
Geographic Location: Middle East.
Summary: The An Nafud Desert is a windswept erg (dune sea) in the northern Arabian Peninsula, with limited biodiversity except in its oases.

The An Nafud or Al-Nafud Desert is in the northern Arabian Peninsula and is considered to be part of the Arabian Desert. It comprises a 40,000-square-mile (103,600-square-kilometer) oval depression across northern Saudi Arabia, and is 180 miles (290 kilometers) long and 140 miles (225 kilometers) wide.

Rainfall is extremely variable and unpredictable, with total annual precipitation typically less than 2 inches (50 millimeters), though as in all deserts, the climate is predominantly hot and arid, turning cold at night because there is so little vegetation to retain heat. Winter temperatures see the greatest diurnal difference, ranging from just above 32 degrees F (0 degrees C) at night to about 70 degrees F (21 degrees C) during the day. Summers often have lows of 80 to 89 degrees F (27 to 32 degrees C) at night and exceed 100 degrees F (38 degrees C) during the day.

Like the Arabian Desert's other major components, the Dahna and the Rub' al Khali to the south, the Nafud is an *erg*, a flat or depressed area with scant vegetation and particularly identified as a sea of dunes or sand sea. The term *erg* is from the Arabic *'arq* or dune field. Along with those of central Asia, the Nafud is one of the oldest ergs in the world, formed by the wind over a period of at least 1 million years. The sand has a red or rusty hue; the wind forms wavelike undulations across it that each may stretch for miles.

Like much of the Arabian Peninsula, the Nafud lies atop a limestone floor riddled with caves and passageways formed by the ancient and ongoing percolation of rainwater, which has absorbed carbon dioxide and formed a weak carbonic acid capable of eating away at the limestone in minuscule amounts at a time.

Though quicksand is not as common as in Rub' al Khali, it does exist in the Nafud, though it is not the suffocation hazard that it is portrayed to be in motion pictures. Quicksand forms when an area of loose sand becomes saturated with water (such as from an underground spring), creating a non-Newtonian fluid that appears to be solid until a change in stress (such as being walked upon) decreases its viscosity. Animals do not become submerged in quicksand, but attempting to withdraw from it at anything but incredibly slow speed requires force comparable to that necessary to lift a horse.

Red Locusts

Red locusts, long relied upon by the Bedouins as a food staple, live in the An Nafud Desert. Entire tribes of Bedouins harvest a flight of locusts, removing the wings, drying the bodies, and grinding them into a protein-rich flour that is used to feed both humans and horses. Locusts are also eaten boiled and are said to be comparable in flavor to green wheat. Many other animals of the desert feed on them as an important protein source.

Desert Vegetation and Wildlife

Springtime in the Nafud, after the extremely light seasonal rains, turns portions of the desert into grassy areas sufficient for pasturing by the nomadic Nejd Bedouins and provides foodstuffs for the indigenous wildlife. Saltbush (*Cornulaca arabica*), *Calligonum crinitum,* and *Cyperus conglomeratus* are among the most common plant species in the Nafud. *Dipterygium glaucum, Limeum arabicum,* and *Zygophylum mandavillei* sometimes occur. Certain local herbs, such as *Danthonia forskallii,* also grow after a rainfall. Sand, however, dominates the landscape. The scarcity of vegetation in the Nafud diminishes potential biodiversity, so there are no formal protected areas in the desert.

Wildlife includes gazelles, which according to popular belief never need to drink, though in fact they are simply desert-adapted and highly efficient in their water use. Ostriches, rodents, spiders, hares, sand cats, spiny-tailed lizards, and oryxes are also well adapted to the extreme environment of the Arabian deserts and appear all over the peninsula. Other species once common here—such as the honey badger, the Arabian ostrich, and the jackal—have become rare or extinct due to hunting and habitat destruction. An endemic (existing here only) species is the *bakkar vanash* (wild cow), a species of humped antelope.

The harvester termite is widely distributed throughout the Arabian Peninsula. Although the

Nafud's sparse vegetation makes it less hospitable to the termite, the insect nevertheless lives here, though with few human settlements to threaten, it is less of a pest. The termite can feed not only on dry weeds and tree stumps, but also on the cellulose in (herbivorous) animal droppings, windblown accumulations of plant debris, and other samples of decomposed flora.

Western Oases

Near the Al Hejaz Mountains in the west are low-lying oases; here some crops are cultivated including fruits like dates and grain such as barley. Wildlife is also more plentiful here, and the climate is less arid. A single oasis has considerably more diversity than the rest of the Nafud. Migratory birds are more common in the mountains, which are relatively cool, yet many of the swifts, larks, wheatears, and other species stop in the desert oases to drink and forage. Raptors also visit sometimes, as well as bustards, sandgrouses, and chukar partridges. Reptiles in the oases include spiny-tailed lizards (*Uromastyx thomasi*) and monitor lizards (*Varanus grisens*).

The oases are home to many of the acacia trees that grow elsewhere in the Arabian Desert, including *Acacia tortilis, A. raddiana, A. gerrardii,* and *A. ehrenbergiana. Ziziyphus spinachristi, Balanites aegyptiaca, Salvadora persica, Moringa peregrina, Capparis decidua, Cordia gharaf, Calotropis procera, Lavandula nubica,* and *Ephedra foliata* all occur in the oasis ecosystems, as do grasses of the *Stipograstis* species. Also growing in the western oases is bdellium (*Commiphora wightii*), a relative of myrrh that exudes an aromatic gum with some value as a trade good.

Bill Kte'pi

Further Reading

Barth, Hans-Jorg and Benno Boer. *Sabkha Ecosystems.* Vol. 2. New York: Springer, 2002.

Edgell, H. Stewart. *Arabian Deserts: Nature, Origin, and Evolution.* New York: Springer, 2006.

Perry, R. A. *Arid Land Ecosystems.* New York: Cambridge University Press, 1979.

Antarctic (Southern) Ocean

Category: Marine and Oceanic Biomes.
Geographic Location: Antarctica.
Summary: The Southern Ocean and its surrounding, mostly ice-covered land area are home to an abundance of marine life that thrives on the cold, oxygen-rich waters.

The Antarctic Ocean—or the Southern Ocean, as it is usually known—is approximately 13.7 million square miles (35.5 million square kilometers) in area, or twice the size of the United States. It is the world's fourth-largest ocean. Because of its unique position and characteristics, the Southern Ocean plays a unique role in regulating the global climate.

Encircling the continent of Antarctica, the Southern Ocean is formed by the convergence of the Pacific, Atlantic, and Indian Oceans as they flow together around Antarctica—the coldest, windiest, and driest of all the world's continents. The Southern Ocean encompasses the Amundsen Sea, the Bellinghausen Sea, part of the Drake Passage, the Ross Sea, a small section of the Scotia Sea, and the Weddell Sea. The deepest point of the Southern Ocean is 27,737 feet (8,454 meters) below sea level, at the South Sandwich Trench.

The waters of the Southern Ocean are characterized by the Antarctic Circumpolar Current (ACC), also known as the Wind Drift Current. As it travels eastward at about 5.4 million cubic feet (153 million cubic meters) per second, this current transports a higher volume of water than any other current in the world, amounting to about 100 times that transported by all the world's rivers. The ACC is the only current that encircles the globe. The Antarctic Convergence, the globe-circling region where the ACC waters blend with the somewhat warmer waters of the other oceans, is a band about 20 to 30 miles (32 to 48 kilometers) wide, ranging between 48 degrees south and 61 degrees south.

Fish Species

Even though the land area of the Antarctic is covered with snow and ice, the biome of the Southern

Ocean supports a wealth of plant and invertebrate animal life. The most abundant groups of fish in the Southern waters are the Antarctic cod and the ice fish, and research tends to concentrate on these two species.

Within the Southern Ocean, upwelling currents cause rich nutrients to be pulled up from the seabed to the surface to provide nourishment to the microscopic algae (plankton) that live near the surface water layers. Krill, small shrimplike crustaceans, feed on the plankton and in turn are eaten by the fish, whales, seals, and birds that live in the Southern Ocean. These cold waters are considered to be as much as four times more productive per acre (hectare) than that of any other ocean in the world. The abundant krill that teem in the surface waters are responsible for the red color sometimes observed in the waters here. Squid and octopus are also important links in the ocean's food chain because they serve as nourishment for sperm whales, seals, penguins, seabirds, and fish. Estimates place the number of squid consumed by whales alone at some 55 million tons annually.

Penguins may be found in colonies of as many as 180,000 birds in Antarctica. Emperor penguins like this one are one of the most common species. (Thinkstock)

Because krill are so plentiful and high in protein, efforts have been made to use them to solve some food shortages in some of the poorer nations of the world. The idea has proved to be largely impractical, however, because of the need for rapid processing and distribution. However, both Japan and Russia have begun to perfect methods of using krill for other commercial purposes.

Biodiversity

Marine mammals that live in the Southern Ocean are no strangers to human activity. During the 18th and 19th centuries, the seals of the Southern Ocean were exploited for their furs, resulting in their ultimately being classified as protected. By the 21st century, these seals have begun a comeback. These seals cannot breed in the water; thus, they head toward land or pack ice during mating seasons. Antarctic seals are larger than those that live in the Arctic, in part due to the easy availability of food and the presence of fewer predators in the area.

The polar winter of the Antarctic generally lasts from November to March and is a period in which violent storms occur. Daylight is absent, and the coastal waters become icebound. When the polar winter ends, more than 100 million seabirds arrive at the Southern Ocean to breed on the mainland or on offshore islands. The albatross, largest of all flying seabirds, is one of the most distinctive. Other species include fulmers, prions, petrels, and shearwaters; more widespread species include shore birds, skuas, gulls, terns, and penguins. Because ice-free land is scant along the shores of the Southern Ocean, these birds tend to breed in large concentrations. When they sense the return of polar winter, the seabirds move to the open sea, sometimes temporarily populating the pack ice.

The penguin may be the most emblematic animal in the Southern Ocean. Much celebrated popularly, the various species are under considerable environmental pressures, however. The Emperor penguin colony of Terre Adélie in east Antarctica, for example, shrank by 50 percent in the 1970s and has yet to recover from that loss. The direct cause was the diminishing acreage of sea ice—as these birds breed and nurture their young on the sea ice; the problem is exacerbated because much of their diet is also dependent on the ice. The fish, squid, and krill that Emperor penguins consume in turn eat the specialized plankton that grows on the underside of sea ice floes. Incremental rise in sea temperature has been enough to upset the balance of this food web.

Seven species of penguins are classified as Antarctic penguins, and more penguins than any other bird species live on this continent. Unlike most birds, penguins do not fly; they use their feet as propellers to move themselves through the waters. The most common species is the Adélie, which may be 24 to 28 inches (60 to 70 centimeters) tall; and the Emperor, which grows to four feet (1.2 meters) in height and can weigh up to 95 pounds (43 kilograms). Some concentrations of penguins may include as many as 180,000 birds. The nests of the Adélie are built of stone; just one or two eggs are incubated in these nests at a time.

Environmental Stresses

Despite its isolation, the Southern Ocean is not immune to some of the same environmental stresses that are threatening ecosystems around the world. The Southern Ocean serves as a massive storehouse for carbon dioxide (CO_2), with up to one-third of all CO_2 stored by the world's oceans present here, according to some estimates. However, scientists are warning that the Southern Ocean is losing its capability to serve as a buffer for the rest of the world because global warming has accelerated the production and accumulation of CO_2 while simultaneously reducing the ability of the Southern Ocean to absorb it, due to altered temperature and salinity regimes.

Human industrial activity has also contributed to the hole in the atmospheric ozone layer over the Antarctic, which tends to add to a warming in its regional climate. This depleted-ozone area allows more solar ultraviolet radiation to penetrate to the surface waters, which some research suggests has reduced the amount of plankton here by as much as 15 percent, and likely has altered the DNA of some fish species.

Exploitation of fishing grounds has become a major problem. Australian scientists have sent out an international call for help because of the increased levels of fishing piracy in the Southern Ocean. One such casualty is the highly prized Patagonian toothfish (*Dissostichus eleginoides*) and its relative the Antarctic cod (*Dissostichus mawsoni),* which are considered a delicacy in both Japan and the United States—where they are marketed as "Chilean sea bass." These fish, known to commercial fishers as the "white gold of Antarctica," have both been placed on the seafood red list of fish to avoid, posted by Greenpeace International. Armed surveillance ships are in some cases being used to protect Patagonian toothfish, which are an integral part of the diets of sperm whales and elephant seals.

A "collateral" casualty of the practice of using long-line baited hooks that stretch out for miles over the surface of the Southern Ocean are some 100,000 seabirds a year. Attracted to the bait, the birds go after the hooks and are trapped as they are pulled below the surface. The great albatross and petrels, according to some sources, may have lately joined the ranks of the most endangered birds on the planet as a result of such practices.

Elizabeth Rholetter Purdy

Further Reading

El-Sayed, Sayed Z. *Southern Ocean Ecology: The Biomass Perspective.* New York: Cambridge University Press, 1994.

Forcada, Jaume and Philip N. Trathana. "Penguin Responses to Climate Change in the Southern Ocean. *Global Change Biology* 15, no. 7 (2009).

Lowen, James. *Antarctic Wildlife: A Visitor's Guide.* Princeton, NJ: Princeton University Press, 2011.

Roberts, Leslie Carol. *The Entire Earth and Sky: Views on Antarctica.* Lincoln: University of Nebraska Press, 2008.

Turner, John. *Climate Change in the Polar Regions.* New York: Cambridge University Press, 2011.

Zarembo, Alan. "Southern Ocean Study Sounds More Global-Warming Alarms." *Los Angeles Times,* May 17, 2007.

Apalachicola Bay

Category: Marine and Oceanic Biomes.
Geographic Location: North America.
Summary: Apalachicola Bay is an ecologically intact bay and estuary system that produces 13 percent of the oysters consumed in the United States. A major issue is reduced freshwater inflow due to urbanization to the north.

Apalachicola Bay is a large estuary and bay located in the region of northern Florida known as the Panhandle, a coastal segment of the Gulf of Mexico. Apalachicola Bay is formed mainly by freshwater input from the Apalachicola River, itself part of a large river system known as the Apalachicola-Chattahoochee-Flint (ACF) and governed by a compact among Florida, Alabama, Georgia, and the U.S. Army Corps of Engineers. The watershed of the bay and 107-mile-long (172-kilometer-long) Apalachicola River encompasses some 20,000 square miles (52,000 square kilometers). The floodplain and freshwater discharge are the largest in the state of Florida. It is bordered in part by the Apalachicola National Forest, one of the largest contiguous blocks of public land east of the Mississippi River.

In 1969, Apalachicola Bay was designated a Florida State Aquatic Preserve, and in 1979, it was designated a National Estuarine Research Reserve by the federal National Oceanic and Atmospheric Administration, one of only 27 in the country. This designation is reserved for those areas deemed most significant for ecological and hydrological factors, and creates protection for long-term research, water-quality monitoring, education and coastal stewardship. Additionally, it was designated a Biosphere Reserve by the United Nations Man and the Biosphere program in 1984, demonstrating its international significance.

Rich Marine Resource

The bay is considered to be one of the least developed and least polluted natural systems remaining in the United States. There are nearly 120 known plant species; more than one-fourth of them are threatened or rare. The largest stand of tupelo trees (*Nyssa* spp.) in the world is located in the region.

A total of 86 freshwater fish species, the largest number in Florida, are found within the basin, and 131 freshwater and estuarine fish species are found within the bay and river. The bay once supported a gulf sturgeon commercial fishing industry until the early part of the 20th century; now, fewer than 400 individuals are left in the bay. In addition, the watershed contains the highest density of diverse amphibians (40 species) and reptiles (80 species) of any zone within North America, apart from Mexico, and the area has significant populations of migratory birds.

The bay is particularly renowned for its production of eastern oysters (*Crassotrea virginica*), producing 90 percent of all the oysters consumed in Florida and 13 percent of the entire consumption of the United States. In recent years, the state has commercially promoted Apalachicola oysters as having a notable flavor and composition that merits distinction alongside oysters from other regions that have a notable provenience (or *merroir*).

By virtue of proximity to the Apalachicola National Forest, much of the watershed is protected to some degree. In addition, the bay's beaches and barrier islands contain two state parks: St. Joseph Peninsula and St. George Island. The area also boasts a national wildlife refuge, St. Vincent, and one additional barrier island, Dog Island. These barrier islands provide significant protection from hurricanes and tropical storms that could otherwise proceed unimpeded into the bay, perhaps permanently altering its vegetation structure and the salt content of its waters. The last major catastrophic storm in the area was Hurricane Elena in 1985.

Human Pressures

The major human settlement in the area is the town of Apalachicola, population 2,231 according to the 2010 U.S. Census. Apalachicola was once the third-largest port on the Gulf of Mexico due to the shipment of cotton that was brought down the Apalachicola River. This 19th-century economy was severely disrupted due to a blockade of the harbor by the Union Navy during the American Civil War, and the area has never recovered its former economic stature. The town today has become largely reliant on tourism. Across the bay is Eastpoint, a more industrially oriented small town where the major economic activity is oyster harvesting. The bay is bisected along its far southern portion by a bridge on Highway 90. This is the only major road development in or along the bay.

Aside from the oyster business, the bay economy is mainly generated by nature- and culture-based tourism and recreational fishing. Apalachicola Bay was not directly affected by the 2010 British Petroleum/Deepwater Horizon oil disaster in the Gulf of Mexico, but the aftermath of that event combined with the 2008 Florida real estate crash have reduced speculative pressures around the bay, with many large planned developments since scaled back or canceled. It is thought the biome is the beneficiary of this slowdown.

Litigation has been ongoing since the 1970s regarding allocation and distribution of the freshwater among Florida, Alabama, and Georgia. Rapid and widespread development in the Atlanta, Georgia metropolitan area has had particularly significant negative effects on the bay. The primary water supply for Atlanta comes from the Chattahoochee River, where it is impounded in the Lake Lanier reservoir. This water supply was never envisioned to support a growing population of more than 5.2 million around Atlanta. The situation has been exacerbated in recent years due to prolonged drought. These pressures have meant that Apalachicola Bay is perpetually underfed by freshwater inputs—there is also concern over the added inorganic nitrogen, dissolved ammonia and nitrate, and phosphorus concentrations due to municipal discharges.

Global warming is being monitored and its looming effects projected. The Apalachicola biome "works" due to its long-term balance of warm temperatures, high humidity, and moderate precipitation levels. Changes in this balance will exert subtle yet perhaps far-reaching effects on the bay's biota, for example its unique array of amphibians and reptiles. To date, the Apalachicola Bay biome is home to far greater populations of turtles, frogs, salamanders, and snakes than to lizards; this mixture has evolved over thousands of years with relatively stable, if steadily somewhat drier, local climate.

Changes in water levels or precipitation patterns could force some altered dimensions in the animals' integrated behavior with the plant kingdom here as well. The more aquatic reptile and amphibian species tend to thrive in tupelo-cypress and tupelo-cypress-hardwood stands, while those that are less water-tolerant prefer the pine and mixed hardwood areas. Rising water levels, temperatures, and/or humidity will most certainly put pressure on the range and success of many such species.

John Richard Stepp

Further Reading

Edminston, H. Lee, ed. *A River Meets the Bay: The Apalachicola Estuarine System*. Tallahassee: Florida Department of Environmental Protection, 2003.

Eidse, Faith, ed. *Voices of the Apalachicola*. Gainesville, FL: University Press of Florida, 2006.

McCarthy, Kevin. *Apalachicola Bay*. Sarasota, FL: Pineapple Press, 2004.

Livingston, R. J., et al. "Modelling Oyster Population Response to Variation in Freshwater Input." *Estuarine, Coastal and Shelf Science* 50, no. 5 (2000).

Apo Coral Reef

Category: Marine and Oceanic Biomes.
Geographic Location: Asia.
Summary: The Apo Coral Reef is an atoll-like platform reef considered to be the second-largest contiguous coral reef in the world and the largest in the Philippines.

The Apo Reef is situated approximately 15 nautical miles (28 kilometers) west of Mindoro, the seventh-largest island in the Philippines. The reef forms the core of the Apor Reef Natural Park, which is comprised of a series of islands, coral reefs, and adjacent waters, all of which aid in conserving a vast diversity of regionally and internationally important species. The park encompasses approximately 61 square miles (158 square kilometers), surrounded by an additional 45-square-mile (117-square-kilometer) buffer zone. Apo Coral Reef itself consists of a series of coral reefs within the park, covering approximately 13 square miles (34 square kilometers).

Apo Reef, the primary geographic feature of Apo Island, is the largest atoll-like reef in the Philippines. Its two platform reefs encompass approximately 34 square miles (88 square kilometers). Apo Reef is located at the northern tip of the Coral Triangle, an area of 2.2 million square miles (5.7 million square kilometers) that contains the seas of six countries, including the Philippines, Indonesia, Malaysia, Papua New Guinea, the Solomon Islands, and Timor-Leste. It has been estimated that approximately 25 percent of the world's islands lie within the triangle and that this area harbors at least 500 species of reef-building corals.

The Apo Reef Natural Park includes Apo Island as well as Apo Menor (Binangaan) and Cayos del Bajo (Tinangkapaang). Apo Island, the largest of the islands in the reef conglomeration at 54 acres (22 hectares), contains a shallow lagoon 6.6 to 32.8 feet (2 to 10 meters) deep, a mangrove forest, and beach vegetation. By contrast, Binangaan is a rocky limestone island and Tinangkapaang is mainly coralline rock; neither of these two has much vegetation.

Apo Island provides a variety of habitat types and ecosystems, accounting for the high density and diversity of species. The lagoon is an important source of food, and is a nursery and spawning ground for a variety of coastal and marine species. The mangroves serve as a vital nursery area for juvenile fish and play an important role in nutrient production that contributes to the aquatic food chain. The mangroves in combination with the lagoon are home to several species of fish, stingrays, jellyfish, marine plants, and other organisms. Sandy beaches, beach forest, and karstic rock habitats (geological formations created by the dissolution of soluble bedrock) serve as transitional zones between marine and terrestrial habitats. The beaches serve as nesting habitats for the endangered green (*Chelonia mydas*) and hawksbill (*Eretmochelys imbricata*) sea turtles. Also dependent on these transitional habitats are a variety of crustaceans, mollusks, and polychaetes.

This 2010 photograph shows a marine plant called a yellow crinoid growing on Apo Reef in Occidental Mindoro in the Philippine Islands.(NOAA/Dwayne Meadows)

Aside from the coral reefs, the marine habitat includes extensive seagrass and macroalgae beds on or adjacent to the reef proper; this zone is home to diverse corals, fish, marine plants, and other marine organisms. The marine biodiversity is impressive, with surveys estimating as many as 385 species of fish, 190 coral species, 26 algae species, and seven species of seagrass. Seabirds are also well represented, with as many as 46 migratory and resident species roosting regularly on Apo's three islands.

Environmental Stresses

Due in part to the vast diversity of marine organisms, Apo Reef was viewed as one of the world's premiere diving destinations 30 or 40 years ago. In the 1970s, however, destructive fishing practices such as dynamite, cyanide, muro-ami (a local net-and-crush technique), and strobe-fishing were implemented on Apo Reef, all which had devastating effects on the marine environment. People from as far away as Cebu, Navotas, and other distant Philippines cities came to fish at Apo Reef to reap the rewards. These negative effects became so pronounced that in the 1980s, the international diving community lost interest in the area. A survey conducted in 1994 concluded that the remaining coral covered only 33 percent of the entire reef.

To add insult to injury, an El Niño event in 1998 raised ocean temperatures in the area, which produced a massive coral-bleaching event and killed countless corals. Additionally, the reef is subject to the same threat as low-lying islets and lands around the world: the sea-level rise attributed to man-made global warming, and the attendant higher intensity of storms and cyclones.

Alternative modes of fishing are slowly being developed and implemented in the area. The use of fish aggregation devices (FADs) in the area, for example, has demonstrated that a single FAD can

Crown-of-Thorns Starfish

The latest immediate threat to Apo Reef is the coral-eating crown-of-thorns starfish (*Acanthaster planci*), which can consume up to 30 square feet (78 square kilometers) of live coral annually. These starfish, which plague Apo Reef by the millions, may flourish along the reef due to the absence of predators such as the giant triton (*Charonia tritonis*), Napoleon wrasse (*Cheilinus undulatus*), and harlequin shrimp (*Hymenocera elegans*).

Crown-of-thorns starfish like this one photographed in 2004 consume live coral and are one of several serious threats to the coral of the Apo Reef. (NOAA/Dwayne Meadows)

yield a minimum of 33 pounds (15 kilograms) of fish daily. The FAD technique is a passive way to attract fish that voids the use of destructive nets, draglines, and other previously very prevalent methods. Some locals are resistant to change, but in the long term, both the fishers and the marine environment will benefit.

Conservation Efforts

Updated surveys conducted in 2003 and 2006 concluded that coral coverage had increased to 43 percent and 52 percent, respectively. In addition to the 19-percentage-point increase in coral coverage in a period of 13 years, larger fish have begun to return. Schools of hammerhead sharks, manta and eagle rays, whale sharks, and sperm whales are being sighted regularly, which is a sign that Apo Reef's biodiversity is rebounding. As the reefs recover and the fish stocks increase, it is hoped that the outlying waters will be seeded sufficiently for sustainability.

Apo Reef lies in waters that are under the jurisdiction of the province of Occidental Mindoro in region IV-B of the Philippines; it is administered by the local government of the municipality of Sablayan. The reef was originally declared a marine park by the Philippines president in 1980. Three years later, the local government of Sablayan declared the reef a special Tourism Zone and Marine Reserve. On September 6, 1996, the entire reef was proclaimed a Protected Area under the category of Natural Park, and its surrounding waters were made a buffer zone by virtue of Presidential Proclamation No. 868. In 2006, the Protected Areas and Wildlife Bureau of the Philippine Department of Environment and Natural Resources submitted Apo Reef to the United Nations Educational, Scientific and Cultural Organization (UNESCO) for consideration as a World Heritage Site; it is currently designated on the UNESCO "tentative" list.

In 2007, fishing, collection, and harvesting of any life form within the park was banned by the Philippines government. The entire park was declared a "no take" zone to allow the reef time to recover from decades of overfishing. Instead, the park will be accessible mainly to scientists and to tourists, who in return will assist in generating funds for the protection of the reef, as well as providing an alternative livelihood for area fishers.

JEFFREY C. HOWE

Further Reading

Jennings, Simon, Michael Kaiser, and John D. Reynolds. *Marine Fisheries Ecology.* Hoboken, NJ: Wiley-Blackwell, 2001.

Sheppard, Charles R. C., Simon K. Davy, and Graham M. Pilling. *The Biology of Coral Reefs.* New York: Oxford University Press, 2009.

UNESCO World Heritage Centre Global Strategy. "Apo Reef Natural Park." 2012. http://whc.unesco.org/en/tentativelists/5033.

Appalachian-Blue Ridge Forests

Category: Forest Biomes.
Geographic Location: North America.
Summary: A stretch of forest covering much of the Appalachian mountain range, this highly diverse and old ecosystem has been largely altered by human use.

Covering about 61,500 square miles (159,284 square kilometers) of northeastern Alabama, Georgia, eastern Tennessee, western North Carolina, Virginia, Maryland, and central Pennsylvania, with small extensions in West Virginia, Kentucky, New Jersey, and New York, the Appalachian-Blue Ridge forests cover the highest parts of the central and southern Appalachian mountain range. These forests are delineated as an ecoregion by the World Wildlife Fund, meaning that it is a large stretch of land that contains characteristic, geographically distinct assemblages of natural communities and species. These forests derive their name from the Appalachian-Blue Ridge Mountains, so called because of their bluish color when seen from a distance. The trees on the mountains release isoprene gas into the atmosphere, giving them their characteristic smoky haze and namesake blue color.

The mountains that these forests adorn began forming more than 400 million years ago and were pushed higher when North America and Europe collided about 320 million years ago. The Appalachian Mountains have been heavily eroded over time, giving them a softer, more rolling look than younger, taller mountain ranges. This long evolutionary history, coupled with geologic stability, has helped these forests become some of the world's richest areas of biodiversity. The habitat is important for many endemic species (those evolved locally and found nowhere else in the world). The area is made up of a large variety of land

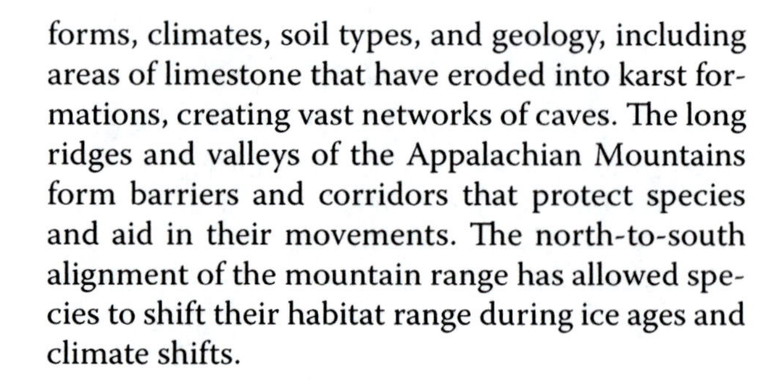

forms, climates, soil types, and geology, including areas of limestone that have eroded into karst formations, creating vast networks of caves. The long ridges and valleys of the Appalachian Mountains form barriers and corridors that protect species and aid in their movements. The north-to-south alignment of the mountain range has allowed species to shift their habitat range during ice ages and climate shifts.

Tree Species

The Appalachian-Blue Ridge forests are home to 158 species of trees—more than any other region in North America. At low elevations, the flora are dominated by deciduous oak forests, transitioning to coniferous spruce fir forests above about 1,350 feet (411 meters). Stunted oak and hickory forests account for most of the Appalachian slope forests. Also common are grasses, shrubs, hemlocks, and mixed oak and pine forests. These forests cover the highest summits in the eastern United States, but despite this elevation, the mountaintop climate is still too warm to create an alpine zone; therefore, no tree line is present on the mountaintops. Instead, the forests generally extend all the way to the mountain peaks. The densest areas of forests tend to be stands of spruce and fir trees.

Animal Species

Many large animals call these forests home, including white-tailed deer (*Odocoileus virginianus*), American black bears (*Ursus americanus*), bobcats (*Lynx rufus*), coyotes (*Canis latrans*), and wild turkeys (*Meleagris gallopavo*). The forests also harbor many threatened and endangered species, including land snails, snakes, spiders,

*The range of the Shenandoah salamander (*Plethodon shenandoah*) is so restricted that the only place it can be found in the world is on three mountains in the Shenandoah National Park. (Flickr/Brian Gratwicke)*

birds, the Virginia big-eared bat (*Corynorhinus townsendii*), and the red wolf (*Canis lupus rufus*), as well as 34 species of lungless *Plethodontid* salamanders that breathe solely through their skin in the moist soils of the mountains.

In fact, the Shenandoah salamander (*Plethodon shenandoah*) has the most restricted range of any terrestrial vertebrate in North America, living solely on three mountaintops in Shenandoah National Park. Salamanders are likely the most abundant vertebrates in the region and account for the highest level of animal biomass in any given patch of forest; therefore, they are extremely important for ecosystem function and nutrient cycling. The Appalachian-Blue Ridge forests have more species of salamanders than anywhere else in the world, an indication that this hot spot could well be the evolutionary origin of this unique order of amphibians.

Threats From Human Activity

Many factors have made the Appalachian-Blue Ridge forests a unique ecosystem, but one that is increasingly threatened with alteration by logging, development, and mining. An estimated 83 percent of the Appalachian-Blue Ridge forests habitat has been altered. Much of it has been logged; cleared for agricultural, urban, and industrial development; or affected by mineral extraction. Valleys and low-elevation areas have been hardest hit, as they are the most accessible and desirable areas for agriculture and residential development.

Only small patches of the original forest biome remain. The largest remnant is in the Great Smoky Mountains National Park, which straddles the border of Tennessee and North Carolina. Many previously altered areas have regrown, such as Shenandoah National Park, which was once almost completely deforested and has since been allowed to regenerate in a largely undisturbed manner. Although these new-growth forests lack many of the large old trees and the historical species diversity of the original forests, they are still important areas for wildlife. Severe fragmentation of habitat in the region threatens species with isolation and the process known as inbreeding depression, which can lead to extinction. Spruce and fir forests were especially heavily logged historically here and have not regenerated in many places. As a result, many mountain summits are covered with low, dense covers of evergreen shrubs; these are called Appalachian bald heaths.

Effects of Introduced Species

Introduced species have also altered the landscape of this forest ecosystem. Lower-elevation forests were historically dominated by a mix of oak and American chestnut trees (*Castanea dentate*). In the early 1900s, the chestnut blight—a disease caused by a pathogenic fungus (*Cryphonectria parasitica)* that was accidentally introduced through importation of Asian chestnut trees—almost wiped out American chestnut populations. This disease changed the makeup of the forest forever, removing the chestnut as an important food source for wildlife and a valuable timber resource for humans.

Other introduced species—such as gypsy moths (*Lymantria dispar*), a foliage-eating pest, and the hemlock woolly adelgid (*Adelges tsugae*), a sap-sucking insect—threaten current forest stands.

Removal of large predators such as wolves and mountain lions has allowed overpopulation of deer and rodents, placing further pressure on the forest through overgrazing. Currently, there is great concern about the overconsumption of acorns and oak seedlings by white-tailed-deer populations, which are preventing the recruitment of new oak trees in forests. This selective overconsumption may cause a shift in the makeup of affected forests from oak-dominated to maple-dominated within the next few decades.

Other Environmental Stresses

Higher-elevation trees are more greatly exposed to the effects of climate change, acid precipitation, and air pollution. Mountain summits are also vulnerable to mountaintop-removal mining—a process in which the soil crests of mountains are blasted off to allow access to the underlying seams of coal. The overburden or excess removed soil is then either replaced, in an attempt to restore the original contours of the mountain, or dumped into neighboring valleys. Although this procedure is perhaps an effort toward developing an economical source of energy that is an alternative to oil, its environmental effects have raised many concerns.

Soil that is dumped into neighboring valleys chokes streams and destroys sensitive ecosystems. Even when soil is replaced on the summit, it is often taken over by nonnative vegetation, and the original biodiversity is unlikely to be restored. There is also concern about the health effects of contaminated stream water, airborne toxins, and dust on humans and wildlife alike.

Conservation Efforts

Much of the remaining land in the Appalachian-Blue Ridge forests has been purchased by state and federal governments, and set aside as protected forests and parks. Most of these lands are mandated to be managed under a multiple-use management plan, mixing timber-harvest regimes with recreation and wildlife purposes. Many sustainable timber-harvest programs help encourage healthy succession in forests. Seasonal hunting in forests and parks is also used to control populations of white-tailed deer and to prevent overgrazing and population-density pressures.

The Appalachian-Blue Ridge forests are traversed by the Appalachian Trail, which begins in Georgia and ends in Maine, giving hikers and campers access to these beautiful forests. Future changes in public demand may steer forest management away from consumptive uses and toward view sheds, camping, hiking, and biodiversity protection.

Hannah Bement

Further Reading

Frick-Ruppert, Jennifer. *Mountain Nature: A Seasonal Natural History of the Southern Appalachians.* Chapel Hill: University of North Carolina Press, 2010.

Hoekstra, J. M., J. L. Molnar, M. Jennings, C. Revenga, M. D. Spalding, T. M. Boucher, J. C. Robertson, and T. J. Heibel, eds. *The Atlas of Global Conservation: Changes, Challenges, and Opportunities to Make a Difference.* Berkeley: University of California Press, 2010.

Ricketts, T. H., E. Dinerstein, D. M. Olson, C. J. Loucks, et al. *Terrestrial Ecoregions of North America: A Conservation Assessment.* Washington, DC: Island Press, 1999.

Spira, Timothy. *Wildflowers and Plant Communities of the Southern Appalachian Mountains and Piedmont.* Chapel Hill: University of North Carolina Press, 2011.

Appalachian Mixed Mesophytic Forests

Category: Forest Biomes.
Geographic Location: North America.
Summary: This deciduous forest community, with rich soils, is in the Appalachian Mountains of the southeastern United States.

The Appalachian mixed mesophytic forests are characterized as moist deciduous forests with nutrient-rich mesic (moderately moist) soils, found throughout the southeastern United States from Tennessee to Pennsylvania. These temperate forest communities have well-developed soil profiles and provide a home for a suite of endemic (adapted specifically for, and unique to, a particular habitat) species including birds, amphibians, and other freshwater-friendly organisms. The forest is composed largely of deciduous tree species; it provides a home for a wide variety of shrubs, ferns, and herbaceous vegetation.

While related to the Appalachian-Blue Ridge forests biome, this biome is spread along the slopes and valleys to the west of that zone—thereby receiving more precipitation from the prevailing west-to-east weather patterns—and exists at markedly lower elevations as well. The Appalachian mixed mesophytic forests are considered to be among the most diverse temperate forest ecosystems on the globe today. These biota-rich forests provide a glimpse into a forest cover type that previously dominated much of the Earth.

Geology and Climate

The Appalachian mixed mesophytic forests are located in two physiographic provinces: the Appalachian Plateau, Ridge, and Valley; and the Blue Ridge provinces. The forest sits upon sandstone, siltstone, shale, and coal, which are the parent material for the rich soils that support a great diversity of plants and animals. Soils here are characterized as mesic loams, with high pH, high nutrient-holding capacity, and rich organic layers. These soils drive the region's rich diversity in vegetation but are less common on the steep hillsides and the very mountainous terrain within this region.

The region is characterized by its warm summers and cool winters, and is classified as a temperature humid continental climate. This mild climate is another support for the biological diversity to thrive here.

Vegetation Community

These forests can be subdivided into lower-elevation and higher-elevation communities, both of which generally are composed of deciduous tree species. Furthermore, the Appalachian mixed mesophytic forests host a series of unique habitats, including relict populations from when the climate was much cooler, so species in these habitats are more similar to northeastern plants and animals than to their southeastern counterparts.

Within the lower-vegetation forests, the tree canopy is dominated by an oak-hickory forest type, which is composed of various oak species (*Quercus* spp.) and hickories (*Carya* spp.), with lower densities of walnuts (*Juglans* spp.), elms (*Ulmus* spp.), birches (*Betula* spp.), ashes (*Fraxinus* spp.), basswoods (*Tilia* spp.), maples (*Acer* spp.), locusts (*Robinia* spp.), and pines (*Pinus* spp.). A second very common forest type within the Appalachian mixed mesophytic lower-elevation forests is composed of tulip poplar (*Liriodendron tulipifera*), blackgum (*Nyssa sylvatica*), Eastern hemlock (*Tsuga canadensis*), black cherry (*Prunus serotina*), sweetgum (*Liquidambar styraciflua*), American beech (*Fagus grandifolia*), and yellow buckeye (*Aesculus octandra*).

Before the decimation of the American chestnut (*Castanea dentata*) by the fungal pathogen chestnut blight (*Cryphonectria parasitica*), chestnuts were dominant features in this landscape. Similarly, the Eastern hemlock today is threatened by an introduced pest, the hemlock woolly adelgid, that has led to dramatic losses of hemlocks throughout this region and the entire eastern United States.

The higher-elevation forests are composed largely of yellow birch (*Betula alleghieniensis),* mountain maple (*Acer spicatum),* sugar maple (*Acer saccharum),* American beech, and Eastern hemlock, with associated understory shrubs such as mountain laurel (*Kalmia latifolia*) and rhododendron (*Rhododendron* spp.).

Additionally, these forests are home to a variety of smaller ecosystems, including bogs dominated by cranberry, blueberry, and sphagnum moss, which provide important habitat for species (such as fisher martens, Northern goshawk, and parasitic plants) that thrive in the acidic environment. This type of bog is more typical of environments such as northern New England. Their presence in the Appalachians provides insights into a cooler climate regime. Other tree species in the mesophytic forests that are reminiscent of cooler past climates include the Canada yew (*Taxus canadensis*), Eastern larch (*Larix laricina*), red pine (*Pinus resinosa*), and balsam fir (*Abies balsamea*).

Important Fauna

The region is especially noted for its high diversity of freshwater organisms, especially fish such as the river trout, which not only provide recreational opportunities for fishing, but also perform important ecosystem services such as nutrient cycling within these freshwater communities. Stream macroinvertebrates are also important players within freshwater communities in the Appalachian mixed mesophytic forests. Many studies have researched the importance of these small organisms in nutrient cycling and fine woody debris decomposition. These organisms act as indicators of stream health (with the loss of pollution-sensitive invertebrates as stream quality declines). A variety of endemic fish species can also be found within the Appalachian mixed mesophytic forests, serving as further indicators of water quality and stream health.

The forests are home to mammals such as the big brown bat, martens, weasels, and black bears, as well as a variety of shrew species. They provide critical habitat for amphibians such as frogs, toads, and salamanders. The Appalachian mixed mesophytic forests are known as the global center of diversity for wood salamanders, which prefer the mild climate and moist conditions. A great diversity of land snails can also be found within these forests, which are considered to be a biodiversity center with claims for the highest endemism for land snails in localized environments.

This biome provides habitat for avian species including warblers (*Dendroica* spp.), vireos (*Vireo* spp.), Northern goshawks (*Accipiter gentilis*), and wood thrushes (*Catharus* spp.). Several reptiles of concern include the scarlet king snake (*Lampropeltis elapsoides*) and timber rattlesnake (*Crotalus horridus*), but a great diversity of reptiles exist and thrive throughout the ecoregion.

The Appalachian mixed mesophytic forest supports additional endemic species, including the Allegheny plum tree (*Prunus alleghaniensis*), the flattened musk turtle (*Sternotherus depressus*), and the Alabama map turtle (*Graptemys pulchra*). Endemic salamanders include the Black Mountain salamander (*Desmognathus welten*), Southern dusky salamander (*Desmognathus auriculatus*), Jordan's salamander (*Plethodon jordani*), and Cheat Mountain salamander (*Plethodon nettingi*).

Land Management

With urbanization occurring throughout the United States, a common policy of not burning forests and grasslands was enacted for forest managers: the so-called no-burn policy. Although this policy has many benefits for humans inhabiting the area, it can cause detrimental effects on the ecosystems that require fire to sustain their natural biologic rhythms. In the mixed mesophytic forests, this policy was especially harmful and disruptive because it reduced and even eliminated naturally occurring fires that, while frequent, usually occur only over a small portion of land and are not catastrophic to the entire ecosystem. Under a natural-burning fire regime, fuel loads do not build up. Conversely, no-burn leads to such buildup, which often allows forest fires to escape to the canopy or tree crowns and destroy virtually all the vegetation within a local community.

Frequent low-intensity fires provide numerous benefits to the ecosystem, including large inputs of nutrient from burned wood, which in turn jump-starts soil fungal and microbial communities to assimilate these new nutrients back into biologically available forms for new understory growth. Frequent low-intensity fires also self-regulate the density of vegetation within forests and can clear out competitors, leading to enhanced growth for the trees and plants that remain after the fires. The Appalachian mixed mesophytic forests experienced these types of low-intensity and frequent surface fires, which promoted the growth and establishment of oak species within this forest type, where oaks make up a large portion of the tree species.

As the United States was industrialized, these practices were switched to high-intensity, catastrophic, stand-replacing fires, which destroyed much of the forest. In more recent times, the no-burn policy has eliminated fire altogether, which, while initially helpful for forest regeneration, now encourages replacement of species such as oaks, which will in turn change the composition of these unique forest communities.

Threats to Community

A variety of forces are threatening the Appalachian mixed mesophytic forests, chief among them anthropogenic factors such as global climate change and pollution, and lesser but sometimes more direct forces such as plant invasions and introduced pests and pathogens. Global climate change poses perhaps the greatest threat, as higher average temperatures, for example, add stress to key tree species by enabling the arrival and penetration of bark-boring pests and their accompanying fungal parasites. Some forms of air and water pollution can greatly diminish the diversity of sensitive species such as stream macroinvertebrates and upset the balance of supportive fungal–plant associations like lichens. Habitat fragmentation is another threat; already it has dramatically reduced the connectivity of the landscape, affecting various species in unpredictable ways.

Regional threats to these forests include invasive animal pests such as the hemlock wooly adelgid; invasive plant species such as Japanese stiltgrass; and habitat fragmentation due to development, logging, mining, and such disruptive add-ons as pipeline construction. Although this forest community is a gem among forested habitats, these regional threats combined with global climate change can profoundly shape the trajectory of future forests, especially the majestic Appalachian mixed mesophytic forests.

RELENA R. RIBBONS

Further Reading

Braun, E. Lucy. *Deciduous Forests of Eastern North America.* Caldwell, NJ: Blackburn Press, 1950.

Eyre, F. H., ed. *Forest Cover Types of the United States.* Bethesda, MD: Society of American Foresters, 1980.

Muller, Robert N. "Vegetation Patterns in the Mixed Mesophytic Forest of Eastern Kentucky." *Ecology* 63, no. 6 (1982).

Arabian Sea

Category: Marine and Oceanic Biomes.
Geographic Location: Northwestern part of the Indian Ocean.
Summary: A large marine ecosystem notably enriched with high productivity supporting fisheries; also a place of great seasonality in weather patterns: an engine of biodiversity.

The Arabian Sea composes the northwestern part of the Indian Ocean, lying west of India and covering a total area of approximately 1,491,000 square miles (3,862,000 square kilometers), and has formed part of the principal sea route between Europe and India over centuries. It is bounded to the west by the Horn of Africa and the Arabian Peninsula, with entrances to the Gulf of Aden/Red Sea and Gulf of Oman/Persian Gulf respectively; to the north by Iran and Pakistan; to the east by India; and to the south by the remainder of the Indian Ocean. The largest river flowing into the Arabian Sea is the Indus River; others include the Netravathi, Sharavathi, Narmada, Tapti, Mahi, and the numerous rivers of the Indian state of Kerala.

The Arabian Sea is a Large Marine Ecosystem (LME), as defined by the U.S. National Oceanic and Atmospheric Administration (NOAA): "relatively large areas of ocean of ocean space of approximately 200,000 square kilometers [77,000 square miles] or greater, adjacent to the continents in coastal waters where primary productivity is generally higher than in open ocean areas." The Arabian Sea LME is considered a Class I, highly productive ecosystem based on SeaWiFS global primary productivity estimates. It has an average depth of 8,970 feet (2,734 meters).

The Arabian Sea has some of the most extreme climatic regimes, due to seasonal fluctuations in air and water temperatures. The sea's average temperature, which used to be 72 to 81 degrees F (22 to 27 degrees C) until the 1980s, is now 81 to 90 degrees F (27 to 32 degrees C), resulting in occurrence of some stronger cyclones. An example is Cyclone Gonu, occurring in 2007 and the strongest cyclone ever recorded in the northern Indian Ocean; it attained a peak wind intensity of 165 miles per hour (270 kilometers per hour). Recent increases in the average intensity of tropical cyclones in the Arabian Sea may be a side effect of increasing air pollution over the Indian subcontinent as well as the higher water temperatures in the sea itself.

The Arabian Sea contributes to a monsoon climate in the surrounding region by providing the water necessary for wet storms. In summer, strong winds blow from the southwest to the northeast, bringing rain to the Indian subcontinent. During the winter, the winds are milder and blow in the opposite direction, from the northeast to the southwest.

During the monsoon season, the generally southwest winds are particularly cold, and are so strong that they succeed in sweeping away some of the salt content of the upper levels of seawater. In monsoon season, the upper waters are less then 35 practical salinity units (psu), while in the non-monsoon season salinity is over 36 psu. Monsoons are characteristic of the Arabian Sea and are responsible for the yearly cycling of its waters.

Species and Economy

Scientists have found that because the Arabian Sea has naturally high productivity, it also has one of the thickest zones of oxygen-depleted water below 330-feet (100-meters) deep. Depending on monsoon winds, local topography, the width and depth of the continental shelf, and drainage of coastal areas, there are three coastal ecosystems in the Arabian Sea, each characterized by its own productivity and species distribution. The Arabian Sea's predictable, seasonally reversing monsoons drive one of the most energetic current systems in the world. The strength of the monsoon winds is regulated; during the monsoon season (May to October), the southwesterly winds in this region are from the southwest, inducing a great deal of evaporation from the warm waters of the Arabian Sea, and heavy rainfall along the coy, a thermal gradient that develops from different heating of land and sea. The winds blow toward India and cause upwelling of low-oxygen waters. There is a concentration of fish in nearshore areas at that time. During the other half of the year, the winds blow in the opposite direction, and not as strongly.

The LMEs produce about 80 percent of the annual world's marine fisheries catch. The United Nations Food and Agriculture Organization (FAO) 10-year trend shows an increase in capture trends; from 1.9 million tons in 1990 to 2.2 million tons in 1999. There are catches of herrings, sardines, anchovies and crustaceans. India's southwest coast is fished for oil sardines (*Sardinella longiceps*), mackerels (*Rastrelliger kanagurta*), and tunas (*Euthynnus affinis* and *Auxis thazard*). Most of this catch comes from a narrow six- to 10-mile (10- to 15-kilometer) coastal belt, and accounts for 23.6 percent of India's fish catch.

The dominant fish species off India's central west coast ecosystem are Sciaenids (*Pseudosciaena diacanthus*), Carangidae (*Caranx* spp.), and Engraulidae (anchovies). The dominant species off India's northwest coast are prawn, Sciaenids (*Pseudosciaena diacanthus*), and Carangidae. Small tuna (*Euthynnus affinis*) migrates to this area to breed. There are also catches of herrings, sardines, anchovies, crustaceans and spiny lobster (*P. homarus megasculptus*), with zonal abundance in relation to surface circulation patterns.

A wide variety of invertebrates and algae exist, including *Sargassopsis zanardinii*, a marine species that is endemic to the Arabian coastline. The endemic (locally evolved and unique to a particular biome) fishes include certain members of the barracuda (*Sphyraena*), wrass (*Labridae*), and damselfish (*Pomacentridae*) families. Also found in the Arabian Sea are dugong (*Dugong dugon:* order *Sirenia* vulnerable International Union for Conservation of Nature [IUCN] 2.3) and several species of turtles, including the green turtle (*Chelonia mydas:* IUCN-listed endangered species), hawksbill turtle (*Eretmochelys imbricata,* critically endangered IUCN), and olive Ridley turtle (*Lepidochelys olivacea* vulnerable, IUCN 2.3).

Of the baleen whales, the following species have been recorded: Bryde's whales (*Balaenoptera edeni,* data deficient IUCN 3.1), minke whales (*B. acutorostrata,* least concern IUCN 3.1), fin whales (*B. physalus,* endangered IUCN 3.1), blue whales (*B. musculus,* endangered IUCN 3.1), and humpback whales (*Megaptera novaeangliae,* least concern IUCN 3.1). Toothed whales include sperm (*Physeter macrocephalus:* vulnerable A1d), orca (*Orcinus orca,* data deficient IUCN 3.1), and false killer (*Pseudorca crassidens,* data deficient IUCN 3.1). At least a dozen species of dolphins (family *Delphinidae*) as well as the finless porpoise (*Neophocaena phocaenoides,* vulnerable IUCN 3.1) exist in the Arabian Sea.

Pressures on Biota

The Arabian Sea, due to its seasonal weather fluctuations, offers excellent examples of biological adaptation to environment. However, the diverse aquatic habitat is currently under threat from the oil industry, which uses the sea as a shipping lane. As a result, oil spills, anchor damage, sedimentation, and other pollution effects are severe threats and have long-standing effects. Mining operations, fishing pressures, destructive fish collecting practices (e.g., dynamiting), residential and commercial development, and effluent discharge have resulted in altered species composition in many areas. Recreation and tourism also contribute to eutrophication and reef degradation. War-related activities provide another source of environmental damage.

Oil from accidents and bilge washings can reach the coast and impact coastal ecosystems. On August 7, 2010, a spill of about 88,040 tons of lubricant oil was poured in the Arabian Sea due to a collision off two cargo ships; this was identified as a cause of tangible damage to mangrove plants and other pelagic organisms. Current fishing methods have resulted in the overexploitation of coastal fishery resources such as prawns, sardines, pomfrets and mackerel. However, most of the fish stocks breed in deeper offshore waters so there is an opportunity to rebuild stocks. The over-exploitation is mostly because of large fishing vessels that fish illegally near the coast.

Population expansion in India will continue to put enormous pressure on the living and nonliving coastal resources. This population pressure also creates major pollution problems, like untreated organic waste as well as sewage, which contribute to the nutrient loading of nearshore areas. Nutrient loading influences productivity cycles and depletes dissolved oxygen supplies. Rapid industrialization is also contributing pollution from industrial wastes and industrial effluents. Heavy metals such as cadmium, lead, manganese, and zinc have been found in core sediment samples to depths up to 18 inches (45 centimeters). Large amounts of pesticides are also deposited in the coastal areas, particularly off the Mumbai coast. This situation would require the dredging and removal of contaminated sediments for proper mitigation. Currently, plans are being considered that would protect the delicate wildlife of the Arabian Sea, particularly the turtle and coral populations.

People fishing from small nonpowered boats with hand nets while cargo ships unload in the background on the coast of India. As much as 65 percent of the fish landed from the Arabian Sea comes from small boats like these. (World Bank)

Sixty-five percent of fish landings in the Arabian Sea derive from artisanal (small scale) fisheries. Coastal populations have traditionally relied on nonmotorized boats. Along its Arabian Sea coast, India operates about 180,000 country crafts, 26,000 motored traditional vessels, 34,000 mechanized boats, and a few large boats. The most important export is prawn. It is uneconomical to harvest the deepwater fisheries here because of low market price, except for tuna, which commands a good price.

Climate change is an interconnected event based on long-term change of regional weather that is thought to be caused by an increase in atmospheric carbon. It is thought that climate change is causing global warming. In the Arabian Sea, it has been recently suggested that climate change is also causing intensified monsoon winds and, in turn, intensified upwellings. This results in huge algal blooms in the western half of the Arabian Sea causing serious changes in productivity. This has detrimental effects by causing oxygen depletion at depth. In addition, incidence of fish mortalities is recorded in association with algal blooms.

Santosh Kumar Sarkar

Further Reading

Dwivedi, S. N. and A.K. Choubey. "Indian Ocean Large Marine Ecosystems: Need for National and Regional Framework for Conservation and Sustainable Development." In Kenneth Sherman, E. Okemwa, and M. Ntiba, eds. *Large Marine Ecosystems of the Indian Ocean: Assessment, Sustainability, and Management.* Cambridge, MA: Blackwell Science, 1998.

Food and Agriculture Organization (FAO) of the United Nations. *Trends in Oceanic Captures and Clustering of Large Marine Ecosystems—Two Studies Based on the FAO Capture Database.* Rome: FAO, 2003.

Goes, Joaquim I., Helga Gomes, Prasad Thoppil, Prabhu Matondkar, and Adnan Al Azri. *Eurasian Warming—Hydrography and Biological Productivity in the Arabian Sea.* East Boothbay, ME: Bigelow Laboratory for Ocean Sciences, 2007.

Pollock, David E. "Spiny Lobsters in the Indian Ocean: Speciation in Relation to Oceanographic Ecosystems." In Kenneth Sherman, E. Okemwa, and M. Ntiba, eds. *Large Marine Ecosystems of the Indian Ocean: Assessment, Sustainability, and Management.* Cambridge, MA: Blackwell Science, 1998.

Arafura Sea

Category: Marine and Oceanic Biomes.
Geographic Location: Pacific Ocean.
Summary: This shallow tropical sea, located in the southwestern Pacific Ocean between Australia and New Guinea, has been threatened by overfishing.

Situated between the Timor and Coral Seas in the southwestern Pacific Ocean, the Arafura Sea is a shallow tropical, semi-enclosed, continental-shelf basin spanning 251,000 square miles (650,087 square kilometers) in the East Indian archipelago. The name *Arafura* is derived from the inhabitants of the Maluku Islands of Indonesia, known as Alforas or Haraforas to 19th-century English anthropologists.

Straddling the Indian Ocean–Australian continental shelves, the Arafura overlays a section of the Sahul Shelf known as the Arafura Shelf, part of a prehistoric land bridge that once connected present-day Australia and New Guinea and was key to the intercontinental migration of humans from Asia into Australia.

The Arafura is bordered by the eastern islands of Indonesia, the Ceram and Banda Seas to the northwest, the southern coast of New Guinea to the north and northeast, the Torres Strait to the east, the Gulf of Carpentaria to the southeast, the northern coast of Australia to the south, and the Timor Sea to the west. It is approximately 800 miles (1,287 kilometers) long and 350 miles (563 kilometers) wide, with depths of 165 to 265 feet (50 to 81 meters) increasing to the west, with depths up to 2,000 feet (610 meters) along its western edge. For most of the year, the Arafura Sea experiences stable trade winds, with monsoons appearing during the summer months.

Marking the boundary between the Indian and Pacific Oceans, the Arafura Sea is an important part of thermohaline circulation, the global heat conveyor belt that is critical to the regulation of the Earth's climate. During the southeast Asian summer monsoon system, a four-month period of extensive convective thunderstorms, the warm waters of the Arafura are pulled westward into the Indian Ocean by the South Equatorial Current. During the Indian winter, conversely, the Equatorial Counter Current reverses the flow, thus known as the Indonesian Throughflow.

Marine Biodiversity

Along with the neighboring Timor Sea, the Arafura Sea is considered to be one of the last remaining areas of tropical marine biodiversity in the Southeast Asia–Australia region, with an estimated 840 to 1,000 species, including corals, marine worms, crustaceans, echinoderms, mollusks, pelagic fish, demersal fish, and benthic fish. Biologically speaking, however, the Arafura remains relatively unknown, with the majority of the studies being done in respect to its commercial fisheries.

The first biological survey of the Arafura's benthic fauna—the animals that live on or near the seabed—was conducted by the Australian Museum in 2005 and revealed several taxa new to science, including the polychaete families *Hartmaniellidae* and *Longosomatidae.*

Threats from Overfishing

The rich marine life of the Arafura has been threatened by overfishing and by unreported and illegal fishing. Between 1992 and 2000, the number of shrimp trawlers operating in the Arafura doubled, while the number of fish trawlers more than tripled. The sustainable shrimp catch limit for the Arafura has been estimated at 21,700 tons (19,686 metric tons) per year. By 2000, however, that limit had already been exceeded even without taking into account illegal fishing. Overfishing pearls, which are harvested in the clear, protected waters surrounding the Aru Islands in the north, has also been reported. A study conducted between 1999 and 2003 found that the Arafura Sea red snapper stock was undergoing a collapse.

In recent years, conservationists have called for tighter fishing regulations and even a fishing moratorium in the Arafura Sea's Exclusive Economic Zones to allow the depleted fish stocks to recuperate. The Arafura Sea and Timor Sea Expert Forum was established to promote the sustainable development of this region to protect not only the area's rich biodiversity, but also the livelihoods of the local communities that depend on a healthy ecosytem.

One of these communities is the Asmat, a group of some 70,000 indigenous people known for head-hunting and wood carving who live along the narrow, muddy rivers that flow into the Arafura Sea on the southern coast of New Guinea. The Asmat use dugout canoes and large hoop nets to catch fish and shrimp.

Hydrocarbon Deposits and Oil Production

Geologically, the Arafura Sea is bordered by the Australian craton to the south, the Tertiary collision zone to the south, the Banda arc collision zone to the west and northwest, and the western limit of late Paleozoic granites to the east. The sediment of the Arafura is rich in calcium carbonate and ranges in age from 540 million to 65 million years, covering the late Paleozoic, Mesozoic, and Cenozoic eras. The seabed of the Arafura has been the focus of energy exploration in the form of hydrocarbons such as oil and natural gas.

Crude oil and natural gas were first discovered here in 2000 by the Abadi-1 exploratory well, operated by the Tokyo-based oil company INPEX. In 2010, the Indonesian government approved a development plan to produce 2.5 million tons a year in the Abadi gas field. In 2011, the Indonesian government awarded the British energy company BP two energy-production-sharing contracts covering more than 6,000 square miles (15,540 square kilometers) of the Arafura Sea.

Reynard Loki

Further Reading

Earl, George Windsor, Dirk Hendrik Kolff, and Justin Modera. *Sailing Directions for the Arafura Sea*. London: Hydrographic Office, 1837.

Institute of Southeast Asian Studies. *Working With Nature Against Poverty: Development, Resources and the Environment in Eastern Indonesia*. Canberra: Australian National University, 2009.

International Hydrographic Organization. *Limits of Oceans and Seas.* Special Publication 23, 3rd ed. Monte Carlo: International Hydrographic Organization, 1953.

Katili, John A. *Future Petroleum Provinces of the World M 40: Geology and Hydrocarbon Potential of the Arafura Sea.* Tulsa, OK: American Association of Petroleum Geologists, 1986.

Rockefeller, Michael. *The Asmat of New Guinea*. Greenwich, CT: New York Graphic Society, 1967.

Wilson, George D. F. *Arafura Sea Biological Survey: Draft Report on Benthic Fauna Collected During RV Southern Surveyor Voyage 05-2005*. Sydney: Australian Museum, 2010.

Arafura Swamp

Category: Inland Aquatic Biomes.
Geographic Location: Australia.
Summary: One of Australia's largest freshwater wetlands, the Arafura Swamp is among the most undisturbed ecosystems in the world.

The Arafura Swamp is a large freshwater wetland situated at the northern coast of the Arnhem Land region in the Northern Territory of Australia. This biodiverse and virtually undisturbed tropical floodplain is home to a wide variety of species, some of which are endangered or vulnerable. The Arafura Swamp is an ecologically and culturally important region to the indigenous Yolngu people. In addition to monitoring the situation of threatened species, conservation efforts have centered on the negative effects caused by fires, feral species, weeds, and the intrusion of saltwater. Australia lists the Arafura Swamp as a wetland of national significance.

Occupying the expansive floodplain of the Goyder–Glyde river system in the Top End of Australia's Northern Territory, and adjoining the coastal plain and tidal waterways of Castlereagh Bay to the north, the pristine Arafura Swamp spans an area of some 300 square miles (777 square kilometers) in Arnhem Land, expanding up to 500 square miles (1,295 square kilometers) by the end of the wet season. It is the largest freshwater basin in East Arnhem Land, one of the largest wooded swamps in the Northern Territory, and one of the most undisturbed ecosystems in the world. Vast, open, and mostly inaccessible, the Arafura receives surface-water inflow from, and provides a flood-control basin for, the Goyder and Gulbuwangay rivers. Additionally, it is fed from below by the numerous springs that line the Goyder River. The water disperses through an irregularly shaped floodplain, discharging northward through the saline flats of the tidally affected Glyde River before emptying into the Arafura Sea through Castlereagh Bay along the northern coast.

Unlike other large coastal swamps in the Northern Territory that open up at their downstream limit toward the coast, the Arafura Swamp is surrounded by a low plateau that rises in parts to 300 feet (91 meters), dominated by scarps to the east and west. It is believed that this bedrock, which constricts the swamp at the northern end, forms a partial barrier preventing the influx of tidal water, creating Arafura's uniquely demarcated wetland basin. In the Top End, the Arafura Swamp is also unique because of the perennial inflow from the springs that line the Goyder River, as well as its lack of a continuous river channel to the Arafura Sea—conditions that result in some portion of the basin being flooded at all times throughout the year. The region has a tropical savanna climate, with average annual precipitation of more than 40 inches (102 centimeters). Rain falls primarily from December to April.

Biodiversity

The Arafura represents a complex, variegated network of landforms and wetland habitats, including virtually undisturbed grasslands, woodlands, coastal plains, sandstone hills, billabongs, lagoons, sinkholes, and drainage channels. Scattered along the edges of the basin are some 7,413 acres (3,000 hectares) of monsoon rainforest that have remained undisturbed by buffalo or pigs. Featuring a wide array of grasses, herbs, sedges, aquatic plants, trees, and floating mat communities, the swamp is dominated by two paperbark species: *Melaleuca cajuputi* and *M. leucadendra,* constituting Australia's largest contiguous forest of *Melaleuca,* a genus of the myrtle family primarily endemic to Australia and commonly known as paperbark because it sheds its bark in flat, paper-like sheets.

Several rare and notable examples of flora thrive in the Arafura Swamp, such as the Capentaria palm (*Carpentaria acuminata*); the rare talipot palm (*Corypha elata*); the fan palm (*Livistona rigida*); the commelinid *Hanguana malayana*, uncommon to other wetlands; the taro (*Colocasia esculenta*); and almost the entire Northern Territory population of the large and imposing Gebang Palm (*Corypha utan*), known for its spectacular inflorescence, which features up to one million individual flowers. The swamp is also home to at least two endangered plant species: the Australian arenga palm (*Arenga australasica*) and *Freycinetia percostata.*

The Arafura supports diverse fauna, including more than 50 plant and five vertebrate species endemic to the Northern Territory, and three recorded plant species known only to the Arnhem Coast within the Northern Territory, though they are also found in other Australian states. The swamp is home to at least five known threatened animal species: the Australian bustard (*Ardeotis australis*), partridge pigeon (*Geophaps smithii*), Merten's water monitor (*Varanus mertensi*), and yellow-spotted monitor (*Varanus panoptes*), all of which are classified as vulnerable; and the critically endangered Northern quoll (*Dasyurus hallucatus*), a native carnivorous marsupial. The only significant breeding population of the endangered hooded parrot outside the Katherine area lives in land adjacent to the swamp. The threadfin rainbowfish (*Iriatherina werneri*), once known only from the Cape York Peninsula in Queensland and in neighboring New Guinea, has been discovered in the Arafura.

Bird Species

A total of 17 migratory species recorded from the Arafura Swamp are protected under bilateral agreements or international conventions. Classified as an Important Bird Area by the Australian government, the Arafura Swamp supports abundant populations, including the red-collared lorikeet (*Trichoglossus rubritorquis*); the brown honeyeater (*Lichmera indistincta*); the silver-crowned friarbird (*Philemon argenticeps*); the white-bellied cuckoo-shrike (*Coracina papuensis*); and Australia's smallest bird, the weebill (*Smicrornis brevirostris*).

The wetlands are also an important refuge and breeding ground for significant populations of several species of waterbirds, most commonly the magpie goose (*Anseranas semipalmata*) and various egret species (*Egretta*). Other waterbirds found in substantial numbers are the green pygmy goose (*Nettapus pulchellus*), the Pacific black duck (*Anas superciliosa*), the little black cormorant (*Phalacrocorax sulcirostris*), the glossy ibis (*Plegadis falcinellus*), the Australian white ibis (*Threskiornis molucca*), the straw-necked ibis (*Threskiornis spinicollis*), the brolga (*Grus rubicunda*), the radjah shelduck (*Tadorna radjah*), the royal spoonbill (*Platalea regia*), the Australian wandering whistling duck (*Dendrocygna arcuata australis*), and darters (*Anhingidae*).

Reptile and Mammalian Species

Reptiles include the freshwater crocodile (*Crocodylus johnstoni*) and saltwater crocodiles (*Crocodylus porosus*).

*Wandering whistling ducks (*Dendrocygna arcuata australis*) in a wetland environment in Australia's Northern Territory. The Arafura Swamp has been named an Important Bird Area by the Australian government and provides a refuge for at least 17 migratory bird species. It is considered to be among the most undisturbed ecosystems on Earth. (Thinkstock)*

Mammalian species include the brushtail possum (*Trichosurus vulpecula*), the northern brown bandicoot (*Isoodon macrourus*), the agile wallaby (*Macropus agilis*), and the delicate mouse (*Pseudomys delicatulus*). The area is also home to a significant number of two species of bats of the genus *Pteropus* (also known as fruit bats or flying foxes), which roost and feed in the Melaleuca forests: the little red flying fox (*P. scapulatus*) and the black flying fox (*P. alecto*).

Environmental Threats

Listed on the Register of the National Estate and part of the Arnhem Land Aboriginal Land Trust, the Arafura Swamp is Aboriginal freehold land, one of the few remaining tropical wetlands managed by Aborigines who employ traditional land management methods such as forest burning regimes.

Though it is one of the most undisturbed wetlands in Australia, the Arafura Swamp is threatened on several fronts. The ecological effects of grazing buffalo; the spread of weeds and invasive exotic pasture grasses; the recent arrivals of feral pigs, feral cats, and cane toads (a poisonous amphibian that has killed many endangered northern quoll); poorly planned fire regimes; and saltwater intrusion due to the effects of climate change have all been causes of concern for conservationists, some of whom have partnered with the Aborigines to develop sustainable land management policies.

Reynard Loki

Further Reading

Australian Nature Conservation Agency. *A Directory of Important Wetlands in Australia*. Canberra: Australian Nature Conservation Agency, 1996.

Brennan, Kim, John Woinarski, Craig Hempel, Ian Cowie, and Clyde Dunlop. *Biological Inventory of the Arafura Swamp and Catchment*. Darwin, Australia: Department of Infrastructure, Planning and Environment, 2003.

Brocklehurst, P. S. and B. A. Brocklehurst. *Vegetation Communities of Arafura Swamp*. Palmerston, Australia: Conservation Commission of the Northern Territory, 1996.

Chatto, R. *The Distribution and Status of Waterbirds Around the Coast and Coastal Wetlands of the Northern Territory—Technical Report 76*. Palmerston, Australia: Parks and Wildlife Commission of the Northern Territory, 2006.

Aral Sea

Category: Inland Aquatic Biomes.
Geographic Location: Central Asia.
Summary: The Aral Sea, once one of the largest lakes in the world, has became a large-scale environmental tragedy, although it now shows some signs of partial recovery.

The Aral Sea was once known as the fourth-largest inland sea of the world by area, with a surface area of 25,500 square miles (66,100 square kilometers) and mean depth of 53 feet (16.1 meters), reaching up to 221 feet (68 meters) in the deepest portions. This inland lake is located in central Asia and it is bordered by the countries of to the north and by Uzbekistan to the south, both previously members of the Soviet Union. It was characterized by once having more than 1,000 islands; its name stems from that fact.

The Aral Sea region is a semi-desert, with annual rainfall of merely eight inches (20 centimeters) and a continental climate (one marked by strong demarkations in seasonal temperatures). The Aral Sea receives most of its water from the Amu Darya River in the south and Syr Darya River in the north, keeping its salinity content at roughly one percent. Geographic conditions of the area have strongly influenced human activities; people in the northern portion have long raised livestock while the ones in the southern region depended on vegetative agriculture. The Aral Sea was of great importance for the region because besides being an abundant fish resource, it facilitated transportation by boat throughout the provinces.

The story of prosperity ended when in the 1960s the Soviet Union diverted the rivers that fed the Aral Sea in order to serve an ambitious plan to

This satellite image shows the two divided lobes of the eastern and western basins of the South Aral Sea in 2010. That year, the eastern sea had recovered slightly due to heavy snowmelt brought by the Amu Darya River. (NASA)

irrigate thirsty crops like rice, cotton, and cereals in the desert. Irrigation channels were grossly inefficient, with an estimated 30 to 75 percent of the water evaporating or leaking away. The farmers did not practice crop rotation, and the extensive monoculture demanded ever-higher quantities of fertilizer and pesticides. The runoff of these chemicals into the shrinking Aral Sea became a big pollution issue. Even today only 12 percent of Uzbekistan's irrigation channel network has antifiltration linings, and although Uzbekistan became a large exporter of cotton, the Aral Sea has contracted ever since, becoming a great calamity to the region.

In 1987 the continued dwindling away of the water level left the Aral Sea split in two, as the North Aral Sea and the South Aral Sea. In 1998 the sea was just the eighth-largest inland sea, with a water surface area reduced by about 60 percent to 11,000 square miles (29,000 square kilometers). Salinity increased from 10 parts per thousand (ppt) to about 45 ppt, making a substantial negative impact in the region. (It would later increase to 100 ppt, about three times as salty as seawater). The decrease in water depth caused surface water temperature to increase during summer, which promoted a still greater evaporation rate. In 2007, the steady decline of the water volume caused the South Aral Sea to be divided into western and eastern basins. By 2009, the North Aral Sea still was alive, but the eastern lake in South Aral disappeared while the western lake shrank to a thin strip of water.

Although the Aral Sea also receives about 5,200 cubic yards (4,000 cubic meters) of water per year from groundwater discharge that finds its way from the Tian Shan and Pamir mountain ranges, this is insufficient to replenish the shrinking sea and stop desiccation or salt intrusion.

Extended Effects

As a result of the Aral Sea shrinkage in recent decades, the diversity and amount of fish radically dropped because they could not adapt to the far higher salinity levels. From 32 species of fish, only six were known to have survived. Thus the fishing industry that was once prosperous collapsed. Muskrat trapping, which once yielded a half-million pelts per year from the wetlands around the Amu Darya River and Syr Darya River deltas, also dwindled drastically.

Shipping trade suffered because the water receded far away from the ports of Moynak to the south and Aralsk in the north. Sixty thousand jobs were lost by the mid-1980s, impacting such water-hungry industries as paper manufacturing. To survive, many men migrated to other regions but left their wives and children behind.

Beyond losing the use of abundant water, the plants, animals, and people of the Aral Sea region also suffered secondary effects. When the sea level dropped by 45 feet (14 meters), some 200,000 tons of salt and sand were exposed on the seabed. The scorching heat dried the debris, and wind carried this salt and sand as far as 190 miles (300 kilometers) every day. The whitish, high-salt layer covering the land surface destroyed many agricultural fields and pastures. Nowadays, before the sowing season, the salt on the land surface must be washed away, which in turns requires more water. Crop yields have diminished; the shortage of forage also has a negative effect on cattle and domestic animals.

The constant use of pesticides and fertilizers in the cotton and rice fields has been a public health concern. The heavily contaminated agricultural waters go back to the rivers and eventually to the Aral Sea. Since most people obtain their drinking water directly from the Syr Darya River, they are highly exposed to contaminants, heavy metals, increased salinity, and pesticides. Health effects are noticeable in the population, in particular in children. Diseases like tuberculosis, viral hepatitis, eye problems, throat cancer, anemia, allergies, digestive disorders, kidney and liver ailments, and birth defects have increased. And the windblown dust seems responsible for many respiratory illnesses.

Climate itself has also changed with the wasting of the Aral Sea. Winters are colder while summers are hotter. Precipitation has decreased, extending drought periods and shortening the growing seasons. Biodiversity has been affected; about half of the native bird and mammal species have withdrawn from the region. Vegetation has also shifted from trees and shrubs to xerophytes and halophytes, plants better tuned to true desert conditions.

The former major island of Vozrozhdeniya in the Aral Sea became a peninsula in 2001, and in 2008 the island merged with the mainland. In 1948, the Soviet Union had a bioweapons laboratory there, but in 1992, the lab was abandoned. Surveys of the area showed that the site was used for production, testing, and disposal of biological weapons like typhus, botulinum toxin, plague, tularemia, and brucellosis. The health risk is that since Vozrozhdeniya is no longer an island, weaponized organisms such as fleas and rodents might reach the cities and spread. Thus, in 2002, the United States and the local government put in action a plan to decontaminate 10 anthrax burial sites, and according to the Kazakh Scientific Center for Quarantine and Zoonotic Infections, all were decontaminated.

In spite of the health, social, and economic repercussions of diverting water from the rivers that feed the Aral Sea into water-intensive crops, the governments of Uzbekistan and neighboring Turkmenistan still plan to enlarge the cotton and rice production in the area. Furthermore, in 2005 they signed an agreement to explore the seabed of the South Aral Sea for gas and oil. As of 2010 they had extracted 654,000 cubic yards (500,000 cubic meters) of gas. Operating under separate priorities, Kazakhstan has tried to recover the North Aral Sea and in 2005 joined efforts with the World Bank and built a gated concrete dam, Dike Kokaral. Since then, salinity has decreased noticeably in the northern zone of the sea, some quantities of fish have returned to the region, and the water level is on the rise again.

Rocio R. Duchesne

Further Reading

Aladin, Nickolay V., et al. "Aral Sea: Water Level, Salinity and Long–Term Changes in Biological Communities of an Endangered Ecosystem—Past, Present and Future." *Natural Resources and Environmental Issues* 15, no. 36 (2009).

Micklin Philip and Aladin Nikolay. "Reclaiming the Aral Sea." *Scientific American* 164, no. 4 (2008).

Okda, Ali. "Aral Sea." 2001. http://nailaokda.8m.com.

Walters, Pat. "Aral Sea Recovery?" *National Geographic Magazine,* April 2, 2010. http://news.nationalgeographic.com/news/2010/04/100402-aral-sea-story.

Araya and Paria Xeric Scrub

Category: Desert Biomes.
Geographic Location: South America.
Summary: The Araya and Paria xeric scrub is a dry coastal shrubland in northern Venezuela that is the last refuge of several rare species.

Bounded on three sides by the Caribbean Sea, the Araya and Paria xeric (arid-adapted) scrub ecoregion is primarily comprised of a 2,000-square-mile (5,180-square-kilometer) peninsula of desert and shrubland located on the northernmost tip of Venezuela, occupying the arid zones of the Araya and Paria Peninsulas of the Cordillera de la Costa, Venezuela's coastal mountain range. This neotropic ecoregion also extends south to Cumaná, the capital of the Venezuelan state of Sucre, and stretches north of the mainland into the Caribbean Sea to include the islands of Margarita, Coche, and Cubagua, all of which are part of the Venezuelan state of Nueva Esparta. The Araya Peninsula juts eastward into the Caribbean Sea from Venezuela's northern coast, and the Paria Peninsula juts westward toward the island of Trinidad and separates the Gulf of Paria from the Caribbean Sea. The combined Araya-Paria Peninsula is connected to the Venezuelan mainland by the Cariaco–Yaguaraparo Isthmus.

Located 23 miles (37 kilometers) north of the Araya Peninsula, and with an area of 360 square miles (932 square kilometers), Margarita Island is comprised of east and west arms joined by an isthmus. To the south lie the smaller islands of Coche and Cubagua. The larger, eastern part of Margarita Island features a mountainous central region of peaks and valleys. The tallest peak, Cerro Copey, reaches a height of 3,000 feet (914 meters). The smaller, western arm also features a mountainous landscape on its Macanao Peninsula, reaching almost 2,500 feet (762 meters) at the top of Cerro Macanao. The large coastal plain stretching between the mountain ranges features white-sand beaches, dunes, and salt marshes.

Mangroves and coconut palms line the coastal areas of the Araya-Paria region. Its inland landscapes feature a mix of shrubland; woodlands of deciduous thorns; and land converted for agricultural use, including farms and pastures. In the east, the Paria Peninsula features the littoral and interior mountain range that forms the northeastern section of the Cordillera de la Costa, a forest and montane ecoregion that is an extension of the Andes Mountains, also known as the Maritime Andes. To the east, a stretch of lowlands populated by bamboo forests and shrubland separates these mountains from a mountainous region in the center of the peninsula dominated by Cerro San José, which reaches an altitude of more than 3,600 feet (1,097 meters). In the west, the Araya Peninsula features a long, sandy, arid plain covered in salinas, or salt marshes, due to its frequent inundation by seawater. The Gulf of Cariaco is bordered by the coastline from the tip of Araya south to Cumaná.

Biodiversity

The biodiversity of this region is similar to that of other xerophytic (characterized by extremely low moisture) regions of the Caribbean, and though it does not have a wide diversity of species, the region is notable for being the home of several rare and restricted-range species. At least 340 bird species have been recorded, including more than 12 hawk, 11 heron, eight dove, five sandpiper, and three parrot species, as well as the endangered red siskin (*Carduelis cucullata*). There are also three invasive bird species: the cattle egret (*Bubulcus ibis*), shiny cowbird (*Molothrus bonariensis*), and great kiskadee (*Pitangus sulphuratus*). The Laguna of Chacopata, located on the Caribbean coast of the Araya Peninsula, is a refuge for Venezuela's largest pelican colony.

The area has more than 140 recorded mammal species, including more than 90 bat, six opossum, and five mouse species. Specific mammals include the vampire bat (*Desmodus rotundus*), agouti (*Dasyprocta leporina*), Venezuelan red howler monkey (*Alouatta seniculus*), crab-eating fox (*Cerdocyon thous*), southern naked-tailed armadillo (*Cabassous unicinctus*), striped hog-nosed skunk (*Conepatus semistriatus*), greater grison (*Galictis vittata*), and silky anteater (*Cyclopes didactylus*). The white-tailed deer (*Odocoileus virginianus*) also is an invasive species.

Vegetation

The coastlines of this region are populated by xerophytic thorn scrubs that thrive in low-water conditions, as well as halophytic coastal herbs that thrive in depressions that are regularly flooded with saltwater. These low-height types of vegetation are well adapted to the saline, semi-desert environment of the Araya-Paria littoral range. Plant species common to this region include perennial herbs that tolerate high-salinity soil, such as glasswort (*Salicornia fruticosa*) and crested saltbush (*Atriplex pentandra*), and coastal herbs such as gullfeed (*Scaevola plumieri*) and salt couch grass (*Sporobolus virginicus*).

Environmental Threats

The Araya-Paria xeric ecosystem faces several threats. Deforestation has affected the Macanao Peninsula lowlands. Vehicles driven on beaches, primarily to support the area's thriving tourism sector, have been found to destroy turtle nests. The xerophytic plants throughout the region are threatened by overgrazing by goats. Threats associated with human expansion—such as urbanization, pollution, and the introduction of nonnative species—also degrade the natural environment. In addition, the pet trade in the blue-crowned parakeet (*Aratinga acuticaudata neoxena*) and yellow-shouldered Amazon parrot (*Amazona barbadensis*), as well as habitat destruction, are pushing these bird species toward extinction.

Protected areas on Margarita Island include Cerro Copey National Park, Guayamurí Natural Monument, and Cerro Matasiete, though the majority of the Macanao Peninsula remains under private ownership and is unprotected. There have been efforts to create a protected fauna refuge in the Macanao Peninsula, though the government has not approved any proposal.

Reynard Loki

Further Reading

Larsen, K. and L. B. Holm-Nielsen, eds. *Tropical Botany*. London: Academic Press, 1979.

Padron, Victor, Jordi Marinell, and Rosa Domenech. *The Marine Neogene of Eastern Venezuela—A Preliminary Report*. Caracas: Universidad Central de Venezuela, 1993.

Stattersfield, A. J., M. J. Crosby, A. J. Long, and D. C. Wege. *Endemic Bird Areas of the World: Priorities for Biodiversity Conservation*. Birdlife Conservation Series No. 7. Cambridge, UK: BirdLife International, 1998.

Arctic Desert

Category: Desert Biomes.
Geographic Location: Arctic.
Summary: This northern desert has more life than its Antarctic counterpart but is almost perpetually covered in ice and snow.

The ecosystem of the Arctic Desert is largely in the northern Arctic region and does not necessarily correspond to the Arctic Circle (65.5 degrees north latitude) or any political boundaries. The Arctic Desert is defined by a climate with an average temperature below 50 degrees F (10 degrees C) in the warmest month. This region qualifies as a desert because of its extremely low precipitation, measuring 4 to 6 inches (100 to 150 millimeters) per year—about the same amount of precipitation as in the Sahara Desert. This is drier than the typical Arctic tundra zones, where precipitation averages up to 10 inches (254 millimeters) annually. The small amount of precipitation in the Arctic Desert falls as snow. The temperature is so cold that the snow does not melt, so the land is covered in snow and ice. The region is not considered to be arid because of its low rates of evaporation. Even in the deserts, air humidity is high and soils are moist during the summer growing period.

Species richness in the Arctic Desert is low and decreases toward the North Pole. Net primary production, decomposition, and net ecosystem production rates are low. Food chains are short, and there are few species at each level within the chain.

The Arctic Desert has only summer and winter seasons. In the summer, the sun is in the sky up to 24 hours a day, yet the temperature stays below 50 degrees F (10 degrees C). In the winter, there are several weeks when the sun never rises and temperatures get very cold—down to minus 94 degrees F (minus 70 degrees C). Snow can fall even at these incredibly cold temperatures as long as a source of moisture exists. The heaviest snows occur in near-freezing temperatures, or 14 to 32 degrees F (minus 10 to 0 degrees C). Typically, the air is so cold that it can hold only very small amounts of water, so snow does not fall often. In the winter, precipitation is very light and can be "diamond dust," a ground-level cloud of tiny ice crystals. Diamond dust can occur even on clear-sky days. Diamond-dust precipitation events can last for several days.

Temperatures within the northern Arctic vary by location. In Greenland, cold water flowing south in the ocean creates cooler temperatures than in the northern European areas, where warm waters flowing in the Gulf Stream warm the land. Eastern Canada (51 degrees north) has polar bears and tundra, for example, and Norway (69 degrees north) supports agriculture.

Terrain and Ecosystems

Seas and mountains, such as the Brooks Range in Alaska, fragment the Arctic land masses. Seas separate large Arctic islands (including Svalbard, Novaya Zemlya, Severnaya Zemlya, New Siberian Islands, and Wrangel Island) and the land masses of the Canadian archipelago and Greenland. Similarly, the Bering Strait separates the Arctic lands of the Old and New worlds.

Soils form slowly in the Arctic Desert; they typically are shallow and underdeveloped. Low nutrient concentrations in the soils correlate to low vegetation productivity. A layer of permanently frozen subsoil called permafrost exists, consisting mostly of gravel and finer material. Permafrost can extend downward for up to 0.6 mile (1 kilometer). Water in the Arctic Desert is permanently frozen, although waters in the southern Arctic will melt in the summer.

Ecosystems in the Arctic Desert are disturbed through mechanical and biological processes that affect organism colonization and survival. Mechanical disturbances include cycles of freezing and thawing in the soils, wind, seasonal changes in ice, flooding, and erosion. Biological disturbances include outbreaks of insect pests, high variance in the numbers of grazing animals from year to year, and fire.

Animal Life

The extreme conditions of the Arctic Desert limit the life found there. The desert has no reptiles or amphibians because it is too cold for them to survive. The native animals have adapted to handle long, cold winters and to breed and raise young quickly in the summer. Animals such as mammals and birds also have additional insulation from fat. Many animals hibernate during the winter because food is not abundant. Another alternative is to migrate south in the winter, as birds do. Due to constant immigration and emigration, the population oscillates.

The top carnivores are polar bears and Arctic wolves. Polar bears are the largest land-based carnivores in the world. Male polar bears weigh 772 to 1,499 pounds (350 to 680 kilograms), and females weigh 331 to 549 pounds (150 to 249 kilograms). Their thick white fur provides camouflage in the snow and ice and is thick enough to keep them warm. Their primary food source is seals, which they hunt from sea ice. Their large, wide paws help the bears walk on the snow; they can kill a seal with just one swipe. The bears roam the shores and swim far out to pack ice; their nostrils can close when swimming underwater. Polar bears

A polar bear eating a seal on the Arctic island of Svalbard. Polar bears are the world's largest carnivores on land, and 20,000 to 25,000 of them survive in the extreme conditions of the Arctic Desert through hunting seals. (Thinkstock)

are considered to be a threatened species. Polar bears play an important role in controlling seal populations, while the seal population limits the bear population. Scientists estimate that 20,000 to 25,000 polar bears live in the Arctic.

Other carnivores are Asian snow leopards, Arctic foxes, ermines, and wolverines. Large herbivores include caribou and muskox; smaller herbivores include hares and lemmings. Offshore are many aquatic mammals, such as walrus, ringed seals, bearded seals, and whales including the beluga and narwhal. Birds found in the northern Arctic generally migrate there in early summer and leave by the end of the season. Birds of prey include owls and hawks; waterfowl include geese, ducks, and loons.

Plant Life

Vegetation in the Arctic Desert covers 5 percent or less of the ground due to a very short, dry growing season; dry air; permafrost; poor soils; and lack of pollinating insects. The entirety of the northern Arctic lies above the tree line, so no full-size tree species grow here. Very few plant species can survive in these conditions. Plants are generally stunted and become more so to the north. These herbs, along with mosses and lichens, stand less than four inches (10 centimeters) high.

The plants as a whole are specially adapted to the harsh winds, low water availability, short growing season, and poor soils. Arctic plants grow in clumps for additional protection from snow and wind. They are able to carry out photosynthesis even in low light conditions and extreme cold, which most other plants could not withstand. Due to the short growing season, these plants reproduce by budding and division.

Environmental Changes

The Arctic is experiencing dramatic environmental changes, which are likely to have profound effects on Arctic ecosystems for many reasons. Temperatures have already increased in some parts of the Arctic, and future global climate changes are expected to be greatest in the northern regions. The Arctic Desert has a longer snow-free period than other parts of the Arctic, and that period has increased five to six days per decade.

Stratospheric ozone has been depleted over recent decades, by a maximum of 45 percent below normal in the high Arctic in spring. This has probably led to an increase in UV-B radiation reaching the Arctic's surface. Arctic plants may be particularly sensitive to increases in UV-B radiation because the low temperatures limit repair to DNA damage. In the future, temperatures could increase by more than 2 to 5 degrees C (3.6 to 9 degrees F) in the Arctic.

Climate change is the primary factor threatening biodiversity here. Current and predicted environmental changes are likely to add more stresses and decrease the potential for ecosystem recovery from natural disturbances while providing thresholds for shifts to new states—when disturbance opens gaps for invasion of species new to the Arctic, for example. Many species may not be able to thrive in their current environment under different climate conditions. Relocation is the most likely response for most species. Some species will not be able to relocate, such as polar bears. Rapid changes in climate beyond organisms' ability to relocate could lead to increased fires, disease, and pests.

Warming in the high Arctic could change the amount of vegetation cover, biomass, and types of reproduction. There could be an increase in sexual proliferation instead of colonial reproduction. As permafrost melts with warming temperatures, decomposition in soils increases, often releasing methane and related greenhouse gases. A lack of soil nutrients will likely limit major vegetation changes.

Human Activities

Approximately 15,000 people live in the northern Arctic. Most of the population are Inuits who maintain subsistence lifestyles, existing on the food they can hunt, catch, or fish. Oil and gas exploration and tourism are other sources of income. One of the world's largest oil fields, located at Prudhoe Bay, Alaska, accounts for roughly 25 percent of U.S. oil production.

Climate change in this region could alter people's food, fuel, and culture. It could have a global economic effect due to changes in oil and gas exploration as more land becomes available with melting ice. Future effects on the moisture content in the Arctic Desert are extremely difficult to predict.

GILLIAN GALFORD

Further Reading

McCord, Howard. *The Arctic Desert.* Berkeley, CA: Stooge Editions, 1975.
Pielou, Evelyn. *A Naturalist's Guide to the Arctic.* Chicago: University of Chicago Press, 1994.
Woodford, Chris. *Arctic Tundra and Polar Deserts.* Chicago: Heinemann Raintree, 2010.

Arctic Ocean

Category: Marine and Oceanic Biomes.
Geographic Location: Arctic.
Summary: The Arctic Ocean holds a multitude of unique life forms that are highly adapted in their life history, ecology, and physiology to the extreme and seasonal conditions of this most-northern of environments.

The Arctic Ocean is unique among the world's oceans for many reasons. It is the most extreme ocean in regard to the seasonality of light and its year-round existing ice cover. Knowledge of what lives in the Arctic Ocean is limited due to the logistical challenges imposed by its many years of ice buildup and its inhospitable climate. Covering about 3 percent of the Earth's total surface area, the Arctic Ocean is the smallest of the world's five ocean regions (after the Pacific, Atlantic, Indian, and the recently delimited Southern Ocean). Most of this nearly landlocked ocean region is north of the Arctic Circle. The Arctic Ocean is connected to the Atlantic Ocean by the Greenland Sea, and to the Pacific Ocean via the Bering Strait. The Northwest Passage (between the United States and Canada) and Northern Sea Route (between Norway and Russia) are two important seasonal waterways.

The center of the Arctic Ocean is covered by a drifting, persistent ice pack that has an average thickness of about 10 feet (three meters). During the winter months, this sea ice covers much of the Arctic Ocean's surface. Higher temperatures in the summer months cause the ice pack to shrink in extent by about 50 percent. In recent years, this polar ice pack has thinned, and its summer extent has contracted, allowing for increased navigation and raising the possibility of future undersea resource claims by neighboring countries and shipping disputes among countries bordering the region.

The Arctic Ocean has the widest continental shelf of any ocean, extending 750 miles (1,207 kilometers) from the coast of Siberia, but also has areas that are quite deep. The shallowest ocean region overall, it has an average depth of 3,450 feet (1,052 meters). The maximum depth is 17,850 feet (5,450 meters). The Chukchi Sea provides a connection with the Pacific Ocean via the Bering Strait, but this channel is very narrow and shallow, so most water exchange is with the Atlantic Ocean via the Greenland Sea. The Arctic Ocean connects to or envelops numerous other smaller seas and bays, including Baffin Bay, Barents Sea, Beaufort Sea, Chukchi Sea, East Siberian Sea, Hudson Bay, Hudson Strait, Kara Sea, and the Laptev Sea.

The polar climate is characterized by persistent cold and relatively narrow annual temperature ranges. The winters are characterized by near-continuous darkness, cold and stable weather conditions, and clear skies; the summers by continuous daylight, damp and foggy weather, and weak cyclones with rain or snow.

Seamounts and Submarine Ridges

The floor of the Arctic Ocean is divided by three submarine ridges—Alpha Ridge, Lomonosov Ridge, and Gakkel Ridge—of which the Lomonosov separates the two great basin deeps, the Amerasian and Eurasian. A further division of the Amerasian deep is the relatively isolated Canadian Basin; this area is particularly interesting to scientists because its isolation could mean that it contains life forms found nowhere else on Earth. The Lomonsov Ridge is an underwater mountain chain that averages 10,000 feet (3,048 meters) above the seafloor plain and in places comes to within 3,000 feet (914 meters) of the sea's surface.

Biodiversity

Researchers have only begun to scratch the surface of biodiversity in the Arctic Ocean, which has at least three distinct biological communities. The Sea-Ice Realm includes plants and animals that live on, in, and just under the surface ice. Because only 50 percent of this ice melts in the summer, ice flows can exist for many years and can reach a thickness of more than 6 feet (2 meters). Sea ice usually is not solid like an ice cube, but is riddled with a network of tunnels called brine channels that range in size from microscopic to more than one inch (25 millimeters) in diameter. Diatoms and algae inhabit these channels and obtain energy from sunlight to produce biological material through photosynthesis. Bacteria, viruses, and fungi also inhabit the channels, and together with diatoms and algae, they provide food for flatworms, crustaceans, and other animals. This community of organisms is called *sympagic*, which means "ice-associated."

Scientists exploring the Arctic Ocean's Canadian Basin, which is so isolated that it is thought to be a possible home for unique life forms. This July 2005 photo shows a summertime mix of meltwater, ice, and open sea. (NOAA)

Partial melting of sea ice during the summer months produces ponds on the ice surface that contain their own communities of organisms. Melting ice also releases organisms and nutrients that interact with the ocean water below the ice. Also, the ice in the Arctic can contain natural-history information about what the world was like thousands of years ago. Much history must be hidden in the Arctic ice, including information regarding historical weather patterns, carbon dioxide levels from the Stone Age, and what organisms were alive thousands of years ago.

The Pelagic Realm includes organisms that live in the water column between the ocean surface and the bottom. Melting sea ice allows more light to enter the sea, and algae grow rapidly because the sun shines 24 hours a day during the summer. These algae and phytoplankton provide energy for a variety of floating animals (zooplankton), including crustaceans and jellyfish. Zooplankton in turn are the main energy source for larger pelagic animals, including fishes, squids, seals, and whales.

When pelagic organisms die, they settle to the ocean bottom and become food for inhabitants of the Benthic Realm. A variety of sponges, bivalves, crustaceans, polychaete worms, sea anemones, bryozoans, tunicates, and ascidians are common members of Arctic benthic (floor-dwelling) communities. These animals provide energy for bottom-feeding fishes, whales, and seals.

Seawater Density and Salinity

Ocean currents are caused by winds and changes in seawater density. Density can be changed by evaporation or freshwater input (which raises or lowers the salinity), as well as by temperature changes. Near the equator, for example, evaporation causes seawater salinity and temperature to increase. The density of seawater increases as salinity rises and decreases as temperature rises.

So even though the water is saltier, the higher temperature keeps the surface water from sinking. But as the water flows toward the poles (driven in part by wind), it becomes cooler and sinks due to the increased density.

Currents

The main ocean current flowing into the Arctic Ocean is the West Spitsbergen Current, a northward extension of the Norwegian Atlantic Current, passing through the Fram Strait and then following a deep trench leading to the Arctic Ocean. The access via the Barents Sea is partially obstructed by shallows. The water in this current is warm, above 37 degrees F (3 degrees C); and relatively salty, with a salinity greater than 34.9 parts per million (ppm). The amount of water transported is uncertain, but it is possible that as much as half of it circulates in the vicinity of the Fram Strait without entering the Arctic Ocean.

Alongside the West Spitsbergen Current is the East Greenland Current. This is the main current out of the Arctic Ocean, carrying cold—below 32 degrees F (0 degree C)—and relatively fresh (less than 34.4 ppm salinity) water southward. The Bering Strait is narrow at 53 miles (85 kilometers) and shallow, reaching just 164 feet (50 meters) in depth, allowing only a small northerly flow.

The Transpolar Drift, a surface current, flows from the Siberian to the Greenland side of the Arctic Ocean, where it feeds into the East Greenland Current. Explorer Fridtjof Nansen trusted this current to carry his ship, the *Fram,* across the Arctic Ocean in the 1890s into the vicinity of the North Pole. A less heroic demonstration of the Transpolar Drift is being provided by a consignment of 29,000 floating plastic bath toys (including ducks, frogs, and turtles) that were lost from a container ship in the North Pacific in October 1992. It was proposed that the floating toys would be transported by the pack ice across the Arctic Ocean into the North Atlantic.

The pack ice, up to 39 feet (12 meters) thick, plays an important part in conserving heat. The seawater, with a mean depth of about 3,937 feet (1,200 meters) and a volume of about four million cubic miles (17 million cubic kilometers), provides an immense heat reservoir. The ice cover reduces heat transfer to the atmosphere by one or two orders of magnitude compared with that from open water. Furthermore, largely because of the inflow of river waters and Bering Strait waters, a layer of low-salinity water floats on top of the denser water in the Arctic basin, producing a marked halocline (a vertical zone in the oceanic water column in which salinity changes rapidly with depth) at 98 to 197 feet (30 to 60 meters), which limits the convection that would otherwise mix the whole water column and promote heat loss. The deepwater consequently remains at 30 to 31 degrees F (minus 0.5 to minus 0.9 degree C), above its freezing point of 28 degrees F (minus 2 degrees C). These various factors contribute to the generally higher temperatures of the Arctic Ocean compared with the Antarctic Ocean.

Plans to divert the southward part of the flow of some Siberian rivers to alleviate water shortages further inland perhaps need not create too much alarm. On present evidence, it seems unlikely that such diversions would have major effects on circulation or sea-ice distribution in the Arctic Ocean.

This complex circulation system in the Arctic, which affects the entire food web, is however, in a delicate balance. In recent years scientists have documented changes in the Arctic systems, including a dramatic reduction in sea-ice cover and a weakening of some of the circulation systems that are attributed to—and perhaps accelerating—climate change. The Arctic Ocean affects not only the Arctic native peoples, but also those living farther south in Europe and North America. As a result, the Arctic Ocean and the effects of changes that are taking place here are the focus of intense study by many nations.

Alexandra M. Avila

Further Reading

Dunbar, M. J., ed. *Polar Ocean.* Calgary, Canada: Arctic Institute of North America, 1977.

Smith, W. O., ed. *Polar Oceanography. Part A, Physical Science.* San Diego, CA: Academic Press, 1990.

Thomas, D. N., G. E. Fogg, and P. Convey. *The Biology of Polar Regions.* New York: Oxford University Press, 2008.

Arctic Tundra, High

Category: Grassland, Tundra, and Human Biomes.
Geographic Location: Arctic.
Summary: At the top of the world rests the high-Arctic tundra. This fortress of snow and ice is home to only the most resilient species.

The high-Arctic tundra is the northernmost terrain in the world. It is characterized by a harsh polar climate, permafrost, and little vegetation, but is home to an array of herbivores, carnivores, and migratory birds. Moss and lichen speckle the desolate landscape, where snowy owls and polar bears roam. Long winters and fleeting summers mark the year. Much of this ecosystem is being lost, however. Global climate change is altering the Arctic, and its species are faced with the loss of their homes and their lives. The high Arctic is not a contiguous land mass; rather, it is composed of islands with similar environmental and biological features. All lands above the 75-degrees north latitude line are high-Arctic tundra, symbolized by places like Spitsbergen, situated in the Arctic Ocean between Norway and the North Pole, and a few southward locations, such as the Canadian Prince of Wales Island.

These northernmost places are not only characteristically cold, but also have an extraordinarily dry climate. Interestingly, the high Arctic has as little precipitation as the Sonoran and Chihuahuan deserts. Due to this dry climate and lack of vegetation, much of the high-Arctic tundra can technically be considered a polar desert. As such, the zone experiences large swings in temperature. The winters are long and devoid of sun, in contrast to the conspicuously short summers with continuous daylight. The temperature can remain below 32 degrees F (0 degrees C) for 9 months, and the freeze subsides only for two to eight weeks in the summer. Unlike temperate ecoregions, the high-Arctic tundra does not have a spring or autumn. Instead, the transition from summer to winter is punctuated by the freezing of ice on coastal shores. Likewise, the period between winter and summer is aptly named "the thaw" to portray the ice retreat.

Flora and Fauna

The austere environment dictates which organisms can survive: only the hardiest plants and animals. The flora and fauna of the high-Arctic tundra differ considerably from that of its southern counterparts. There are no trees and relatively few plants; those plants that have adapted to this environment generally are no more than a few centimeters tall. There is little plant diversity. Patches of cushion plants, sedges, grasses, moss, and lichen are interlaced among large areas with little to no flora. Vegetative areas are dominated by moss and lichen, with the former thriving in moist settings and the latter being the only vegetation in dry tracts of land.

One condition limiting which plants survive in the Arctic is the ever-present frozen ground: permafrost. The permafrost can reach its deepest extent, as much as 984 feet (300 meters) thick in the high-Arctic tundra; it also contains massive pieces of ice here. A slice of the top layer melts in the summer, allowing phototrophic plants to establish shallow roots while taking advantage of the perpetual summer sunlight. The surge of new greenery during the summer growing season is the sustenance for herbivore populations. Herbivores residing in the high tundra include lemmings, Arctic hare, caribou, and musk oxen. These land mammals must deal with two unique situations: Their food sources are available only during the summer, and they must cope with temperatures as low as minus 58 degrees F (minus 50 degrees C) in winter.

None of the high-Arctic herbivores hibernate. They have behavioral, physiological, and physical adaptations that allow them to handle severe conditions. One common physiological adaptation is the storage of brown fat. This type of fat is the metabolic equivalent of a furnace, so energy-rich that it generates sufficient heat to keep the animal warm in the harshest external conditions. In addition to brown fat, many Arctic mammals use snow for warmth. Temperatures under the snow are much higher than those in the open air, and resourceful Arctic residents take advantage of this fact. Lemmings, for example, use snow to insulate their underground burrows and shelter themselves from icy winds.

In addition to the mammalian herbivores, the tundra is home to several carnivores, including polar bears, Arctic foxes, and wolverines. Polar bears are unique cases because they don't spend much time on land. They occupy much of the winter hunting seals while wandering atop the ice covering the Arctic Ocean. When the ice melts in the summer, they move onto land, where they forage on berries. In addition to taking advantage of the bountiful summer plant life, pregnant females den on land to hibernate and give birth.

Like the polar bear, the Arctic fox is a definitive Arctic species. The primary prey of these foxes are lemmings; they also scavenge marine-mammal carrion and actively hunt Arctic hares and ptarmigans.

The avian inhabitants of the high-Arctic tundra are adorned with snowy-white plumage. They include the snowy owl; rock ptarmigan; snow goose; snow bunting; and the largest falcon in the world, the gyrfalcon. Some of these birds are year-round residents; others call the Arctic their summer home. Like all high-Arctic fauna, the birds have notably differentiated characteristics. Ptarmigan, for example, grow feathers on the soles of their feet and their toes. These feathers effectively act as snowshoes and allow them to scamper upon deep snowdrifts. How much longer deep winter snowdrifts will occur in the Arctic is an open question, however.

Climate Change

Arctic snows are melting earlier each year, and the accumulation of snow is delayed by increasingly late freezes. Global climate change is taking a great toll on the Arctic; the fragility of this wilderness is a glaring issue that has sparked international concern. The seriousness of the threat of climate change was exemplified by the 2008 listing of polar bears as a threatened species. The reduction of snow cover, sea ice extent, and ice thickness, along with the melting of glaciers and permafrost, are leading to the degradation of the Arctic ecosystem. Only time will tell whether the Arctic of the future bears in-depth resemblance to the ecosystem we know today.

Micaela E. Martinez-Bakker

Further Reading

Bliss, Lawrence, et al. "Arctic Tundra Ecosystems." *Annual Review of Ecology and Systematics* 4 (1973).

Pielou, Evelyn. *A Naturalist's Guide to the Arctic.* Chicago: University of Chicago Press, 1994.

Reynolds, James F. and John D. Tenhunen, eds. *Landscape Function and Disturbance in Arctic Tundra*. New York: Springer, 1996.

Strain, Daniel. "Collapsing Coastlines: How Arctic Shores are Pulled A-Sea." *Science News* 108, no. 2 (2011).

Arctic Tundra, Low

Category: Grassland, Tundra, and Human Biomes.

Geographic Location: Arctic.

Summary: The Arctic tundra is perhaps the most pristine human-inhabited biome. The low-Arctic tundra is home to hundreds of species, from inconspicuous reindeer lichen to majestic polar bears.

The Arctic tundra is one of the world's northernmost ecosystems. Low-Arctic tundra circles the globe, nestled between boreal forests to the south and mid-Arctic tundra to the north. Low-Arctic tundra makes up the northern mainland of North America, Scandinavia, and the Russian Federation. Indigenous people have inhabited the Arctic for thousands of years, and those living there today rely on subsistence harvests and traditional knowledge to survive. Though trees are absent from the tundra, it is rich with life and is home to a plethora of plants, rodents, caribou, birds, and carnivores.

Denizens of this ecosystem are specifically adapted to handle the harsh Arctic climate, but the fate of this region is unknown. Global climate change is altering the Arctic landscape and threatening all who reside in this wilderness. Indeed, the consensus among most scientific authorities is that the Arctic is affected more rapidly and to greater extremes than other areas of the Earth by the pressures—warmer water and air temperatures, altered salinity and thermohaline circulation—of global warming. In part this is due to the accelerated positive feedback loops between Arctic ice cover and albedo (reflectivity index), ice thickness, and age of the ice pack.

The climate of the low Arctic consists of long, bitter winters and little year-round precipitation. Arctic summers are short and chilly. Air temperature remains below freezing during most of the year, dipping below minus 22 degrees F (minus 30 degrees C) in February; highs in July exceed 41 degrees F (5 degrees C). Due to the cold climate, the soils of the Arctic are perennially frozen permafrost. The permafrost is about 656 feet (200 meters) deep, but the topmost layer melts seasonally, with soil reaching depths up to 31 inches (80 centimeters). Even with the low levels of precipitation, a mere four inches (100 millimeters) of summer rains can saturate the shallow soil and facilitate explosive plant growth. Annual precipitation can average up to 8 to 10 inches (200 to 254 millimeters) here; further north, where the annual rate is often less than 6 inches (150 millimeters), the area transitions to the Arctic Desert biome.

Plant Life

Hundreds of species of vascular plants, moss, and lichen richly carpet the tundra. Characteristic plants in this largely treeless biome are the low-lying perennials that live for several years, buried alive under the snows of winter and in full bloom under the 24-hour sunlight of high summer. Plant communities are delimited by soil microclimate. Sedge meadows dominate wet soils; diverse flowering plants occupy moist environments; dry areas are paved with tussocks of cottongrass. The vascular plants are far too many to enumerate, but some of the most alluring are the Arctic poppy, saxifrage, and Arctic lupine.

In addition to carpets of flowers, an abundance of bryophytes and lichen are distinguishing landscape features. Many of these plants have astonishing colors and are important components of herbivore diets. Red moss, among its shamrock-green counterparts, for example, creates a striking landscape accentuated by caribou grazing on reindeer lichen.

Mammalian Life

The tundra is home to herbivores large and small, including musk oxen, caribou, Arctic hares, ground squirrels, lemmings, and voles. Each species is adapted to the harsh conditions. Arctic herbivores are very resilient. Caribou, for example, travel up to 621 miles (1,000 kilometers) between sub-Arctic wintering grounds and calving sites on the tundra. Caribou, like other tundra mammals, have physical and physiological features that ameliorate the bitter winter conditions; they grow specialized winter coats, for instance, to shield them from the elements.

Only the Arctic ground squirrel escapes the frost by entering hibernation. Nonetheless, to stay alive, herbivores must not only prevail against the environment, but also avoid predation. The Arctic is the citadel of carnivores, the most remarkable of which is the polar bear. Both polar and brown bears inhabit the tundra, and though they are closely related, only the polar bear is confined to the Arctic. Neither Arctic bear is strictly carnivorous: Polar bears prey upon seals on the coast,

An Arctic fox (Alopex lagopus) pup with its brown summer coat, which will change to all white in the winter, among tundra grasses on St. Paul Island in Alaska. (NOAA)

and grizzlies take down caribou, but both species enjoy feasting on berries and lush tundra grasses.

The Arctic is also home to red and Arctic foxes, wolves, wolverines, and weasels. Like brown bears, the wolves and red foxes are also found in forest ecosystems, yet they take advantage of the tundra where they can escape the human hunting pressures they are subject to elsewhere.

Bird Species

Mammals are not the lone fauna of the low Arctic tundra; the zone also provides breeding sites for nearly 100 bird species. Temperate-climate, predominantly wetland migrants such as cranes, swans, loons, and sandpipers use the tundra as breeding grounds. In addition to these visitors, inimitable species are permanent residents here, such as the snowy owl and the rock ptarmigan.

Because breeding in the Arctic is such a ubiquitous feature of wetland and shore birds, the survival of many populations is dependent on the security of Arctic habitat. Their security, like that of other species, is threatened by global climate change.

Conservation Efforts

The Arctic National Wildlife Refuge is a conservation region comprising 19.3 million acres (7.8 million hectares) in the Alaskan Arctic, the sole purpose of which is to conserve this one-of-a-kind wilderness and its wildlife. The largest hurdle to management of this ecosystem is global climate change. In relation to the rest of the world, the rate of climate change has ratcheted up in the Arctic. Warming temperatures have led to tundra erosion; rising sea levels; and the melting of permafrost, sea ice, and glaciers. All these events, taken together, are resulting in loss and deterioration of tundra habitats. Conservation initiatives are concerned with the preservation not only of wilderness, but also of the indigenous culture that is deeply rooted in the wilds of the north. In the face of habitat destruction in ecosystems around the world, the hinterlands of the Arctic tundra have become a key remaining pristine and unbroken habitat for humans—but the fate of this vulnerable ecosystem is yet unknown.

MICAELA E. MARTINEZ-BAKKER

Further Reading

Bliss, Lawrence, et al. "Arctic Tundra Ecosystems." *Annual Review of Ecology and Systematics* 4 (1973).

Pielou, Evelyn. *A Naturalist's Guide to the Arctic.* Chicago: University of Chicago Press, 1994.

Reynolds, James F. and John D. Tenhunen, eds. *Landscape Function and Disturbance in Arctic Tundra.* New York: Springer, 1996.

Strain, Daniel. "Collapsing Coastlines: How Arctic Shores are Pulled A-Sea." *Science News* 108, no. 2 (2011).

Argentine Monte

Category: Grassland, Tundra, and Human Biomes.

Geographic Location: South America.

Summary: Temperate-arid band of scrub-dotted steppe, home to many unique plant and animal species, many under threat from human activities and climate change.

The Argentine Monte corresponds to the most arid region of Argentina. It has a landscape of extensive plains and plateaus. The climate is temperate-arid, with rainfall of four to eight inches (10 to 20 centimeters) per year and marked annual temperature ranges. The vegetation is less diverse toward the south, with disappearing cactus and mesquite. The fauna is similar to that of the Patagonian steppe.

The Monte biogeographical province occupies a vast territory of more than 114 million acres (46 million hectares) that runs east of the Andes and toward the south. It is shaped like a sweeping band that starts in the northwest, in the province of Salta in the eastern foothills of the Andes, and becomes wider as it passes through the Patagonian steppe, going as far south as the province of Chubut, where it extends outward to the Atlantic Ocean. Despite its large extension, its physiognomy, and floristic composition, the climatic conditions are fairly homogeneous.

The Monte goes from an elevation of 9,186 feet (2,800 meters) in the Andes foothills to sea level at the Atlantic coast. The westernmost part of the Monte transitions into southern Andean steppe as elevation increases. To the east of the Rio Colorado, the Monte gradually becomes pampas. The fauna and flora in the Monte area are closely related to those of the Chaco biogeograhical province, with those in the central and southwestern area showing similarities to the fauna and flora of Patagonia.

The climate is temperate-arid; the general area receives very little rainfall, although there is a marked and highly variable rainfall gradient from east to west, ranging from only three inches (7.6 centimeters) per year in the east to 10 inches (25.4 centimeters) of annual rainfall in the west. It rains primarily during the summer in the northern and central regions of the Monte, but in the colder south, the rainfall is distributed throughout the year, particularly south of the Rio Diamante. Another notable feature of the Monte is its fairly uniform temperature (isothermal), with average temperatures varying only from 55 degrees F to 64 degrees F (13 degrees C to 18 degrees C).

Scrublands

The shrub steppe occupies most of the Monte. The province is characterized by the presence of thorn scrub and dry grasslands; of shrub steppes of the *Zygophyllaceae* family; and with sparse gallery forests scattered throughout those areas; all are fed by a continuous if scant supply of water from the river systems that wind through this area or wherever underground water is available. The Argentine Monte is therefore considered to be a warm scrub desert.

The most widespread and characteristic type of steppe is the *jarillal* or *larrea* (creosote bush) steppe. The *jarillal* tends to grow in pockets on sandy or rocky-sandy soil. The scrubs are 5 to 8 feet (1.5 to 2.5 meters) in height, not exceeding 10 feet (three meters). There also are cactus (in northern parts) and low trees and shrubs of medium size, such as *Bulnesia* spp., *Monthea aphylla, Bougainvillea spinosa, Cassia aphylla, Cercidium praecox, Chuquiraga erinacea, Prosopis alpataco,* and *Zuccagnia punctata.* The herbaceous cover is spatially variable, depending on rainfall and the effect of livestock.

In the piedmont area, thorny steppes can be found, mainly featuring *Plectrocarpa* species. Other local types of shrubs are jumeales (*Suaeda divaricata*) and zampales (*Atriplex* spp.), which are associated with saline conditions; cardonales, which are giant cactus found on the northern rocky slopes; and the chilcales (*Baccharis salicifolia*), which are associated with wet soils. Other plant communities are made up of *Prosopis* scrubs, of *Baccharis salicifolia* (chilca), and *Tessaria dodonaefolia* (pájaro bobo) in humid places; *Atriplex* scrubs in clayish soils; and *Suaeda divaricata* and *Allenrolfea vaginata* in salty soils.

These scrublands are, for the most part, wide and open tracts of land. The hotter and drier areas are colonized by *Larrea cuneifolia;* near the rivers, the *L. divaricata* can be found. *L. nitida* grows in the cold areas of the Monte as well on the slopes of the Andes.

Endemic Plants and Animals

Since the Oligocene, the Monte area has played an important role in the evolution of temperate biota on the continent and has led to a great level of endemism in the area. This region has several endemic species of flora and fauna. Among the endemic (unique to the area) plants are *Romorinoa girolae* (chica); *Gomprhena colosacana* var. *andersonii* from Sierra de las Quijadas National Park; and species with limited distribution in Argentina, such as verdolaga (*Halophytum ameghinoi*), a fleshy grass that grows in bogs.

One characteristic mammal fauna is the edentates quirquincho chico or piche llorón (*Chaetophractus vellerosus*), which is a species of armadillo. Others include the pichiciego (*Chlamyphorus tuncatus*); carnivores such as puma *(Felis concolor),* zorro gris chico *(Pseudalopex griseus*), zorrino chico *(Conepatus castaneus*), huroncito or hurón chico (*Lyncodon patagonicus*); the ungulate guanaco (*Lama guanicoe*); and the rodents cuis chico (*Microcavia australis*) and mara (*Dolichotis* spp.).

The red viscacha rat (*Tympanoctomys barrerae*) and the pichiciego (*Chlamyphorus truncatus*) are endemic to this biome; they are also listed as vulnerable according to International Union for Conservation of Nature (IUCN) categories. The critically endangered rodent *Ctenomys validus* (tuco-tuco de Guaymallén) makes the Monte its home; vulnerable rodent species such as *Octomys mimax, Andalgalomys roigi,* and *Salinomys delicatus* also abound here, along with the mara (*Dolichotis patagonum*); and in some parts of the region the guanaco (*Lama guanicoe*) can be found.

The avian fauna of the Monte include the copetona or martineta común (*Eudromia elegans*), monterita canela (*Poospiza ornata*), inambú pálido or petiso (*Nothrura darwinii*), loro barranquero (*Cyanoliseus patagonus*), and others. Among the threatened birds are the Falco peregrinus (*halcón peregrino*) and the águila coronada (*Harpyhaliaetus coronatus*). There are also threatened reptiles species that inhabit the Monte such as the terrestrial turtle (*Chelonoidis chilensis*) and lampalagua (*Boa constrictor*). Various species of reptiles can be found here like the iguana colorada (*Tupinambis rufescens*) and snakes (ophidia), such as the falsa yarará (*Pseusotomodon trigonatus*), yarará ñata (*Bothrops ammodytoides*), lampalagua (*Boa constrictor occidentalis*) and falsa coral *(Lystrophis semininctus*). There is less variety of amphibians found in the Monte, but one of importance is the *Pleurodema nebulosa* which is endemic to Argentina.

Protected Areas and Damage

There are many national and provincial protected areas within the central and southern parts of the Argentina Monte region. These include Los Cardones National Park (Salta province), Sierra de las Quijadas National Park (San Luis province), Talampaya National Park (La Rioja province), Valle Fértil Provincial Reserve (partially within the ecoregion), San Guillermo Provincial Reserve, Ichigualasto Provincial Park (San Juan province), Telteca Provincial Reserve, Nacuñán Provincial and Biosphere Reserve, Laguna de Llancanelo Provincial Reserve, Divisadero Largo Provincial Reserve (Mendoza province), Lihué Calel Provincial and National Park, La Humada Provincial Reserve, La Reforma Provincial Reserve, Salitral Levalle Provincial Reserve (La Pampa province), Cinco Chañares Provincial Reserve, Complejo Islote Lobos Provincial Reserve, Caleta de los Loros Provincial Reserve (Rio Negro province), El Mangrullo Provincial Reserve (Neuquén province), and Península de Valdés provincial Reserve (Chubut province). Despite the list being extensive, none of these protected areas are in the northern parts of the Monte biome and the total area of protection actually represents less than 2 percent of the Monte's surface area. This means that the large tracts of land needed by many species are not available for them to be able to complete their life cycles.

Serious damage is occurring and has occurred in the Monte, as well as the Chaco and Patagonia lowland, due to human (anthropogenic) activities—in particular, overgrazing by goats, sheep, and cattle; clear-cutting for fuel; and land clearing for agriculture, mining, and oil exploration. Human populations are found mainly in the oases in valleys and other locations close to rivers that make irrigation possible. This is why some sections of the Monte have been intensively altered while others are fairly untouched.

The forest also underwent significant depredation as man occupied patches and used wood for vineyards, mining, furniture making, construction, and fuel. Overgrazing and deforestation has caused erosion that has affected millions of hectares of the Monte. The deforestation and selective extraction of hardwood and clear-cutting of mesquite forests began in the late 19th and early 20th centuries and continues today. These activities have caused pronounced desertification, disruption of habitats, and changes in the biodiversity and geographic ranges of many species. The effects of global warming, too, must be added to the list. A region of delicate moisture balance, the Argentine Monte harbors many species that depend on pockets of land and water where seasonal conditions are now moving toward climate extremes on a long-term basis.

ALEXANDRA M. AVILA

Further Reading

Ezcurra, E., C. Montaña, and S. Arizaga. "Architecture, Light Interception, and Distribution of Larrea Species in the Monte Desert, Argentina." *Ecology* 72, no. 1 (1991).

Ojeda, R. A., C. M. Campos, J. M. Gonnet, C. E. Borghi, and V. G. Roig. "The MaB Reserve of Ñacuñán, Argentina: Its Role in Understanding the Monte Desert Biome." *Journal of Arid Environments* 39, no. 2 (1998).

Roig, V. G. "Desertification and Distribution of Mammals in the Southern Cone in South America." In M. A. Mares and D. Schmidly, eds. *Latin America Mammalogy: History, Biodiversity, and Conservation.* Norman: University of Oklahoma Press, 1991.

Roig-Juñent, S., G. Flores, S. Claver, G. Debandi, and A. Marvaldi. "Monte Desert (Argentina): Insect Biodiversity and Natural Areas." *Journal of Arid Environments* 471 (2001).

Arizona Mountains Forests

Category: Forest Biomes.
Geographic Location: North America.
Summary: Arizona's geology, elevation, and climatic diversity produces a landscape rich in montane habitats.

Arizona, the sixth-largest state in the United States with an area of 114,000 square miles (295,259 square kilometers), offers one of the most spectacular and diverse landscapes in North America. It spans nearly six degrees of latitude and longitude, with elevations ranging from 72 feet (22 meters) to 12,637 feet (3,852 meters). Arizona comprises two physiographic provinces: the Basin and Range in the south and west, and the Colorado Plateau, including the Grand Canyon, in the north and east. In the center is a transition zone whose main feature is the Mogollon Rim, which marks the southern edge of the Colorado Plateau. Portions of four deserts—Chihuahuan, Great Basin, Mojave, and Sonoran—fall within its borders.

The incredible geologic and climatic range produces great biotic diversity. The primary gradients influencing the composition of Arizona mountain forests include latitude, elevation, and moisture. Other factors that influence forest competition are disturbance (such as fire) and underlying geology. Predominant climate types are hot desert, hot steppe, and cold steppe, but a variety of warm and cold mid-latitude climate types are found at higher elevations. There is rather striking differentiation in forest composition along elevation gradients. In arid environments with large topographic diversity here, one may notice two tree lines on mountain slopes. A lower tree line occurs where lack of water limits tree growth. An upper tree line occurs where low temperatures and short growing seasons likewise limit tree growth.

Upper-Tree-Line Vegetation

Moisture patterns add another layer of complexity to vegetation zonation in these mountains. Different exposures—the direction that a slope faces—are exposed to different patterns of shading, heating, and cooling, which in turn influence moisture supply. Southwest-facing slopes tend to be driest; northeast-facing slopes wettest. The effect is that boundaries between vegetation belts tend to be lower on the wetter slopes and higher on the drier ones.

Many of Arizona's mountain ranges rise from desert- or steppe-like environments where trees are scattered—often confined to ravines—and shrubs, grasses, or succulents like cacti are the dominant residents of the landscape. The dominant plant in parts of the Sonoran Desert, for example, is the giant saguaro (*Carnegiea gigantea*), a cactus that may reach more than 40 feet (12 meters) in height—twice as high as the palo verde (*Parkinsonia microphylla*), a tree with which the saguaro is commonly associated.

Lower-Tree-Line Vegetation

Trees are uncommon at lower elevations because of the lack of precipitation, coupled with high evapotranspiration demand—loss of water through evaporation and transpiration, and loss of water through stomates (openings) in their leaves. Higher up the mountain slopes, as water becomes less limiting, woodlands begin to replace desert scrub or grassland as the dominant vegetation type.

The photo shows two distinct types of vegetation growing at lower and higher elevations below a mountain in Sedona, Arizona. A greater variety of tree species may be found by rivers and in canyons like this. (Thinkstock)

The type of woodland found at lower elevations varies according to location. In the southeast, oak–pine woodland is typically found from 4,000 to 7,000 feet (1,219 to 2,134 meters). Emory oak (*Quercus emoryi*), Arizona oak (*Q. arizonica*), and blue oak (*Q. oblongifolia*) dominate, with alligator juniper (*Juniperus deppeana*) and Mexican piñon pine (*Pinus cembroides*) being common associates. The open canopy of these woodlands leaves room for succulent and semi-succulent species, including *Agave*, *Opuntia*, and *Yucca*, as well as for small trees and shrubs such as acacias (*Mimosa*), manzanita (*Arctostaphylos*), mountain mahogany (*Cercocarpus*), and sumac (*Rhus*).

In the northern part of the state, the oak–pine woodland is replaced by piñon–juniper woodland; oaks are a minor component here. The dominant species are piñon pines and junipers. Mexican piñon pine, more abundant in the south, is replaced by Colorado piñon (*Pinus edulis*) in the north. Likewise, alligator juniper is replaced by other juniper species, such as oneseed juniper (*Juniperus monosperma*), Rocky Mountain juniper (*J. scopulorum*), and Utah juniper (*J. osteosperma*).

Along the Mogollon Rim in the central part of the state, chaparral is the primary type of woodland at lower elevations. Sonoran scrub oak (*Quercus turbinella*) is the dominant tree species. Other important species of the chaparral include pointleaf manzanita (*Arctostaphylos pungens*), hairy and birchleaf mountain mahogany (*Cercocarpus breviflorus* and *C. betuloides*), skunkbush sumac (*Rhus trilobata*), desert ceanothus (*Ceanothus greggii*), Wright's silktassel (*Garrya wrightii*), and Apache plume (*Fallugia paradoxa*).

Pine and Spruce–Fir Forests

Above these low-elevation woodlands is a pine-forest belt from 6,000 to 9,000 feet (1,829 to 2,743 meters). Ponderosa pine (*Pinus ponderosa*) is the dominant species in the north, and Arizona pine (*P. arizonica*) is dominant in the south. Associated species include Apache pine (*P. engelmannii*), Chihuahua pine (*P. leiophylla*), Mexican white pine (*P. strobiformis*), alligator juniper, Gambel's oak (*Quercus gambelii*), silverleaf oak (*Q. hypoleucoides*), and madroño (*Arbutus arizonica*) in the south. In the north, Gambel's oak, quaking aspen (*Populus tremuloides*), and Douglas fir (*Pseudotsuga menziesii*) are the most common associates of the pines.

Spruce–fir forests occur above the pine-forest zone at elevations ranging from about 7,000 feet (2,134 meters) to the upper tree line, approaching 11,400 feet (3,475 meters). Douglas fir and white fir (*Abies concolor*) are the dominant species in this belt south of the Mogollon Rim, with Mexican white pine being a common associate. Engelmann spruce (*Picea engelmannii*) and corkbark fir (*Abies lasiocarpa* var. *arizonica*) dominate stands at the upper limits of the spruce–fir zone in the south.

North of the Mogollon Rim, the spruce–fir zone is more diverse, with Douglas fir, white fir, subalpine fir (*Abies lasiocarpa* var. *lasiocarpa*), blue spruce (*Picea pungens*), and Engelmann spruce being important members of the community. The spruces are typically found at higher elevations. They are joined by limber pine (*Pinus flexilis*) in isolated locations, such as the San Francisco Peaks.

On the Kaibab Plateau, which forms part of the North Rim of the Grand Canyon in northern Arizona, blue and Engelmann spruce dominate. Quaking aspen occurs in scattered stands. Other common trees include Gambel's oak, water birch (*Betula occidentalis*), and box elder (*Acer negundo*).

Effects of Substrate and Slope

Substrate (geology and soils) modifies these general patterns. Soils derived from granite, for example, support vegetation characteristic of more mesic (wet) conditions, whereas soils derived from limestone support vegetation characteristic of more xeric (dry) conditions.

Likewise, slope position has a significant effect. The riparian zone supports a richer assortment of species because of the increased water supply and sometimes greater shelter. Species found in canyons and along streams in Arizona's mountains include blue spruce, box elder, white fir, bigtooth maple (*Acer grandidentatum*), narrowleaf cottonwood (*Populus angustifolia*), Arizona walnut (*Juglans major*), and several alder (*Alnus*) and willow (*Salix*) species.

Fauna

Apex predators across each of these zones include grizzly and black bears, cougars, gray wolves, and raptors. Their prey consists of a wide range of mammals—including both wild species such as antelope, black-tailed deer, bighorn sheep and various domesticated herd animals—as well as reptiles such as the docile western box turtle and the considerably more feisty rattlesnake. These in turn are supported by a full panoply of smaller mammals such as the kangaroo rat, white mouse, and squirrels, which depend ultimately on a diet of invertebrates and vegetation.

David M. Lawrence

Further Reading

Peet, Robert K. "Forests and Meadows of the Rocky Mountains." In Michael G. Barbour and William Dwight Billings, eds. *North American Terrestrial Vegetation.* New York: Cambridge University Press, 2000.

Shreve, Forrest. "The Vegetation of Arizona." In Thomas H. Kearny and Robert H. Peebles, eds. *Flowering Plants and Ferns of Arizona.* Washington, DC: Government Printing Office, 1942.

Wentworth, Thomas R. "Vegetation on Limestone and Granite in the Mule Mountains, Arizona." *Ecology* 62, no. 2 (1981).

West, Neil E. and James A. Young. "Intermountain Valleys and Lower Mountain Slopes." In Michael G. Barbour and William Dwight Billings, eds. *North American Terrestrial Vegetation.* New York: Cambridge University Press, 2000.

Whittaker, R. H. and W. A. Niering. "Vegetation of the Santa Catalina Mountains, Arizona: A Gradient Analysis of the South Slope." *Ecology* 46, no. 4 (1965).

Arkansas River

Category: Inland Aquatic Biomes.
Geographic Location: Western region of North America.
Summary: The Arkansas River watershed is the largest river basin in the lower Great Plains region of the United States, providing habitats for a large number of algal, plant, and animal species, as well as human settlement and use.

The Arkansas River basin and the Red River basin collectively drain a large portion of the southern Great Plains region of the United States. The region drained by the Arkansas River is the western Mississippi River basin, south of the area drained by the Missouri River and its tributaries. This region includes all of Oklahoma, most of western and central Arkansas, parts of northeastern New Mexico, Colorado, Kansas, northern and western Texas, and western and central Louisiana.

The Arkansas River begins in central Colorado in extremely rugged, high-elevation terrain, dropping almost exactly 2 miles (3.2 kilometers) from there to the point where the Arkansas River joins the Mississippi River. All the major rivers of the southern Great Plains region drain primarily from northwest to southeast and are considered to be tributaries of the Mississippi River. Along its path, the Arkansas River passes through many types of terrestrial habitats, climates, and types of human land use.

Vegetation

The river passes through six terrestrial ecoregions: the Western Short Grasslands, the Central and Southern Mixed Grasslands, the Central Forest Grassland Transition Zone, the Ozark Mountain Forests, and the Mississippi Lowland Forests.

Vegetation across the Arkansas River basin varies tremendously. There are coniferous forests in the higher elevations of the Rocky Mountains; short native grasslands in the western regions; mixed to tallgrass prairies farther to the east of the basin; deciduous forests on the slopes of the Ozark Mountains; and relatively dense Coastal Plain forest in eastern Arkansas close to the Mississippi River.

Climate

Climate varies tremendously along the Arkansas River, which is not surprising, since it moves from high elevations in the Rocky Mountains to the lowland areas of eastern Arkansas. At its headwaters in central Colorado, winters are cold and summers are cool with low humidity—while in the eastern regions of the basin, the climate is warm-subtropical, with hot, humid summers, and mild winters with rare extreme cold events. From Colorado through western Kansas, there is usually less than 20 inches (50 centimeters) of precipitation per year, but as the Arkansas River flows through Kansas and Oklahoma, it crosses into areas that experience much greater amounts of precipitation, passing 40 inches (100 centimeters) per year in northeastern Oklahoma. In the extreme eastern end of the basin in Arkansas, rainfall is greater than 50 inches (125 centimeters) per year.

The highly variable climate exhibited in the Arkansas River basin inevitably influences the type of land use that can be practiced in any given area of the basin. Human land uses are governed principally by the amount of annual precipitation, which varies considerably from west to east across the basin. In the west, Colorado grasslands dominate and are used as grazing lands for cattle, being too dry for most other types of agriculture. Moving east into Kansas, increasing annual precipitation permits wheat to be grown; this crop dominates the landscapes of Kansas. In Oklahoma and Arkansas, row crops and livestock pasture are widespread in the river valleys, with the surrounding hills and mountains covered by forests.

The entire region, with the exception of a handful of major cities (Pine Bluff, Little Rock, and Fort Smith, Arkansas; Muskogee and Tulsa, Oklahoma; and Wichita, Kansas), is primarily rural with very little heavy industry and only relatively localized mining operations. The U.S. Department of Agriculture (USDA) estimates land use outside of urban areas in this region as 50 percent rangeland and 50 percent cropland in the western areas of the Arkansas River basin; and 50 percent forest, 15 percent cropland, and 25 percent pasture in the eastern parts of the basin.

The Arkansas River has a high mean discharge of 1,313 cubic feet per second (1,004 cubic meters per second); however, the flow is highly variable along the length of the river. At certain times, the upper main stem, characterized by rocky bottoms and woody snags, lacks flow completely, and the mid- and downstream regions are wide, shallow, and have mud or sand bottoms. When exposed to full sun during low water flow conditions, water temperatures may reach 104 degrees F (40 degrees C). Thus, the Arkansas River and other rivers in the Southern Great Plains are excellent examples of extremely hot and harsh aquatic habitats.

Aquatic Life

Because the Arkansas changes significantly along its length in terms of flow, chemistry, and type of bottom substrates, it traverses four major freshwater ecoregions: Southern Plains, Central Prairie,

Ozark Highlands, and Mississippi Embayment. Consequently, the aquatic communities vary greatly throughout the reaches of the Arkansas River, depending on the depth, bottom type, chemistry, flow rate, and temperature. In the headwaters of the river, the productivity of the system is likely dominated by the availability of rocky substrates, and the relatively clear water here may be nutrient-limited. In contrast, much farther downstream the river is larger and more turbid, which likely limits photosynthesis to the upper region of the water column and processes associated with hard bottom substrates are not very important. Insect, crustacean, and mussel assemblages vary depending on the part of the river examined.

The nonnative zebra mussel has invaded the river via barges in the navigation channel of the lower part of the river. Another threat is nutrient runoff from farm fertilizer; the USDA, through its Natural Resources Conservation Service, has launched a series of efforts to help farmers reduce soil erosion, trap nutrient runoff, improve water quality, and protect wildlife species and habitats.

The Arkansas River has 141 native fish species and approximately 30 nonnative fish species. Other vertebrates include several frog species; a variety of turtle species (e.g., common snapping turtle, false map turtle, yellow mud turtle); northern water snake; and numerous riparian bird and mammal species including herons, kingfishers, beaver, muskrat, and the North American river otter. The river otter is an Oklahoma state species of special concern; its numbers are on the rise, possibly because of stocking programs. The nonnative nutria is increasing in Arkansas; this species has been introduced and its numbers are growing in this region. The interior least tern, a bird species listed on the Federally Endangered Species list, has breeding populations on nonvegetated beaches and sandbars of the Arkansas River.

Human Populations

Humans were living in the Arkansas River basin by 11,500 to 10,000 years ago; these people were hunter-gatherers, pursuing the large mammals of the last ice age. By 900 years ago, major human settlements arose in the fertile river valleys in eastern Oklahoma and neighboring Arkansas. Major mound-building cities also arose in this region, such as Spiro Mound near the Arkansas River in eastern Oklahoma. In the western regions, humans hunted bison and farmed along several tributaries of the Arkansas and Red Rivers. By the mid-1500s, Spanish explorers such as Francisco Vásquez de Coronado and Hernando de Soto claimed the region for the Spanish empire. These Europeans reported large populations of Native Americans along the Arkansas River at this time, and the villages were fortified. In the western regions of the Arkansas River basin, nomadic tribes such as the Kiowa, Comanche, Arapaho, and Cheyenne lived on the plains. The release of horses by the Spanish gave these native human populations increased mobility and changed their way of life substantially.

By the early 1700s, many French explorers, trappers, and traders arrived in the Southern Great Plains region, and made contact with the native peoples in the region. Spain was acknowledged as the owner of the region by treaties in 1762-63, but Spain transferred the ownership of "Louisiana" to France in 1800–02. In the Louisiana Purchase of 1803, the United States acquired most of the Arkansas and Red River basins, together with the Missouri and western Mississippi River basins.

Among the major waterways of the southern Great Plains, the Arkansas River is unique because it is most influenced by snowmelt from the Rocky Mountains, and it traverses perhaps more diverse landscapes than any other river in this region. During most of the time of European settlement along the Arkansas River, the main river remained mostly untouched. However, in the latter half of the 1900s, major changes took place. Numerous large dams were constructed, and greater water withdrawals occurred in the western half of the Arkansas River basin. Depletion of water in the main river channel in western Kansas has been severe; this loss of water undoubtedly has had undocumented negative impacts on the animals and plants of the Arkansas River in that region. The effects may be exacerbated by drought and by more extreme seasonal temperatures as part of climate change.

Daniel M. Pavuk

Further Reading
Benke, Arthur C. and Colbert E. Cushing, eds. *Field Guide to Rivers of North America.* Boston: Elsevier/ Academic Press, 2010.
Brown, Arthur V. and William J. Matthews. "Stream Ecosystems of the Central United States." In Colbert E. Cushing, Kenneth W. Cummins, and G. Wayne Minshall, eds. *River and Stream Ecosystems of the World.* Berkeley: University of California Press, 2006.
Matthews, William J., Caryn C. Vaughn, Keith B. Gido, and Edie-Marsh-Matthews. "Southern Plains Rivers." In Benke, Arthur C. and Colbert E. Cushing, eds. *Rivers of North America.* Boston: Elsevier/Academic Press, 2005.
Pierson, S. T., et al. "Phosphorus and Ammonium Concentrations in Surface Runoff From Grasslands Fertilized with Broiler Litter." *Journal of Environmental Quality* 30, no. 5 (2001).

Arnhem Land Tropical Savanna

Category: Grassland, Tundra, and Human Biomes.
Geographic Location: Australia.
Summary: The Arnhem Land tropical savanna is a vast, relatively intact expanse of tropical savanna embedded with monsoon rainforests, large river systems, extensive freshwater floodplains, and rocky heathland communities.

Arnhem Land is a large region in Australia's Northern Territory; not sharply defined, Arnhem Land is variously considered to range from 37,000 square miles (95,000 square kilometers) to 56,000 square miles (145,000 square kilometers). It is dominated by vast tracts of tropical savanna—a discontinuous layer of trees over a continuous layer of grasses—across the central northern coast of Australia. Embedded within the savanna matrix are small patches of monsoon rainforest, large river systems with extensive freshwater floodplains, and rocky heathland communities.

The climate of the region is classically monsoon tropical, with year-round warm temperatures and extreme rainfall seasonality; abundant rain falls during the austral summer monsoon season (December to April) with very little during the winter dry season (May to November). Rapid grass growth in the wet season and grass curing in the dry season promote high frequencies of low-intensity grass fires, with around 40 percent of the region burning in a typical year, making it one of the most frequently burned regions on Earth. Although the majority of fires are lit by humans, natural fire frequencies would most likely have been similarly high, though concentrated later in the year following ignition by lightning strikes during the buildup (premonsoon) season. Only a small proportion of the region—around 3 percent—has been cleared for agriculture or other high-impact developments.

Fire has played a prominent role in shaping the biota of the region. The tree layer of the savannas is strongly dominated by eucalypts (genera *Eucalyptus* and *Corymbia*), which are extremely fire-tolerant and able to resprout prolifically

Grasses burning in the Kakadu National Park in Australia's Northern Territory. As much as 40 percent of the Arnhem Land burns every year, one of the highest rates anywhere. (Thinkstock)

from burned stems. Eucalypt densities are typically high, though with open canopies that allow a dense grass layer to form underneath and provide fuel for savanna fires. Despite the fire-prone nature of the savannas, fire-sensitive plant communities such as monsoon rainforests are able to persist, mostly on perennially wet sites or those with some degree of topographic fire protection, such as rocky cliffs and gorges.

Topography

The region is dominated by three broad topographic units, each with distinct biota: undulating lowlands, rocky sandstone plateaus, and freshwater floodplains. The lowlands are almost entirely dominated by savannas, an important habitat for a range of species, most of which are widespread across northern Australia.

The heavily dissected Arnhem Plateau, in the center of the region, contains several fire-sensitive plant communities, including large tracts of rainforest along its western margin and extensive areas of heathland. It is an important center of plant and animal endemism, most likely due to its high level of isolation from similar habitats, long-term geomorphological stability, and high topographic complexity. Notable endemic species include the dominant rainforest tree *Allosyncarpia ternata*, the black wallaroo (*Macropus bernardus*, a small kangaroo), the Oenpelli python (*Morelia oenpelliensis*), and the black-banded fruit-dove (*Ptilonopus alligator*).

Extensive floodplains occur where the major rivers approach the sea and support grasslands, sedgelands, and *Melaleuca* woodlands. The floodplains support massive numbers of a range of waterbirds, most notably the magpie goose (*Anseranus semipalmata*); for this reason, the floodplains of Kakadu National Park are on the Ramsar List of Wetlands of International Importance.

Aboriginal Land Use

Australia has a very long history of human occupation, and well-dated human remains from Kakadu National Park suggest that humans have occupied the Arnhem Land region for at least 50,000 years. Despite European colonization in the 19th century and the consequent widespread and sometimes catastrophic disruption of Aboriginal society, parts of the region remain strongholds of traditional Aboriginal culture. Perhaps most notable is the Arnhem Land Aboriginal Land Trust, a vast, 37,000-square-mile (95,000-square-kilometer) area of Aboriginal freehold land in the northeastern part of the region.

More than 80 percent of the region is Aboriginal land, the majority administered by local land councils, but with large parts also incorporated into national parks jointly managed by local Aboriginal people and government agencies. The largest, at 7,645 square miles (19,800 square kilometers), and most notable of such parks is Kakadu National Park, established in 1979 and inscribed on the United Nations Educational, Scientific and Cultural Organization (UNESCO) World Heritage List in 1981 for both its outstanding natural and cultural values.

Other than traditional Aboriginal land uses and conservation, the predominant land use in the region is cattle-based pastoralism. Although pastoralism typically leaves the woody vegetation largely intact, there is clear evidence that high stocking rates can have negative impacts on biodiversity, as can exotic grasses deliberately introduced as fodder for cattle.

Biodiversity Decline

Despite the absence of large-scale land clearing, a significant biodiversity decline is occurring in the region and across the wider savanna landscapes of northern Australia. Most notable is the dramatic decline in the abundance of small to medium-size mammals in the past few decades, even in relatively intensively managed national parks such as Kakadu. Species such as the golden bandicoot (*Isoodon auratus*), black-footed tree rat (*Mesmbriomys goudlii*), and common brushtail possum (*Trichosaurus vulpecula*) have experienced large range contractions and are now rarely encountered in most areas. Alarmingly, this pattern appears to mirror the catastrophic loss of a diverse suite of mammals in central Australia in the early to middle 20th century.

A similar decline, though much less pronounced, is also occurring in granivorous bird species across northern Australia. The causes of these declines remain unclear, but ecologists strongly suspect two events: changes in fire regimes in recent decades (specifically, the increasing frequency, intensity, and size of vegetation fires), and predation by house cats, introduced during the 19th century. It is now well understood that increases in the frequency and intensity of vegetation fires are having negative effects on components of the region's vegetation. Well-documented examples include the northern cypress pine (*Callitris intratropica*), a fire-sensitive conifer that has declined substantially over the past century, and the sandstone heath communities that contain a high proportion of shrub species that are killed by fire and need relatively long fire-free periods to regenerate.

The most severe threat to the region's biodiversity in coming decades is likely to be the spread of the exotic gamba grass (*Andropogon gayanus*). This species rapidly invades savanna, resulting in fuel loads more than four times that observed in non-invaded areas. Such fuel loads produce extremely intense fires that can kill mature savanna trees. Recent research has shown that gamba can convert a savanna woodland to grassland monoculture within a decade or two of initial invasion. Gamba is predicted to spread throughout the region in the coming decades, accompanied by severe effects on savanna biodiversity. Given that removal of gamba, once established, is effectively impossible at large scales, preventing further spread must be a key objective of efforts to effectively conserve the biodiversity values of the region.

Fire Management

The need to better manage destructive fire regimes on the Arnhem Plateau has led to the establishment of an innovative fire management project known as the Western Arnhem Land Fire Abatement (WALFA) project. Operating since 2005, WALFA covers 9,266 square miles (24,000 square kilometers) of mostly savanna woodland and heathlands, and aims to improve fire management on the plateau. The project uses income from greenhouse-gas abatement based on fire management. Recent research has demonstrated that savanna fire regimes that are destructive to biodiversity (such as frequent, intense fires) are also major contributors to greenhouse-gas emissions.

WALFA is the world's first major carbon-abatement project based on fire management and was developed by a group comprising government agencies, Aboriginal Traditional Owners, and land management groups. Funding for the WALFA project is provided by Darwin Liquefied Natural Gas (part of one of the world's largest energy companies, ConocoPhillips), which is providing about $1 million annually for 17 years in return for an annual abatement credit of greenhouse gases equivalent to 100,000 tons of carbon dioxide. This approach to funding improved fire management in tropical savanna landscapes, with accompanying biodiversity benefits, is being rapidly adopted in other regions of northern Australia, including several national parks. It is hoped to be one element in a broader effort to sustain resources such as those found in Arnhem Land against the encroachments fed by climate change and global warming.

Brett P. Murphy

Further Reading

Andersen, A. N., G. D. Cook, and R. J. Williams. *Fire in Tropical Savannas: The Kapalga Fire Experiment.* New York: Springer, 2003.

Fitzsimons, J., S. Legge, B. Traill, and J. Woinarski. *Into Oblivion? The Disappearing Native Mammals of Northern Australia*. Melbourne, Australia: The Nature Conservancy, 2010.

Press, A. J., D. Lea, A. Webb, and A. Graham. *Kakadu: Natural and Cultural Heritage and Management.* Canberra: Australian Nature Conservation Agency, 1995.

Russell-Smith, J., P. J. Whitehead, and P. Cooke. *Culture, Ecology and Economy of Fire Management in North Australian Savannas: Rekindling the Wurrk Tradition*. Collingwood, Australia: CSIRO Publishing, 2009.

Woinarski, J., B. Mackey, H. Nix, and B. Traill. *The Nature of Northern Australia: Natural Values, Ecological Processes, and Future Prospects*. Canberra: Australian National University Press, 2007.

Ascension Scrub and Grasslands

Category: Desert Biomes.
Geographic Location: Atlantic Ocean.
Summary: First discovered in 1501, Ascension was a true desert island. Nearly 200 years of settlement have produced a landscape in rapid transition.

On first stepping out of an airplane at Ascension Island's Wideawake Field, a visitor cannot avoid being thunderstruck by the desolation of the landscape, dominated by vast expanses of jagged, gray volcanic rock with occasional rust-red volcanic cones. A glimpse to the east, however, reveals something different: a tall greenish peak obscured by clouds. That peak, Green Mountain, is the highest point on Ascension. It is also the focus of a two-centuries-long effort to transform—terraform—the island's landscape.

Ascension Island, just south of the equator, is the tip of a large volcano about 60 miles (97 kilometers) west of the mid-Atlantic Ridge. This volcano rises from the ocean floor, about 10,000 feet (3,048 meters) deep, ascending to 2,817 feet (859 meters) above sea level on the summit of Green Mountain. The tip of the volcano forms a roughly arrowhead-shaped island, roughly as broad as it is long, with the tip pointing to the east. The island—only 34 square miles (88 square kilometers) in area—features more than 60 volcanic vents, ranging from small craters to the summit of Green Mountain.

More than half the surface consists of lava flows; most of the rest, with the exception of storm-deposited material along the island's beaches, consists of pyroclastic deposits. The oldest rocks on the surface range from 0.6 to 1.5 million years old. The most recent eruption may have taken place a few hundred years ago, but no eruptions have occurred there since the island's discovery by Portuguese admiral João da Nova in 1501.

The climate at sea level is a hot desert climate—high temperatures with little precipitation. In the island's main settlement of Georgetown, the mean annual temperature is about 84 degrees F (29 degrees C), and the mean annual precipitation is about 5 inches (127 millimeters). At higher elevations on Green Mountain, the desert climate grades into a hot steppe climate and ultimately into a mild, humid climate with no dry season. Mean annual temperature on Green Mountain is about 66 degrees F (19 degrees C); mean annual precipitation there is about 26 inches (661 millimeters).

Vegetation

Before its discovery, the island was sparsely vegetated, the only hint of visible greenery being at higher-elevation slopes. The island's known vegetation consisted of only one shrub, *Oldenlandia adscenionis,* scattered around on some slopes, along with *Portulaca oleracea,* grasses including *Aristida adscensionis,* and ferns and club mosses including *Marrattia purpurascens* and *Lycopodium cernuum.* On lava flows and other barren areas, *Euphorbia origanoides* and other grasses could be found, whereas *A. adscensionis* and *Cyperus appendiculatus* could be found in gullies. *P. oleracea* and *Ipomoea pes-caprae* were scattered along the coast. A variety of ferns dominated the summit of Green Mountain.

Besides all these plants, *O. adscensionis* was endemic to the island (found only here; evolved uniquely to fit this biome) but is now believed to be extinct. Extant endemic species include the grasses *Sporobolus caespitosus* and *S. durus* and the ferns *Anogramma ascensionis, Asplenium ascensionis, Dryopteris ascensionis, Marratia purpurascens, Pteris adscensionis,* and *Xiphopteris ascensionse.*

Bird Life

By contrast with the paucity of vegetative flora at the time of discovery, Ascension was rich in bird life, as it was one of the few large sites in this marine region suitable to oceanic birds for breeding and rearing young. The island's signature avian residents were arguably the endemic Ascension frigate birds (*Fregata aquila*). Another notable resident included the sooty tern, or Wideawake (*Sterna fuscata*).

Further common bird species were the masked, brown, and red-footed boobies (*Sula dactylatra, S. leucogaster,* and *S. sula*); the brown and black noddies (*Anous stolidus* and *A. minutus*); the red-billed and white-tailed tropicbirds (*Phaethon aethereus* and *P. lepturus*); the Madeiran storm petrel (*Oceanodroma castro*); the white tern (*Gygis alba*); Audubon's shearwater (*Puffinus lherminieri*); the Ascension night heron (*Nycticorax* sp.); and the Ascension rail (*Atlantisia elpenor*). The latter two species are now extinct. Although the Ascension rail was noted (and sketched) by the 17th-century explorer Peter Mundy, the Ascension night heron is known only from fossils.

Animal Life

Ascension had only one large land animal: a crab, *Johngarthia lagostoma*, that can still be found from sea level to the summit of Green Mountain. Its most prized land visitor, however, was its breeding population of green turtles (*Chelonia mydas*). For several months of the year, the females of the species climb out of the sea to lay eggs on the thin ribbons of sandy beach scattered along the coast. Sailors frequently stocked their holds with adult turtles, which they used for fresh meat during long sea voyages.

Effects of Human Activities

Ascension's terrestrial life changed radically after its discovery. Introduced animals, such as rats, goats, and cats, devastated the island's native plant life and drove many of its seabird colonies to rookeries on adjacent Boatswainbird Island. But the island's greatest transformation begin after Britain's Royal Navy occupied the island in 1815.

Determined to make it more hospitable to humans, the British sought the advice of Sir Joseph Hooker, who prepared a plan to terraform the island by planting species new to the area. Hooker's proposal included planting trees at higher elevations to increase rainfall, planting valley slopes with species that promote soil formation, planting dry areas with drought-tolerant species, and adding a mix of European and tropical species to gardens on the upper slopes of Green Mountain.

Because Ascension Island provides a rare site for ocean birds to breed, the island is home to a variety of sea birds, including the white tern (Gygis alba), shown here. (Wikimedia/ Drew Avery)

The results of this effort are novel terrestrial ecosystems. A lush cloud forest consisting of a mixture of tropical and subtropical species—including bamboo (*Bambusa* spp.), banana (*Musa* spp.), *Casuarina equisetifolia, Eucalyptus camaldulensis,* fig (*Ficus carica*), ginger (*Alpinia speciosa*), mulberry (*Morus* spp.), Norfolk Island pine (*Araucaria excelsa*), and screwpine (*Pandanus* spp.)—occupies the upper slopes of Green Mountain. Vast expanses of greasy grass (*Melinis minutiflora*), guava (*Psidium guajava*), and other introduced grasses and shrubs occupy the middle and lower slopes of Green Mountain. Introduced desert species, such as mesquite (*Prosopis juliflora*) and prickly pear cactus (*Opuntia cochinillifera* and *O. vulgaris*), are spreading across the volcanic flats.

David M. Lawrence

Further Reading

Ashmole, Philip and Myrtle Ashmole. *St. Helena and Ascension Island: A Natural History.* Oswestry, Shropshire, UK: Anthony Nelson, 2000.

Duffey, E. "The Terrestrial Ecology of Ascension Island." *The Journal of Applied Ecology* 1, no. 2 (1964).

Grazier, Kevin R., ed. *The Science of Dune: An Unauthorized Exploration Into the Real Science Behind Frank Herbert's Fictional Universe.* Dallas, TX: BenBella Books, 2008.

Hart-Davis, Duff. *Ascension: The Story of a South Atlantic Island.* Garden City, NY: Doubleday, 1973.

Lawrence, David M. "The Shade of Uliet: Musings on the Ecology of Dune." In Kevin R. Grazier, ed. *The Science of Dune: An Unauthorized Exploration into the Real Science Behind Frank Herbert's Fictional Universe.* Dallas, TX: BenBella Books, 2008.

Ashmore Reef

Category: Marine and Oceanic Biomes.
Geographic Location: Indian Ocean.
Summary: Ashmore Reef, a large platform reef in the eastern Indian Ocean, is uniquely positioned to provide vital habitat for a vast range of marine and avian fauna.

Ashmore Reef is a large platform reef consisting of an atoll-like structure with three low-lying, vegetated islands; several shifting sand cays; and two lagoons. The Ashmore Reef National Nature Reserve and the Cartier Island Marine Reserve aid in conserving regionally and internationally important sea turtles, dugongs (*Dugong dugon*), and birds (both migratory and seabird species), as well as an unusually high diversity and abundance of sea snakes. The Ashmore Reserve, located off the coast of northwestern Australia, encompasses approximately 225 square miles (583 square kilometers). About 25 nautical miles (46 kilometers) southeast of Ashmore lies Cartier Island Marine Reserve, which covers approximately 64 square miles (167 square kilometers).

The surface waters around the reserves, which are warm and generally nutrient-poor, are dominated by the Indonesian Throughflow Current. During the summer months, the current weakens, resulting in a northward movement of surface water due to the Eastern Gyre Current. At the same time, cooler, nutrient-rich water, thought to be associated with the northward-flowing West Australian Current, is flowing just beneath the surface. Mixing of both bodies of water is thought to occur along shelf-break areas adjacent to the reef systems, which results in localized upwelling and increased nutrient availability.

The Indonesian Throughflow Current is responsible for transporting genetic material southward from the biologically diverse reefs of the Philippines and Indonesia. Consequently, the Australian North West Shelf reefs and shoals play primary roles in maintaining the biodiversity in reefs to the south. In addition, the Leeuwin Current, which is the major southern-flowing current originating in the reserves' area, and the other currents with which it interacts play important roles in maintaining coral reef and algal communities to the south. As a result, the Cartier and Ashmore reefs facilitate the transport of biological material to the reef systems along the western Australian coast.

Vital Species Base

Ashmore Reef National Nature Reserve encompasses two lagoons, several sand flats, shifting sand cays, an extensive reef flat, and the three vegetated islands simply called East, Middle, and West Islands. Together these are the base for a variety of marine habitats. The reef front and crest are composed of hard and soft corals, gorgonians, sponges, and encrusting organisms. In addition, this area provides abundant niches for fish, crustaceans, and echinoderms. Based on recent research, the total number of fish species at Ashmore and Cartier reefs may be as high as 747; such high density and diversity is attributed to the vast diversity of habitat types.

The reef flats are composed of seagrasses, which provide critical habitat for numerous species, including dugongs and sea turtles. The sand flats support species including foraging sea turtles, sting rays, echinoderms, mollusks, and crustaceans. In addition, the sand flats are important grounds for migratory wading and shorebirds. Both lagoons support a variety of fish, as well as apex predators such as sharks and sea snakes, and provide additional feeding areas for both dugongs and sea turtles.

Due to seasonally shifting marine conditions and diverse benthic (seafloor) habitats, Ashmore Reef supports distinct benthic and pelagic communities. Also unique to Ashmore Reef is the fact that it resides in a transitional area between algal-dominated communities to the north and coral communities in the south. The reef is recognized for its high biological diversity, supporting the greatest number of reef-building and non-reef-building corals on any reef off the western Australian coast. As a critical habitat for sea snakes, the reserves are internationally significant. A total of 17 species of sea snakes in the subfamily *Aipysurini* have been recorded in the reserves—the greatest number of the species recorded at any location worldwide. Three of these species are endemic (uniquely evolved to fit a specific biome niche) to Australia's North West Shelf. Some sea-snake species appear to have very small, restricted ranges within the reef.

Providing a base for both foraging and breeding, Ashmore Reef is crucial for sea turtles, and supports a large and significant population of endangered loggerhead (*Caretta caretta*), green (*Celonia mydas*), and hawksbill (*Eretmochelys imbricate*) sea turtles. These individuals, present year-round, may spend decades foraging in the reserves before migrating up to 1,243 miles (2,000 kilometers) to other reefs for nesting. Adult sea turtles that feed elsewhere as juveniles return to the reserves for breeding and nesting. Undisturbed reef flats and sandy beaches are critical in supporting these populations and ensuring reproductive success in perpetuity. The small population of dugongs on Ashmore Reef is thought to have a range that extends to Cartier Island and to some of the submerged shoals in the region that support seagrass beds.

The Ashmore islands provide important nesting habitat for seabirds and migratory shorebirds. Despite the smallness of the islands, Ashmore supports some of the most important seabird rookeries on the Australian North West Shelf; it is a staging point for migratory wetland birds, especially waders. A total of 78 species have been recorded at Ashmore, of which 17 have been recorded breeding. Among these are 35 species that are noted in international agreements between the Australian government and the governments of China, Japan, and the Republic of Korea concerning the conservation of migratory birds and their habitats.

Fisheries and Stresses

The reserves are within an area subject to a Memorandum of Understanding (MOU) between Indonesia and Australia, which allows for continued Indonesian traditional fishing activities in an area referred to as the MOU Box. Fishing on this scale has been of mixed impact to the environment of Ashmore and Cartier, but it is seen as a component in the sustainable future of the species that depend on the habitat here.

The reserves were created by the Commonwealth of Australia to protect biologically diverse marine ecosystems within Australia's North West Shelf and to facilitate scientific research. Established in 1983 and 2000 respectively, the Ashmore and the Cartier Reserves protect an important genetic resource for the region, while the MOU underpins vital economic needs of many Indonesian subsistence and small-scale commercial fishers.

Global climate change is considered a factor in future planning for the protection of these habitats, and is being analyzed along with El Niño events for effects on the reef systems here. Two bleaching events, in 1998 and 2003, affected the coral reefs at Ashmore by the action of higher-than-normal water temperatures. However, the reefs rapidly recovered, with a tripling of hard corals and a doubling of soft coral covering during the 2005–09 survey period, according to an Australian government team. The survey cited the relatively light impact of human activities in the reserves, and the particular success of the fast-growing *Acropora* and *Pocillopora* coral genera. While the survey found a shift in dominance among the coral species, the habitat base seemed to have been quite well preserved over the course of these events.

Jeffrey C. Howe

Further Reading

Berry, P. F., ed. *Marine Faunal Surveys of Ashmore Reef and Cartier Island.* Perth: Western Australia Museum, 1993.

Ceccarelli, Daniela, et al. "Rapid Increase in Coral Cover on an Isolated Coral Reef, the Ashmore Reef, North-Western Australia." *Marine and Freshwater Research* 62, no. 10 (2011).

Commonwealth of Australia. *Ashmore Reef National Nature Reserve and Cartier Island Marine Reserve (Commonwealth Waters) Management Plans.* Canberra: Environment Australia, 2002.

Smith, L., M. Rees, A Heyward, and J. Cloquhoun. *Survey 2000: Beche-de-mer and Trochus Populations at Ashmore Reef.* Casuarina: Australian Institute of Marine Science, 2000.

Atacama Desert

Category: Desert Biomes.
Geographic Location: Northwest region of South America.
Summary: This extremely arid coastal and mountain desert is located in western South America, in the main area of the cold Humboldt Current influence.

The Atacama Desert stretches along the western Pacific margin of South America from southwestern Peru to northwestern Chile. The desert includes not only the basal coastal strip below 6,560 feet (2,000 meters), called the "Atacama hyperdesert," but also the adjacent cold and very arid western mountain slopes of the Andes, rising to more than 13,125 feet (4,000 meters), an area called Puna de Atacama.

The coastal zone of the Atacama Desert, located at 6,500 to 9,840 feet (2,000 to 3,000 meters) altitude depending on latitude, is considered to be among the driest deserts on Earth. In fact, it has been described as an absolute desert or hyperdesert, with 0 to 0.2 inch (0 to 5 millimeters) of average annual precipitation. The extreme drought is due mainly to the strong influence in this area of the cold Humboldt coastal marine current, which flows south to north, and to the rain-shadow effect caused by the high Andes, which intercept the arrival of the moist trade winds from the Atlantic Ocean to the coastal Pacific. The Humboldt Current creates a temperature inversion in the atmosphere due to the cooling of the layers of air in contact with the ocean. Cold air cannot ascend enough to cause cloudiness and rainfall, so it originates dense, almost-permanent coastal fogs. These fogs are the main sources of moisture in the coastal Atacama Desert, supporting a specialized ecosystem called the "Atacama lomas formation" that develops inland on west-exposed hillsides, mainly from 2,300 to 3,930 feet (700 to 1,200 meters).

Tourists at the Tatio geothermal geyser field outside of San Pedro de Atacama in the Atacama Desert in northern Chile. Chile's possession of nearby salt flats is the result of a five-year war with Bolivia and Peru in the late 1880s. (Thinkstock)

Vegetation

The Atacama lomas formation is dominated largely by several species of terrestrial *Tillandsia* (*Bromeliaceae*), called "airplants" because they depend exclusively on atmospheric moisture for their nourishment. Another type of lomas is dominated by arborescent and shrubby cacti, mainly endemic species of *Haageocereus* and *Eulychnia*, which form rather small, isolated populations on steep slopes exposed to the west. *Tillandsia* and the cacti lomas ecosystem have a patchy and fragmentary geographic natural distribution related to the presence of the main areas of coastal fog's inland penetration. These are highly specialized biological communities, threatened by impacts such as the construction of coastal roads and power lines. They are also vulnerable to the effects of climate variability associated with the El Niño Southern Oscillation (ENSO) that periodically causes changes in the normal regime and intensity of the Humboldt Current, which can produce anomalies in the distribution and duration of the coastal fogs.

Outside lomas, the vegetation is almost totally absent; the barren landscape consists of large rocky low mountainous areas (the Pacific Coastal Ranges), often with an aeolian (wind-driven) sandy coverage, interspersed with rocky plateaus and some salt flats. Due to the relative proximity of the Andes, there are several seasonal or temporal allochthonous (from elsewhere) rivers (locally called *quebradas* or *arroyos)* that originate in the eastern high mountains and reach the coast by crossing the desert. It is in these arid valleys that human populations have concentrated, both the present settlements and historically the diverse indigenous cultural groups that historically inhabited the Atacama Desert, including the Nazca, Atacameño, and Inca cultures. In the valleys some agriculture can be practiced, due to moisture and fertility periodically provided by the muddy sediments deposited by rivers in flood times. Along with agriculture, fishing in the sea provides the main traditional source of livelihood for Atacama coastal human populations.

The original natural vegetation of the Atacama Desert valleys are low and open forests and shrublands, which are dominated mostly by several species of *Prosopis* (mesquite tree) specially adapted to live by exploiting the groundwater levels in the valleys. This natural vegetation has been largely replaced by crops and pastures, but there remain several managed populations of mesquite because of its traditional use as firewood, timber, and for support of livestock.

Salt Lakes and Plateaus

Salt deposits are very characteristic of the Atacama Desert, where they are widespread and represent a key mineral resource of great economic value. In Atacama there are two main types of salt deposits: those in which precipitate a remarkable diversity of evaporite salts, mainly chlorides, iodides, carbonates, and sulphates of sodium, potassium, boron, and lithium (*salares*); and the salt flats characterized by nitrate deposits and locally called *salitreras. Salares* are distributed both in coastal areas as well in the high Andean plateaus of the Puna de Atacama, while *salitreras* are restricted to basal and pre-Andean regions of the Atacama Desert in Chile. The dispute over the possession of the valuable nitrate deposits was a contributing factor in a five-year war between Bolivia, Peru, and Chile in 1879, as a result of which Chile took possession of the *salitreras* and the desert region where they are located, which previously belonged to Bolivia and Peru.

On the western slopes of the Andes, above 6,560 to 9,840 feet (2,000 to 3,000 meters) altitude, the dense fog layers typical of the basal Atacama Desert disappear, and some rains occur seasonally across the high mountains from the east during the warmest time of year (December to March), creating a tropical rainfall regime. These scarce rains bring only small amounts of rainfall, less than 3 inches (80 millimeters) annually, and characterize the cold, high-Andean Atacama Desert or Puna de Atacama. These rains are distributed in northeastern Chile (Antofagasta and Atacama regions), the northwest corner of Argentina (Jujuy), and adjacent areas in far southwestern Bolivia (Lípez).

The topography of the Puna de Atacama is characterized by high and large stony desert plateaus (*altiplano*) and mountain slopes of volcanic origin, with the presence of numerous volcanoes both dormant and semi-active, many of them reaching altitudes above 16,400 feet (5,000 meters). At the foot of volcanoes and plateaus, diverse salt flats, salt lakes, and ponds occur, mostly located above 13,125 feet (4,000 meters). Natural vegetation is sparse, consisting mainly of open bunch-grassland (*pajonal*) and widely scattered low shrubby vegetation adapted to the strong diurnal alternation of ice and thawing which is typical of the high tropical mountains. Moreover, in small humid valleys and in the margins of saline lakes and ponds, a peat-swamp Andean vegetation (*bofedal*) locally develops, mainly conditioned by the presence of underground geothermal water springs related to the regional volcanism.

The human resident population in the Puna de Atacama is extremely scarce, mainly due to the harsh climatic conditions and high altitudes. However, the region has a great scenic value, with extraordinarily luminous and unique landscapes, annually attracting a significant flow of tourists from around the world to visit the protected areas.

GONZALO NAVARRO

Further Reading

Chile, Ministerio Del Interior. *Nitrate and Guano Deposits in the Desert of Atacama: An Account of the Measures Taken by the Government of Chile to Facilitate the Development Thereof.* Charleston, SC: Nabu Press, 2010.

Gajardo, Rodolfo. *La Vegetación Natural de Chile: Clasificación y Distribución Geográfica.* Santiago de Chile: CONAF-Editorial Universitaria, 1994.

Gutiérrez, J., F. López-Cortes, and P. Marquet. "Vegetation in an Altitudinal Gradient Along the Rio Loa in the Atacama Desert of Northern Chile." *Journal of Arid Environment* 40, no. 4 (1998).

Lautaro Núñez, L., Martin Grosjean, and Isabel Cartagena. *Ocupaciones Humanas y Paleoambientes en la Puna de Atacama.* San Pedro de Atacama: Universidad Católica del Norte, 2005.

Navarro, G. and S. Rivas-Martínez. "Datos Sobre la Fitosociología del Norte de Chile: La Vegetación en un Transecto Desde San Pedro de Atacama al Volcán Licancabur (Antofagasta II Región)." *Chloris Chilensis* 8, no. 2 (2005).

Rundel, P. W., M .O. Dillon, B. Palma, H. A. Mooney, S. L. Gulmon, and J. R. Ehleringer. "The Phytogeography and Ecology of the Coastal Atacama and Peruvian Deserts." *Aliso* 13, no. 1 (1991).

Atchafalaya Swamp

Category: Inland Aquatic Biomes.
Geographic Location: North America.
Summary: The largest river swamp in the United States, the Atchafalaya Swamp is one of the last great wildernesses and contains a rich variety of plant and animal species.

The Atchafalaya is the largest river swamp in the United States and is contained within the extensive Atchafalaya Basin—3,222 square miles (8,345 square kilometers) of the Mississippi River deltaic plain in south-central Louisiana. The basin begins near Simmesport, Louisiana, where the Mississippi, Red, and Atchafalaya Rivers meet, and extends 137 miles (220 kilometers) south to the Gulf of Mexico. The Atchafalaya River is the fifth-largest by discharge in North America and the largest distributary, or outlet stream, of the Mississippi River, which is its greatest water source.

A navigable river, the Atchafalaya provides a significant industrial shipping channel. Since 1928, the Atchafalaya Basin has also been part of the Mississippi River & Tributaries Project (MR&T), a program of the U.S. federal government that provides flood protection for millions of people. Flood-protection levees confine the Atchafalaya's original floodplain to three floodways that are just half of the basin's original size. The Atchafalaya remains one of the last great wilderness areas in the United States, however, with marshes, cypress swamps, and bottomland hardwoods interspersed with bayous and a rich variety of animal species.

Delta-Building Activity

For the past 10,000 years, delta-building in the Mississippi River system has consisted of cyclic construction and abandonment of delta lobes, with each major lobe being active for 1,000 to 1,500 years. The Mississippi River built southern Louisiana, depositing sediment from upstream and changing river course periodically, looking for a shorter route to the Gulf of Mexico. Large basins such as the Atchafalaya Basin were formed between current and old Mississippi River channels and their natural levees. A dynamic system, the Atchafalaya swamp was shaped by this systematic series of delta-building events. The basin contains one of the few remaining largely natural deltaic ecosystems in the world.

With the Atchafalaya River's progressive capture of more flow of the Mississippi River since the 1500s, the Atchafalaya is currently in a delta-building phase. The Atchafalaya River provides a much more efficient route (137 miles or 220 kilometers) for water and sediment to reach the Gulf of Mexico than the current Mississippi River course (323 miles or 520 kilometers). The natural boundaries of the Atchafalaya Basin are the modern Mississippi River levee to the east and the levees of the Teche (an old course of the Mississippi River) to the west.

From the 1500s to the early 1900s, the large Atchafalaya Basin filled mainly with swamp deposits and from the Atchafalaya River sediment filling in lakes. The Atchafalaya River's increasing capture of flow prompted the U.S. Army Corps of Engineers to build the Old River Control Structure, another feature of the MR&T. Completed in 1963, this floodgate system was built at the meeting point of the two rivers in an attempt to prevent total capture and to maintain the inflow of the Mississippi to the Atchafalaya at 30 percent. Starting in the 1950s, the Atchafalaya River increasingly deposited sediment at the coast, bypassing the basin, which had infilled. After a large flood in 1973, both the Atchafalaya and Wax Lake outlets developed sand-rich bars that were exposed during low tide and rapidly developed into two new deltas, some of the newest land in the world.

Forest and Wildlife

Bottomland hardwood forest grows at the highest elevations in the northern part of the basin. Extensive swamps made primarily of bald cypress and water tupelo are found in the middle of the basin, which is regularly flooded. Freshwater and brackish marshes occur in the lower region, where the Atchafalaya River meets the Gulf of Mexico, and include the habitats of the newly formed Wax Lake and Atchafalaya.

The geological processes of delta formation are still transforming the region. Representing the largest contiguous bottomland hardwood forest and overflow alluvial swamp remaining in the United States, the Atchafalaya provides habitat for wildlife species such as alligators, bald eagles, and crawfish, as well as some endangered species, such as the Louisiana black bear and pallid sturgeon. Designated an Important Bird Area of global significance by the National Audubon Society, the Atchafalaya is a key component of the Central and South American flyway for migratory birds.

Effects of Human Activity

The bountiful resources of the Atchafalaya provided sustenance for Native Americans and, in the 18th century, for thousands of Cajuns who settled in the basin. Inhabitants adapted their lives to the changing conditions of water levels, often living in houseboats. The large 1927 flood devastated many communities and prompted people to migrate from the basin. Almost half of the Atchafalaya Basin Floodway is publicly owned. The rest is privately owned land that includes upland forests and deepwater swamps. The basin is now sparsely populated, but loggers and fishermen still use the basin's resources for their livelihoods. The rich biodiversity also makes the Atchafalaya highly valued by hunters and other sportsmen, photographers, and birders.

Extensive logging has depleted the old-growth cypress trees in the basin, and oil and gas activities have created numerous canals that cut through the swamp, bringing in saltwater. Freshwater flow from the Atchafalaya River keeps Atchafalaya Bay almost salt-free (salinity below 0.5 ppt) most of the year. The system has salinity pulses during low-flow conditions, driven by onshore winds associated with tropical storms and cold fronts.

The basin also has compromised water quality from pollutants, again due to the changes in natural water-flow patterns, which, coupled with changes in sediment deposition, have led to the loss of bald cypress. Restoration plans are underway to restore natural hydrology and to improve the ecological health of the Atchafalaya Basin and the Mississippi River Delta Plain. Some of these measures incorporate the secondary threats of climate warming, including sea-level rise and greater storm surges—another salt intrusion factor. Another element is the vast increase in nitrogen deposition throughout the delta from upstream farmland runoff, itself exacerbated by documented heavier summer precipitation in the Upper Mississippi River Basin, which in turn is in part accelerated by higher global temperatures (which cause more water to evaporate from ocean surfaces, thus contributing to greater precipitation). Along with other strategies, the Atchafalaya National Wildlife Refuge Comprehensive Conservation Plan of 2011 outlines mitigation efforts to reduce and clean up pollutants from the oil industry, agricultural sources, and other human activities.

Angelina M. Freeman
Juanita Constible

Further Reading

Couvillion, Brady, et al. *Land Area Change in Coastal Louisiana From 1932 to 2010.* Reston, VA: U.S. Geological Survey, 2011.

Louisiana Department of Natural Resources (LDNR). *Atchafalaya Basin FY2012 Annual Plan, Atchafalaya Basin Program.* Baton Rouge, LA: LDNR, 2011.

Roberts, Harry. "Delta Switching: Early Responses to the Atchafalaya River Diversion." *Journal of Coastal Research* 14, no. 3 (1998).

Athabaska, Lake

Category: Inland Aquatic Biomes.
Geographic Location: North America.
Summary: This large freshwater lake supporting multiple species of fish and several mammal and bird species on its shores has considerable amounts of oil deposited in its nearby tar sands.

Canada's Lake Athabasca lies in northeastern Alberta and in northwestern Saskatchewan. The Athabasca River flows from Jasper National Park northward, where it meets two other large rivers at the Peace-Athabasca Delta on the western shore of the lake. The lake has a surface area of 3,010 square miles (7,800 square kilometers) and an estimated water volume of 49 cubic miles (204 cubic kilometers). Outflow is northward, through Great Slave Lake, then eventually joining the Mackenzie River and thereby the Arctic Ocean.

The area of land that collects water draining into the Athabasca River and Lake Athabasca is known as the Athabasca watershed; its extent of 100,000 square miles (260,000 square kilometers) can be subdivided into multiple regions such as Rocky Mountain forests, temperate grasslands, wetlands, boreal forest, taiga, and tundra. Together, these biomes support a diversity of terrestrial and freshwater ecosystems.

The lake itself is home to multiple aquatic species, including large fish such as lake trout, northern pike, and Arctic grayling. The rivers feeding the lake are rich with nutrients, supporting the baitfish that the larger fish feed on. The banks of the lake's tributaries support mixed-wood forests and other riparian vegetation, which draws in mammal and amphibian species and serves as a staging site for migratory birds

The plant makeup of the Peace-Athabasca Delta was largely affected by the building of the hydroelectric Bennett Dam on the Peace River in the 1960s. Originally a marsh, the area experienced significant water loss after the dam was built, leading to the exposure of mud flats, which were quickly colonized by tundra-subarctic forest transition species such as spike rush, slough grass, common great bulrush, smartweed, sedges, reed grass and willows. These are plants that are able to survive low water levels.

Invertebrates and Fish

The Peace-Athabasca Delta and the channels associated with it are important because they act as feeding, spawning, and nursing areas for fish in Lake Athabasca. The considerable phytoplankton and aquatic macrophyte communities in the lake serve as a food source for smaller fish and invertebrates. The lakes and channels contain large daphnids and copepods; notably, the copepod *Diacyclops bicuspidatus thomasi* acts as an intermediate host for the tapeworm species *Triaenophorus crassus*, which infects lake whitefish and cisco in Lake Athabasca.

The dominant zooplankton species in the benthic (deepwater) region of the lake is the amphi-

Athabaska Sand Dunes

The Athabasca Sand Dunes, which are located along the southern shore of Lake Athabasca, have National Park status. Conditions in the area are essentially desertlike, and there is little vegetation in active dunes, but rare and endemic (evolved and native only to a specific biome) plant species such as field chickweed, felt-leaved willow, Mackenzie hairgrass, Tyrrell's willow and floccose tansy are known to thrive. Inactive and stable dunes are held together by stands of jack pine, low shrubs, herbs, and grasses.

pod *Monoporeia affinis* (formerly known as *Pontoporeiea affinis*), and fingernail clams; midge larvae and snails are also present. The standing stock of benthic invertebrates tends to be low, however, because many areas of the delta freeze to the bottom.

Lake Athabasca is home to many species of fish and supports commercial, domestic, and sport fishing industries. Many of the sport and commercial fish use the delta as a spawning and nursery area. The species found in the lake include cisco, Arctic grayling, longnose sucker, white sucker, lake chub, spottail shiner, ninespine stickleback, trout perch, yellow perch, burbot, and longnose dace.

Lake trout, lake whitefish, walleye, and northern pike populations support the domestic and commercial fisheries. The latter set of fish are largely omnivorous, mostly feeding on benthic plankton, aquatic insects, crustaceans, and other invertebrates in their juvenile stages; as they increase in size, they prey on smaller fish, frogs, small mammals, and small waterfowl. Fish such as northern pike that feed on small land mammals seek their prey at the edge of weed beds, where they can attack unsuspecting animals.

Birds and Land Animals

The riparian areas of the Athabasca region draw in the most bird and mammal species as the increased vegetation provides a more hospitable environment than the harsh climate and sandy, shallow soils of the surrounding areas. The Peace-Athabasca Delta terrain is the most important to birds because it is relatively undisturbed and is positioned on several of North America's major migratory flight paths, or flyways. Whistling swans, geese, and the endangered whooping crane all use it for breeding, staging, and molting, as well as for nesting.

Notably, the peregrine falcon, which became a threatened species because of organochlorine pesticide poisoning in the 1950s and 1960s, is known to nest in the region. Waterfowl, including mallard, common merganser, and the common loon, are also present along the lake's shores and in vegetation along adjoining rivers. All of these birds feeds on small invertebrates, such as crustaceans, insects, and worms, but mergansers and loons also feed on small fish.

The delta region is also a habitat for mammals such as caribou, which feed mostly on lichens, and moose, which are drawn to areas of dense shrub or wetlands where vegetation is rich in nutrients. Several species of hares, voles, shrews, mice, and porcupine serve as important food sources for larger predators. Also present are muskrat, beaver, marten, and mink; the sale of their furs is an important industry for First Nations groups.

The muskrats, which are semi-aquatic rodents, thrive in perched basins (small lakes that are refilled only by flooding) in the delta, and their numbers have decreased as a result of the less frequent flooding in the area due to the building of the Bennett Dam. Importantly, the sedge and grass meadows of Wood Buffalo National Park, which surrounds the delta, is home to the world's largest herd of roaming wood buffalo, a threatened species. The wood buffalo are preyed upon by wolves, the dominant carnivore in the area, although black bears, wolverines, weasel, marten, and lynx are also present.

Human Incursions

The health of the lake and its surrounding lands has been threatened by the extraction of uranium and tar sands, a major source of oil, in nearby areas. At Fort McMurray, Alberta, south and west of Lake Athabasca, lie the Athabasca Tar Sands, the world's largest known deposit of crude bitumen. The oil sands are thought to contain over 1.75 trillion barrels of oil. Water from the Athabasca River must be diverted to extract crude oil from the tar sands, and there are grave concerns about the effects of toxicants, especially heavy metals, polycyclic aromatic hydrocarbons, and solvents released or required for extraction, on the ecological and human communities in the area.

The results of recent studies on the effects of heavy metal poisoning on fish have so far been inconclusive or conflicting. In 2012, First Nations fishermen and ecologists working upstream of the sands, however, reported an increase in tumors, disease, and deformity in river fish, including whitefish, sucker, burbot, and northern pike. A compromise in the safety of the fish in the area may not only have serious effects on the viability of the Lake Athabasca fishing industry but also on the larger organisms that feed on them.

The Lake Athabasca region has been exposed to previous chemical threats due to uranium mining; the basin is among the world's largest known uranium resources. The search for uranium in the area began in 1942 for military reasons, and the wartime ban on private prospecting was lifted in 1947, leading to the discovery of major deposits on the northern shore of Lake Athabasca. In the 1950s, the provincial government established Uranium City to house men working at nearby mines; when a major mining company officially encouraged its employees to live in the city, the population began to boom, eventually reaching numbers close to 10,000. With the Cold War came increasing public anxiety about nuclear power, and when uranium's value dropped and the mines were shut down in 1983, the city collapsed. Though most of the area's former inhabitants have left, fears of contamination in the lake due to mine tailings and infrastructure dislocation persist.

Global warming may affect the lake in various ways. In close proximity, mainly to the north, lie vast stretches of peatlands, a major carbon sink. As temperatures rise, portions of this carbon, in the form of methane, are known to gas out, engendering a positive feedback loop by increasing concentrations of this potent atmospheric greenhouse gas. Lake Athabasca is expected to receive greater precipitation as ambient atmospheric temperatures rise, and winters are likely to grow somewhat milder. Earlier spring snowpack melts upstream are projected to yield higher water levels in the lake at that time of year. However, studies are unclear on how average water levels will play out over the other seasons, as factors such as smaller snowpack, increased evaporation, and greater available inflow due to expanded deforestation for agricultural purposes must be taken into consideration. The situation remains a dynamic, and complex one.

Yasmin M. Tayag

Further Reading

Athabasca Land Use Planning Interim Advisory Panel. "Athabasca Land Use Plan." Government of Saskatchewan, 2003. http://www.environment.gov.sk.ca/Default.aspx?DN=77e08791-38ff-4b6c-bbd3-79c2af8320cc.

Mitchell, P. and E. Prepas, eds. *Atlas of Alberta Lakes.* Edmonton: University of Alberta Press, 1990.

Piper, Liza. *The Industrial Transformation of Subarctic Canada.* Vancouver: University of British Columbia Press, 2009.

Schindler, D. W. "Sustaining Aquatic Ecosystems in Boreal Regions." *Conservation Ecology* 2, no. 2 (1998).

Atlantic Coastal Forests, Middle

Category: Forest Biomes.
Geographic Location: North America.
Summary: The Middle Atlantic coastal forest is among the most diverse landscapes in North America, harboring unique and rare ecosystems that are under increasing pressure from a growing human population.

Stretching from the eastern shore of Delaware and Maryland to Georgia, the Middle Atlantic coastal forest is sandwiched between the Atlantic Ocean to the east and the uplands of the Piedmont region to the west. This region cannot be defined by a single land use type or ecosystem: It is a patchwork of wild lands mixed with various forms of human development. Unique in many ways, this landscape harbors some of the richest ecosystems in North America, in terms of number of species and biological productivity. Forbidding swamps are among its most well-known habitats, although various types of forests and wetlands cover this vast coastal lowland. At the same time, humans have altered this landscape in a variety of ways, fragmenting much of the original habitat into isolated blocks of wilderness. As a result, many of the species and ecosystems characteristic of this region are now considered to be endangered.

Water is a defining feature of this biome, fostering a diverse array of habitats dependent on the amount of water and the timing of flows. Slow-moving rivers flowing from the Appalachian highlands create vast bottomland swamps before they reach the Atlantic Ocean. Rivers such as Roanoke, Cape Fear, James, and Savannah generate expansive floodplains that have mostly avoided human development. Deep swamps dominated by bald cypress, water tupelo, and Atlantic white cedar trees flank the riverbanks. Thick mats of Spanish moss blanket many of these trees, creating an image often associated with the southern United States. Many of these bottomland forests are constantly covered in water, although there is seasonal variability. The flow of water brings with it sediments and organic matter from the upper regions of the watershed, which feeds plant communities. Even areas away from rivers are governed by water, for this region harbors the greatest diversity of wetland communities in North America.

The push and pull of fire and water reveal two of the forces most responsible for shaping the ecology of this region. The Atlantic Ocean provides a temperature buffer across the region, generating a warm, temperate climate. Winters are mild and are followed by wet, warm springs that send a flood of water into coastal areas. Summers are hot and humid, conditions that can extend into the fall. Fire frequency is at its greatest during late summer and early fall. Species are adapted to the effects of both flooding and fire. Hurricanes are another important force in this coastal region, capable of causing widespread disturbance and dispensing drenching rains.

Spanish moss hanging from a tree in a swampy area near open water on Bulls Island in the Cape Romain National Wildlife Refuge in South Carolina in 2011. (U.S. Fish and Wildlife Service/Steve Hillebrand)

Wetland Vegetation

Pine trees dominate in drier, sandy areas, whereas hardwood tree species are found in the moister environments. Unique ecosystems include the longleaf pine savannas and pocosin wetlands. Longleaf pine savannas consist of open expanses of grasses sparsely populated by longleaf pine trees. This habitat harbors an incredible array of flora, including several species of carnivorous plants, such as Venus flytraps, sundew, and pitcher plants. These ecosystems would not exist if not for fire: Periodic burning reduces the density of trees and shrubs, allowing grasses and other low-growing plants to thrive. Fire also returns nutrients to the soil. Some plants' seeds will not germinate until they have been exposed to high temperatures. Full-grown longleaf pines are mostly resistant to small fires, allowing them to compose the upper canopy of these savannas.

Pocosin is an Algonquin word for "swamp on a hill," which reflects the unusual elevated nature of these thick evergreen wetlands. Occurring on damp soils, pocosin habitats develop in poorly drained locations away from rivers in which water is provided from underground seeps. Vegetation in these areas is very thick, usually dominated by pond pine and evergreen shrubs such as gallberry. Again, this ecosystem is dependent on fire: Pond pines are serotinous, meaning that their cones will not open and disperse seeds unless they are burned. Given the thick vegetation and dense organic soil, fires in pocosins can burn underground, potentially for long periods.

Wildlife Species

Due to the foreboding nature of this swampy biome, much of the region has escaped human development, allowing it to serve as a sanctuary for robust wildlife populations such as white-tailed deer, coyotes, foxes, bobcats, and black bears. Aquatic areas are patrolled by beavers, muskrats, otters, and invasive nutria, which were introduced to the southeastern United States for the fur trade. Because of its remoteness, this region was selected as a reintroduction site for the critically endangered red wolf, first introduced in 1987. The red wolf population in northeastern North Carolina currently numbers about 100 to 120 individuals.

Wetlands sustain tremendous levels of bird diversity, and areas such as northeastern North Carolina support large concentrations of migratory geese and swans. Alligators roam many of the waterways, along with an assortment of turtles and frogs. Venomous snakes are common, including copperheads, water moccasins, and rattlesnakes. Many of the larger rivers are important spawning areas for anadromous marine fish, such as striped bass, American shad, shortnose sturgeon, and Atlantic sturgeon. Efforts are underway to increase the populations of these migratory species. Warm-water game fish such as largemouth bass, catfish, panfish, and gar are popular species sought by anglers and are plentiful in many of the waterways.

Protected Areas

Many of the areas spared from human development are now protected, creating a patchwork of protected lands mixed throughout the landscape. The Roanoke River National Wildlife Refuge shelters one of the largest remaining bottomland swamps in the eastern United States. Alligator River and Pocosin Lakes National Wildlife Refuges in eastern North Carolina harbor remaining pockets of pocosin habitat and concentrations of migratory waterfowl and red wolves. The Great Dismal Swamp along the Virginia–North Carolina border is rumored to have been a stopover area for the notorious pirate captain Blackbeard. Croatoan National Forest, Holly Shelter Game Lands, and Green Swamp in eastern North Carolina contain some of region's remaining longleaf pine savannas and are sanctuaries for carnivorous plants. Francis Marion National Forest in South Carolina contains diverse blackwater swamps. The Savannah River, although developed in several areas, still contains several isolated pockets of natural habitat. All these areas are within a few hours' drive of millions of people.

Ecoregion Threats

Despite the presence of these remaining wildernesses, this ecoregion is considered by many conservation groups to be endangered. Development has expanded over the past several decades, pushing people closer to wild areas. Many of the swamps have been drained to make way for farming operations. The area is crisscrossed by canals that shuttle water area from natural wetlands, creating drier areas suitable for farming. Commercial timber harvesting is also prevalent, with monotypic stands of loblolly pine covering thousands of acres. Pockets of urbanization continue to expand and encroach on natural areas.

The resulting landscape is a diverse mix of cities, suburbs, timber plantations, farms, woodlands, wetlands, and protected areas. Many ecosystems—such as longleaf pine savannas, cypress forests, and cedar swamps—have been reduced to a fraction of their former range. Both forest cover and wetland area have been reduced over the past 40 years, whereas the amount of developed land has increased.

Although forestry and agriculture have converted much of the native habitat, many generalist wildlife species manage to thrive in these areas. Habitat destruction has led to the endangerment of several species, however, especially those that are dependent on specific habitat types. Destruction of river environments and construction of dams have reduced the populations of many anadromous fish species and altered natural water flows that feed coastal swamps.

Climate change may pose additional risks in the future, not only to wild species, but also to human populations, given that much of this area lies at or near sea level. More violent storms and hurricanes can lead to stream dislocation and advanced erosion. Higher temperatures and drier conditions that studies suggest may result here from global warming will also open the door to population booms among tree-damaging insects. These threats will be met in some cases by determined human foes, as efforts are underway to restore native habitats return to natural processes, and otherwise defend this diverse and unique system.

JUSTIN BOHLING

Further Reading

Conner, William H., Thomas W. Doyle, and Ken W. Krauss. *Ecology of Tidal Freshwater Forested Wetlands of the Southeastern United States.* Dordrecht, Netherlands: Springer, 2007.

Davis, Donald E. *Southern United States: An Environmental History.* Santa Barbara, CA: ABC-CLIO, 2006.

Dickson, James G. *Wildlife of Southern Forests: Habitat and Management.* Blaine, WA: Hancock House Publishers, 2006.

Jose, Shibu, Eric J. Jokela, and Deborah L. Miller. *The Longleaf Pine Ecosystem: Ecology, Silviculture, and Restoration.* New York: Springer Science+Business Media, 2006.

Messina, Michael G. and William H. Conner. *Southern Forested Wetlands: Ecology and Management.* Boca Raton, FL: CRC Press, 1998.

Atlantic Coastal Forests, Southern

Category: Forest Biomes.
Geographic Location: South America.
Summary: An extensive biome extending along the Brazilian coast and reaching into inner plains and even mountain areas, this high-diversity biome is also a highly threatened biodiversity hot spot.

The southern region of the Atlantic coastal forests was originally one of the largest rainforests of the Americas, extending over 580,000 square miles (150 million hectares) and covering greatly heterogeneous environmental conditions, which contributed to its exceptionally high biodiversity. With a latitudinal span of 38 degrees, it extended into tropical and subtropical regions of South America, north to south along today's Brazilian Atlantic coast. Its wide longitudinal range, extending inland toward plains and mountains—with altitudes ranging from sea level to over 9,500 feet (2,900 meters) reaching to what is currently Argentina and Paraguay, also contributed to its high biodiversity, together with a variety of soil types and variation in rain patterns: Areas closer to the coast receive up to 160 inches (400 centimeters) of rain annually, while inland areas get in the range of 40 inches (100 centimeters).

These diverse environmental conditions have resulted in very different formations within the biome, comprising both coastal rainforests and seasonal mixed coniferous-deciduous forests further inland. Mixed *Araucaria* pine forests and distinct *Lauraceae*-dominated forests are found in the south; deciduous and semi-deciduous forests are located inland. A number of associated formations include mangroves, *restingas* (coastal forest and scrub on sandy soils), high-elevation grasslands (*campo rupestre*), and *brejos* (humid forests resulting from orographic rainfall in otherwise semi-desert scrub in the northeast of Brazil).

Unique Species

The rich biodiversity of this biome has made it one of the 25 recognized biodiversity hot spots of the world, which together account for over 60 percent of all terrestrial species. As with other biodiversity hot spots, the Southern Atlantic coastal forest has experienced significant environmental loss. It is in fact considered one of the most devastated and highly threatened biomes of the planet—over 93 percent of its original cover has disappeared. Overall, the recognized biodiversity hotspots have lost three-fourths of their original landscapes.

Even though only 7 percent of the original Southern Atlantic coastal forest is intact, it still hosts more than 8,000 endemic species (native and unique to a biome) of vascular plants, amphibians, reptiles, birds, and mammals. Surprisingly enough, most of the original species that were thought to integrate the original biome can still be found, although most often in small areas and highly fragmented landscapes. Around 3,000 species of plants, 104 birds, and 35 mammals are currently threatened to some degree.

Agriculture and Other Stresses

The drivers of biodiversity loss in this forest are extremely complex, as they stem from the varying socioeconomic conditions of the different regions it comprises, north to south and east to west, today and dynamically back through history. Since the colonization by the Portuguese and the Spanish, the biome has experienced severe transformations, mostly related to intensive land use related to the production of commodities, from brazilwood to sugarcane, coffee, and cocoa plantations, and cattle ranching, and more recently, the expansion of soy fields and pine and eucalyptus plantations.

High urbanization rates in recent decades are also related to environmental loss in the biome. The population in areas of the Southern Atlantic Coastal Forests biome has kept increasing sharply in the three countries over which the biome extends—Brazil, Paraguay, and Argentina. Three of the largest cities of South America, including its biggest urban center, Sao Paulo, are located within the area originally occupied by the biome. As much as 70 percent of the Brazilian population, or some 100 million people, live here.

Yet human influences in these woodlands started millennia ago. Records of human presence indicate that the first humans inhabited the biome 11,000 years ago, with evidence of agricultural activity dating to 3,900 years ago. This implies previous ecological transformations, especially from the introduction of slash-and-burn agriculture, which uses fire to manage plant succession and has a reductive effect on the complexity and biomass of the forest.

Indigenous populations dispersed cultivated plants, which ultimately affected natural selection

Canoes against a background of tropical rainforest on the Amazon River in Brazil. The population of original indigenous people in the Southern Atlantic Coastal Forests region has declined to less than 150,000. (Thinkstock)

and the hybridization of wild species. Yet indigenous groups here maintained low levels of trade, which resulted in a low-intensity land use pattern, as opposed to the agricultural intensification that followed European colonization. This factor helped prevent the Southern Atlantic coastal forests from becoming an entirely secondary formation.

Indigenous groups in the region gathered more than 100 forest fruit species and hunted deer, marmosets, turtles, crocodiles, monkeys, sloths, peccaries, agoutis, armadillos, capybaras, tapirs, and otters among larger animals. Coastal groups intensively used various fish and shellfish species, including mullet and 23 other species of saltwater fish, eight species of freshwater fish, as well as crabs, cockles, shrimp, and manatee. Biocultural diversity has been eroded among the descendants of these ethnic precursor groups, with fewer than 150,000 people belonging to some of the original indigenous groups currently living there.

Recent Developments

Global climate change will be a factor in the sustainability of the remnants of the Southern Atlantic coastal forests. The challenges include increased saltwater intrusion, altered precipitation patterns, invasive species expansion, and rapidly changing nutrition and mineral availability to plants and animals due to the combined effects of all of these factors.

Against these negative vectors, the activities of human populations can have mitigating effects, if they are applied with intelligence and determination. The Southern Atlantic Coastal Forests biome has become a top-priority hotspot for biodiversity conservation at the local, regional, national, and international scale. Over 40 protected areas have been created in recent years, with roughly 20 percent of the current forest remnants under firm protection measures. Increasing environmental awareness regarding the value of the forest resources among the general public, together with the involvement of conservationist groups, the academic community, farmers, private business, and governments have resulted in numerous high-quality studies, conservation measures, enforcement practices, and sustainable development initiatives. Yet the extent to which this recent interest in recovering part of the splendor of the biome will be successful still remains to be seen.

Raquel Moreno-Peñaranda

Further Reading

Dean, Warren. *With Broadax and Firebrand: The Destruction of the Brazilian Atlantic Forest.* Berkeley: University of California Press, 1997.

Galindo Leal, Carlos and Ibsen de Gusmão Câmara, eds. *The Atlantic Forest of South America: Biodiversity Status, Threats, and Outlook.* Washington, DC: Island Press, 2003.

Ribeiro, Milton Cezar, et al. "The Brazilian Atlantic Forest: How Much is Left, and How is the Remaining Forest Distributed? Implications for Conservation." *Biological Conservation* 142, no. 6 (2009).

Tabarelli, M., et al. "Challenges and Opportunities for Biodiversity Conservation in the Brazilian Atlantic Forest." *Conservation Biology* 19, no. 3 (2005).

Atlantic Equatorial Coastal Forests

Category: Forest Biomes.
Geographic Location: Africa.
Summary: The Atlantic Equatorial coastal forests are critical for the conservation of large forest mammals of Africa that are under pressure from hunting, poaching, and indiscriminate land uses.

Covering about 73,244 square miles (189,700 square kilometers) and extending along the Atlantic coast from low, undulating hills in the north to mountains farther south and east, the Atlantic Equatorial coastal forests biome is a tropical moist broadleaf forest zone of central Africa, stretching over the Atlantic coasts of Cameroon, Equatorial Guinea, Gabon, Republic of Congo, Angola, and the Democratic Republic of Congo. It forms the southernmost part of the Lower Guinea forests complex in the west, transits into the North Congolian lowland forests, is bounded to the southeast by the Western Congolian forest–savanna mosaic, and lies to the east of the Atlantic Ocean.

The Atlantic Equatorial coastal forests zone consists of Precambrian metamorphic rocks such as schists, amphibolites, quartzites, and gneiss. The southern boundary is marked by the limits of these Precambrian rock outcroppings, with younger rocks on either side. Several important river systems crisscross the region. The northern limit is the Sanaga River; farther south are the Ogooué and Nyanga Rivers, which form extensive coastal deltas that are prone to significant flooding, and the Kouilou River, just north of the Congo River.

Biodiversity

The Atlantic equatorial coastal forests are located in the tropics and receive high rainfall throughout the year. Therefore, the diversity of life here is immense. In fact, the region holds about 50 percent of the endemic (evolved specifically in and unique to a biome) plant life of tropical west Africa. The southern Cristal Mountains contain more than 3,000 species of plants. The Mayombe area—on the borders of the Democratic Republic of Congo, Angola, and the Republic of Congo—is home to the sun-tailed monkey, long-footed shrew, lesser Angolan epauletted fruit bat, and African smoky mouse. The forests are also very rich in bird life. Prevalent reptiles include the Ogowe river frog, Gabon dwarf-clawed frog, Apouh night frog, and Perret's snout-burrower.

The region is most globally known for its richness of forest mammals, from lowland gorillas and chimpanzees to forest elephants and African buffalo. Most parts of these forests are used as areas for conservation of large mammals. They are also good habitats for small mammals, birds, amphibians, reptiles, and invertebrates.

Effects of Human Activity

Humans living in and near the area have long relied on hunting the animals for bushmeat. They also hunt because certain parts of some animals are believed to bring good health or luck. Most of the animals are relatively easy to hunt, and their populations are slow to recover when many individual members of the population are killed. Furthermore, there is little or no protection for these animals in these forests, making them easy prey for hunters and poachers. Other threats to fauna and flora come from extensive logging and native agricultural activities. The area overall, however, is very thinly populated by humans, which can be attributed to the density of the forests and the belief in some quarters that such forests are the abode of supernatural powers and spirits, and not to be ventured into.

The southern region (in Cameroon) is highly forested and contains the Kribi Coastal Resort and the Campo Ma'an National Park. The entire mainland of Equatorial Guinea is in this region. In Gabon, the forests, including logging camps, are inhabited by small groups of agricultural and fishing people, mainly traditional forest dwellers such as the Bakola and Bagyeli. The Gabonese towns within the coastal forests include Lambaréné, the logging base of Ndjolé; Fougamou, a base for visiting Waka National Park; and Gamba, an oil hub and base for visiting Loango National Park.

With the onset of global climate change, the Atlantic Equatorial coastal forests biome has been recognized as a key tropical ecosystem for its capacities to act as a carbon sink, absorbing quantities of carbon dioxide from the atmosphere. While future precipitation and temperature regimes in the area are subject to uncertainties as climate change accelerates, it is clear that species here and abroad will benefit to the degree that sustainable forestry practices are encoded and enforced in this biome.

Peter Elias

Further Reading

Barnes, R. F. W., K. Beardsley, F. Michelmore, K. L. Barnes, M. P. T. Alers, and A. Blom. "Estimating Forest Elephant Numbers With Dung Counts and a Geographic Information System." *Journal of Wildlife Management* 61 (1997).

Blaney, S. and M. Thibault. *Utilisation des Ressources Naturelles Pour la Sécurité Alimentaire et Nutritionnelle des Populations d'une Aire Protégée du Sud-ouest du Gabon.* Libreville, Gabon: World Wildlife Fund Central Africa Regional Program Office, 2001.

Hilton-Taylor, C. *The IUCN 2000 Red List of Threatened Species.* Cambridge, UK: International Union for Conservation of Nature, 2000.

Kamdem Toham, A., J. D'Amico, et al., eds. *Biological Priorities for Conservation in the Guinean-Congolian Forest and Freshwater Region.* Libreville, Gabon: World Wildlife Fund, 2000.

Kingdon, J. *The Kingdon Field Guide to African Mammals.* London: Associated Press, 1997.

Atlantic Ocean, North

Category: Marine and Oceanic Biomes.
Geographic Location: Atlantic Ocean north of the equator.
Summary: The complex environment of the North Atlantic has supported human populations for centuries, but the seemingly inexhaustible fisheries of the past are threatened by human activity.

The North Atlantic Ocean, the portion of the Atlantic that lies north of the equator, is a vast and growing expanse of water with complex ecology due to numerous connections with adjoining oceans, seas, gulfs, bays, and currents that help set the pace of the Earth's thermohaline circulation—a temperature- and salinity-driven system that exerts a strong influence on the planet's climate.

As a whole, the Atlantic is the Earth's second-largest ocean, but it is expanding while the largest ocean (the Pacific) is shrinking as a result of plate tectonics. On the west, the North Atlantic is bordered by the continents of South and North America and by the Caribbean Sea, Gulf of Mexico, Bay of Fundy, and Gulf of St. Lawrence. On the north, it is bounded by the Earth's largest island, Greenland, as well as Iceland and the Greenland and Norwegian Seas. On the east, it is bounded by the continents of Europe and Africa; the British Isles; the North, Scottish, and Irish Seas; the Bristol and English Channels; the Bay of Biscay, Mediterranean Sea, and Gulf of Guinea.

The North Atlantic is slightly larger in area than the South Atlantic, 15,930,000 square miles (41,259,000 square kilometers) to 15,617,000 square miles (40,447,000 square kilometers). But the North Atlantic has a lower average depth, 11,180 feet (3,408 meters) compared to 13,020 feet (3,967 meters), and smaller volume at 35,049,000 cubic miles (146,090,000 cubic kilometers) compared to 38,826,000 cubic miles (161,833,000 cubic kilometers).

Zonal Characteristics

There are numerous ways to classify the environmental zones of the oceans. One is by type of environment, whether pelagic (open water) or benthic (sea bottom). The dominant residents of the pelagic environment are either plankton, free-floating organisms that are usually microscopic in size; or nekton, free-swimming organisms that are typically fish, but include sea mammals such as whales, and reptiles such as sea turtles. The dominant residents of the benthic zone are invertebrates such as hard and soft corals, polychaetes, crustaceans, and mollusks.

The pelagic environment can be subdivided into two provinces: the neritic, which extends from the high tide line out to the edge of the continental shelves at about 660 feet (200 meters) deep, and which is subject to influence from adjoining terrestrial environments; and the oceanic, which encompasses the deepwater environment beyond the edge of the continental shelves and is less influenced by the input of nutrients, sediment, and other pollutants from land.

The pelagic environment can be further subdivided into zones based on depth. The epipelagic zone is the surface layer down to 660 feet (200 meters). By definition, all of the neritic province would fall into the epipelagic zone. The mesopelagic zone extends from 660 feet down to 3,300 feet (1,000 meters). The bathypelagic zone extends from 3,300 feet down to 13,000 feet (4,000 meters). Below 13,000 feet lies the abyssopelagic zone.

The benthic or sea-bottom environment can likewise be subdivided into two provinces: the subneritic, which encompasses the continental shelves down to about 660 feet; and the suboceanic, which encompasses the sea bottom in the deeper waters beyond the edges of the continental shelves. As with the pelagic environment, the benthic environment can be divided into zones based on depth. The littoral zone corresponds with the edge of the sea—the area between the high- and low-tide lines. The sublittoral zone encompasses the rest of the seafloor of the subneritic province, from the low-tide line down to about 660 feet.

The Sunlight Factor

This littoral zone can be further subdivided into inner and outer sublittoral zones. The inner sublittoral is defined as the portion where sufficient light reaches the ocean floor to permit the growth of photosynthetic marine algae (such as kelp) attached to the bottom. The maximum depth of the inner sublittoral varies based on the amount of solar radiation reaching the ocean surface as well as the turbidity (cloudiness) of the water, but is usually about 160 feet (50 meters).

The bathyal zone corresponds with the location of the continental slope—a drop-off from the edge of the continental shelf down to the abyssal plains of the deep ocean—from 660 feet down to 13,000 feet. The abyssal zone is the seafloor in waters between 13,000 feet and 20,000 feet (6,000 meters) in depth. Below 20,000 feet lies the hadal zone.

Another classification system is based on what is arguably the single most important variable influencing the distribution of marine life—sunlight. The euphotic zone has sufficient light for photosynthesis to occur. It extends from the surface down to about 330 feet (100 meters). The primary photosynthetic organisms are phytoplankton (microscopic organisms such as diatoms, coccolithophores, dinoflagellates, and cyanobacteria) along with fixed algae, attached to the bottom in shallower waters. Because of the incredible productivity made possible by the energy in sunlight, the euphotic zone is home to the vast majority of all marine life.

Below the euphotic zone is the disphotic zone, which extends to about 3,300 feet in depth. While not sufficient to permit photosynthesis, the disphotic zone has measurable quantities of light which may have significant ecological effects, such as in concealing prey from or revealing prey to predators. The aphotic zone lies below the disphotic zone. No sunlight penetrates those depths. Any light that exists in the aphotic zone is the product of bioluminescence—light produced by the organisms themselves.

North Atlantic Gyre

Centuries ago, oceanographers realized that a series of currents flowed through and subdivided the North Atlantic. The first of these currents to be discovered was the Gulf Stream, known to European explorers such as Juan Ponce de León as early as the beginning of the 16th century and named and mapped by Benjamin Franklin in 1770.

The Gulf Stream originates off the tip of Florida where the Caribbean Current and Antilles Current—both offshoots of the Atlantic North Equatorial Current—rejoin one another. Atmospheric circulation drives the current northward along the east coast of North America and east into the Atlantic south of Greenland. There it splits into two currents: the North Atlantic Drift, which flows northeastward along the coast of Scandinavia into the Arctic Ocean; and the Canary Current, which flows east and south along the western coasts of the Iberian Peninsula (Spain and Portugal) and North Africa. The Canary Current turns west off the coast of Africa, becoming the Atlantic North Equatorial Current and completing the final leg of a clockwise-rotating system known as the North Atlantic Gyre.

By bringing warm tropical water into the higher latitudes, the Gulf Stream helps make northern and western Europe warmer than they would otherwise be. It also permits the development of an extensive coral reef system around Bermuda, which at 32 degrees north latitude is the most northerly expanse of coral reefs in the entire Atlantic.

Where the Gulf Stream comes into contact with the colder Labrador Current, which flows southward between Canada and Greenland into the northwestern North Atlantic, it creates potentially stormy conditions, but it also creates some of the richest fisheries on the planet. One such contact is off the Grand Banks, an undersea rise southeast of Newfoundland; another is off the Outer Banks barrier island system of North Carolina.

The wind-driven surface currents of the North Atlantic Gyre are connected with the density-driven, deep currents of the thermohaline circulation—where colder and saltier, hence denser, water sinks to form strong currents that circulate through the Earth's ocean basins—to ultimately form the global oceanic "conveyor belt." The conveyor begins in the North Atlantic off Greenland and Iceland where cold water sinks to form North Atlantic Deep Water, which bends southward along the continental slopes of the Americas to join with Antarctic waters that in turn spread into the Indian and Pacific Ocean basins.

By its interactions with the atmosphere as well as all the Earth's seas, the global ocean conveyor helps act as a thermostat on the Earth's climate, keeping global temperatures relatively stable. Disruptions in the North American portion of the conveyor—such as a massive influx of cold freshwater from rapidly melting glaciers in Greenland—could trigger sudden and devastating climate change.

A North Atlantic right whale and her calf surface at sea level. An endangered species, the majority ranges from wintering off the southeastern United States to summer feeding and nursery grounds in New England waters and the Bay of Fundy and Scotian Shelf. (NOAA)

Fauna Profile

The rich fisheries off the east coast of North America have had significant impact on the history of Europe and the Americas. Arguably the most important was that of the North Atlantic cod, which drew European sailors across the Atlantic—hundreds of years before Columbus's 1492 voyage—and which fueled 16th-century Europe's ambitions for colonizing the northeastern United States and Canada. Cod survived intense fishing pressure for 400 years until technological innovations allowed commercial fishers to so thoroughly deplete the population that the fishery has collapsed throughout the species' range.

While abundant fisheries have historically been found off the east coast of North America and on the northwestern fringes of Europe, inside the great loop of the North Atlantic Gyre it is something of a biological wasteland. The nutrient-poor waters have relatively depauperate animal populations, except where upwelling currents bring cold, nutrient-laden water up from the depths.

What is found in abundance inside the gyre is floating plastic debris. Plastic refuse, discarded on land, washes into coastal waters and is picked up by the currents circulating around the edges of the gyre. As a result of the wind effect known as Ekman transport, surface waters—and the pollution they carry—are deflected to the right of the direction of the prevailing winds that drive the gyre's currents in the Northern Hemisphere (in the Southern Hemisphere, the deflection is to the left). This transport concentrates plastic debris in the center of the gyre, where the debris frequently enters marine food webs after ingestion by marine animals.

In addition to cod, the cooler waters of the western North Atlantic are home to Atlantic herring, whose populations have also suffered from overfishing in recent decades. Among notable mammalian species are beluga whale, a formerly heavily hunted species now suffering from the effects of water pollution and diseases; the endangered northern right whale; fin whale; humpback whale; and grey seal.

The warmer waters of the western North Atlantic, under the influence of the Gulf Stream, support a rich temperate fish fauna—which includes snappers, groupers, grunts, and porgies—with a significant number of residents more typical of tropical water, such as wrasses and damselfish.

Development, pollution, and introduced species and diseases are significant environmental concerns in this region.

The northern European waters offer largely shallow water environments with areas of upwelling that bring cool, nutrient-rich waters to the surface that have traditionally supported robust fisheries in species such as Arctic cod, Arctic char, and capelin. They are also home to porbeagle shark; several cetacean species, including the blue whale, minke, killer, beluga, narwhal, bowhead, and northern bottlenose; and the grey and harbor seal. Overfishing and pollution are the primary environmental threats to the region.

Tropical Populations

While the coral reefs of the tropical North Atlantic are less diverse than their Indo-Pacific counterparts, they are diverse enough to support major feeding grounds that draw a wide range of species from the temperate North Atlantic. Among the more important residents are parrotfish, wrasses, Cherub fish, large grouper including Nassau grouper, drum, and conch. The region serves as a breeding haven for the humpback whale, offers crucial habitat for such littoral fauna as the American crocodile and manatees, and hosts a robust array of sea turtles including loggerhead, green, hawksbill, leatherback, olive ridley and Kemp's ridley. Loss of habitat from development and disturbance, epidemic disease, coral bleaching, and pollution threaten many reefs in the region.

The waters of the northeast Brazil shelf offer a diversity of habitats, including mangroves and coral reefs, that are home to a number of sea turtle species as well as breeding grounds for humpback whales and smalleye hammerhead sharks. Development, pollution, and destructive fishing practices are significant environmental concerns here.

Several areas of upwelling occur where the Canary Current approaches the coast of southern Europe and northwestern Africa. These areas support significant fisheries for deep-sea fish and lobsters. The Canary Current region is also home to several economically important sardine species (sardine, Spanish sardine, and pilchard), horse mackerel, hake, green and olive ridley sea turtles, and West African manatee; and serves as a breeding ground for humpback whale. As with other marine provinces near areas of dense human habitation, development and pollution are significant concerns.

David M. Lawrence

Further Reading

Earle, Sylvia A. and Linda K. Glover. *Ocean: An Illustrated Atlas.* Washington, DC: National Geographic Books, 2008.

Hutchinson, S., J. R. E. Lutjeharms, B. McMillan, J. Musick, B. Stonehouse, and M. Tomczak. *Illustrated Atlas of the Sea.* Sydney, Australia: Weldon Owen, 2010.

International Hydrographic Organization (IHO). *Limits of Oceans and Seas.* 3rd ed. Special Publication No. 23. Monte Carlo: IHO, 1953.

Kurlansky, Mark. *Cod: A Biography of the Fish That Changed the World.* New York: Walker, 2007.

Law, Kara Lavender, Skye Morét-Ferguson, Nikolai A. Maximenko, Giora Proskurowski, Emily E. Peacock, Jan Hafner, and Christopher M. Reddy. "Plastic Accumulation in the North Atlantic Subtropical Gyre." *Science* 329, no. 5996 (2010).

Roberts, Callum. *The Unnatural History of the Sea.* Washington, DC: Island Press, 2007.

Spalding, Mark D., et al. "Marine Ecoregions of the World: A Bioregionalization of Coastal and Shelf Areas." *BioScience* 57, no. 7 (2007).

Thurman, Harold V. and Alan P. Trujillo. *Introductory Oceanography.* 10th ed. Upper Saddle River, NJ: Prentice Hall, 2004.

Atlantic Ocean, South

Category: Marine and Oceanic Biomes.
Geographic Location: Atlantic Ocean south of the equator.
Summary: The waters of the South Atlantic Ocean encompass a range of environments from the harsh conditions of the near-Antarctic region to lush coral reefs in the tropical waters of Brazil.

The South Atlantic Ocean, the portion of the Atlantic that lies south of the equator, has historically been relatively untouched in terms of economic exploitation, but humanity's search for natural resources is increasingly focused on its waters.

The generally accepted boundaries of the South Atlantic are rather fluid, changing several times over the course of the 20th century as experts debated the existence and boundaries of an ocean further south variously named the Southern, Austral, or Antarctic Ocean. Early in the 20th century, the Antarctic Ocean was defined as lying below the Antarctic Convergence, an area where cold waters from Antarctica sink below warmer subtropical waters from the north. For most of the latter part of the 20th century, the International Hydrographic Office failed to recognize such an ocean, but in 2000 proposed resurrecting the Antarctic Ocean in the form of the Southern Ocean, whose northern boundary would be 60 degrees south latitude. The proposal had not been ratified as of 2012.

Southern boundaries aside, the western boundary of the South Atlantic is defined as the east coast of South America and the great Rio de La Plata estuary. On the north, it is bounded by the equator and the Gulf of Guinea. On the east, it is bounded by the west coast of Africa. If the proposed boundary of the Southern Ocean is accepted, then the South Atlantic would be slightly smaller in area than the North Atlantic, but it is deeper and has a greater volume than its northern counterpart.

The South Atlantic Ocean, like other large ocean basins, has a gyre—the South Atlantic Gyre—at its heart. Here, it circulates in counterclockwise manner, bringing warm water poleward and cold water toward the equator. Its western arm, the Brazil Current, is a weaker, southern counterpart to the North Atlantic's Gulf Stream, flowing south toward Antarctica. Off the tip of South America it collides with the Malvinas (Falklands) Current, itself a north-flowing offshoot of the eastward-flowing Antarctic Circumpolar Current. Part of the Brazil Current breaks away in the vicinity of this confluence to form the eastward-flowing South Atlantic Current, which, upon reaching Africa, splits, with a northward-flowing branch forming the Benguela Current.

The Benguela Current follows the west coastline of Africa. Ekman transport (a wind-driven force) triggers upwelling of cold, nutrient-rich waters inshore of this current, creating conditions for traditionally rich fisheries in the upwelling zone; these are found off the southwestern coast of Africa and the Cape of Good Hope. As the Benguela Current flows northward, its waters warm. It turns westward near the equator, becoming part of the South Atlantic Equatorial Current—the northern part of the South Atlantic Gyre. In the vicinity of Brazil's Cabo de Sao Roque, the South Atlantic Equatorial Current splits, one portion flowing north as the North Brazil Current, the other portion flowing south, becoming the Brazil Current and completing the circulation. The North Brazil Current is much the stronger; most of it flows northwestward into the Northern Hemisphere, but part of it splits off, flowing eastward as the North Equatorial Counter Current.

Marine and Coastal Life

The upwelling zone of the Benguela has created conditions for rich fisheries of species such as rock lobster, pilchard, anchovy, Cape horse mackerel, and hake. Overharvesting, however, has depleted the stocks of rock lobster and pilchard, among other species. Other notable species include the jackass penguin and a number of sea mammals, such as the Cape fur seal, southern right whale, and Heaviside's dolphin. Kelp forests, similar to those that occur off the coast of California, are locally important.

The cold, stormy waters of the southwest Atlantic have been havens for a wide variety of marine life, offering breeding areas for marine birds and mammals—including Magellanic penguin, imperial cormorant, pale-faced sheathbill, southern right whale, humpback whale, Commerson's dolphin, southern elephant seal, southern sea lion, and South American fur seal. The whaling and sealing industries began taking a heavy toll on many marine mammal species in the 19th century. Currently, the region supports economically important fisheries for hake, common squid, shortfin squid, and other species.

Macaroni penguins on mountainous South Georgia Island, which provides breeding grounds for large numbers of seabirds in the South Atlantic Ocean. (Thinkstock)

In subtropical and tropical waters farther north, collisions of warm currents with upwelling zones support a diverse environment, including mangroves and coral reefs. The beaches are important nesting areas for communities of leatherback, green, hawksbill, olive ridley, and loggerhead turtles. The waters are important tropical breeding areas for humpback whales and the smalleye hammerhead.

The South Atlantic features a number of isolated islands or island groups that encompass a significant amount of environmental and biological diversity. South Georgia and the South Sandwich Islands feature mountain ranges covered by snow and ice; their coastlines are breeding grounds and havens for marine birds and mammals.

Near the equator, Ascension Island—roughly midway between Africa and South America—appears to be the definitive desert island, both on land and at sea. The large expanses of rust-brown volcanic cinder cones and grayish-white lava flows were once breeding grounds for a number of seabirds, including the wideawake tern and Ascension frigatebird. The island's beaches are important nesting sites for green and hawksbill turtles. The low-nutrient marine environment has no coral reefs and lacks the diversity of reef ecosystems at similar latitudes. Ascension's nearshore waters are noted for abundance of black durgon and ocean triggerfish; they also host a number of endemic (adapted and native uniquely to a biome) species such as resplendent angelfish and Lubbock's yellowtail damselfish.

Impact of Mankind

Many of the ecosystems of the South Atlantic had long been relatively sheltered from human activity, but in the 20th century, a burgeoning human population began putting greater and greater demands on the environment. Petroleum exploration and development, for example, is bringing much-needed cash to developing nations of Africa and South America, but at the cost of environmental disruption because of construction and accidents, including oil spills. Additionally, introduced species ravage once-isolated regions, and coastal construction—such as for resorts in tropical areas—leads to increasing loss of natural coastal areas and seascapes. Newly discovered resources, such as deep-sea fisheries, are depleted before there is sufficient scientific understanding of the resource to guide the development of sound management policies.

Much work remains to be done to identify and understand the ecosystems of the South Atlantic. A further overriding concern, of course, is the oncoming suite of changes being wrought by global climate change. A heavier atmospheric carbon dioxide load migrating into seawater causes its acidification—a process with often striking negative effects on corals, crustaceans, mollusks, and other organisms whose ability to create their calcium structures is blocked by the unbalanced seawater chemistry. Ocean warming (and further south, melting sea ice) has been shown to rather dramatically reduce the supply of Antarctic krill in the South Atlantic; the depletion of this bottom-of-the-pyramid food source directly affects the population stability of such higher animals as penguins and baleen whales. Increased temperatures in the major surface currents along the southern African coast, several studies suggest, will likely cause increased rainfall in coastal areas, which in turn would alter the mix and augment the amount of land-based pollutants and nutrients flowing into the South Atlantic.

David M. Lawrence

Further Reading

Ashmole, Philip and Myrtle Ashmole. *St. Helena and Ascension Island: A Natural History.* Oswestry, UK: Anthony Nelson, 2000.

Earle, Sylvia A. and Linda K. Glover. *Ocean: An Illustrated Atlas.* Washington, DC: National Geographic Books, 2008.

Hutchinson, Stephen, Johann R. E. Lutjeharms, Beverly McMillan, John Musick, Bernard Stonehouse, and Matthias Tomczak. *The Illustrated Atlas of the Sea.* Sydney, Australia: Weldon Owen, 2010.

International Hydrographic Organization (IHO). *Limits of Oceans and Seas.* 3rd ed. Special Publication No. 23. Monte Carlo: IHO, 1953.

Spalding, Mark D., et al. "Marine Ecoregions of the World: A Bioregionalization of Coastal and Shelf Areas." *BioScience* 57, no. 7 (2007).

Thurman, Harold V. and Alan P. Trujillo. *Introductory Oceanography.* 10th ed. Upper Saddle River, NJ: Prentice Hall, 2004.

Australian Alps Montane Grasslands

Category: Grassland, Tundra, and Human Biomes.
Geographic Location: Australia.

Summary: This mountainous ecoregion, which is threatened by periodic bushfires, is home to numerous unique species.

The Australian Alps montane grasslands ecoregion consists of the high-elevation lands in the Australian Alps mountain ranges in southeastern mainland Australia. The highest point in the ecosystem is the peak of Mount Kosciuszko in the Snowy Mountains, at 7,310 feet (2,228 meters). The highest mountains in Australia, the Australian Alps are the southeastern section of the continent's Great Dividing Range, running for 373 miles (600 kilometers) from the Brindabella Ranges near Canberra along the borders of the Australian Capital Territory, New South Wales, and Victoria. Ranges in the Australian Alps include the Snowy Mountains in New South Wales, the Victorian Alps in Victoria, and the Brindabella Ranges in the Australian Capital Territory. The higher mountains are especially cold; Kosciuszko is snow-covered most of the year. Outside of Tasmania, the montane region is the only part of Australia where deep snow occurs regularly. The montane grasslands are surrounded by the Australian temperate forests at lower elevations, called the tableland, below 3,609 feet (1,100 meters).

Despite the name, the Australian Alps are neither as high nor nearly as steep as the European Alps, and most peaks can be reached on foot without specialized mountaineering or rock-climbing equipment. There are few permanent human settlements in the Alps, apart from the ski resorts and the town of Cabramurra in the Snowy Mountains Scheme. Though the Alps constitute less than 1 percent of Australia's landmass, they receive about a quarter of the country's rainfall and supply the water for nearly half of the population. The eastern slopes' runoff is diverted into the Murray and Murrumbidgee Rivers, through the Snowy Mountains Scheme.

Bushfires

Australia is predominantly hot and dry, and though the Alps are less so, they are still prone to frequent bushfires. The Victorian Alps have been almost completely burned by bushfires on several occasions, and the fires of 1851 ("Black Thursday"), 1939 ("Black Friday"), 2003, 2006 to 2007, and 2009 ("Black Saturday") are infamous. Several eucalypt and banksia species depend on bushfire events, however, to cause their seed pods to open before germination. The fires also encourage the growth of new plants throughout the grasslands.

But the fires can often be catastrophic. The 2003 fires destroyed two-thirds of the pasture, forest, and nature parks in the Australian Capital Territory and spread into the suburbs of Canberra, destroying 500 homes and killing four people. The Black Saturday bushfires were the deadliest in Australian history, killing 173 people, with hundreds more sustaining serious burns and related injuries. More than one million acres (404,686 hectares) of land were burned and 12,000 head of livestock killed. More than a million wild animals were estimated to have been killed, and the only known habitat of Leadbeater's possum was burned. A great many surviving animals were seriously injured and may have died later as a result; large numbers of kangaroos, for example, were burned while traveling through smoldering grasslands.

With the acceleration of global warming, conservationists have posited whether the greater frequency and intensity of bushfires in the region in recent years are in part fueled by the drying effects of various climate change factors. As the toll of such events includes flora, wild fauna, and increasingly human lives and property, action to protect the biological infrastructure of the Australian Alps is a focus. It is noted, for example, that reduction of plant root systems can contribute to looser soil structure and thus exacerbate erosion and intensify flood damage, yet another concern.

Flora

The ecoregion is home to 1,400 higher plant species, 66 of which are endemic (uniquely evolved to fit a biome niche) and 26 of which are endemic specifically to the subalpine and alpine bands. The ecoregion is a mix of grassland, heath, and bog, divided into montane, subalpine, and alpine bands in ascending order of elevation. Different trees flourish at different elevations: the mountain ash (*Eucalyptus regnans*) and alpine ash (*E. delegatensis*) in the montane band, and the snow gum (*E. pauciflora*) and black sallee (*E. stellulatea*) in the subalpine. Above the tree line in the alpine band, the dominant species is a species of snow grass (*Poa*), growing along with shrubs like orites, *Grevillea, Prostanthera,* and *Hobea.*

Hikers in the grasslands on Mount Kosciuszko in the Snowy Mountains of the Australian Alps. While they only make up under 1 percent of Australia's landmass, the Alps receive about 25 percent of the country's rainfall and provide water for almost 50 percent of the population. (Thinkstock)

Alpine vegetation grows no more than 3 feet (1 meter) high, with tall herbs growing in rich humus soil. The only alpine conifer in Australia is the mountain plum pine (*Podocarpus lawrencei*). Stream beds are populated by sphagnum bog communities. Little light reaches the forest floor, so the understory is sparser than in other montane forests. The alpine plants are well adapted to the cold climate and winter drought. *Caltha introloba* flowers under the snow; others form floral buds during the fall and flower as soon as the snow has melted. Few alpine plants produce seeds, germination and taking root being too difficult in the cold, hard soil; instead, they grow from rhizomes, root nodes, or bulbs.

Fauna

Endemic wildlife in the region includes the mountain pygmy possum (*Burramys parvus*); the corroboree frog (*Pseudophryne corroboree*); and the alpine thermocolor grasshopper, which changes hue like a chameleon. Both the possum and the corroboree frog, which are two of the very few alpine endemics, are found strictly in the alpine areas. The possum is also the only marsupial to hibernate for long periods, hibernation being its method of dealing with the long winter.

The Baw Baw Plateau in the Victorian Alps is home to the Baw Baw frog (*Philoria frosti*), which is found nowhere else. Other notable species in the high elevations include the red cryptic treefrog (*Litoria paraewingi*), Spencer's treefrog (*L. spenceri*), dendy toadlet (*Pseudophryne dendyi*), and Maccoy's elf skink (*Nannoscincus maccoyi*). In the montane bands, there are red-necked wallabies (*Macropus rufogriseus*), swamp wallabies (*Wallabia bicolor*), tiger quolls (*Dasyurus maculatus*), platypuses (*Ornithorhynchus anatinus*), and various species of wombats (*Vombatus*). The montane woodlands support large breeding populations of flame robins and pilotbirds, and have been classified by BirdLife International as an Important Bird Area.

Bill Kte'pi

Further Reading

Costin, A. B. "The High Mountain Vegetation of Australia." *Australian Journal of Botany* 5, no. 2 (1957).

Gibson, David J. *Grasses and Grassland Ecology.* New York: Oxford University Press, 2009.

Johnston, Frances M. and Catherine Pickering. "Alien Plants in the Australian Alps." *Mountain Research and Development* 21, no. 3 (2001).

Australian Temperate Forests, Eastern

Category: Forest Biomes.
Geographic Location: Australia.
Summary: This geologically complex and biologically diverse region, covered by eucalypt forests and patches of rainforests, faces threats from deforestation and other human activities.

The eastern Australian temperate forest region is located between Australia's east coast and the Great Dividing Range (also known as the Eastern Highlands mountain range), extending from the central coast of New South Wales near Sydney into southeast Queensland. This area supports a large variety of animals and plants, some of which are endemic (uniquely evolved to fit a biome) to this region and others that are globally threatened. High diversity exists here, in part due to the changes that occur along the landscape, characterized by dramatically different substrates and microclimates.

Temperate eucalyptus forests are dominant, with patches of rainforest dispersed throughout. Four types of rainforests exist in this area: subtropical, dry, warm temperate, and cool temperate. In the coastal area, heath, shrubs, and other sand-dune vegetation flourish. Specific areas of the eastern Australian temperate forest region are recognized internationally as biodiversity hot spots and regions of high plant endemism. Human civilization, however, has had a significant impact on this region's rich biota through deforestation and development.

The Australian temperate forest region is geologically complex, and a wide variety of substrates underlies these distinct regions of rich vegetation.

Coastal Zone

In the coastal zone, extensive sand deposits make up the large sand mass of Fraser Island and Cooloola in Queensland. Together, these sand masses are named the Great Sandy Region. Fraser Island is one of the largest sand islands in the world; it is included on the World Heritage List because of its unique landscape. It is the only place in the world where tall subtropical rainforests, known as littoral rainforests, grow amid giant sand dunes. The large and extensive sand deposits are the result of climatic and sea-level changes over the past 700,000 years. The sand dunes on Fraser Island can reach elevations of more than 656 feet (200 meters).

Coastal vegetation includes shrublands, heaths, satinays, kauri pines, and other salt-tolerant vegetation. Mammals in this area include wallabies, dingoes, bandicoots, and potoroos. The Great Sandy Region also provides critical habitat for birds; more than 350 species of birds have been recorded, including sea eagles, osprey, kites, and pelicans.

Border Ranges

Forming the boundary between Queensland and New South Wales, the Border Ranges are remnants of 20-million-year-old shield volcanoes. Volcanic activity during the Tertiary Period formed large areas of basalt and elevated landscapes such as these. The Border Ranges—including the McPherson Range, Tweed Range, and some notable plateaus—are recognized as biodiversity hot spots. The area contains more than 1,000 vascular plants, including several that are endemic, and some threatened species. This area is considered to be a center for vascular-plant species richness and endemism. More than 100 genera are Gondwanan in origin and include both rainforest and nonrainforest genera.

The Border Ranges are also similarly diverse in fauna, with a variety of mammals, reptiles, amphibians, birds, and invertebrates. Volcanic activity during the Tertiary Period created the distinct landscapes and large areas of basalt. The central ancient shield volcano, Tweed Volcano, formed Mount Warning, which dominates the landscape of far northeastern New South Wales. About 20 million years ago, this area covered 1,544 square miles (4,000 square kilometers) and was nearly twice its present height. Throughout time, erosion has produced unique landforms, including an erosion caldera known as Tweed Valley. Mount Warning, also called Wollumbin and Cloud-Catcher by the Aboriginal inhabitants, is the first place on Australia's mainland to be touched by the morning sun.

Greater Blue Mountain Area

The Blue Mountains form the central part of the Great Dividing Range in New South Wales, which parallels much of the eastern margin of Australia. The Greater Blue Mountain area is approximately 2.4 million acres (1 million hectares) of well-forested sandstone plateaus and cliffs. Much of this area has protected status, to preserve the high diversity of scleromorphic species. It is a center for plant endemism.

The high diversity of scleromorphic taxa consists of 20 plant families, including *Fabaceae, Myrtaceae,* and *Orchideae.* The diverse and endemic biota of this region exists partly due to the region's geologic stability and inaccessibility. This area contains 92 species of eucalyptus, 10 percent of Australia's vascular flora, and 13 percent of the global total. Many of these species are rare or threatened, and some are endemic species, such as the Wollemi pine. Rainforest communities are interspersed throughout the region. The high diversity of plants create equally diverse habitats for fauna. The region has more than 400 vertebrate taxa, including native mammals, birds, reptiles, and frogs. Many of these species are rare, endemic, or threatened.

Megan Machmuller

Further Reading

Australian Department of Sustainability, Environment, Water, Population and Communities. "Australian Places on the World Heritage List." 2011. http://www.environment.gov.au/heritage/places/world.

Crisp, M. D., et al. "Endemism in Australian Flora. Special Issue: Historical Biogeography and Patterns of Diversity." *Journal of Biogeography* 28, no. 2 (2001).

Mansergh, I. and D. Cheal. "Protected Area Planning and Management for Eastern Australian Temperate Forests and Woodland Ecosystems Under Climate Change—A Landscape Approach." In M. Taylor and P. Figgis, eds. *Protected Areas: Buffering Nature Against Climate Change.* Sydney: World Wildlife Fund-Australia, 2007.

World Conservation Monitoring Centre. "World Heritage Sites." 2012, United Nations Environmental Programme. http://www.unep-wcmc.org/world-heritage-sites_189.html (Accessed July 2012).

Australia Woodlands, Southwest

Category: Forest Biomes.
Geographic Location: Australia.
Summary: This biodiverse region is the last refuge for some of Australia's most threatened marsupials.

Rising from the Swan Coastal Plain via the Darling Scarp, the Yilgarn Block is an inland plateau 919 to 1,115 feet (280 to 340 meters) above sea level that is dominated by the giant jarrah (*Eucalyptus marginata*) and marri (*Eucalyptus calophylla*) trees of Australia's southwestern woodlands. These hardwoods reach their peak in the high-rainfall (55 inches or 1,400 millimeters annually) western edge of the region but slowly decline to stunted mallees at the opposite edge of the rainfall gradient, where only 25 inches (635 millimeters) of rain falls per year and wandoo (*E. wandoo*) begins to dominate. Plant-species richness follows this gradient.

Wildlife

This region is the last refuge of a multitude of Australian marsupials that were decimated following the arrival of the European red fox (*Vulpes vulpes*) to the region in the early 1930s. While the fox drove up to 22 Australian-native mammals weighing 0.07 to 12 pounds (35 grams to 5.5 kilograms) to extinction throughout the rest of the continent, several of these species persisted in the southwestern woodlands, in places such as the Dryandra Woodland and Perup Forest. These two conservation areas have high densities of native shrubs of the genus *Gastrolobium,* and these plants contain naturally high levels of the toxic substance sodium monofluoroacetate, or 1080. The native fauna have evolved to cope with this poison, and the herbivores that regularly eat the plants have a high tolerance to it. Introduced species, such as foxes and cats, have little tolerance to 1080, so their densities have been kept at low levels through secondary poisoning by feeding on the native species. This low density of predators has allowed species like numbats (*Myrmecobius fasciatus*) and brush-tailed bettongs (*Bettongia penicillata*) to persist.

Numbats are small, obligate termite feeders that are the sole members of their families. Numbats are active during the day, when their russet coat with white stripes acts as wonderful camouflage from aerial predators like wedge-tailed eagles (*Aquila audax*) and other raptors. Brush-tailed bettongs are small kangaroos, weighing two to four pounds (one to two kilograms), with a prehensile tail that carries nesting material, which they deposit in a small depression beneath a shrub or log.

In the trees, herbivores like western ringtail possums (*Pseudochirus occidentalis*) and insectivores like red-tailed phascogales (*Phascogale calura*) roam. Native predators like western quolls and chuditch (*Dasyurus geoffroii*) preyed on these species but were generalist predators, like foxes or cats, so the system was in relative equilibrium.

The quokka (*Setonix brachyurus*) is another native kangaroo that survives in the upper reaches of watercourses in the higher-rainfall sections of the southwestern woodlands. Quokkas are also threatened by foxes, and their catastrophic decline with the arrival of the European red fox attests to this fact. The threat from foxes is exacerbated by the alteration of their habitat due to the change in fire regimes since European colonization of Australia. For almost 50,000 years, quokkas were able to cope with the cool, high-frequency fires employed by the local Nyoongar Aboriginal people on a four- to eight-year rotation, reaching their peak when their habitat went 10 to 15 years without a fire. Beyond this period, however, the swampy habitat favored by the quokkas became too open, as it was dominated by a few plant species that offered little refuge from predation or diversity of forage.

Ecosystem Threats

As these examples of the plight of the southwestern woodland fauna illustrate, this ecosystem is threatened by several factors. The 1930s catastrophic species declines, pegged to the arrival of the fox, was certainly exacerbated by changes in fire regimes as Aboriginal people left the land and fire management entered a phase of control that reflected the prehuman fire regimes of large, hot, lightning-ignited bushfires. This phase lasted until the 1960s, when more intensive fire management was employed to reduce the effect on humans. Now fire managers seek to benefit biodiversity through their activities, while also reducing the risk to people and infrastructure.

Fire is a natural process throughout most of Australia and is not a threat to most species, as numerous plants have evolved to resprout via epicormic growth or from lignotubers. Some species need fire to open up the canopy and allow their seeds to drop and regenerate. Other species with short-lived seeds may be threatened if the fire return is too rapid to allow them to reach seed-bearing age. Yet these threats are not occurring in isolation. A total 90 percent of the eucalypt woodlands in Australia have been cleared, and the vast majority of the western side of the southwestern woodlands—an area known as the Wheatbelt—has been cleared for agriculture.

The Jarrah Forest is a multiple-use forest that has been extensively logged and that contains several bauxite mines and dams that supply water to the state capital, Perth. Many of the most important and common plant species in the southwestern woodlands are also threatened by root rot or dieback caused by *Phytophthora cinnomoni,* which was introduced from eastern Australia. The spores of this fungus are transported in moist soil, so large areas of vegetation in the southwestern woodlands, particularly along tracks, are dying.

Conservation Efforts

The southwestern woodlands of Australia have several protected areas, including Dryandra Woodland Nature Reserve at 108 square miles (280 square kilometers), Perup Nature Reserve at 201 square miles (520 square kilometers), and Lane Poole Reserve at 212 square miles (550 square kilometers). Nevertheless, simply designating land as a conservation estate rarely solves conservation problems, unless land clearing is the sole threatening process.

In the 1990s, the Western Australian Department of Environment and Conservation instigated a broad-scale fox control project called Western Shield. This project involved aerially spreading dried meat baits injected with 1080 poison across large swaths of western Australia. This resulted in initial improvement in the status of numerous species, including the numbat, brush-tailed bettong, southern brown bandicoot (*Isoodon obesulus*), and quokka. More recently, however, mesopredators such as cats have increased in abundance and are now having as much of an effect as foxes did previously. An effective method of controlling feral cats is urgently needed, but it appears to be a long way off, given that the problem has been researched for several decades.

MATT W. HAYWARD

Further Reading

Burbidge, A. A. and N. L. McKenzie. "Patterns in the Modern Decline of Western Australia's Vertebrate Fauna: Causes and Conservation Implications." *Biological Conservation* 50, no. 1–4 (1989).

De Tores, P. J. and N. J. Marlow. "A Review of the Relative Merits of Predator Exclusion Fencing and Repeated 1080 Fox Baiting for Protection of Native Fauna: Five Case Studies From Western Australia." In M. J. Somers and M. W. Hayward, eds. *Fencing for Conservation.* Upper Saddle River, NJ: Springer, 2011.

Friend, J. A. "The Numbat 'Myrmecobius fasciatus' (Myrmecobiidae): History of Decline and Potential for Recovery." *Proceedings of the Ecological Society of Australia* 16, no. 1 (1990).

Hayward M. W., P. J. de Tores, and P. B. Banks. "Habitat Use of the Quokka 'Setonix Brachyurus' (Macropodidae: Marsupialia) in the Northern Jarrah Forest of Australia." *Journal of Mammalogy* 86, no. 4 (2005).

Short, J. and A. P. Smith. "Mammal Decline and Recovery in Australia." *Journal of Mammalogy* 75, no. 2 (1994).

Azores Temperate Mixed Forests

Category: Forest Biomes.
Geographic Location: Atlantic Ocean.
Summary: The evergreen broadleaved forests, or laurisilva, of the Azores archipelago are remnants of a broader historical forest ecosystem that once spanned parts of Europe and Africa.

Before human settlement, forests covered 90 percent of the surface of the Azores volcanic archipelago. The chain of nine islands contains unique endemic (uniquely evolved to fit a local biome niche) species, but deforestation and the introduction of foreign biota prompted degradation of native flora and fauna.

Stretching across approximately 373 miles (600 kilometers) of the Atlantic Ocean, the Azores are part of the Macaronesian region—a collection of archipelagos including Madeira, the Savage Islands, the Canary Islands, and the Cape Verde Islands. Situated 2,423 miles (3,900 kilometers) from North America and 932 miles (1,500 kilometers) from Portugal, the Azores are considered to be the westernmost point of Europe, although two of the islands (Flores and Corvo) actually rest on the North American plate.

Laurisilva

The evergreen forest, or laurisilva, is regarded as the most characteristic indigenous flora of the Azores. The native evergreens were once part of a much larger forest biome that covered portions of Europe and Africa nearly 2.75 million years ago, during the late Tertiary era. Scientists estimate that the Azores' cool, oceanic climate significantly contributed to laurisilva survival on the islands, in contrast to the continental climate extremes that fueled the extinction of European and African counterparts. Settlement, agriculture, and livestock domestication led to deforestation of the original laurel forest cover—a humid, evergreen broadleaf

Thick vegetation on the steep cliffs of São Jorge Island in the central group of the Azores Islands. A small settlement on a fajã, or low coastal land formed by past rockfalls, can be seen at left. Azores evergreen forests contain about 30 known endemic plants that grow only within the Azores laurisilva. Several of these endemisms are considered living fossils. (Thinkstock)

variety specific to the Azores—although somewhat similar species have been found on Madeira and the Canary Islands. Close to 98 percent of the native forests were decimated by human activity after initial settlement in the 15th century.

As a result of 20th-century conservation enlightenment and in response to the potential of tourism to drive sustainable economic growth, the Azorean government simultaneously helped and hindered the laurisilva. The majority of the native forests are now protected in 26 Recreational Forest Reserves; the Regional Directorate for Forest Resources established reserves on eight of the nine islands. To satisfy an agenda of Azorean beautification for tourism, however, and also to reduce erosion, produce timber, and minimize water loss, exotic species were imported. Subsequently, a series of plantations now grow foreign biota to replace the diminished native broadleaved evergreens. Two species in particular—the coniferous Japanese cedar (*Cryptomeria japonica*) and the Australian cheesewood (*Pittosporum undulatum*—dominate the Azorean landscape. Although Azorean heath (*Erica azorica*) and fire tree (*Myrica faya*), both endemic species, have flourished where agricultural cultivation has been abandoned, the highly invasive *Pittosporum undulatum,* introduced from Australia during the 19th century, threatens the more fragile endemic flora.

Unprotected remnants of indigenous evergreens have survived at elevations above 1,640 feet (500 meters). At this elevation, Azores laurel (*Laurus azorica),* Azores juniper (*Juniperus brevifolia),* and Azores heath compose the dense canopy of native evergreen forest.

Endemic Vegetation

Despite remaining portions of laurisilva at high altitudes and those found in reserves, the endemic Azores evergreen forests remain relatively unexplored and marginally researched. There are, however, some 30 known endemic vegetation that appear exclusively within the Azores laurisilva. Several of these endemisms derive substantial significance as survivors of geologic time, or living fossils. They are peculiar to the Azores and offer evidence of vegetation that once existed on the continents, as well.

Bird Life

Faunal biodiversity, primarily a bird population consisting of 36 reported species, has been historically minimal due to the geographic isolation of the Azores. In particular, the Azores bullfinch (*Pyrrhula murina*) that resides in high-altitude forests is listed as an endangered species by the International Union for Conservation of Nature. A mere 250 birds remain in the bullfinch population, which mainly inhabits the Azorean island São Miguel.

Conservation and Preservation

With its history of relatively dramatic ecosystem changes that have resulted from human activity and invasive species, the Azores and its mother country, Portugal, host a scientific community that is keen to establish protections against a range of contemporary threats that include ocean temperature warming, seawater acidification, climate change, and species genetic jeopardy. These often interconnected vectors have been under study for years; various mitigation initiatives are in the planning stages. With the unifying goal of encouraging and sustaining biological diversity, the Azores is seen as an ideal location for habitat reserves, ecotourism, and related efforts in this campaign.

Matthew Alexander

Further Reading

Morton, Brian and Joseph C. Britton. "The Origins of the Coastal and Marine Flora and Fauna of the Azores." *Oceanography and Marine Biology* 38, no. 1 (2000).

Networking Tropical and Subtropical Biodiversity Research in Outermost Regions and Territories of Europe in Support of Sustainable Development (NetBiome). "Azores." 2012. http://www.netbiome.org/index.php?option=com_content&view=article&id=65%3Aazores&catid=51%3Amonographies&Itemid=1 (Accessed July 2012).

Santos Guerra, Arnoldo. *Evergreen Forests in the Macaronesian Region.* Strasbourg, France: Council of Europe, 1990.

Baffin Bay

Category: Marine and Oceanic Biomes.
Geographic Location: North America.
Summary: Bordered by Canada's Nunavut Province on the west and Kalaallit Nunaat (Greenland) on the east, Baffin Bay is known for its marine mammals, seabirds, and polar bears.

Baffin Bay is positioned to the northwest of the North Atlantic Ocean, straddles the Arctic Circle, and features a surface-frozen channel that connects to the Arctic Sea. The bay, with a surface area of approximately 300,000 square miles (775,000 square kilometers), is cradled between two giant boreal islands: to the west, Baffin Island, within the Canadian province of Nunavut; to the east, Greenland. To the north is a third major island, Ellesmere; between it and Greenland the relatively constricted chain of Smith Sound–Kane Basin–Kennedy Channel–Hall Basin–Robeson Channel, together known as Nares Strait, forms a roughly 350-mile-long (565-kilometer-long) link to the Arctic Sea. To the south, the broader Davis Strait opens upon the Labrador Sea, itself a lobe of the North Atlantic.

Much of Baffin Bay is jammed with sea ice more than nine months of the year, but the biome features some polynyas, or areas of seasonal open water surrounded by pack ice, that provide oases for life adapted to the Arctic cold. The most important of these is called the North Water. A mighty polynya considered one of the world's largest, the North Water, extends across as much as 33,000 square miles (85,000 square kilometers) of northern Baffin Bay and Smith Sound late each spring. Heavy ice floes from the Arctic are blocked at this time of year by an ice arch that forms across the southern aperture of the Kane Basin.

The West Greenland Current, sweeping northward out of the Atlantic and along the Greenland coast until it enters the North Water, is somewhat saltier and warmer than the prevailing Baffin Bay waters. This helps make North Water a haven for marine life in the bay. Currents generally arc east to west through North Water, then the Baffin Island Current heads south to help generate the southeasterly Labrador Current into the North Atlantic. Thus the macrocirculation picture in Baffin Bay is counterclockwise: north along Greenland and south along Baffin Island.

Biota

Arctic tundra borders the bay in narrow coastal plains and rockbound pockets on Baffin Island

and stretches of the Greenland coast. Characteristic vegetation runs from stunted stands of salt-tolerant willow and alder to a relative abundance of flowering plants such as woodrush (Juncaceae spp.), harebell (Campanula rotundifolia), and Arctic poppy (Papaver radicatum). The waters of the bay support vital algae colonies, especially pronounced in areas like Lancaster Sound (between Baffin and Ellesmere Islands) and North Water.

Arctic cod is a vital component of the ecosystem in Baffin Bay, as a prime food source for the vast marine mammal population. Humans, too, have long relied on the cod fishery of Baffin Bay and its associated inlets, sounds, and channels. Lancaster Sound alone is home to cod schools that have been estimated at up to 30,000 tons per school.

Birds resting among the rocks below the steep cliffs of Frobisher Bay on the southeast of Baffin Island near Iqaluit. Five wildlife refuges protect birds and other life in and around Baffin Bay. (Thinkstock)

Protected Areas

Canada's Sirmilik National Park includes the Bylot Island Migratory Bird Sanctuary, adjoining Lancaster Sound near the northeastern tip of Baffin Island. Park authorities estimate up to 320,000 thick-billed murres and 50,000 black-legged kittiwakes, seabirds hailing from the North Atlantic, make use of the cliffs in spring, while nearby polynyas offer good foraging for both the seabirds and marine mammals. Another seabird, the northern fulmar, makes perhaps its largest Canadian colony at the Ellwin Inlet and Baillarge Bay areas of the park, nesting on cliffs 1,970 feet (600 meters) high. Eclipse Sound, another area of the park, is covered by ice until midsummer but offers prime habitat for narwhals and polar bears.

Besides the sanctuary on Bylot, there are five other reserves offering refuge to fowl, mammals, and other fauna around Baffin Bay. The northernmost is Nirjutiqavvik National Wildlife Area on Coburg Island, a nesting area for 385,000 birds including black-legged kittiwakes, thick-billed murre, northern fulmar, and black guillemot. Also present are polar bears, ringed seals, walruses, and narwhal and beluga whales. Cambridge Point on the island overlooks a polynya. To the southwest, on the western edge of Lancaster Sound, is Prince Leopold Island Migratory Bird Sanctuary; its vertical cliffs of sandstone and limestone rise to 820 feet (250 meters), providing additional nesting sites. Niginganiq (Iqaliqtuuq) National Wildlife Area includes summer habitat for a population of bowhead whales. Adults and adolescent whales frequent this area on the east coast of Baffin Island.

Akpait National Wildlife Area, located in southern Baffin Bay at the easternmost point of Baffin Island, includes Canada's largest thick-billed murre colonies, nesting black-legged kittiwakes, glaucous gulls, and black guillemots. Ocean cliffs here rise to 3,002 feet (915 meters), with polar bears and marine mammals active below. At nearby Qaqulluit National Wildlife Area, two rock towers rise 1,411 feet (430 meters) out of the ocean and host nesting northern fulmars. The surrounding waters are important for marine mammals.

Development and Sustainability

Despite the harsh climate, development is proposed in some areas of Baffin Bay. At the Mary River Project on Baffin Island, the Baffinland Iron Mines Corporation plans to develop an iron-ore mine. The minerals would be shipped via a new railroad south to the Foxe Basin. Seasonal access is also proposed at Milne Inlet, a southern extension of Baffin Bay known as Eclipse Sound, via a 62-mile mile (99-kilometer) haul road. The project is awaiting the approval of the Nunavut government.

In its Canadian territorial waters, Baffin Bay falls mainly within that government's Eastern Arctic bioregion and is part of one or more planned marine protected areas (MPA). The internationally recognized MPA designation attempts to frame and encourage such goals and measures as biological diversity, conservation, restoration, and integrated ocean management.

In 2009, authorities from Canada, Nunavut, and Greenland agreed on ordinances and enforcement aimed at protecting the polar bear population of Baffin Bay and Kane Basin; the bears are subject to pressures from human hunting, pollution, prey depletion, and global warming.

Global warming is already affecting Baffin Bay, with major effects manifesting in obvious ways such as accelerated glacial melt from Greenland—adding to freshwater volume and altering the salinity balance—and in less predictable ways such as potential disruptions of the flow of higher-saline seawater from the Atlantic; greater algal blooms yielded by more open water; and increasing human industrial activity as ice-free sea channels expand to new areas and for longer periods each year, making oil drilling, shipping, and port facility development more feasible.

HAROLD DRAPER

Further Reading

DeVilliers, Marq. *Guide to America's Outdoors: Eastern Canada*. Washington, DC: National Geographic Society, 2001.

Krajick, Kevin. "In Search of the Ivory Gull." *Science* 301, no. 5641 (2003).

McKinnon, L., et al. "Lower Predation Risk for Migratory Birds at High Latitudes." *Science* 327, no. 5963 (2010).

Shubin, Neil. *Your Inner Fish*. New York: Random House, 2008.

Tang, Charles C. L., Charles K. Ross, Tom Yao, Brian Petrie, Brendan M. DeTracey, and Ewa Dunlap. *The Circulation, Water Masses, and Sea-Ice of Baffin Bay*. Dartmouth, Canada: Bedford Institute of Oceanography, 2004.

Baffin Coastal Tundra

Category: Grassland, Tundra, and Human Biomes.
Geographic Location: North America.
Summary: The Baffin coastal tundra, one of the coldest year-round lands on Earth, is subtly beautiful and home to various low-growing plants and northern wildlife.

The Baffin coastal tundra is located along the northeastern coast of Baffin Island and corresponds with the Baffin Island coastal lowlands ecoregion of the Arctic Cordillera ecozone. Clyde River or Kanngiqtugaapik (in Inuktitut) is the largest village, with a population of approximately 800 of mostly Nunatsiarmuit ("People of the Beautiful Land"), an Inuit subgroup. The Baffin Coastal Tundra biome spans some 3,500 square miles (9,065 square kilometers). The climate of the region, among the world's coldest, is classified as high Arctic. Humid, with long, cold winters and very short, cold, foggy summers, the Baffin version of this climate features an average annual temperature of 11.3 degrees F (minus 11.5 degrees C). The average winter temperature is minus 8.5 degrees F (minus 22.5 degrees C), while the summer average is 33.8 degrees F (1 degree C). Annual precipitation is 8 to 12 inches (20 to 30 centimeters).

This coastal plain is approximately 25 miles (40 kilometers) wide. It is mostly treeless, with a

A caribou antler lies on a bed of low-growing plants on the tundra on Baffin Island in Canada. Vegetation in wet areas of the Baffin coastal tundra consists of low plants such as mosses, rushes, and saxifrages, and may cover as much as 60 percent of the land. (Thinkstock)

frozen subsurface. It is dissected by many fjords, such as Clark, Gibbs, Eglinton, and McBeth; most fjords extend landward to the Davis Highlands. There are several inlets (e.g., Scott and Clyde), capes (e.g., Hewett and Adair), and bays (e.g., Hamilton and Isabella) along the shore. The gently rolling landscape has been sculpted and warped by frost weathering and ice wedging.

Surface deposits of alluvial, residual, and/or glacial deposits appear, but bare bedrock and small bedrock outcrops are the dominant features. The coastal areas are predominately rocky, with low bluffs. The beaches are made of sand and gravel, and there are marine terraces. The Baffin coastal tundra has deep, continuous, low-ice-content permafrost.

Flora

The coastal plain is sparsely covered with vegetation, due to poor soil conditions, low precipitation, cold temperatures, and the limited growing season. Low biotic diversity and simple vegetation structure are characteristic; much of it consists of prostrate and low-lying dwarf shrubs, mosses, low-growing forbs, sedges, shrubs, rushes, and lichens. In the wet areas, vegetation coverage can reach 60 percent, and is mostly made up of various mosses, rushes (e.g., wire rush and Arctic and northern woodrush), and saxifrages.

Resident forbs include alpine mountain sorrel, alpine bistort, and moss campion. Typical mosses are turgid aulacomnium, tomentypnum, juniper polytrichum, and splendid feather moss. Lichen species include Cetrariella deliseii, felt, snow, chocolate chip, and whiteworm. Other vegetation includes tundra grass, Arctic and polar willows, white Arctic mountain heather, Lapland rosebay, mountain avens, various berries, mushrooms, river beauty, purple saxifrage, kobresia, sedge, Arctic poppies, and seaweed from the ocean. All vegetation on the coastal plain is vulnerable to frost at any time.

Fauna

The Baffin coastal tundra supports a limited number of mammalian species. Common year-round mammals include barren-ground caribou (Rangifer tarandus groenlandicus); polar bears; Arctic wolves, foxes, and hares; short-tailed weasels (stoat or ermine); and American brown and Peary Land collared lemmings. Marine mammals include walrus; harp, ringed, and bearded seals; as well as several whales, such as beluga, killer, bowhead, and narwhal. Common fish species just off the coast in Baffin Bay include Arctic char, polar cod, Arctic flounder, halibut, turbot, four-horned sculpins, capelin, and herring. Baffin Bay and Davis Strait are home to large cold-water reefs, sponges, clams, shrimp, and echinoderm species.

Birds present year-round are the gyrfalcon, snowy owl, common raven, and rock and willow ptarmigan. Most birds of the Baffin coastal tundra are migratory. Common such fowl are snow, Ross, Brant, and Canadian geese; tundra swans; king and common eiders; surf scooters; and long-tailed ducks. There are several loon species, including red-throated, Pacific, common, and yellow-billed, as well as plovers (white-faced, American golden-plover, and semi-palmated). Birds of prey include the rough-legged hawk, peregrine falcon, and short-eared owl.

Other common migratory birds include sandhill cranes, Arctic terns, thick-billed murres, black guillemots, black-legged kittiwakes, horned larks, Wheater thrush, Savannah sparrows, Lapland longspurs, snow buntings, and redpolls (common, Greenland, and Arctic). There are several species of sandpipers (semi-palmated, pectoral, Baird's, least, and buff-breasted) and gulls (ivory, sabine, Ross's herring, and Iceland). The ecoregion has no reptiles or amphibians.

Land Use and Conservation

Land use consists mostly of subsistence use. The area is mostly intact, but the coastal line, naturally fragmented by fjords, is vulnerable to human land use and activity. Other concerns are transboundary heavy metals and pesticide pollution. There is the possibility of oil spills, which would be disruptive to the tundra environment. In cold temperatures, oil has a low evaporation rate, and spilled oil would persist for many years, disturbing the thin layer of fragile vegetation and resulting in thermokarst erosion.

In 2008, after a 26-year Inuit-initiated effort, the first National Marine Wildlife Area was established in Canada. Niginganiq (formerly called Igaliqtuuq) National Wildlife Area (Isabella Bay), simply called Niginganiq, is an 830,274-acre (336,000-hectare) conservation and bowhead-whale preserve.

Arctic tundra is a regulator of global climate, in the sense that permafrost is a temperature-sensitive carbon sink; as temperatures rise, methane gas that originated in dead organic matter is released from within the thawing permafrost. This contributes to the atmospheric carbon load in a positive feedback loop as this potent greenhouse gas in turn adds to climate warming trends. Ultimate effects of this trend could mean a shortened snow season, disruption of pollination patterns, altered terrestrial food web across areas of the tundra, and saltwater inundation of coastal areas from sea-level rise.

Andrew Hund

Further Reading

Derocher, Andrew E. and Wayne Lynch. *Polar Bears: A Complete Guide to Their Biology and Behavior.* Baltimore, MD: Johns Hopkins University Press, 2012.

Ricketts, T., E. Dinerstein, D. Olson, C. Loucks, et al. *Terrestrial Ecoregions of North America: A Conservation Assessment. World Wildlife Fund–United States and Canada.* Washington, DC: Island Press, 1999.

Sempels, Jean-Marie. "Coastlines of the Eastern Arctic." *Arctic* 35, no. 1 (1982).

Watson, Adam. *A Zoologist on Baffin Island, 1953.* Northampton, UK: Paragon Publishing, 2011.

Bahia Forests

Category: Forest Biomes.
Geographic Location: South America.
Summary: The Bahia forests, located on the southeastern coast of Brazil, are home to several critically endangered species threatened by agricultural and other land-clearing activities.

The Bahia forests are composed of three main sections: coastal, interior, and southern moist forests. The average annual rainfall is 39 to 71 inches (1,000 to 1,800 millimeters), with a temperature range of 62 to 83 degrees F (17 to 28 degrees C). The Bahia coastal forest covers approximately 38,610 square miles (100,000 square kilometers) and occupies a strip 93 miles (150 kilometers) wide bordering the Atlantic Ocean.

This coastal forest is home to a large number of endemic (evolved in and uniquely native to a specific biome) plants, birds, primates, and butterflies that have become critically endangered, including the three-toed maned sloth (Bradypus torquatus) and the golden-headed lion tamarin (Leontopithecus chrysomelas). The types of vegetation that can be found in the coastal forest are Atlantic moist (evolved to thrive in high-rainfall zones) and semi-deciduous (losing leaves only in the short dry season), often in a setting with up to four strata topped by a 100-foot-high (30-meter-high) canopy.

Researchers have found that almost 30 percent of the vascular plants in these coastal forests are endemic. This segment of the Bahia Forests biome represents a range limit for many Amazonian species of plants and birds. In other words, many such species have a particular habitat niche, food source, or temperature requirement that can be found here, but not north of this region. Diversity in the Bahia forests approaches the level of diversity in the Amazon rainforests, but the species profile is distinctly regional.

Interior Forests

Interior forests occupy 41 percent of the total land area in Bahia and constitute 75 percent of Bahia forests. This area of 88,803 square miles (230,000 square kilometers) contains both semi-deciduous and seasonal deciduous areas, and is home to several critically endangered endemic species. The Coimbra titi monkey (Callicebus coimbrae), for example, was recently found to be an endemic species in danger of extinction. Although the interior is considered to be a species-rich area, the exact amount of biodiversity therein is unknown. In 1993, biologists discovered the presence of two highly specialized endemic birds: the slender antbird (Rhophornis ardesiaca) and the narrow-billed antbird (Formicivora iheringi).

Interior forests are among the most modified and fragmented ecoregions of the Bahia forests. Due to activities such as logging and cocoa plantations, few continuous stretches of interior forest cover more than four square miles (10 square kilometers). Less than 1 percent of the interior forests are protected. The most common method of protection has been the creation of parks, with the most representative park of the ecoregion being the State Park of Rio Doce, comprising about 139 square miles (359 square kilometers) in the middle of the valley of Rio Doce, just south of the city of Ipatinga. This park, small in comparison to the vast forest resources of the area, does not provide adequate protection for these habitats. Though the Brazilian Forestry Code was passed in 1965, with the aim of protecting 247 million acres (100 million hectares), lack of proper enforcement has resulted in approximately 40 percent of the area being deforested.

Southern Moist Forest

The moist forest of southern Bahia extends southward from the Salvador to northern Espirito Santo state. It is approximately 62 to 124 miles (100 to 200 kilometers) inland from the coast, originally covering 27,220 square miles (70,500 square kilometers), manifesting as a half wet and half mesophytic (midrange of moisture) forest. Trees in the region flower throughout the year but usually experience a distinct peak from October to December.

The moist southern forests have epiphytes (plants perched upon other plants) on up to 80 percent of the trees. As many as 40 percent support lianas (large, rooted woody vines). Several important plant families are present in terms of diversity, density, and dominance, including myrtle (Myrtaceae), the evergreen Ericales shrubs and trees (Sapotaceae), Poincianas and other Fabaceae (Caesalpiniaceae), laurels (Lauraceae), and coco plums and related woody evergreens (Chrysobalanaceae).

Effects of Human Activity

Human activities have reduced the forests by more than 35 percent, to a current area of less than 9,527 square miles (24,675 square kilometers). Approximately 494,211 acres (200,000 hectares) are used for pastures, cocoa plantations, and rubber plantations. Attempts are being made to reduce these activities, however, as approximately 98,842 acres (40,000 hectares) of this forest are in the initial stages of regeneration. Despite the Forestry Code of 1965, less than 0.1 percent of wet forests and none of the mesophytic forests are protected.

The Una Biological Reserve is one of the most important forest sanctuaries, preserving one of the largest areas of contiguous forests—as much as 49 percent of the mature forests. Many of the mature forest areas in Una, however, are directly exposed to human activities. Further, most forest reserves are inadequately staffed or patrolled, allowing continued fragmentation and degradation of the habitat.

Approximately 1,931 square miles (5,000 square kilometers) of Atlantic coastal forest is already subject to cocoa planting and cattle ranching, which have increased habitat loss and forest fragmentation, endangering many of the valuable endemic species of the region. Less than 5 percent of the original coastal forest remains, which has caused conservationists to place it at the top of the priority list for biodiversity protection. One measure has been to create biological and forest reserves, such as the Sooretama Biological Reserve and Linhares Forest Reserve, to protect the forests from any potential anthropogenic threat to the habitat and the native species.

The demand for access to land in Bahia threatens all forest resources. The creation of forest reserves and state parks are to date inadequate measures to provide protection for these vast resources. Given limited government intervention and enforcement of the 1965 Forestry Code, there is a need for creative means of protection, such as providing greater enforcement and management autonomy to indigenous populations and enforcing stringent laws that value the ecological resource. Though there is nongovernmental organization work on deforestation, such groups often face opposition from government, which believes that deforestation is necessary for national development.

Global warming presents additional pressures, as rising temperature patterns and altered precipitation regimes can lead to more severe fire damage, erosion threats, and species dislocation through habitat degradation and fragmentation—and through these effects, the loss of a valuable carbon sink that can help mitigate greenhouse gas buildup. But these forests can only perform that ecological service to the extent that they are protected and sustained.

April Karen Baptiste
Yasmin Mannan

Further Reading

Mori, Scott A., Brian M. Boom, Andre M. de Carvalho, and Talmon S. dos Santos. "Southern Bahian Moist Forests." *Botanical Review* 49, no. 2 (1983).

Paciencia, Mateus L., and Jefferson Prado. "Effects of Forest Fragmentation on Pteridophyte Diversity in a Tropical Rain Forest in Brazil." *Vegetation* 180, no. 1 (2005).

Voeks, R. A. "Tropical Forest Healers and Habitat Preference." *Economic Botany* 50, no. 4 (1996).

Baikal, Lake

Category: Inland Aquatic Biomes.
Geographic Location: Southeast Siberia, Russian Federation.

Summary: Lake Baikal is a UNESCO World Heritage Site in Siberia that is among the world's oldest, largest, and deepest freshwater lakes; it is home to many unique species, but under threat from anthropogenic pollution.

At 25 million years old and 5,580 feet (1,700 meters) deep, Lake Baikal is the world's oldest and largest freshwater lake. With a volume of 5,500 cubic miles (23,000 cubic kilometers), this isolated lake in southeast Siberia is estimated to contain one-fifth of the world's unfrozen freshwater, making it the largest single liquid freshwater reservoir. Its isolation and age give Lake Baikal and the surrounding shores a unique ecology with tremendous biodiversity and unique freshwater species. Lake Baikal is designated as a United Nations Educational, Scientific and Cultural Organization (UNESCO) World Heritage Site. Despite its unique environmental qualities and irreplaceable value, the lake is at risk from industrial pollution.

The lake is situated in southeastern Siberia near the Mongolian border, along a continental rift valley. The surrounding terrain is mountainous; some 300 streams and rivers flow into the lake, including these major rivers: the Selenga, the Barguzin, and the Upper Angara.

Drainage is into the Angara, a tributary of the Yenisei, which ultimately flows into the Arctic Ocean. Lake Baikal is approximately 396 miles (640 kilometers) long and 49 miles (80 kilometers) wide. It is considered the deepest lake in the world; in most places, the shoreline plunges deeply into the lake and there are very few coastal plain areas or islands—a striking exception being the 45-mile-long (72-kilometer-long) Olkhon Island. Thermal vents found deep in the lake produce heated, oxygenated water that helps Lake Baikal circulate oxygen more thoroughly than many other deep lakes.

Mosaic of Endemic Species

Many endemic (evolved specifically to a biome and not found elsewhere) species of plants and animals are found in Lake Baikal. Of more than 1,500 plant and animal species living in the lake and its environs, as many as two-thirds are thought to be endemic. From phytoplankton, diatoms, and protozoa to bacterial mats, sponges, worms, shrimp, fish, seals, chipmunks, moose, elk, and bears, the lake area teems with a rich mosaic of biodiversity.

Baikal sturgeon (Acipenser baerii baicalensis) are one of the endangered species endemic to the lake. The Baikal sturgeon preys upon mollusks—there are more than 100 species here—as well as sponges, which live at depths up to 3,280 feet (1,000 meters).

The transparent, tiny, shrimplike epischura (Epischura baikalensis) enjoys the dominant position among all zooplankton in the lake; its feeding upon microscopic life and algae helps filter and purify the lake's water. Much of the cold-water bacteria in the lake also helps break down organic matter and decomposing entities. This filtering action accounts for the extreme clarity of Baikal waters.

Among and across the rugged mountains that surround Lake Baikal, a rich biota inhabits the areas of light coniferous forests, pine forests, and deciduous forests. Characteristic plants include Siberian fir (Abies sibirica), alpine bearberry (Arctostaphylos alpina), dwarf birch (Betula divaricata), and Siberian larch (Larix sibirica), as well as mountain pine, aspen, and Adam's rhododendron (Rhododendron Adamsii).

Baikal Seals

The Baikal seal (*Phoca sibirica*) is a unique freshwater species that is endemic to the lake. Also called nerpa, it is one of only a handful of species of completely freshwater seals in the world. Also among the smallest species of seals, the Baikal seal dwells hundreds of miles away from the nearest ocean. Some scientists suggest the species may date to 2 million years ago.

The population of Baikal seals numbers around 60,000. The lake is largely covered with ice in the winter; female Baikal seals build ice dens in which to birth and raise their young, typically a single pup.

The seals' diet is quite dependent upon golomyankas or Baikal oilfish (*Comephorus baikalensis, Comephorus dybowkii*), another endemic species. This fish is viviparous, hatching live young rather than laying eggs, a relative rarity among piscine species. Golomyankas migrate vertically daily, coming closer to the surface at night.

The nerpas also prey upon gobi, of which at least two dozen species are endemic in Lake Baikal. Also found in the lake is the endemic omul (*Coregonus migratorius*), a salmon-type fish that is considered a delicacy in the region and subject to commercial fishing.

Lake Baikal's freshwater seals, or nerpas, are one of the few species of exclusively freshwater seals in the world. They live hundreds of miles away from the ocean and some scientists think that they may date back as far as 2 million years. (Thinkstock)

Signature mammal species include European roe deer (Capreolus capreolus), red deer (Cervus elaphus), wolverine (Gulo gulo), Siberian chipmunk (Tamias sibiricus), brown bear (Ursus arctos). Eurasian lynx (Lynx lynx), and sable (Martes zibellina). Among the avian notables are black stork (Ciconia nigra), red grouse (Lagopus lagopus), spotted nutcracker (Nucifraga caryocatactes), grey partridge (Perdix perdix), and the peregrine falcon (Falco peregrinus).

Protection and Threats

Lake Baikal's UNESCO World Heritage Site designation came in 1996 due to its valuable and unique natural qualities according to four UNESCO criteria: (1) The lake is an exceptional natural phenomenon with aesthetic value for its natural beauty, (2) the age of the lake places it as a valuable example of Earth's history, (3) the isolation and biodiversity of the lake make it an example of significant ecological and biological processes in the evolution of ecosystems and plant and animal life, and (4) Lake Baikal and the surrounding area also contain important and significant natural habitats needed for conservation of biodiversity, including threatened species. In 2010, the organization considered removing Lake Baikal from the World Heritage list because of concerns about the adverse environmental effects from industrial pollution tied to a pulp and paper mill; the issue has remained unresolved.

The most significant threat to the health of Lake Baikal is industrial development, including two pulp mills, as well as logging and a hydroelectric dam. One contentious source of pollution has long been the Baikalsk Pulp and Paper Mill on the lake's southwestern tip. The mill has been under continued pressure from Russian ecologists and environmental activists to close operations since its opening in 1966. Ecologists have argued that the plant pollutes Lake Baikal through its discharge of wastewater into the lake; mill operators argue that the wastewater is adequately treated before discharge and poses no harm.

Biologist and ecologist Marina Rikhvanova received a Goldman Environmental Prize in 2008 for her efforts to preserve Lake Baikal from such threats. Her research showed that the water near the mill is polluted, resulting in decreasing populations of crustaceans, genetic mutations, and disoriented fish unable to find food or spawning grounds. Among the mill's wastes are dioxins, extremely toxic and known to affect hormones and weaken immune systems. The people and animals who eat fish with dioxin contamination can suffer from ill health effects. The plant was closed in 2008 due to environmental contamination concerns, and had gone bankrupt, but was allowed to reopen in 2010 with permission from Russian prime minister Vladimir Putin and with the Russian government owning 49 percent of the company. As of summer 2012, the mill's license to release waste into Lake Baikal was under regulatory review. Despite assurances of minimal or no environmental effect, local fisherman must move further away from the mill to find their catch as they claim that fish have become scarce closer to the mill.

Lake Baikal is unique in another way, related to global climate change. Sampling of sediment cores has the potential to provide a detailed historical record of climate variation. Most long-term historical climate data is taken from oceanic marine environments; there is little comparable data available from continental interiors. Core sampling from Lake Baikal can provide such data. Scientists can use that core sampling to provide climatic, environmental, and geological history reaching back millions of years.

Shaunna Barnhart

Further Reading

Hutchinson, D. and S. Colman. "Lake Baikal—A Touchstone for Global Change and Rift Studies." U.S. Geological Survey, July 1993. http://marine.usgs.gov/fact-sheets/baikal.

Kozhova, O. M. and L. R. Izmest'eva. *Lake Baikal: Evolution and Biodiversity.* 2nd ed. Kerkwerve, Netherlands: Backhuys Publishers, 1998.

Minoura, Koji. *Lake Baikal: A Mirror in Time and Space for Understanding Global Change Processes.* Amsterdam, Netherlands: Elsevier Science 2000.

Moskvitch, Katia. "UN May Strike Baikal off World Heritage List." *BBC News.* July 23, 2010.

Baja California Desert

Category: Desert Biomes.
Geographic Location: North America.
Summary: The only desert in the world surrounded by two seas, this geologically isolated peninsula sets the stage for a myriad of remarkable plants and animals exemplifying adaptations to an isolated and arid environment.

The Baja California Desert is the peninsular arm of the mainland Sonoran Desert, and although closely related to each other, they contain dramatically different evolutionary histories. While the mainland Sonoran Desert biota evolved connected to both northern temperate biomes and southern tropical forests, the evolution of the Baja California peninsula took a different trajectory due to its long history of isolation. The Baja California Desert is a paradigm of the importance of geography and time, the two axes along which life develops its variations in shaping the natural world.

Distributed through the major part of the second-longest peninsula in the world at 800 miles (1,300 kilometers), the Baja California Desert is a peninsular desert. This fundamental fact implies two straightforward but decisive ecological consequences: isolation, and the climatic influence of the surrounding seas. Also to consider are four more geographical attributes: location at mid-latitudes (23 degrees 30 minutes north to 30 degrees north); almost north-south orientation; very narrow width of 31 miles (50 kilometers) on average; and an intermittent mountain range along its length. With these few, but determinant, attributes one can outline the basis for an understanding of the ecological singularity of the Baja California Desert.

Cleaving this arid region into two divergent but fundamentally united biomes (the Baja California and Sonoran Deserts) is the Gulf of California. The gulf is the greatest physical barrier in northwest Mexico. It opened at least 5.5 million years ago, splitting Baja California from the mainland. Since then, this barrier has impeded dispersal by many plant and animal species from the mainland to the peninsula, and vice versa. This near-insularity of the Baja California Peninsula has been the most crucial factor to determine the uniqueness of this desert.

In addition to the isolated nature of the peninsula as a whole, scattered and sequestered habitats at different scales are superimposed on the desert along its length. Sea islands of various sizes are present along the Pacific coast; they are especially abundant throughout the Gulf of California. The highest tips of the mountain ranges that form the backbone of the peninsula contain small sky islands of relict temperate ecosystems.

Scattered palm oases in deep and sheltered disjunct canyons represent mesic (moderately moist) refuges within a landscape of dry rock and sand. Both seacoasts are dotted with coastal lagoons often harboring mangroves, here at their northernmost occurrence in the Pacific Coast of North America, that constitute critically important wetland ecosystems. This mosaic of insularity at different temporal and spatial scales constitutes the driving force of biological speciation: adaptation to local, isolated microenvironments with distinct microclimates.

In contrast to other coastal deserts in the world, the Baja California Desert is a bicoastal desert. On the western side of the peninsula is the cold Pacific coast, its chilled waters of the California Current coming from polar latitudes. Meanwhile, the eastern coast is warmed by the Gulf of California, which has been considered the only large evaporation basin of the Pacific Ocean due to the high temperature in the region. The contrast of climatic influences between two coast lines, separated by the narrow width of the peninsula, combined with the presence of transpeninsular mountain ranges acting as barriers between the two climatic influences, establishes a sharp west-east climatic gradient. At the same time, there is a long ecological transition between the northern temperate region showered by winter rains, and the southern dry tropical forest soaked by summer storms and hurricanes. Consequently, the central part of the Baja California Desert contains a biseasonal and unreliable precipitation regime.

Throughout this climatic background of scarce and unpredictable precipitation, there are anomalous events of abundant precipitation. These are caused by two principal climatic sources: tropical cyclones from the south in summer months, or El Niño Southern Oscillation (ENSO) years that bring above-average winter rain. These pulses of abundant resources are crucial periods in the ecological organization and dynamics of the Baja California Desert. In these brief periods of intense activity, the desert becomes renewed and prepares to tenaciously face the next years of hardship.

Biota of the Desert

The existence of recurrent insularity during millions of years and different climatic influences along both north-south and east-west axes has stimulated the forces of evolution and generated a plethora of singular life forms of desert plants and animals. Many of these are endemic, evolved and found here and nowhere else. About 20 percent of vascular plant species here are endemic, whereas within animals the level of endemism is particularly high in invertebrates (scorpions, at least 40 percent, and tenebrionid beetles, 45 percent) and mammals (about 45 percent, at the subspecies level), followed by amphibians and reptiles (30 percent).

Some of the most striking vegetation gradients in the world are seen in Baja California. If one heads south from the U.S.–Mexico border in Tijuana, the trip begins in the Mediterranean region where rain only comes in winter months. Around 30 degrees latitude, a dramatic change in the landscape occurs, a transition from the mediterranean coast to the desert. Suddenly one sees giant columnar cacti called cardones, rosy boa (Lichanura trivirgata), and a bizarre, massive carrot-shaped plant called the boojum tree (Fouquieria columnaris).

Further south, cold marine air and fog from the Pacific Ocean along the western side of the peninsula stimulate the growth of plants with much reduced stems that instead have succulent leaves to allow them to persist through the long dry summer months; these include agaves, yuccas, and dudleyas. Air plants and lichens are seen clustered on the stems of many shrubs, able to exist detached from the soil due to the cool Pacific fog from which they absorb moisture.

However, a few kilometers to the east, on the other side of the mountains, along the warm-water Gulf of California coast, there is a flora dominated by woody legumes and trees with gigantic and fleshy stems, such as copalquín (Pachycormus discolor)the elephant tree (Bursera microphylla), the copal (Bursera hindsiana), and various Jatropha species.

The Baja California Desert during a lush spring in March 2010 with cirio and cardón plants rising above the various other desert scrub species. About 20 percent of vascular plant species in the Baja California desert biome are endemic. (Benjamin Theodore Wilder)

Further south still, these give way to dense and diverse vegetation where cacti, trees, shrubs, and other succulents intermingle. The southernmost part of the peninsula is rich in legume tree species and columnar cacti of tropical origin. These life forms singularize the physiognomy (the morphological appearance) of the vegetation of this desert, which give it the unexpected appearance of a strangely arborescent arid wilderness, compared with other nearly barren deserts of similar latitude in other parts of the world. The voyage from the mediterranean region to the tropics has concluded.

However, that is not all. If one travels through the Baja California Desert in one of those rare years when unpredictable and large rain events occur, one can observe a hidden treasure. Ephemerals, plants that complete their life cycle in one year, are seen in abundance during these periods of plenty, making the desert come alive in color while they replenish their seed banks. Fields of yellow, purple, violet, and orange run the length of the peninsula during such times. Meanwhile, long-lived desert plants establish during these brief intervals of bonanza, and initiate growth and resource accumulation in order to resist the long droughts soon to come. Usually unseen frogs and desert toads emerge full of activity and sound-looking, to close the magic circle of reproduction and leave a set of descendants that then await the next precipitation pulse.

Hallmark animal species of the Baja California Desert range from reptiles like the Baja California rattlesnake (Crotalus enyo), peninsular leaf-toed gecko (Phyllodactylus xanti), and the endangered Santa Catalina Island rattlesnake (Crotalus catalinensis) to mammals such as the peninsular pronghorn (Antilocarpa americana peninsularis), coyote (Canis latrans peninsulae), and the endangered black jackrabbit (Lepus insularis).

More than 500 species of birds inhabit the biome, thanks in part to the ample watering areas on either flank. Endemic birds include Xantus's hummingbird (Hylocharis xantusii) and Gray Thrasher (Toxostoma cinerum); that haunt the periphery are Craveri's murrelet (Synthliboramphus craveri), yellow-footed gull (Larus livens), and Townsend's shearwater (Puffinus auricularis).

Climate and Habitat Change

The isolation of the Baja California Desert and its extreme environmental conditions have maintained low levels of human presence through time. Even after European colonization in the 17th century introduced agriculture and livestock to the peninsula, human populations remained relatively low until the 1950s. The consequently minor level of anthropogenic activities allowed the peninsular desert to remain largely undisturbed, preserving the majority of its wilderness.

However, the last decades have brought rapid population growth in urban centers to the north and south regions bordering the desert. In 1950 there were 288,000 inhabitants in the peninsula; in 2010 there were 3,792,000 inhabitants, a population increase of 1,200 percent. The pressure of natural resource use has also been slowly increasing and spreading through the desert lands. Some areas along the Pacific coast that were developed for intensive agriculture production in the mid-20th century are now suffering from depletion of underground aquifers. Free-roaming cattle that have been long established in many desert areas throughout the peninsula have affected vegetation dynamics in a still unknown way.

Tourism-based urban sprawl, adventure-tourism, (e.g., off-road vehicle use in the open desert), and clearance of mangroves for coastal developments have made large negative impacts on peninsular ecosystems, especially coastal areas. Additionally, invasive plant and animal species are becoming widely established in the Baja, representing a growing threat to the native species.

Additional potential pressure can result from global warming. The Baja Peninsula Desert is a transition region between two different climatic regimes that indeed are apparently changing. On the one hand, drier and warmer winters with a delayed onset of winter rains have been observed in the continental Sonoran Desert. On the other hand, climate change is predicted to increase the frequency and intensity of tropical hurricanes that cross the peninsula from the south. A shift northward in species distribution and an expansion of the desert is one possible future scenario. Furthermore, it is not yet clear how the two seas surrounding the Baja Desert—the cold Pacific on the west and the warm Gulf of California on the east—are responding to global warming. Therefore, it is still largely unknown, how global climate change is affecting the Baja Californian Desert

Acknowledgement of the natural treasures of this unique land has grown among the local population, those that visit, and within public administrations during recent decades. This awareness has led to the declaration of more than three million hectares of land placed under legal conservation status, and more importantly, to a growing consciousness that the future of the people is inextricably linked to the future of desert.

Pedro P. Garcillán
Brigitte Marazzi
Benjamin Theodore Wilder

Further Reading

Case, Ted J., Martin L. Cody, and Exequiel Ezcurra. *A New Island Biogeography of the Sea of Cortés.* New York: Oxford University Press, 2002.

Grismer, L. Lee. *Amphibians and Reptiles of Baja California, Including its Pacific Islands and the Islands in the Sea of Cortés.* Berkeley: University of California Press, 2002.

Shreve, Forrest and Ira L. Wiggins. *Vegetation and Flora of the Sonoran Desert.* Palo Alto, CA: Stanford University Press, 1964.

Balkhash, Lake

Category: Inland Aquatic Biomes.
Geographical Location: Central Asia.
Summary: Lake Balkhash, one of the largest lakes in Asia and unique in having both freshwater and saltwater tributary rivers, is in danger from diversion of water and from overusage.

Lake Balkhash is located in east-central Kazakhstan in central Asia, in the deepest part of the Balkhash-Alakol depression, currently at about 1,222 feet (342 meters) above sea level. The Balkhash-Alakol depression was formed by the sloping trough of the Turan Plate during the Neogene-Quaternary Period, between 23 and 2.6 million years ago. Lake Balkhash is located within an endorheic basin, a closed drainage catchment that retains the water that flows into it without releasing any water through other rivers, lakes, or oceans.

Some 375 miles (605 kilometers) long from east to west, the surface area of Lake Balkhash varies from about 6,000 square miles (15,500 square kilometers) in dry years, to as much as 7,300 square miles (19,000 square kilometers) during peak inflow years. Changes in the total area are accompanied by about a 10-foot (3-meter) change in the water level, depending on how much water is flowing into the lake.

Lake Balkhash contains both freshwater—mainly from the Ili River flowing in from the western end—and saltwater from the Karatal, Aqsu, and Lepsi Rivers flowing in from the east and southeast. Ili River water originates from melted snow coming off the glaciated Tien Shan mountains of China's Xinjiang region. Before the late 20th century, the Ili River had provided Lake Balkhash with as much as 90 percent of its total inflow; this proportion has decreased, in part due to municipal and industrial use, but the Ili still provides up to three-fourths of the inflow. From scouring minerals in the arid Balkhash-Alakol basin, the rivers to the east bring in sufficient saline content to equate to salinity levels in eastern Lake Balkhash that reach up to eight times that of the western area of the lake.

The western, freshwater part of Lake Balkhash is wide and shallow, with its depth not reaching further than 36 feet (11 meters). The saline, eastern part of the lake is more narrow and relatively deep, reaching up to 85 feet (26 meters). A narrow, shallow segment, Uzynaral Strait, and an associated peninsula help keep east-west exchange of water to a minimum. The depth here is approximately 21 feet (6.4 meters).

Winters tend to be harsh in Kazakhstan and around Lake Balkhash, with the whole lake actually freezing over from the end of November to the beginning of April each year. The region receives about 17 inches (43 centimeters) of precipitation per year, yielding an arid grassy plain or steppe, in the surrounding area, as well as some semi-desert and desert stretches.

Species and Human Impacts

Vegetation around the lake—much of it salt-tolerant—features such forest species as Schrenk's spruce (Picea schrenkiana), Manchurian ash (Fraxinus sogdiana), and Semenov's maple (Acer semenovii); scrub plants like honeysuckle (Lonicera hispida), barberry (Berberis heteropoda), sea-buckthorn (Hippophae rhamnoides), salt cedar or tamarisk (Tamarix ramosissima), and saxaul or suo suo (Haloxylon ammodendon); and herbaceous flora such as the Chinese medicinal herb Anabasis (Anabasis salsa), amaranths (Nanophyton erinaceum, Halocnemum strobilaceum), and sagewort (Artemisia terrae alba).

There is a relative abundance, especially around the southwestern shores of Lake Balkhash, of common species of reeds (Phragmites spp.), cattails (Typha spp.), and Russian thistle (Salsola spp.). Such stands provide a refuge for fish, land animals, and birds, although the marshlands in general have receded as water flow and levels have receded over the past several decades.

Before the quality of the lake declined in the late 20th century—due mainly to spreading human agricultural, industrial, and climate-change impact—dozens of fish species thrived here, of which a handful were peculiar to the lake itself (endemic). Among the surviving endemic species are Balkhash perch (Perca schrenki), Balkhash marinka (Schizothorax argentatus), and plain stone loach (Nemacheilus labiatus). All three are economically important to the human fishery in Lake Balkhash. Other popular species have been introduced to the lake, including sturgeon, pike, eastern bream, and Aral barbell (Barbus brachycephalus).

In the vicinity of the lake, Canadian muskrat were introduced in the 1940s to spur the fur trade. This has since come to a halt due to the downsizing of the muskrat habitat areas. Another animal that used to live in the region along the southern shore was the Caspian tiger (Panthera tigris virgata), now extinct.

Reptiles and amphibians are represented here by the toad agama (Phrynocephalus guttatus), steppe agama (Agama sanguinolenta), gray monitor (Varanus griseus), central Asiatic frog (Rana asiatica balchaschensis), and semirechensk salamander (Ranodon sibiricus).

Mammals include several considered rare or endangered, such as the marbled polecat (Vormela peregusna), a rodent known as the five-toed dwarf jerboa (Cardiocranius paradoxus), and the goitered gazelle (Gazella subgutturosa).

The Lake Balkhash biome still supports many species of birds, ranging from the great egret (Casmerodius albus) to cormorants and pheasants. Some birds, however, are in danger, such as the Dalmatian pelican (Pelecanus crispus), whooper swan (Cygnus cygnus), and Eurasian spoonbill (Platalea leucorodia).

Along with municipal and agricultural uses including hydroelectric dams and irrigation networks, Lake Balkhash waters are also consumed, diverted, or polluted by a copper mining and smelting industry that deposits lead and zinc into the lake, as well as copper residues. Dust storms are an increasing problem, as global warming seems to be decreasing regional rain averages.

The Lake Balkhash biome has some protections, including a Ramsar-listed Wetland of International Importance: the Ili River Delta and South Lake Balkhash wetlands of some 3,735 square miles (968,000 hectares) in area. Considered vital for fish, birds, desert flora, and other wildlife, this Ramsar site is within three State Nature Reserves—the Balkhash, Karroy, and Kukan—and is administered by a state national nature park agency. This area is downstream from the Alakol-Sasykkol Lakes System wetlands, another Ramsar site in Kazakhstan that is also crucial for nesting and migratory birds and several vulnerable mammal species. This upstream wetland area comprises some 3,530 square miles (915,000 hectares).

William Forbes
Katherine Ruminer

Further Reading

McGivering, Jill. "Kazakh 'National Treasure' Under Threat." *BBC News*, September 28, 2005.

http://news.bbc.co.uk/2/hi/asia-pacific/4286916 .stm.

Mitrofanov, V. P. and T. Petr. *Fish and Fisheries in the Altai, Northern Tien Shan, and Lake Balkhash (Kazakhstan).* Rome: Food and Agriculture Organization of the United Nations, 1999.

Zhatkanbayev, Altay Zhumakan-Uly. "Creating Protected Areas on Lake Balkhash and Ile River Delta in the Kazakhstan Republic." In G. C. Boere, C. A. Galbraith, and D. A. Stroud, eds. *Waterbirds Around the World.* Edinburgh, UK: The Stationery Office, 2006.

A Baltic cod caught among fishing lines and net. Stocks of cod and salmon in the Baltic Sea have been overfished for centuries, but fishing moratoriums have helped the eastern stock of cod recover to a sustainable level. (Thinkstock)

Baltic Sea

Category: Marine and Oceanic Biomes.
Geographic Location: Northern Europe.
Summary: The Baltic Sea is an enclosed brackish-water sea with low species diversity, threatened by eutrophication and pollution from its catchment area.

The Baltic Sea, a mostly enclosed sea in northern Europe, comprises a number of branches and basins which include, from northeast to southwest: the Gulf of Bothnia, Gulf of Finland, and Gulf of Riga; the central-southern area known as the Baltic Proper; and the Danish straits of Kattegat and Skagerrak, which connect to the North Sea, an arm of the Atlanic Ocean. To the east, the White Sea-Baltic Canal, opened by Russia in 1993, allows traffic from the Baltic—via the Gulf of Finland and a chain of rivers and major lakes—to the White Sea, itself an arm of the Arctic Ocean. To the southwest, the Kiel Canal, opened in 1895, cuts through the Jutland peninsula to connect the Baltic with the North Sea.

Total surface area of the Baltic is approximately 163,000 square miles (422,000 square kilometers). Volume is some 5,040 cubic miles (21,000 cubic kilometers). The Baltic is a small sea in comparison with other global seas. However, it is among the world's largest brackish water bodies, in the same league with the Black Sea, Caspian Sea, and Hudson Bay. Connected to the Atlantic Ocean by the narrow Danish straits, the salinity in the Baltic varies between 20 practical salinity unit (psu) in the southern Kattegat to less than 1.0 psu in the innermost areas of the Gulf of Finland, which makes it a unique, but also a very sensitive, environment. The Baltic Sea is bordered by nine countries: Finland, Sweden, Denmark, Germany, Poland, Lithuania, Latvia, Estonia and Russia. More than 85 million people live in its catchment area.

The history of the Baltic Sea is strongly linked with the ice age and the land uplift that took place after the ice had melted. The fossil history clearly demonstrates an alternating dominance by typical freshwater and marine species since the last glaciation period. Approximately 4,000–7,500 years ago, during the so-called Littorina Sea stage, the connection through the Danish straits to the Atlantic was formed.

The average salinity of the Baltic Sea was then about eight times higher than it is today. Consequently, the species that were found then differed drastically from the ones that can be found in the Baltic Sea today. The Baltic Sea reached its current form and salinity level about 3,000 years ago, a very short period in geological terms. Due to its young age, there are still ecological niches available to be filled in the Baltic Sea ecosystem.

Species and Stresses

Only a few marine animals and plants are able to tolerate low salinity. Currently there is a total of 59 species and 16 biotopes that are considered threatened and/or declining in the Baltic Sea. The species distribution depends on the geographic location, with more marine species encountered in the southern Baltic Sea and more brackish and freshwater species in the northern Baltic Sea. There are only 20 zoobenthic animal species in the benthic (deepwater) ecosystem of the Bothnian area, but 500 in the southwestern Baltic Sea, and about 1,600 in the open Skagerrak waters. The number of macroalgae drops from 250 in the Skagerrak area to 40 in the Bay of Bothnia.

The low salinity of the Baltic Sea suppresses the growth of some species. For instance, the blue mussel species Mytilus edulis is, at its longest, only an inch (2.5 centimeters) long in the Baltic Sea, but reaches up to nearly 3 inches (7 centimeters) in length in the oceanic environment of the North Sea. On the other hand, some freshwater fish species such as pike grow faster and larger in the Baltic Sea than they typically do in lakes or rivers.

Apart from its low salinity, the lack of a proper intertidal zone or a deep sea ecosystem are other contributing factors to the low number of species found in the Baltic Sea. Indeed, the Baltic Sea overall is a very shallow sea, with an average depth of only 177 feet (54 meters). The average depth of the world's oceans is 12,465 feet (3,800 meters), about eight times deeper than the maximum depth of some 1,505 feet (459 meters) of the Baltic Sea.

The Baltic Sea is divided into subbasins with different topographic, chemical, and biological traits. The inflow from the catchment area is larger than the inflow of seawater through the Danish straits, which means that for the most part, the water in the Baltic Sea remains in the basins for decades. This leads to a strong stratification of the water column. The saline water is heavy and sinks to the bottom, forming a distinct halocline, a layer of saltier water. Vertical mixing and ventilation of the benthic ecosystem are quite constricted. Hence, some areas on the seafloor are anoxic, or depleted of oxygen.

The inflow of oxygenated Atlantic Ocean seawater into the deep bottom habitats is essential for the maintenance of a healthy benthic ecosystem in the Baltic Sea. These saltwater pulses, however, are relatively infrequent, occurring about every two years. Anoxic areas have increased in recent times in the Baltic Proper and in the Gulf of Finland; in autumn 2006 and spring 2007 they corresponded to half of the surface area of the basins.

Principal negative habitat pressures acting on the Baltic Sea are eutrophication—the overabundance of nutrient inflow, destabilizing ecosystem balance—and loss of biodiversity. The heaviest anthropogenic pressure in the Baltic Sea is experienced in the Gulf of Finland, Gulf of Riga, the southeastern part of the Baltic Proper, and the southern Baltic Sea. Riverine input of nutrients, organic matter, and heavy metals are among the main challenges in the Gulf of Finland and southeastern part of the Baltic Proper. The southern parts of the Baltic host the most dense human population, the heaviest fishing pressure, and large inputs of nitrogen and heavy metals through atmospheric deposition.

Nutrients entering the Baltic Sea environment have altered the food-web structure, changed species composition, and altered population dynamics. Nutrients can increase the growth of phytoplankton, often leading to severe blooms of cyanobacteria. The Baltic Sea is remarkable for these major cyanobacteria blooms in a marine environment; elsewhere in the world such phenomena tend to be found in freshwater ecosystems. Excess biomass of dead blooms sinks onto the seafloor and leads to an increased growth of microbes there, which eventually leads to anoxic conditions on the seafloor and, if prevailing for a prolonged period, may result in die-offs of benthic organisms.

The main sources of nutrients and organic matter in the Baltic Sea are the rivers that flow into it: Vistula from Poland, Neva from Russia, Oder from Germany, Daugava from Latvia, and Nemunas from Lithuania. The most severe nutrient enrichment pressure is found in the Baltic Proper, the Gulf of Finland, and the Gulf of Riga. Sewage discharges from ships account only for 1 percent of the total inputs of nutrients, but this nutrient enrichment may increase the phytoplankton blooms.

The average waterborne inputs of nitrogen and phosphorus have remained roughly steady since 1990, as some coordinated efforts have been made to reduce output from municipal and agricultural sources. However, draining of wetlands in the catchment areas of rivers in the Gulf of Bothnia has increased the flow of organic matter into the Baltic Sea. Natural background input of organic matter is high in the Gulf of Bothnia compared to the other areas of the Baltic. Runoff from farmlands and forests still represents the main input of man-made organic matter into the Baltic Sea.

Physical disturbance to the seafloor can be very destructive on a local scale, but researchers suggest these inputs make the least impact among stressors on the overall health of the Baltic Sea. Nevertheless, dumping dredged material on the seabed can smother benthic communities and lead to significant losses in biodiversity, favoring opportunistic species. Harbors, offshore wind farms, cables, bridges, coastal structures, and oil platforms occur along the Baltic Sea coastline. Additional seafloor areas are likely to be disturbed in the future, since more constructions and windfarms are expected to be built. Construction may alter the water flow and increase erosion in coastal areas.

Fishing Industry

Overfishing is a serious threat to the world's oceans; there is no exception in the Baltic Sea. Industrial fishing has removed large quantities of predatory fishes such as cod, pikeperch, pike, and salmon, as well as the grey seal. The removal or depletion of these top-of-the-chain species has contributed to ecosystem regime shifts, increased nuisance species, and further eutrophication.

The stocks of cod (Gadus morhua) and salmon (Salmo salar) are of particular concern, from both economic and ecological standpoints. The eastern and the western stocks of cod have been overexploited for centuries, while the reproductive capacity of salmon in rivers flowing into the Baltic has also dwindled, even as stocks of Baltic herring (Clupea harengus membras) and sprat (Sprattus sprattus) have been harvested sustainably. The eastern stock of cod has only recently—through various fishing moratorium efforts—returned to what is thought to be a sustainable level; it is anticipated that the western stock can also recover in the near future.

In recent history, two species of fish have become extinct in the Baltic Sea: the Atlantic sturgeon (Acipenser oxyrinchus), which has suffered from overexploitation and by the obstruction of its migratory pathways; and the bluefin tuna (Thunnus thynnus), which was overfished in the Kattegat area in the early 20th century.

Climate change will continue to challenge the health of the Baltic Sea. The average annual sea surface temperature in the southern Baltic has in recent decades increased by approximatively 1.8 degree F (1 degree C), whereas in the northern Baltic Sea the changes are mainly seasonal. The length of the ice season has decreased in the range of 14 to 44 days during the last century. Precipitation is predicted to increase in the future, which may lower salinity levels and increase riverine flow of nutrients that may in turn, together with increasing water temperatures, aggravate the eutrophication and anoxic conditions.

International management of the Baltic Sea in general has the aim to improve its health and support the ecosystem services the sea can provide to its people in the future. A principal goal for recovery is to reduce the nutrient input sources. Coordinated actions will likely be costly in the near term, but delaying such actions is likely to increase overall costs.

Jessica Haapkylä

Further Reading

Beaugrand, G., M. Edwards, K. Brander, C. Luczak, and F. Ibanez. "Causes and Projections of Abrupt Climate-Driven Ecosystem Shifts in the North Atlantic." *Ecology Letters* 11, no. 11 (2008).

Heck, K. L. Jr. and J. F. Valentine. "The Primacy of Top-Down Effects in Shallow Benthic Ecosystems." *Estuaries and Coasts* 30, no. 3 (2007).

Helsinki Commission (HELCOM). *Ecosystem Health of the Baltic Sea 2003–07: HELCOM Initial Holistic Assessment.* Helsinki, Finland: HELCOM, 2010.

Leppäranta, Matti and Kai Myrberg. *Physical Oceanography of the Baltic Sea.* New York: Springer, 2009.

MacKenzie, B. R. and R. A. Myers. "The Development of the Northern European Fishery for North Atlantic Bluefin Tuna (*Thunnus thynnus*) During 1900–1950." *Fisheries Research* 87, no. 1 (2007).

Paaver, T. "Historic and Recent Records of Native and Exotic Sturgeon Species in Estonia." *Journal of Applied Ichthyology* 15, no. 4–5 (1999).

Powilleit M., J. Kleine, and H. Leuchs. "Impacts of Experimental Dredged Material Disposal on a Shallow, Sublittoral Macrofauna Community in Mecklenburg Bay (Western Baltic Sea)." *Marine Pollution Bulletin* 52, no. 4 (2006).

Rönkkönen, S., E. Ojaveer, T. Raid, and M. Viitasalo. "Long-Term Changes in Baltic Herring (*Clupea harengus membras*) Growth in the Gulf of Finland." *Canadian Journal of Fisheries and Aquatic Sciences* 61, no. 2 (2004).

Bangweulu Wetlands

Category: Inland Aquatic Biomes.
Geographic Location: Africa.
Summary: This wetlands is a biodiverse ecosystem of critical importance to many migratory bird species, but under severe environmental stress from human settlement and agricultural activity.

The Bangweulu Wetlands biome, situated in a headwaters area of the Congo River system in west-central Africa, consists of the Bangweulu swamps, Lake Bangweulu, and its associated floodplain in northern Zambia—a total combined area of about 5,800 square miles (15,000 square kilometers) that drains an extremely flat catchment area of some 42,500 square miles (110,000 square kilometers).

The system is fed by 17 rivers, the largest of which is the Chambeshi, which is the longest tributary of the Congo River; it flows into the lake from the northeast. Rich deltas are found where the rivers Lupososhi, Luena, Lukuto, Chambeshi, and Luansenshi enter the lake or swamps.

The Bangweulu (the name means "where the water sky meets the sky" in Bemba dialect) wetlands and lake system is drained by the Luapula River to its the south; the swamps help prevent the Luapula from flooding by releasing water slowly through channels and lagoons. The area is not easily navigable because of the amount of vegetation, including floating beds of papyrus. For hundreds of years, efforts have been made to improve navigation by cutting channels, but dugout canoes remain the vessels that are easiest to navigate through Bangweulu. Other vessels may be too wide to pass through clogged arteries, and motorboats face the problem of vegetation clogging their propellers.

Lake Bangweulu is about 5,792 square miles (15,000 square kilometers) in surface area in the rainy season and just 1,158 square miles (3,000 square kilometers) in the dry season, with an average depth of about 13 feet (four meters) year-round. Prominent lagoons include Lake Chali, Lake Chaya, Lake Wumba, and Pook Lagoon. The areas is riddled with islands, many of them inundated or entirely submerged during the rainy season.

Vegetation

The Bangweulu wetlands are among the most dilute water bodies in Africa, with a very low level of dissolved solids and a generally low level of nutrients. Phytoplankton is present only in low concentrations. On the other hand, the swamps and the channels that run through them are frequently clogged by heavy stands of papyrus (Cyperus papyrus). Sedges, spearworts, wild rice, and hippo grass can all be found on the firmer ground and the islands, and the permanent water bodies are fringed by Eleocharis dulcis, Phragmitus pungens, and Nymphea. Termite mounds are so numerous and prominent that they act as small islands that become temporary homes to small wildlife during floods, and sometimes preserve tree seedlings from fig trees, sausage trees (Kigella africana), and the water berry (Syzygium cordatum).

Dugout canoes on the shore during the dry season in the Bangweulu wetlands in Zambia in 2006. Hand-powered canoes like these are still the best method of navigating the thick vegetation of the area's swamps. (Wikimedia/Mehmet Karatay)

Animals

The region is one of the only breeding grounds for the endangered shoebill stork (Balaeniceps rex), a bird that nests on the ground and has no more than two offspring per year. This odd-looking bird is well known among the ecotourists who visit the region during the months just before the rainy season, when it dwells on the fringe between the floodplain and the permanent waters. Saddle-billed stork, spur-winged goose, sacred ibis, black-crowned night heron, Denham's bustard, and very numerous waterfowl complement the avian picture.

The region is also home to the black lechwe, an antelope that gathers in herds of up to 10,000 during floods, as well as tsessebe, reedbuck, oribi, sitatunga, African buffalo, crocodile, and elephants.

The wetlands boasts at least 83 species of fish in 13 identified families. Common larger species include the bulldog-nose or elephant-nose (Marcusenius macrolepidotus), redbreast tilapia (Tilapia rendalli), yellow-belly bream (Serranochromis robustus), and African tiger fish (Hydrocynus vittatus). Lower on the food chain are various species of barbus, as well as banded tilapia (Tilapia sparrmanii), Churchill (Petrocephalus catostoma), and African butter catfish (Schilbe mystus).

An African legendary creature called the Emela-ntouka is alleged to live in Lake Bangweulu. It is described by the local Pygmy tribes as being the size of an African elephant, with a body similar to a rhinoceros and a single long horn or tusk on its snout. There is a wide spectrum of thinking about the Emela-ntouka. At one end is the fringe theory that the region is in general home to surviving prehistoric animals, including some dinosaurs, but this theory has no explanation for the lack of a dinosaur like the Emela-ntouka (a ceratopsian, for example) in Africa's fossil record. At the other end of the spectrum is simple disbelief: The Emela-ntouka is a legendary creature like Bigfoot or the Abominable Snowman. Somewhere in the middle is the suggestion that a real animal has taken on the appearance of legend. The ecosystem is far from exhaustively recorded; the discovery of a new species of semiaquatic rhinoceros, for example, may be feasible.

Human Stresses

Despite facing significant environmental stress as a result of nearby human settlement and the use of the area for hunting, fishing, and grazing, the ecoregion lacks a wildlife reserve; neither Bangweulu National Park nor the nearby Isangano National Park feature major conservation management activities, and there are no protected areas within the wetlands. The Zambian government has historically had trouble funding and amply staffing its system of 19 national parks. The human settlements in the Bangweulu wetlands region are largely impoverished, relying on fishing and hunting for subsistence; combatting poaching in parklands is an ongoing challenge.

Bill Kte'pi

Further Reading

Brelsford, W. V. "Making an Outlet From Lake Bangweulu in Northern Rhodesia." *Geographical Journal*, no. 1–2 (1945).

Howard, G. W. and D. R. Aspinwall. "Aerial Censuses of Shoebills, Saddlebilled Storks, and Wattled Cranes at the Bangweulu Swamps and Kafue Flats, Zambia." *Ostrich* 55, no. 4 (1984).

Jeffrey, R. C. V. and P. M. Chooye. "The People's Role in Wetlands Management: The Zambian Initiative." *Landscape and Urban Planning* 20, no. 1–3 (1991).

Kolding, Jeffrey. *A Brief Review of the Bangweulu Fishery Complex.* Langbroek, Zambia: African Parks Bangweulu Wetlands Project, 2011.

Richardo-Bertram. "The Fishes of the Bangweulu Region." *Zoology Journal of the Linnean Society of London* 41, no. 278 (1943).

Smardon, Richard. *Sustaining the World's Wetlands.* New York: Springer, 2009.

Banrock Station Wetlands

Category: Inland Aquatic Biomes.
Geographic Location: Australia.
Summary: The Banrock Station wetlands of Australia's Murray River floodplain are known for their biodiversity within the country's mostly arid ecosystems.

The Banrock Station wetlands are located in the Riverland region of southeastern Australia. These wetlands border the Murray River and are classified as a floodplain wetland. The reserve features a 5.3-square-mile (1,375-hectare) central wetland basin or lagoon, known as Banrock Lagoon, and an 0.5-square-mile (130-hectare) eastern lagoon. The Banrock Station wetlands are known for their biodiversity, and were designated of international importance under the Ramsar Convention in 2001. Banrock Station Wines owns and manages the reserve area.

The Banrock Station wetlands are set within an area that encompasses both freshwater and saltwater habitats and is marked by lagoons, basins, and channels as well as seasonal ponds and marshes recharged by a rainy period in the springtime. Specific habitat types here include the mallee, wetland foreshore, and wetland proper.

Permanent and Migratory Species

Common swamp life flourishes in Banrock Station, including approximately 120 species of plants and 138 species of birds, as well as a number of species of insects, frogs, fish, reptiles, and mammals. The floodplain hosts woodlands of River Red gum (Eucalyptus camaldulensis), coolibah (Eucalyptus coolabah), and black box (Eucalyptus largiflorens), along with shrublands of lignum (Muehlenbeckia florulenta). Sedge (Cyperaceae spp.) is also common, as are an array of herbs.

The wetlands host both permanent and migratory bird and fish populations. Close to 90 bird species have been recorded, including several that are not often found within southern Australia. Migrating fish traverse the Banrock as they travel along the Murray River during the seasonal spring floods, also stopping to spawn and nurse within its waters.

Fourteen species dependent on the wetlands for their survival are among those classified as threatened, including several species of waterbirds. The southern bell frog (litoria raniformis), eastern regent parrot (Polytelis anthopeplus monarchoides), and river snail (Notopala sublineata) are among the regionally and nationally endangered species that reside within the wetlands.

The habitats in the wetlands are shaped by precipitation patterns. During a cyclical dry period, the exposed moist wetland soils mineralize nutrients and absorb oxygen. A wet period characterized by heavy rainfall then marks the beginning of a new cycle of germination, breeding, and spawning.

Ecological Alterations

Wetlands are significant absorbers of greenhouse gases such as carbon dioxide; wetland preservation has emerged as a vital concern in Australia, a country known for its mostly arid ecosystems. More than half of Australia's wetlands have already disappeared and more than 70 percent of those that lie within the Murray River floodplain have suffered environmental degradation. The Banrock Station wetlands area had remained permanently inundated with water throughout most of the 20th century, a condition that had severe environmental consequences. The disruption to the natural hydrological wet and dry cycles was the result of a series of locks and weirs installed along the Murray River for transport and agricultural purposes. The increased salinity of groundwater and the introduction of exotic species are additional key environmental concerns.

Wetland rehabilitation efforts had begun by the late 20th century. In 1992, Wetland Care Australia introduced a management plan designed in part to restore more natural hydrological cycles of flooding and drying. Locks and weirs were removed from the Murray, allowing restoration of a more natural water flow. Large structures are used to limit water flow, inducing dry cycles approximately every other year. Removal of the structures allows for springtime flood cycles. The first significant dry cycle occurred in 2007.

Wetlands management has also included the protection, recovery, and reintroduction of native and threatened plant and animal species as well as the targeted removal and exclusion of exotic species. Threatened species have begun to recover in numbers, while the invasive common carp (Cyprinus carpio) has been largely eliminated from the area through the use of fish screens. Public environmental education is also a primary goal. Severe, persistent droughts have limited ongoing restoration efforts in the early 21st century due to the impact of newly introduced governmental water use restrictions, which limit induced wetlands flooding programs.

Banrock Station Wines, part of the larger Constellation Wines Australia (formerly BRL Hardy), owns and manages the area, operating a commercial winery on the premises. A Wine and Wetland Centre offers public educational experiences dedicated to wine growing and the surrounding habitat. The company also constructed the Banrock Station Boardwalk, a series of nature trails that allow public exploration of the wetlands. Interpretive signs, information huts, story areas, rest areas, and bird blinds are featured. The company also donates parts of its commercial proceeds to global wetland preservation efforts.

Marcella Bush Trevino

Further Reading

Dugan, Patrick. *Guide to Wetlands.* Buffalo, NY: Firefly, 2005.

Keddy, Paul A. *Wetland Ecology: Principles and Conservation.* New York: Cambridge University Press, 2000.

Stuart, M. Whitten and Jeff Bennett. *Managing Wetlands for Private and Social Good: Theory, Policy, and Cases from Australia.* Northampton, MA: Edward Elgar Press, 2005.

Wetherby, K. G. *Banrock Station Walking Trail Guide: The Mallee Meets the Valley Trail and the Boardwalk Trail.* Kingston-on-Murray, Australia: Banrock Station, 2002.

Barataria-Terrebonne Estuary

Category: Marine and Oceanic Biomes.
Geographic Location: North America.
Summary: This estuarine and wetlands habitat on the northern edge of the Gulf of Mexico supports wildlife and human populations—but is rapidly disappearing.

Deep in southeast Louisiana, between the Atchafalaya and Mississippi Rivers, lies more than four million acres (two million hectares) of estuarine and associated wetland habitat. The Barataria-Terrebonne Estuary, part of one of the largest river deltas in the world, supports diverse wildlife and a vibrant human culture that depends on its habitats for food, shelter, jobs, and recreation. The wildlife and human communities are threatened, however, by one of the highest rates of wetland loss in the world.

Barataria-Terrebonne is a mosaic of habitats shaped largely by a transition from saltwater to freshwater that extends more than 25 miles (40 kilometers) inland. Thick seagrass beds grow in the shallow waters of the northernmost part of the Gulf of Mexico. Salt marsh, a habitat dominated by smooth cordgrass (Spartina alterniflora), fringes the coast. As salinity decreases inland, brackish and then intermediate marsh grows. Farthest away from the sea, and most influenced by rainfall and river water, is freshwater marsh.

By one estimate, the diversity of plants in the freshwater marsh here is five times higher than that in the salt marsh. There are also shrubby and forested habitats within and between the broad swaths of marsh. Bottomland hardwood is a type of forested wetland along the rivers' edges, while cypress–tupelo swamp is a nearly perpetually flooded habitat dominated by iconic bald cypress trees (Taxodium distichum).

Wild blue iris growing in a marsh at the Barataria Preserve outside of Marrero, Louisiana, in 2009. Coastal Louisiana has lost about 25 percent of its land area since 1932, and about half of that loss occurred in Barataria-Terrebonne. (U.S. Army Corps of Engineers/Team New Orleans)

Animal Biodiversity

The coastal habitats of Barataria-Terrebonne are critical to both land-dwelling and marine animals. Nearly 700 species of mammals, birds, reptiles, amphibians, and fish live in the Barataria-Terrebonne part or all of the year, along with a vast array of invertebrates like shrimp and oysters. The wetlands are winter habitat for waterfowl and raptors from Canada and the northern United States, and nesting habitat for colonial birds like brown pelican (Pelecanus occidentalis).

Larval fish and shellfish use the wetlands as nursery habitat, feeding in warm, shallow waters and hiding from larger predators. And for migratory songbirds moving from Latin America to Canada's boreal forest, Barataria-Terrebonne's coast is an important resting spot after a grueling flight over the gulf. These lush resources inspired early French settlers to give Terrebonne Parish, the southwest corner of Barataria-Terrebonne, its name: Terre Bonne means "good earth."

Today, the resources of Barataria-Terrebonne still shape the lives of the people who live there. The region is the most densely populated part of Louisiana's coastal zone outside of New Orleans. One of the most important occupations in Barataria-Terrebonne is commercial fishing, and fish landings in this region help make Louisiana the second-largest seafood producer in the United States. Hunting and recreational fishing are important pastimes for residents and visitors, and lucrative sources of income for the tourist industry. Wetland vegetation also provides coastal protection, reducing shoreline erosion and the destructive power of hurricane-driven waves and storm surge.

Wetland Loss

For nearly 80 years, Barataria-Terrebonne's rich landscape has suffered a slow-motion collapse that has inundated wetlands and human communities alike. Coastal Louisiana lost nearly 1,892 square miles (4,900 square kilometers) of land between 1932 and 2010, or about 25 percent of the land area present in 1932. About half that loss occurred in Barataria-Terrebonne. One of the main causes of the wetland loss is the isolation of the Mississippi River from the surrounding wetlands.

The Mississippi River drains a watershed covering 41 percent of the continental United States. Over much of the past 10,000 years, sediments from the river dropped into shallow gulf waters and built land seaward in a series of overlapping delta lobes. Every 1,000 to 1,500 years, the river changed course, building land in a new lobe adjacent to the river and abandoning the old lobe. Barataria-Terrebonne is made of three such lobes, the oldest of which started to form about 5,500 years ago. The elevation stayed above sea level as long as new land was built at the same rate as the old land subsided. When the lobes were abandoned, subsidence and marine processes began to dominate, and the extremely productive Barataria-Terrebonne Estuary biome evolved.

Today, the wetlands are still subsiding—the rate can vary locally from 0.5 feet (0.15 meter) per century to more than 4 feet (1.2 meters) per century—but there is little opportunity for the Mississippi to build new land. Nearly the entire length of the river is lined by human-made levees and dams that disrupt water and sediment flow. The southernmost section of the river is also dredged to keep the channel clear for barges and other ship traffic, and these dredged sediments are discharged to the gulf.

Other Ecosystem Threats

The wetlands are subject to many other challenges. The oil and gas industry has cut more than 9,942 miles (16,000 kilometers) of canals through the marsh, allowing saltwater to flow into normally fresh areas. When canals are dredged, the spoil or sediment is usually deposited along the banks, which blocks natural tidal inflow to the marsh surface beyond. This reduces new sediment from spreading across marshland, which often results in the reversion of the habitat to open water. Large withdrawals of oil and gas from Barataria-Terrebonne have also sped subsidence.

Tropical storms and hurricanes are double edged-swords: Although they can nourish some wetlands by depositing sediment, they can devastate others, ripping up vegetation and pushing saltwater far inland. Nutria (Myocastor coypus), large rodents introduced to Louisiana in the 1930s, can rapidly turn wetlands into open water through their ravenous feeding upon plant stems, roots, and rhizomes, along with their widespread compacting of vegetation to build burrows and dikes, much in the manner of beavers.

The BP Deepwater Horizon, an oil rig 50 miles (80 kilometers) off Louisiana's shores, exploded in 2010, spilling up to 185 million gallons (700 million liters) of petroleum into the gulf. It will be years before scientists know the full impact of the BP oil disaster on Barataria-Terrebonne. Finally, Barataria-Terrebonne's low-lying coast is vulnerable to the effects of global warming—particularly changes in rainfall patterns and global sea-level rise.

Juanita M. Constible
Angelina M. Freeman

Further Reading

Couvillion, Brady, et al. *Land Area Change in Coastal Louisiana From 1932 to 2010.* Reston, VA: U.S. Geological Survey, 2011.

National Commission on the BP Deepwater Horizon Oil Spill and Offshore Drilling. "Deep Water: The Gulf Oil Disaster and the Future of Offshore Drilling—Report to the President." January 2011. http://www.gpo.gov/fdsys/search/pagedetails.action?packageId=GPO-OILCOMMISSION.

Barents Sea

Category: Marine and Oceanic Biomes.
Geographic Location: North of Norway and Russia.
Summary: The Barents Sea hosts large flocks of seabirds and stocks of fish. Although on the

fringe of civilization, it is exposed to overfishing, oil exploration, and nuclear contamination.

A branch of the Arctic Ocean, the Barents Sea ecoregion includes the heavily glaciated arctic archipelagos of Svalbard and Franz Josef Land to the north, and Novaya Zemlya to the east. It extends to mainland Norway and Russia, and abuts the coastal White Sea along northwestern Russia. Bordered to the west by the Greenland Sea and to the southwest by the Norwegian Sea, the Barents Sea receives flow from warm currents through the latter; this North Atlantic Drift current helps keep ports along its southern shores ice-free in winter. The Barents is considered the warmest of the circumpolar seas—although its northern reaches are ice-bound most of the year, and it receives cold polar currents from several directions.

A continental-shelf sea, the Barents has an average depth of 755 feet (230 meters). With a surface area of 540,000 square miles (1.4 million square kilometers), water volume is about 77,250 cubic miles (322,000 cubic kilometers).

Indigenous settlements of the Same people (formerly called Lapps) within the biome are widespread across the northernmost European coastal areas. There is also a sparse presence of the Nenets and the Komi along the Russian northern rim. Traditionally, the peoples have been fishermen, hunters, and reindeer herders. Present-day economic activities extend to mining and some agriculture. Contemporary human pressures affecting the Barents come largely from the major shipping, fishery, and petroleum development activities in Norwegian and Russian territorial waters.

Species Richness

The benthic (deep sea layer) fauna and flora of the Barents Sea is diverse, including approximately 2,500 vertebrate species and up to 300 species of phytoplankton. The mixing of warmer North Atlantic Drift waters with cold Arctic Sea waters, as well as nutrient-rich river inflows from the continent, together provide a fertile basis for vast amounts of krill, copepods, jellyfish, and mollusks. In addition, there are extensive cold water reefs that harbor crustacean populations and anchor kelp forests, making for even greater biological productivity.

Capelin (Mallotus villosus) in great numbers feed upon crustaceans and krill here, forming a vast feast for higher creatures including the economically vital cod, cetaceans, and other marine mammals and seabirds. Besides the cod (Gadus morhua) and polar cod, the Atlantic salmon (Salmo salar) is a major component of the fisheries.

The large stocks of fish and plankton in the Barents Sea support some 20 million seabirds during summer months. Out of the 30 key seabird species in the biome, the little auk (Alle alle), black-legged kittiwake (Rissa tridactyla), common guillemot (Uria aalge), and Atlantic puffin (Fratercula arctica) make up more than half of the breeding population.

Marine Mammals

The Barents Sea ecoregion host 24 species of marine mammals: 12 species of large cetaceans, five of dolphins, and seven pinnipeds. Most of the dolphins and cetaceans are long-distance migrants; the only three permanent residents are the narwhal (Monodon monoceros), white whale or beluga (Delphinapterus leucas), and bowhead whale (Balaena mysticetus). Among the whales, the northern right whale (Eubalaena glacialis) came to the brink of extinction due to industrial whaling pressure, while the blue whale (Balaenoptera musculus) and bowhead whale have not recovered their population size since the major whaling period. The fate of the other cetaceans is at risk; among those listed as endangered, vulnerable, or rare in Russia and/or Norway are the northern bottlenose (Hyperoodon ampullatus), humpback (Megaptera novaeangliae), sei (Balaenoptera borealis), fin whale (Balaenoptera physalus), and narwhal.

The polar bear (Ursus maritimus) was also close to extinction as a result of hunting but in 1973 was protected and its population is recovering; estimated numbers range from 3,000 to 5,000 individuals. Among the pinnipeds, the walrus (Odobenus rosmarus) was close to extinction by the 1950s and is currently listed as protected in Norway and vulnerable in Russia. In contrast, the population of the harp seal (Pagophilus groenlandicus) is about 2 million; it is a dominant species here, spending winters on the ice and the rest of the year in open water.

Economy and Threats

The rich biodiversity of the Barents Sea ecoregion faces great challenges. Among them are the fishing industry, radioactive contamination, shipping activities, petroleum exploration and exploitation, introduced species, and climate change in the near future. Large-scale offshore fisheries came in by the middle of the 20th century, building upon centuries of more artisanal fishing techniques. Among the most devastating industrial techniques are bottom trawling, dredging, and the use of drift nets. The first two damage the ocean bottom and deepwater corals, while drift nets are responsible for drowning adult seabirds. In spite of the large fish stocks in the region, the population of fish diminished greatly; for instance spring-spawning Norwegian herring (Clupea harrengus) stocks crashed in the 1960s and took more than 20 years of concerted efforts to rebuild.

The Barents Sea and its sediments have been impacted by the introduction of man-made nuclear material from a variety of sources. These include atmospheric fallout, including from the Chernobyl disaster; radioactive industrial material discharged by European nations and dumped at sea by the former Soviet Union; and nuclear reactor and weapons testing by the Soviet Union in Novaya Zemlya. Heavy metals like mercury and lead, as well as persistent organic pollutants (POPs) like DDT (dichlorodiphenyltrichloroethane), HCHs (hexachlorocyclohexanes), and PCBs (polychlorinated biphenyls), also find their way to the Barents Sea. Since POPs are fat soluble, they readily bioaccumulate as they move up the food chain toward top predators.

Since 1980, the Barents Sea has seen continually accelerating gas and oil exploration. Potential undersea fields are being actively developed on both the Norwegian and Russian zones. Underwater construction, spills, and increased shipping traffic are ongoing risks. With global warming, it is predicted that the opening of sea lanes across the deicing polar waters north of Europe and Asia will see industrial transport multiply at exponential rates, along with port facility and energy industry development.

On the aquaculture front, economically suitable fish such as various species of salmon and cod, as well as king crab (Lithodes maja, Paralithodes camtschaticus, et al.) are likely to make inroads into existing habitats, competing for food, as well as raising the risk of diluting the genetic defenses of their wild cousins through interbreeding. Introduced species reported in the Barents Sea already include a range of crab species as well as Pacific salmon and several plant types.

Conservation efforts have been made to protect parts of the Barents Sea ecoregion. The total marine protected area is 22,800 square miles (59,000 square kilometers), equal to less than 3 percent of the marine realm, while the protected land area (mostly islands) is 21,000 square miles (54,500 square kilometers).

One obstacle to continued rich biodiversity here is the prevalence of petroleum fields in highly sensitive habitats. The Norwegian Pollution Control Authority, for example, has called for the Nordland VI oil field to be declared a "petroleum-free area," due to its congruence with vital fish and seabird habitats.

Rocio R. Duchesne

Further Reading

Hamre, Johannes. "Biodiversity and Exploitation of the Main Fish Stocks in the Norwegian-Barents Sea Ecosystem." *Biodiversity and Conservation* 3, no. 6 (1994).

Honneland, Geir. *Making Fishery Agreements Work: Post-Agreement Bargaining in the Barents Sea.* Northampton, MA: Edward Elgar, 2012.

Larsen, Tore, Dag Nagoda, and Jon Roar Andersen, eds. *Description of the Barents Sea Ecoregion.* In *The Barents Sea Ecoregion: A Biodiversity Assessment.* Washington, DC: World Wildlife Fund, 2004.

Matishov, Dmitry G. and Gennady G. Matishov. *Radioecology in Northern European Seas.* New York: Springer, 2010.

Belize Barrier Reef Reserve System

Category: Marine and Oceanic Biomes.
Geographic Location: South America.
Summary: This coral reef system, one of the world's most extensive barrier reefs, is recovering from 1990s mass-bleaching events.

The Belize Barrier Reef, a 190-mile (306-kilometer) section of the Mesoamerican Barrier Reef System, hugs the coast of Belize, with a proximity ranging from 980 feet (299 meters) to 25 miles (40 kilometers). Like other large coral reef systems, this western Caribbean Sea reef is highly biodiverse, but it is not clear yet just how biodiverse, as surprisingly little of it has been subject to organized study. Only about 10 percent of the reef has been researched so far, revealing more than 100 species of coral—70 hard and 36 soft—as well as hundreds of invertebrates and fish. As an example of the potential species richness, a single investigation of the reef at a depth of 32 feet (10 meters) in 2003 identified 19 coral taxa.

The climate of the reef system is predominantly hot, although strong winter storms from October to February bring heavy rains, strong winds, and cooler temperatures.

The reef is critical both to Belize's tourism industry and to its commercial fishing; 370 square miles (958 square kilometers) of the reef are protected by the Belize Barrier Reef Reserve System. Tourists often come to visit the Great Blue Hole, made famous by Jacques Cousteau in 1971 when he declared it one of the best scuba-diving sites in the world. A system of karst limestone formations and submerged caves, the hole is home to numerous reef sharks and groupers.

The reserve includes 450 cays (sometimes spelled "caye" or "key," these are small sandy islands found atop coral areas), three atolls, and seven marine reserves; it has been designated a World Heritage Site since 1996. Tourism and fishing, although critical to Belize's economy and to providing the government a practical economic means for caring for the reef, also deplete fish populations and pollute the waters. Climate change, however, is the biggest threat. As much as 40 percent of the reef has been damaged since the 1990s, as rising ocean temperatures increase the risk both of more violent hurricanes and the vulnerability of corals to bleaching events.

Coral Bleaching

Much of the damage observed since the 1990s was the result of two mass-bleaching events: a 1995 bleaching that killed a tenth of the coral and a 1997–98 event in the aftermath of Hurricane Mitch that may have killed much more.

Coral bleaching is not always permanent damage. Coral's hue comes from zooxanthellae endosymbionts, which are photosynthetic unicellular organisms that live in the tissue of the hard corals (Scleractinia spp.) that build and constitute the reef itself. Bleaching is the loss of these zooxanthellae; the endosymbionts are often highly specifically adapted, leaving them vulnerable to changes in environmental factors like water temperature, water chemistry, sedimentation, salinity, and changes in sunlight. They can also be hurt or killed by exposure.

Mass bleachings that follow hurricanes may be caused by any combination of these factors; other mass bleachings are usually triggered by warm waters, bacterial infections, or an overall decline in the coral reef's health that makes the coral unable to provide the zooxanthellae the nutrients needed to perform photosynthesis. Coral depends on biological services provided by its zooxanthelle; when the endosymbionts die, so does much of the coral. What survives can repair the damage, but during that recovery time, the coral is extremely vulnerable to bacterial and other disease, which is more likely to kill the reef than the bleaching event itself.

It has been widely noted that global warming exacerbates the vulnerability of corals worldwide to bleaching events, as such events are generally correlated with rises in seawater temperature. It is not known how or to what extent either the corals themselves, their helpful zooxanthellae, or other antibacterial symbionts as yet unidentified can adapt or adjust as average temperatures rise in various oceans and seas.

Fish Species

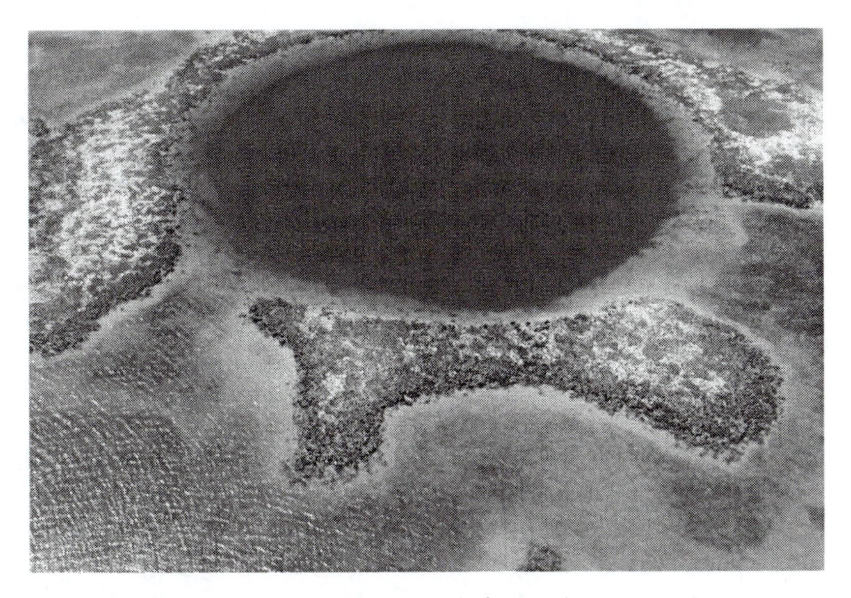

The Great Blue Hole of Belize, a site made famous by Jacques Cousteau on his television show, is one of seven marine reserves in the Belize Barrier Reef Reserve System. The hole is 1,000 feet wide and 412 feet deep. (Thinkstock)

With these two types of life—and many species of each—living in symbiosis, a coral reef is a sophisticated ecosystem in and of itself, but it also attracts numerous other fauna. Many species of reef fish feed on plankton in the surrounding waters, and the sandy floor around much of the reef is home to meadows of seagrass (primarily Thalassia testudinum and Syringodium filiforme), which is foraged upon and within by more reef fish (such as parrotfish and hog fish), as well as numerous types of sea turtles and crustaceans. Many of the seagrass meadows of the biome are in protected areas; if they were to be overfished, the increased amount of decaying seagrass would lead to a greater number of algal blooms, which would in turn jeopardize the oxygen balance of the reef area.

Undersea mangrove forests are home to seahorses (genus Hippocampus), angelfish (family Pomacanthidae), and grunts (family Haemulidae). At the bottom of the reef-coastal channel, spotted eagle rays (Aetobatus narinari) and Southern stingrays (Dasyatis americana) may be found, while moray eels (family Muraenidae) and sea anemones (order Actiniaria) live in the rocky areas. Dolphins like the short-beaked common dolphin (Delphinus delphis) and pantropical spotted dolphin (Stenella attenuata) swim nearby, along with the world's largest population—about 500 individuals—of West Indian manatees (Trichechus manatus).

Marine Reserves

The seven marine reserves of the Reserve System are Bacalar Chico National Park and Marine Reserve, Blue Hole Natural Monument, Half Moon Caye Natural Monument, South Water Caye Marine Reserve, Laughing Bird Caye National Park, Sapodilla Cays Marine Reserve, and Glover's Reef Marine Reserve. Glover's Reef is a partially submerged atoll at the outermost boundary of the barrier reef, with 850 reef patches and pinnacles rising to the surface of its interior lagoon.

Glover's is believed to contain the greatest diversity of reef species in the Belize Barrier Reef network. It is also the largest of only two spawning sites for the endangered Nassau grouper (Epinephelus striatus), which has long been threatened by overfishing. One of the largest fish in the barrier reef ecosystem, it is solitary, feeding in the daytime on small fish and crustaceans, which it gathers up in a large, wide mouth. Its spawning occurs in December and January, always under the full moon—the very predictability of which naturally led to the overfishing that has depleted its population.

Northern Caye, one of 14 major cayes in the Belize Barrier Reef Reserve System, is home to both the snowy egret (Egretta thula) and American saltwater crocodile (Crocodylus acutus), which is one of the largest nonextinct reptiles in the world. Male American saltwater crocodiles reach sexual maturity at about 11 feet (3 meters) and can exceed 20 feet (6 meters) in length, weighing more than 1 ton (0.9 metric ton). The crocodile divides its time between freshwater rivers in the wet season, and estuaries and the sea. It is fiercely territorial, fighting other male crocodiles as well as humans and other predators that enter its territory. Adult crocodiles feed mainly on fish, turtles, and crustaceans, but also will consume monkeys, bats, birds, large mammals, and even sharks. Studies have suggested they have an intelligence comparable to that of rats, and are capable of remembering and tracking the migration patterns of their chosen prey.

Bill Kte'pi

Further Reading

Cho, Leandra. "Marine Protected Areas: A Tool for Integrated Coastal Management in Belize." *Ocean and Coastal Management* 48, no. 11–12 (2005).

Gibson, J., M. McField, and S. Wells. "Coral Reef Management in Belize: An Approach Through Integrated Coastal Zone Management." *Ocean and Coastal Management* 39, no. 3 (1998).

Hartshorn, G., L. Nicolait, L. Hartsthorn, G. Bevier, R. Brightman, J. Cal, et al. *Belize Country Environmental Profile: A Field Study.* Washington DC: United States Agency for International Development, 1994.

McField, M. D., P. Hallock, and W. C. Jaap. "Multivariate Analysis of Reef Community Structure in the Belize Barrier Reef Complex." *Bulletin of Marine Science* 69, no. 2 (2001).

Ruiz-Zarate, M. A. and J. E. Arias-Gonzalez. "Spatial Study of Juvenile Corals in the Northern Region of the Mesoamerican Barrier Reef System." *Coral Reefs* 23, no. 4 (2004).

Self-Sullivan, Caryn, Gregory W. Smith, Jane M. Packard, and Katherine S. LaCommare. "Seasonal Occurrence of Male Antillean Manatees on the Belize Barrier Reef." *Aquatic Mammals*, 29, no. 3 (2003).

Bengal, Bay of

Category: Marine and Oceanic Biomes.
Geographic Location: South Asia.

Summary: This major lobe of the Indian Ocean, along with its coral reef, estuarine, and coastal habitats, supports great numbers and diversity of species.

The major northeastern lobe of the Indian Ocean, the Bay of Bengal encompasses some 850,000 square miles (2.2 million square kilometers) of ocean fed by such major rivers as the Mahanadi-Brahmani, Godavari, Krishna, and Kaveri from India in the west; the Ganges-Brahmaputra-Meghna system from India and Bangladesh in the north; and the Irrawaddy from Burma (Myanmar) in the east. Arcing from Sri Lanka and the east coast of India and across the Ganges Delta to Burma and the Andaman-Nicobar archipelago, the Bay of Bengal has an average depth of 8,500 feet (2,600 meters), a maximum depth of 17,260 feet (5,260 meters), and water volume of some 1.3 million cubic miles (5.6 million cubic kilometers). It is considered the largest bay on Earth.

The floor of the Bay of Bengal is almost entirely made up of the Bengal Fan, a structure first created in the ancient tectonic collision of the Indian Plate and Eurasian Plate, then eroded over eons into a series of underwater canyons by the massive inflow of sediment from the Himalayan headwaters through the Ganges-Brahmaputra-Meghna river system.

Annual continental runoff of freshwater into the bay, estimated at more than 710 cubic miles (3,000 cubic kilometers) is considerable, and peaks during the monsoon season, June through September. The rivers transport minerals and nutrients into the bay, as well as effluents and other pollutants from the cities, towns, and farms they pass through. Total sediment discharge into the Bay of Bengal is estimated at 24 billion tons annually.

Ecology and Species

The Bay of Bengal biome comprises estuaries and wetlands, coral reefs, and such unique zones as the world's largest salt-tolerant tidal mangrove forest—in the Sundarbans, where river and marine environments merge across the Ganges Delta. Upwelling along many of the near-shore areas contributes to the relative ease with which populations of phytoplankton and zooplankton thrive in these seas. Atoll, fringing, and barrier reefs all occur within the Bay of Bengal coral matrix, supporting substantial communities of sponges, mollusks, crustaceans, echinoderms, and fish.

Key fish species in the Bay of Bengal include the commercially important tuna, including striped bonito (Sarda orientalis), eastern little tuna (Euthynnus affinis), bullet tuna (Auxis rochei), and big eye tuna (Thunnus obesus); Indopacific king mackerel (Scomberomorus guttatus), Spanish mackerel (Scomberomorus commerson), and Indian mackerel (Rastrelliger kanagurta); Indian salmon (Eleutheronema tetradactylum); at least 15 species of anchovy, four of sardine, and two herring; and tiger prawn (Penaeus monodon).

Sea turtles include the leatherback (Dermochelys coriacea)—particularly numerous along the island chains of the Andamans and Nicobars, a prime egg-laying region—and the endangered olive ridley sea turtle (Lepidochelys olivacea). The olive ridley has been known to mass on the beaches of the eastern Indian state of Orissa in numbers of more than a half-million during its nesting season. The Gahirmatha Marine Wildlife Sanctuary, with some mangrove wetlands areas first set aside in 1975, protects this overfished reptile and also helps shield such at-risk fauna as the saltwater crocodile (Crocodylus porosus); eight types of at-risk kingfisher, including the black-capped kingfisher (Halcyon pileata) and pied kingfisher (Ceryle rudis); and resident herons such as Asian open bill (Anastomus oscitans), night heron (Nycticorax nycticorax), and white ibis (Threskiornis melanocephalus).

Mammals thriving in the Bay of Bengal include the dugong (Dugong dugon), which favors the seagrass beds found along the island chains; the Irrawaddy River dolphin (Orcaella brevirostris), preferring estuary zones and the Sundarbans; and cetaceans like Bryde's whale (Balaenoptera edeni), which inhabit such areas as the submarine canyon known as the Swatch of No Ground.

Crowning the biological richness of the Bay of Bengal in its northern reaches is the Sundarbans mangrove: Its National Park area in India is a United Nations Educational, Scientific and Cultural Organization (UNESCO) World Heritage Site, and its entire Indian expanse is a UNESCO Global Biosphere Reserve, while the Sundarbans Reserve Forest in Bangladesh is a Ramsar Wetland of International Importance spanning 2,323 square miles (601,700 hectares). The Sundarbans is vital for migratory fowl. Such apex predators as the greater spotted eagle (Aquila clanga) and Pallas's fish-eagle (Haliaeetus leucoryphus) winter here and summer in north-central Asia. Indian skimmer (Rynchops albicollis), spotted greenshank (Tringa guttifer), and spoon-billed sandpiper (Eurynorhynchus pygmeus) also rely on the Sundarbans.

Species as diverse as the Bengal tiger (Panthera tigris) and the rare Ganges shark (Glyphis gangeticus) depend on this vast mangrove, with its teeming food supply, prime breeding grounds, and protection from cyclones. Subsistence fishermen here sometimes train the smooth-coated otter (Lutrogale perspicillata) to drive fish into their nets.

Balancing Human Activities

Despite the protected areas within this vast region, much of the coast habitat is under threat from energy industrial and general urban develop-

Bengal tigers are among the mammals that feed on prey in the mangrove wetlands along the coast of the Bay of Bengal, which is considered the Earth's largest bay. (Thinkstock)

ment, the mangrove timber industry, wet rice cultivation, bird hunting, and aquaculture in the form of shrimp pond farms. Often the reach of these threats extends outward across the Bay of Bengal. For example, transmission of aquaculture effluents or escaped cultivars that can carry toxins and parasites is likely to be a factor in widespread disease impacts such as the as-yet unknown pathogen that in 2009 began affecting fish, mollusks, and crustaceans including such commercially vital species as the tiger prawn and white-leg shrimp (Litopenaeus vannamei). The Network of Agriculture Centres in Asia-Pacific is one of the organizations attempting to move its members toward more sustainable economic pursuits that build habitat protection into their methods and practices.

About 450 million people live in the coastal zone around the Bay of Bengal; annual fishery production here is on the order of three million tons, with a value of approximately $4 billion. Poverty and population pressures, insufficient governmental planning and control, increased tourism, and insufficient attention to waste management are all threats to the ecology of the region. Due to the low-lying nature of much of the land bordering the bay, the region is also extremely sensitive to changes in ocean level; there is growing concern for how global warming may affect the climate and habitats of the region. Salt intrusion, invasive species, and phytoplankton population vector alterations are among the catalog of risks being examined.

One cooperative international group focused on these issues is the Bay of Bengal Large Marine Ecosystem Project (BOBLME), an effort by Bangladesh, India, Indonesia, Malaysia, the Maldives, Myanmar, Sri Lanka, and Thailand. BOBLME aims to halt the ecological decline of the bay, improve the lives of people living on the coastline, prepare for changes expected from climate change, regulate growth and development, and better manage the fishing industry within this biome. The organization is only in the very early stages of formulating ways to combat the effects of global warming across the Bay of Bengal.

Sarah Boslaugh

Further Reading

Kumar, M. Dileep. *Biogeochemistry of the North Indian Ocean*. New Delhi: Indian National Science Academy, 2006.

Quader, O. *Coastal and Marine Biodiversity of Bangladesh (Bay of Bengal)*. Dhaka, India: Space Research and Remote Sensing Organization, 2010.

Sengupta, Debasis, Bharath Raj, and S. S. C. Shenoi. "Surface Freshwater From Bay of Bengal Runoff and Indonesian Throughflow in the Tropical Indian Ocean." *Geophysical Research Letters* 33, no. 1 (2006).

Beni Savanna

Category: Grassland, Tundra, and Human Biomes.

Geographic Location: South America.

Summary: This large, flood-prone tropical savanna in the lowlands of Bolivia constitutes a unique and exceptionally adaptive cultural landscape, with high levels of species and ecosystem biodiversity.

The Beni savanna occupies a large area in the warm tropical floodplains of the Mamoré and Beni Rivers in eastern Bolivia; it constitutes the largest area of flooding savannas in the Amazon drainage basin. It is surpassed in size only by the Llanos del Orinoco in Venezuela and the Pantanal of Mato Grosso in Brazil, which belong to the Orinoco and La Plata River basins, respectively.

Most of this region is flooded each year by overflowing rivers during the rainy season. The flooding extends roughly from December to June. By contrast, the other months are characterized by a marked shortage of rainfall, leading to a long dry season, which has its greatest intensity in the relatively cooler months of the year. The marked seasonal rhythm imposed by the annual flooding and drought is the main factor that influences the functioning of ecosystems in the Beni and the traditional patterns of human settlements, as well as the use of landscape and natural resources.

Overall, the Beni has a flat topography, forming an extensive alluvial plain. This plain has a characteristic microrelief pattern with altitude differences of just a few feet (meters) or less between different landforms. In this sense, the Beni has four main land units that are closely related to the intensity and duration of floods: not-flooded areas (upland, or altura), scarcely flooded areas (wetland, or semialtura), seasonally flooded areas (low and depressional lands, or bajura estacional), and permanently flooded areas (swamps, marshes, and lakes, or bajura permanente).

The Beni generally has nutrient-poor soils, particularly in the more distant, vast plains of the major river channels, which are seasonally flooded both by rainwater and by overflow from rivers and streams in the secondary river system, carrying mostly clear or black nutrient-poor waters. By contrast, the recent alluvial plains close to major rivers have richer soils because they are inundated by white water carrying considerable amounts of suspended sediment and average to moderate concentrations of dissolved mineral salts. This spatial pattern of soil fertility explains the preferential location of traditional human settlements on the banks of the rivers, where growing crops is possible and where transport through the river is easy with canoes and boats. In addition, the banks are topographically higher than the surrounding savanna plains and thus are affected only by floods of low intensity and duration.

Human Populations and Activities

The Beni savanna was originally populated by several indigenous groups, mostly belonging to the Arawak, Movima, Moseten, and Yurakare linguistic families. The largest populations today are the Mojeño and Baure (Arawak), Movima, Tsimane (Moseten), and Yurakare. The first three are the more numerous groups; they still have a traditional ethnic economy based on small farming and ranching, supplemented by fishing, hunting, and gathering in the savannas, forests, and rivers. By contrast, Yurakare and Moseten are eminently fishers, hunters, and gatherers—seminomads who practice cultivation only on a very small scale.

After the Spanish colonization and the later republic, a mixing of peoples and races created a significant Creole population. Creole and indigenous people—mostly Mojeño, Baure, and Movima—burn the savanna annually during the dry season so that after burning, the new grass will provide softer, more palatable forage for livestock. Since the introduction of cattle in the Beni by Jesuit and Franciscan missionaries during the 17th century, the savanna has undergone a long process of co-adaptation to fire and grazing. Thus, the Beni is now a remarkable cultural landscape where high levels of biodiversity remain, together with traditional patterns of adaptive human practices and use of natural resources.

Despite the creation of an extensive state-protected area that is also Indian territory, officially called Isiboro-Secure Indian Territory and National Park (TIPNIS), this area is threatened by the imminent construction of a new highway between Bolivia and Brazil that will cross the park's core area. This development will mean deforestation, loss of critical ecosystems and agricultural frontier, and expansion of the coca monoculture, as well as acculturation and loss of indigenous groups.

Global warming is already forcing changes upon the habitat selection of various species in the Beni savanna. Field studies have revealed, for example, that a number of species have been migrating upslope into the Andes foothills to the west of the savanna, most likely in order to maintain their optimal temperature and moisture conditions.

An area of traditional practice that may have worldwide import as solutions to global warming are sought is the ancient use of pyrolysis or biochar to enrich local soils. Simply put, causing organic matter—wood, grain husks, plant stalks, animal remains—to smolder or carbonize in a low-oxygen trench enclosure was a means of manufacturing a charcoal-like soil enricher. To Portuguese settlers, this native practice was called terra preta do indio; deposits of it are still in use as an extra-fertile soil in many areas of South America. Researchers have in recent years suggested biochar can contribute to carbon sequestration efforts on a global scale, as the process is readily measurable and verifiable—and thus tradable as a carbon emissions footprint-reducer commodity.

Gonzalo Navarro

Further Reading

Hanagarth, Werner and Stephan G. Beck. "Biogeographie der Beni-Savannen (Bolivien)." *Geographische Rundschau* 4, no. 11 (1996).

Navarro, Gonzalo and Mabel Maldonado. *Geografía Ecológica de Bolivia. Vegetación y Ambientes Acuáticos.* Cochabamba, Bolivia: Editorial Centro de Ecología Simón I. Patiño, 2002.

Pouilly, Marc, Stephan G. Beck, Mónica Moraes, and Carla Ibáñez, eds. *Diversidad Biológica en la Llanura de Inundación del Río Mamoré.* Santa Cruz de la Sierra, Bolivia: Editorial Centro de Ecología Simón I. Patiño, 2004.

Thomas, Evert. "New Light on the Floristic Composition and Diversity of Indigenous Territory and National Park Isiboro-Sécure, Bolivia." *Biodiversity and Conservation* 18, no. 7 (2009).

Bering Sea

Category: Marine and Oceanic Biomes.
Geographic Location: Northern Hemisphere.
Summary: A shallow, productive oceanic ecosystem in the sub-Arctic North Pacific, the Bering Sea is a very important fishing ground for finfish (such as cod, haddock, and pollock) and invertebrates (mostly crabs).

The Bering Sea is located in the North Pacific, joining that ocean with the Arctic Ocean by way of the Bering Strait that spans the gap between easternmost Asia and westernmost North America. This Large Marine Ecosystem (a designation of the U.S. National Oceanic and Atmospheric Administration, or NOAA) encompasses in its western areas the deepwater Aleutian Basin, with depths well in excess of 10,000 feet (3,048 meters), as well as a large area of shallow, broad continental shelf in the east, where depths seldom exceed about 200 feet (60 meters). The rapid change in bathymetry (depths) between these two regions brings cold, nutrient-rich bottom water onto the shallow shelf, which thus supports a vast food web.

The Bering Sea is host to a rich and diverse assemblage of marine life, with a large variety of marine mammals, fish, invertebrates, and seabirds living within its bounds. The region also supports very large and productive fisheries for groundfish, salmon, and crabs. Many of these species were severely overfished with the onset of mechanized fishing in the second half of the 20th century, but improved international treaties and domestic management regimes have contributed to stabilization of these high-value fisheries over the past decade.

The borders of the Bering Sea are defined by mainland Alaska on the east, the Kamchatka Peninsula of Russia on the west, Alaska's Aleutian Island chain on the south, and the Bering Strait, which separates the Bering Sea from the Arctic Ocean, on the north. The sea's total area is roughly 849,425 square miles (2.2 million square kilometers). Its borders lie mostly within the 200-mile (322-kilometer) Exclusive Economic Zone (EEZ) of either the United States (in the eastern Bering Sea) or Russia (in the western Bering Sea), largely because of the large number of islands dotting the continental shelf and Aleutian chain, each of which commands a 200-mile (322-kilometer) EEZ radius. A small area of international waters in the center is commonly referred to as the "doughnut hole."

Oceanography and Marine Life

The physical oceanography of this region is particularly important. Sharp changes in depth from the Aleutian Basin to the North American continental shelf create localized upwelling events, which bring nutrient-rich bottom water to the surface and also generate strong currents throughout the shelf portion of the system. Strong winds and frequent storms create hazardous conditions for seagoing vessels, with waves frequently exceeding 30 feet (10 meters) and winds often exceeding 50 miles per hour (80 kilometers per hour), but also create a great deal of mixing of deep ocean nutrients into the well-lit surface layer. These conditions are ideal for the growth of phytoplankton—the tiny marine plants that are the basis of the Bering Sea food web. Seawater stratification from freshwater formation along the edge of the ice front, which advances and retreats seasonally, also creates areas of high primary productivity.

This productivity supports a range of marine life from varied pelagic (surface-layer-dwelling) and demersal (near-bottom-dwelling) fish species, to large populations of benthic (bottom-dwelling) invertebrates such as crabs, which are scavengers that feed on the large flux of organic matter raining down from the richly productive waters above. The western Bering Sea has the largest population of gadoid (cod-like) fishes in the world; the eastern Bering Sea is home to an array of fish species from cod and haddock to flatfish (flounder, sole, and halibut), to pelagic species like salmon. The system also supports an abundance of marine-mammal species, ranging from seals, walruses, and toothed whales (which feed on the abundant fish stocks) to baleen whales (which feed on krill), as well as pelagic invertebrates.

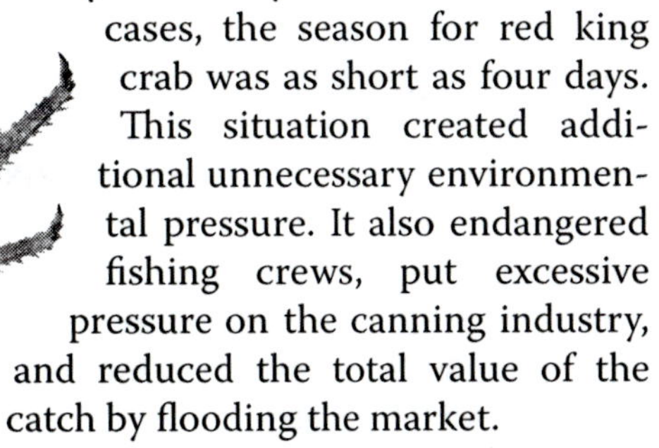

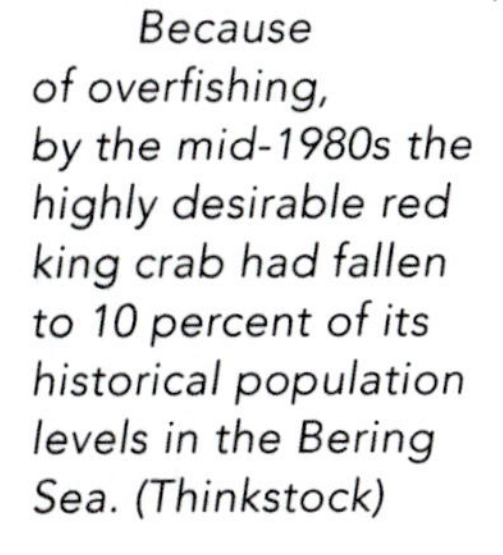

Because of overfishing, by the mid-1980s the highly desirable red king crab had fallen to 10 percent of its historical population levels in the Bering Sea. (Thinkstock)

Coastal regions of the Bering Sea also support large populations (up to 20 million individuals) of predatory fish-eating birds, such as puffins, albatross, kittiwakes, and eagles. Polar bears, too, prey along these shores. All told, more than 15 mammal, 30 bird, and 400 fish species call the Bering Sea home—almost as many fish species as can be found in the Caribbean Sea.

Fisheries and Challenges

With the onset of industrialized fishing in the post–World War II era, large mechanized fishing fleets from various nations (chiefly the United States, Russia, Canada, China, and Japan) rapidly increased fishing pressure in the Bering Sea. A broad range of techniques and target species are employed in this fishery, although the principal fisheries are groundfish (cod, haddock, pollock, and halibut), salmon, and crabs (king, tanner, bairdi, and opilio). Of these species, the most valuable—and therefore most overexploited—is the red king crab, the population of which was reduced to less than 10 percent of historical levels by the mid-1980s.

Early management efforts by the National Marine Fisheries Service were predominantly based on single-species catches, setting size limits and total allowable catches. In large part, these measures created a derby-type fishery, in which large numbers of entrants fished target species extremely heavily for a very short time. In some cases, the season for red king crab was as short as four days. This situation created additional unnecessary environmental pressure. It also endangered fishing crews, put excessive pressure on the canning industry, and reduced the total value of the catch by flooding the market.

Recently, management efforts have shifted to a system dominated by individual transferrable quotas (ITQs), in which each entrant to the fishery is assigned a specific percentage of the total quota and can fish that quota over a longer period. This management strategy has worked very well in this system, improving the biomass of fisheries species and the total value of the catch while dramatically reducing the fatality rate among fishermen. The system has also dramatically reduced the number of entrants into the fishery, however, creating some contention due to the high unemployment rates among local fishers.

Although the Bering Sea is in many respects a "poster child" example of the effectiveness of ITQ fisheries, it is not without major environmental concerns. Availability of food in this system is highly correlated to temperature, with cold years being more favorable for high production than warm ones. Climate change presents a particular concern to the Bering Sea system for this reason, and also because the specialized physical oceanographic features of the system prevent species from simply moving north to remain in their preferred temperature range.

Some scientists estimate that primary productivity in this system has already dropped by 30 to 40 percent, and they have shown that high-energy phytoplankton species (diatoms) are being outcompeted by lower-energy species (coccolithophores) during warm years, which results in less energy moving up the food chain. Loss of sea ice is also problematic to species such as polar bears, which hunt along the ice edge. Although cycles of warm and cold years are part of natural climate variability, a continued warming trend may significantly affect diversity and productivity in this system throughout the 21st century.

Jason Krumholz

Further Reading

Gates, Nancy. *The Alaska Almanac: Facts About Alaska.* 30th Anniversary ed. Anchorage: Alaska Northwest Books, 2006.

MacLeish, Summer. *Seven Words for Wind: Essays and Field Notes From Alaska's Pribilof Islands.* Anchorage: University of Alaska Press, 2008.

North Pacific Research Board and National Science Foundation. "Understanding Ecosystem Processes in the Bering Sea." http://bsierp.nprb.org.

Tang, Q. and K. Sherman, eds. *Large Marine Ecosystems of the Pacific Rim.* Malden, MA: Blackwell Science, 1999.

Bikar Atoll

Category: Marine and Oceanic Biomes.
Geographic Location: Pacific Ocean.
Summary: Bikar is a small, remote coral atoll in the Marshall Islands, known for its relatively undisturbed and pristine flora and fauna.

The Bikar Atoll is among the northernmost of the Marshall Islands, a sparsely inhabited republic of 29 coral atolls and five islands situated in the tropical North Pacific Ocean, approximately 2,000 miles (3,400 kilometers) northeast of Papua New Guinea, 2,800 miles (4,600 kilometers) southeast of Japan, and 2,500 miles (4,100 kilometers) west of Hawaii. Like most atolls, Bikar was formed as the fringing reef of a now-extinct basaltic volcano, which has since subsided into the ocean, leaving a ring of small islets.

The small area, lack of freshwater, and remote location of Bikar Atoll contribute to the lack of any permanent human settlement. Though some fishing pressure exists on and around Bikar, the flora and fauna here have remained relatively pristine and undisturbed throughout history. Even this remote location is not immune to all human influence, however, as the Bikar islets, which are home to a native population of Polynesian rats (Rattus exulans), have been impacted by invasive black rats (Rattus rattus) from fishing vessels; the invasive rat species, which is more aggressive and carnivorous, poses a threat to seabird and turtle nesting sites. Also, the soil of Bikar was contaminated with nuclear fallout from the 1954 Castle Bravo nuclear test conducted by the United States in Bikini Atoll, some 300 miles (485 kilometers) upwind.

Bikar Atoll is one of the smallest atolls in the Republic of the Marshall Islands (RMI), and is located roughly 360 miles (580 kilometers) north of Majuro Island, the capital of the RMI. The diamond-shaped fringing reef of the atoll is continuous except for a single narrow, forked passage. The reef encloses a lagoon covering roughly 14 square miles (37 square kilometers), though the combined land area of its handful of islets is only about 0.2 square mile (0.5 square kilometer). The largest of these, also called Bikar, reaches a height of 20 feet (6 meters) above sea level, common across the RMI. Though direct measurements have never been taken, evidence from nearby similar atolls indicates that the fringing reef may extend as much as 4,600 feet (1,402 meters) beneath the ocean surface and be approximately 55 million years old. The islets and atoll have built up around the central cone of an extinct volcano that rises more than 10,000 feet (3,048 meters) from the seafloor.

Bikar Atoll is relatively warm and semiarid, with an average air temperature of 82 degrees F (28 degrees C). Average annual precipitation is less than 45 inches (114 centimeters); most rain falls during the July-to-October wet season, and little freshwater is available.

Flora and Fauna

Flora on Bikar consists predominantly of atoll forest with scrub understory. The tough, spotty forest is largely made up of low, salt- and drought-resistant species common to the Indo-Pacific realms, such as Mapou (Myrsine australis), grand devil's claws (Pisonia grandis), and scattered screwpine (Pandanus tectorius). Tree heliotrope (Tournefortia argentea) plays an opportunistic role. Herbaceous plants such as spiderlings (Boerhavia spp.) also occasionally make inroads. Some coconut palms originally introduced by visiting islanders are present, though such plants are periodically decimated by cyclone activity.

Bikar is a traditionally important nesting and feeding ground for several dozen species of birds, including frigatebirds such as the great frigatebird (Fregata minor), terns including the sooty tern (Onychoprion fuscatus), and boobies like the red-footed booby (Sula sula). The atoll is especially important as a nesting ground for the endangered green turtle (Chelonia mydas). Bikar is one of just three sites in the RMI deemed significant in this regard, and it is thought at various times to have hosted the largest number of green turtles in the Marshalls.

Beyond the islets and below the waves, the coral reef itself is a secondary harbor of life here, although not as abundantly blessed with species diversity as reefs in other biomes. Dozens of species of coral occur at Bikar, supporting various types of green, blue, and brown algae including the green macroalgae Halimeda taenicola and Halimeda lacunalis, and mollusks like the bear claw clam (Hippopus maculatus). Drawing fishermen are the two-spot red snapper (Lutjanus bohar), humpback red snapper (Lutjanus gibbus), leopard grouper (Mycteroperca rosacea), and humphead parrotfish (Bolmetapan muricatus).

There is among the reef fishes a substantial variety of moray eel species, including the Seychelles moray eel (Anarchias seychellensis), yellow-edged moray (Gymnothorax flavimarginatus), giant moray (Gymnothorax javanicus), hookjaw moray (Enchelycore bayeri), viper moray (Enchelynassa canina), and snowflake moray (Echidna nebulosa).

Human Activity

The United States assumed ownership of the Marshall Islands at the close of World War II, and this area was the site of several nuclear tests during the 1950s, including the 1954 Castle Bravo nuclear test. That test near Bikini Atoll was well west of Bikar, but Bikar was directly in the fallout plume. The explosion left elevated levels of radiation in Bikar's soil for several decades, though a 1981 survey of fish caught in and around the Bikar lagoon showed minimally elevated levels of radiation at that time.

In 1986, the United States ended its territorial trusteeship over the Marshall Islands, transferring power to the RMI. The Marshalls remain under a Compact of Free Association with the United States.

In October 2011, the RMI government dedicated what is one of the world's most extensive shark sanctuaries, comprising 770,000 square miles (two million square kilometers) of ocean surrounding the Marshall Islands. A primary aim of the Marshall Islands National Shark Sanctuary is to protect the health of the reefs, as the shark population in its natural role controls the scope of the schools of other large predator fish, which in turn affect the types of small fish and other organisms that trim plankton blooms, clean corals, and otherwise maintain a balanced, healthy reef ecosystem.

Global climate change presents a fairly uniform threat to many atoll reef systems. The mean level of Bikar Atoll's lagoon is about 1.5 feet (0.5 meter) above the ocean level. Most of the

Bikar Atoll, which is surrounded by a fringing reef that may be as much as 55 million years old, is home to many types of moray eels. At right, an ocellated moray eel or honeycomb moray eel (Gymnothorax saxicola). (NOAA)

coral lies below sea level, but some coral heads are situated above the shallow lagoon level. Complete inundation of these outposts is a very real possibility; however saltwater intrusion is not a factor. A more pervasive effect is likely to be increased storm damage from cyclones, which already wreak considerable destruction upon the forests here. Reef infrastructure will be more directly at risk from water-temperature rise, which has been indicated as a negative factor that has contributed to coral bleaching cases worldwide.

JASON KRUMHOLZ

Further Reading

Maison, Kimberly A., Irene Kinan-Kelly, and Karen P. Frutchey. *Green Turtle Nesting Sites and Sea Turtle Legislation Throughout Oceania.* Honolulu, HI: National Marine Fisheries Service, 2010.

Mohamed, Nizar and Jorelik Tibon. *The Republic of the Marshall Islands Biodiversity Strategy and Action Plan.* Majuro: Republic of the Marshall Islands Department of Resources and Development, 2010.

Paine, James R. *IUCN Directory of Protected Areas in Oceania.* Gland, Switzerland: International Union for Conservation of Nature, 1991.

Black Sea

Category: Marine and Oceanic Biomes.
Geographic Location: Southeastern Europe.
Summary: The Black Sea is a great inland sea characterized by a low proportion of oxygen-rich water; not lacking in biological diversity at the surface and along its coastline habitats, the sea has been under great environmental stress.

The Black Sea, located in southeastern Europe and western Asia, is a largely enclosed sea with a narrow link to the Mediterranean that permits two-way flow of waters. With a surface area of 178,000 square miles (461,000 square kilometers), the Black Sea is bound on the south by Turkey, the west by Bulgaria and Romania, the north by Ukraine, the northeast by Russia, and the east by Georgia. Its drainage basin draws upon 965,000 square miles (2.5 million square kilometers) in this Eurasian region. With its average depth of about 4,185 feet (1,275 meters), the Black Sea encompasses an average water volume of approximately 130 cubic miles (545 cubic kilometers).

Hydrology

Most of the freshwater flows into the Black Sea from the northern half. The Danube River, which drains a catchment of some 309,000 square miles (801,000 square kilometers) as it winds 1,775 miles (2,857 kilometers) across Europe, is the main inflow source of freshwater. Other major sources are the Dniestr, flowing from the northwest through Moldova; Dniepr to the north, the largest river in Ukraine; also to the north the Don, which flows into the Sea of Azov, the northeast lobe of the Black Sea; the Rioni and Chorokhi in Georgia; and several rivers in Turkey, among them the Yesil-Irmak, Filyos, and Sakarya.

The Black Sea receives inflow of 83 cubic miles (348 cubic kilometers) of freshwater annually, with by far the greatest flow from the Danube River, at some 48 cubic miles (200 cubic kilometers). About 1.8 billion cubic feet (52.2 million cubic meters) of sediment are deposited into the sea annually, of which well over half flows from the Danube. An undersea river of saline water flows in from the Mediterranean via the Bosporus Strait. The Black Sea surface-layer salinity is about half that of the average for oceans around the world, thus it is considered brackish.

Due in part to its particularly stable water column layering—there is a vast anoxic level in the depths below about 800 feet (245 meters), accounting for well over 80 percent of all the sea's water—the Black Sea harbors a relatively low number of species for its size. However, its surface waters and coastal realms support great diversity of species and abundance of biota. Still, with changing levels of pollution and nutrient runoff, species population shifts, and other factors that have interacted with its already low-oxygen regime and enclosed nature, the sea has at times sustained very high levels of eutrophication, hypoxia, and anoxia.

Changing episodes of human impact have also altered the mix; it is scientifically notable that the collapse of the Soviet Union in 1991 and the bottoming-out of its intensive agriculture practices led directly to a profound decrease in the inorganic fertilizer-originated nutrient inflow, helping resuscitate the Black Sea from a period of oxygen-choking eutrophication.

Habitats and Vegetation

The Black Sea is ensconced in a wealth of coastal-area habitats that include continental shelf, coastal lagoons, intertidal zones, wetlands, estuaries, dunes, cliffs, grasslands, steppes, rivers, and forests. Among the most richly endowed is the Danube Delta, a large and critical habitat area encompassing some 1,740 square miles (4,500 square kilometers) and one of the largest estuary wetlands in Europe. Great breeding populations of birds nest here among estuarine forests and reeds, such as the white pelican (Pelecanus onocrotalus), glossy ibis (Plegadis falcinellus), pygmy cormorant (Phalacrocorax pygmeus), and marsh harrier (Circus aeruginosus).

Within the deeper waters of the Black Sea, populations of phytoplankton and zooplankton swell and contract in sway with seasonal changes to temperature, salinity, and nutrient inflow. These foundation species support the food web. In turn, oyster beds, marine mammals, and anadromous fish schools thrive or diminish; overflying and nesting bird flocks, too, have their own rhythms.

Some of the Black Sea habitat areas have undergone immense change. The northwestern shelf area, for example, was once dominated by the red seaweed Zernov's Phyllophora, but which, having been nearly eliminated by more aggressive macroalgae, now is sustained in the area mainly by carefully planned and protected human cultivation. Similarly, thickets of marine eelgrass (Zostera spp.) in the same area have also been decimated.

Above the water, the mainly oak forests of the Strandzha region of Bulgaria, host to such types as white oak, Hungarian oak, and turkey oak (Quercus polycarpa, Q. frainetto, and Q. cerris), also feature Western Pontic beech (Fagus orientalis) and understory species such as Pontic rhododendron (Rhododendron ponticum) and twin-flowered daphne (Daphne pontica).

Fauna of the Region

The Black Sea has long supported significant fisheries. European anchovy (Engraulis encrasicolus) and sprat (Sprattus sprattus) are among the dominant commercial types currently, increasing in importance as stocks of predatory fish have been decimated; these have previously included bonito, horse mackerel, and bluefish.

Some 30 species of sturgeon (Acipenser spp.) were historically found in abundance here; favored feeding areas were located all around the coastal waters with the exception of much of the southern shoreline. However, overfishing and pollution have devastated their numbers. Aquaculture solutions have yielded mixed results, particularly in light of the long period of maturation (as much as 12 years) of sturgeon.

Among marine mammals, endemic (found only here) subspecies include the Black Sea bottlenose dolphin (Tursiops truncatus ponticus), Black Sea common dolphin (Delphinus delphis ponticus), and Black Sea harbor porpoise (Phocaena phocaena relicta).

Upwards of 300,000 seabirds and as many as 100,000 raptors rest here each autumn on their way south, part of the second-largest flyway in Europe. Threatened or endangered avian species include Dalmatian pelican (Pelecanus crispus), ferruginous duck (Aythia nyroca), and the slender-billed curlew (Numenius tenuirostris).

Human and Climate Issues

Along with urban, industrial, and agricultural pollution, the Black Sea biome has had to contend with overfishing, coastal habitat destruction, and invasive species. Key species that have been wiped out or severely depleted here by direct human activity include the endangered semi-aquatic European mink (Mustela lutreola) and the endemic Black Sea turbot (Scophtalmus maeotica). The Black Sea oyster (Ostrea edulis) has been a victim of the rapacious whelk (Rapana venosa), and also by an introduced jellyfish.

In the 1980s the warty comb jelly (Mnemiopsis leidyi), a species native to the Atlantic Ocean and with few predators, arrived in the Black Sea. The warty comb jelly gorged on zooplankton as well as the fish eggs and larvae of many species. Its population exploded, precipitating the collapse of anchovy, chub mackerel, and two dozen other fisheries, as well as engendering the rise of red algae blooms. Fortunately, quantities of pink comb jelly (Beroe ovata), a Mediterranean species that is a natural predator of the Mnemiopsis, arrived in the Black Sea in the late 1990s, coming there to feast upon its prey and to somewhat contain the warty population.

Climate change in some cases augments the effects put into action by humans and invasive species. Research has clearly shown, for example, that warmer average temperatures in several periods since the late 1980s likely contributed to the spread of the warty comb jelly. The warmer regime may also be contributing to increased anchovy and sprat catches, a habitat positive, but also results in increased blooms of toxic cyanobacteria, a habitat negative. Greater extremes in precipitation in Ukraine are likely to aggravate problems of coastal erosion and flooding, affecting habitats along the northern Black Sea and Crimean peninsula. On the other hand, the Danube Delta is projected to remain relatively stable, as a balance seems to be struck by consistent sediment flow, very moderate sea-level change locally, and general continuation of current precipitation levels.

James Mammarella

Further Reading

Adams, Terry D., Michael Emerson, Laurence David Mee, and Marius Vahl. *Europe's Black Sea Dimension.* Brussels, Belgium: Centre for European Policy Studies, 2002.

Bat, Levent, et al. "Biological Diversity of the Turkish Black Sea Coast." *Turkish Journal of Fisheries and Aquatic Sciences* 11, no. 4 (2011).

Commission on the Protection of the Black Sea Against Pollution (BSC). "Strategic Action Plan for the Environmental Protection and Rehabilitation of the Black Sea." BSC, 2009. http://www.blacksea-commission.org/_bssap2009.asp#_Toc222222293.

Kideys, Ahmet E. "Fall and Rise of the Black Sea Ecosystem." *Science* 297, no. 5586 (2002).

Blue Mountains Forests

Category: Forest Biomes.
Geographic Location: North America.
Summary: Rising above the American northwestern deserts, the Blue Mountains create a dramatic landscape composed of tall peaks, deep river valleys, and vast forests.

Situated amid the deserts and grasslands of the Columbia Plateau, the Blue Mountains form a verdant sanctuary in this landscape. This mountain range occurs mostly in the northeastern quadrant of the state of Oregon, with parts in neighboring Idaho and Washington. The range, which forms a portion of the Snake-Columbia River watershed, supports forests that form a natural link between the Cascade Range to the west and the Rockies to the east. These mountains harbor robust wildlife populations—but the forests of the Blues, as these mountains are commonly called, have been dramatically affected by human activity.

As in any mountainous environment, the ecology of the Blue Mountains is shaped by topography. Much of the land between the Cascades and Rocky Mountains experiences the rain-shadow effect, creating vast deserts that stretch south through this realm from central Washington to the Great Basin of Nevada and Utah. The Blue Mountains, however, are an exception. Tall peaks drive moist air upward, generating rainfall. Because the Blues lie east of the Cascades, they themselves are subject to a slight rain shadow that causes them to be drier than the very moist coastal areas of Oregon. Nevertheless, rainfall at higher elevations promotes the growth of expansive forests surrounded by an expanse of low-elevation sagebrush and dry steppe environments.

A few of the birds of the Blue Mountains Forests biome are the greater sage-grouse (Centrocercus urophasianus), western tanager (Piranga ludoviciana), and pileated woodpecker (Dryocopus pileatus), seen here. (Thinkstock)

Vegetation Zones

Elevation creates a zonation of plant communities moving up and down the mountains. From the base of a mountain, vegetation shifts from sagebrush (Artemisia tridentata) steppe to juniper (Juniperus spp.) woodlands to ponderosa pine (Pinus ponderosa) forests. These ponderosa forests are often described as parklike, for trees usually are sparse, permitting the growth of understory grasses. Fire maintains these forests by burning understory vegetation and leaving mature, fire-resistant ponderosa pines standing. Above these forests, moister conditions allow the growth of Douglas fir-spruce (Pseudotsuga menziesii) forests. Spread across much of the Blues, these forests generally are more productive due to the increase in rainfall and humidity. Moister conditions allow communities of shrubs to flourish, including the huckleberry (Vaccinium parvifolium), thimbleberry (Rubus parviflorus), and bracken-fern (Pteridium aquilinum).

Still further upslope, lodgepole pine (Pinus contorta) becomes the dominant tree species. It can tolerate the short summers and heavy snowfall of the upper montane zone. Stands of lodgepole are often very dense, prohibiting the development of forest-floor plants. Again, this community is controlled by fire: Dense growth of mature trees encourages intense fires that burn large areas. Fires remove the most mature trees, permitting the next generation to replace them. Lodgepole pine grows fast, but the successional communities that soon follow a burn are productive and important for herbivorous animals.

The highest forest type in the Blues is the subalpine forest, which is composed around several tree species adapted for an extreme environment. The density of these forests varies, with stands becoming thinner higher up the mountains. Subalpine fir (Abies lasiocarpa), western white pine (Pinus monticola), Engelmann spruce (Picea engelmannli), and aspen (Populus tremuloides) are common species in this vegetation zone. Snowfall in these areas can be substantial, and the growing season is short. As a result, many trees are short and stunted due to the harsh environmental conditions. Eventually, forests give way to alpine meadows that are covered by snow for much of the year. Following snowmelt, a rich plant community develops during the short growing season. Grasses and wildflowers abound during these months, many of which are found only on isolated high mountain peaks.

Fauna of the Blues

Wide-ranging herds of mule deer (Odocoileus hemionus), bighorn sheep (Ovis canadensis), and elk (Cervus elephas) share prime mammalian habitats with black bear (Ursus americanus) and cougar (Puma concolor). The gray wolf (Canis lupus) vanished from Oregon—and nearly all the contiguous 48 states—by the 1930s, but has sparsely reappeared in the Blue Mountains since 2000; this apex predator enjoys a protected/endangered status here, as does the Canada lynx (Lynx canadensis).

Among amphibians and reptiles, the Columbia spotted frog (Rana luteiventris) ranges along the rivers while western painted turtle (Chrysemys picta bellii) stalks drier zones.

Avian members of the Blues community include such denizens as the greater sage-grouse (Centrocercus urophasianus), western tanager (Piranga ludoviciana), and pileated woodpecker (Dryocopus pileatus); as well as predators such as the great gray owl (Strix nebulosa), ferruginous hawk (Buteo regalis), and bald eagle (Haliaeetus leucocephalus). Listed as a species of concern is the northern spotted owl (Strix occidentalis caurina).

Human Impact

Forests are defining hallmarks of much of the Blue Mountains, but this ecoregion has other notable features. Between the peaks are deep valleys with rich soils that support grasslands, many of which have been converted for agriculture; crops such as wheat and alfalfa are common here. These valleys have been carved by powerful rivers such as the John Day, Grande Ronde, Powder, and Malheur. Perhaps the most impressive of these areas is Hells Canyon, created by the Snake River as it passes between the Wallowa Mountains in Oregon and the Seven Devils Mountains in Idaho. Many wildlife species are dependent on riparian areas here. In the past, nearly all these rivers supported massive salmon runs, although those runs have now dwindled to a fraction of their former levels.

The rugged nature of the Blue Mountains landscape supports a sparse human population. Most live in the valleys, which can support agriculture, but the entire area has been extensively affected by human activity. A large proportion of forests have been logged in the past, leaving few old-growth forests that are essential for many species. (One protected old-growth stand envelops the Emigrant Springs State Heritage Area, a campground and interpretive site that recalls the Oregon Trail pioneering days.)

Livestock grazing on both forested and open lands has degraded many local ecosystems, especially riparian zones. Outbreaks of pine beetles and other pests have created large swaths of dead trees. Invasive species range from Kentucky bluegrass (Poa pratensis) and tansy ragwort (Senecio jacobaea) to European red fox (Vulpes vulpes) and Virginia opossum (Didelphis virginiana).

Land management programs aim to address these issues while balancing human use of natural resources. A related issue is firefighting policies. Studies have called into question, for example, the suppression of fire in terms of impact on overgrowth of certain evergreen species and consequent loss of elk grazing habitat. Research continues both on the use of prescribed fire and overall fire regime decision making in light of new understanding of the balance of mature trees, understory growth, and the forage needs of a wide range of animals.

Fire policy becomes even more crucial with the progression of global warming, although the effects of climate change in the Blue Mountains Forests biome are unpredictable. The Pacific Northwest as a whole is projected to gain in average annual precipitation due to global warming—but recent years have seen reduced precipitation in parts of the Blue Mountains. It is certain, however, that higher average temperatures will lead to reduced snowpack, earlier snowmelts, and lower water flow in the warm season. Such alterations to the hydrologic cycle will cause sometimes dramatic changes in local habitats, which naturally must give rise to species migration to other temperature zones, behavioral and genetic adaptation, or extinction.

Holding back habitat fragmentation becomes especially critical under such conditions. Federal reserves situated wholly or partly in this biome include the Malheur, Umatilla, and Wallowa-Whitman National Forests. There are many state parks, as well as federal Wilderness Areas. The Blue Mountains Forest Reserve was established on 2.63 million acres (1.06 million hectares) in 1906 following several years of patchwork, informal reserve-creating in Oregon. President Theodore Roosevelt in 1907 signed an expansion of the reserve; much of this area has subsequently been subdivided by the National Forest Service for better regional management.

Justin Bohling

Further Reading

Franklin, Jerry F. and C. Ted Dyrness. *Natural Vegetation of Oregon and Washington.* Corvallis: Oregon State University Press, 1973.

National Research Council. *Environmental Issues in Pacific Northwest Forest Management.* Washington, DC: National Academies Press, 2000.

Quigley, Thomas M. *Forest Health in the Blue Mountains: Science Perspectives.* Portland, OR: USDA, Pacific Northwest Research Station, 1994.

U.S. Forest Service. *Wildlife Habitats in Managed Areas: The Blue Mountains of Oregon and Washington, Agriculture Handbook No. 553.* Washington, DC: U.S. Department of Agriculture, 1979.

Borneo Rainforests

Category: Forest Biomes.
Geographic Location: Asia.
Summary: The third-largest island on Earth contains five major forested ecoregions that exist because of Borneo's unique geomorphology and the climatic and biological characteristics that define the Cool Equatorial Rainforest biome.

Thick, impenetrable jungle sparsely populated by tribal headhunters is a description that could have at one time summarized the island of Borneo, at least for a great many travelers and writers familiar with world geography. This depiction still circulates today, but the reality is that more than 20 million people inhabit the island, mostly along the coastline, and that almost 70 percent of the arboreal descendants of the oldest tropical rainforest in the world have been razed during the past 50 years for the sake of economic development. Much of the physical landscape of Borneo has become in one way or another a harvestable natural resource, and many indigenous people and wage laborers and their cultures have become tourist commodities. What remains of long-standing ecoregions that formed in the relationship between geology and climate are generally traces left mostly unaltered by human beings, save for the occasional wildlife surveillance camera set to capture a portion of the immense biodiversity that the island still generates.

The extraordinary flora and fauna of these regions depend on a continuously warm and moist atmosphere—two climatic conditions that qualify the island as a global biome called Cool Equatorial Rainforest. The life of these ecoregions also depends on environmental policies and initiatives set forth by Brunei, Indonesia, and Malaysia, the three countries that share the island and are responsible for its future.

Geologic Foundation of Biology

At 287,001 square miles (743,330 square kilometers), Borneo is the largest island of the Malay Archipelago. Its geologic origins remain somewhat of a mystery, although most scientists agree that the original surface rose above sea level some 130 million years ago, long after the split of the supercontinent Pangea into the land masses of Gondwanaland and Laurasia. As Gondwanaland's tectonic plate began to fragment and sections gradually moved into the positions they are in today, at times and in places they also converged on the Laurasian plate, causing layers of rock to crunch and then uplift. Around this rock core, sediment and debris carried on ocean currents and forged from the erosion of the core continue to collect and consolidate, thereby shaping the island as it moves, along with the underlying tectonic plate, ever so slowly to the north.

At the geological moment when Borneo rose above sea level, one of the world's oldest rainforest environments began to appear. The island's high level of biodiversity, however, owes much to the dynamics of Pleistocene glaciation.

About 3 million years ago, glaciers began to form in fits and spurts on the surface of the Earth, stretching down from the poles and alpine regions. Some 18,000 years ago, they covered about a quarter of the Earth's surface before they began to melt and recede.

By this time, Borneo had been in place for 50 million years in an area we now call southeast Asia, all the while engaged in the cycle of erosion

A liana, or long woody vine, spiraling up into the forest canopy in the Borneo rainforest, which is one of the oldest tropical rainforests in the world. (Ken Whalen)

with the Eurasian landmass and volcanic islands of the Malay Archipelago that built up the Sunda continental shelf. With the formation of glaciers and concomitant lowering of sea level, the shelf emerged from the sea as a land bridge of tropical savanna vegetation linking Eurasia with insular southeast Asia. Exotic plants, animals, and people were able to colonize Borneo. When the glaciers ebbed, the shelf became the floor of shallow seas, and the colonizers were isolated from the mainland. In isolation, plants and animals on the island evolved uniquely.

Climate and Ecoregions

An important influence on the evolution of Borneo's flora and fauna is the Intertropical Convergence Zone, also known as the Thermal Equator, which is an undulating band of consistently low air pressure, high temperatures, and rainfall that follows the movement of the most intense solar rays hitting Earth. These rays move generally north and south across the surface within the tropical latitudes. The Thermal Equator is never far from Borneo, so convectional rainfall is continuous year-round, save for the barely perceptible less-wet period.

Borneo has five types of rainforest ecoregions spread across Malaysia's states of Sabah and Sarawak, Indonesia's Kalimantan, and the Sultanate of Brunei. They are lowland climax rainforest, southwest Borneo freshwater swamp and peat-swamp forests, the Sundaland heath forests (Kerangas), and the Sunda Shelf mangroves. Together, these ecoregions contain thousands of species of flowering plants and trees, including 267 species of dipterocarps (tropical hardwood family with two-winged fruit), and hundreds of species of terrestrial mammals and birds, amphibians, and reptiles, not to mention a plethora of insects, many of which are endemic (evolved in a local biome and found nowhere else).

Lowland Rainforest

The lowland rainforest once covered most of Borneo. What remains are diminishing patches of canopy that together equal about 165,059 square miles (427,500 square kilometers). Emerging from the rusty-red soil, rich in aluminum and iron oxides, are hardwood evergreen dipterocarps, a family of tropical trees that produce seeds with wings that float down from the highest stories of the canopy. Because the soil is relatively infertile, trees have buttressed root trunks that spread widely along the forest floor, enabling them to absorb organic nutrients at the surface.

The luxuriance of life at midcanopy revolves around myriad species of lianas, orchids, and ferns. Growing in the understory—particularly near rivers where sunlight reaches the forest floor—are shrubs, bamboo, rattan, and ferns. Fungi and moss prefer darker places under the canopy, where they break down organic remains into nutrients for trees.

The most significant natural resource of the lowland rainforest is a particular genus of dipterocarps known as Shorea,of which there are 138 species on the island, 91 of them endemic. Some of these flowering trees can grow more than 200 feet (61 meters) tall. In 1990, 50 percent of the world's tropical-wood exports came from Borneo. Dipterocarps—most of which are now critically endangered—were cut down, made into plywood, and exported to Europe and Japan. Rattan, a kind of palm, is also harvested; its core is used to make furniture, and its outer sheath is stripped to make matting and basketry. Indonesia supplies 70 percent of the world's raw rattan, with much of it coming from Borneo.

Finally, the lowland rainforest is a natural resource that draws tourists from around the world. People come to gaze at some of the most unique wildlife on the planet, such as the Bornean orangutan (Pongo pygmaeus), one of two extant species of great apes in Asia; the endemic proboscis monkey (Nasalis larvatus); and the Bornean pygmy elephant (Elephas maximus borneensis), mostly found in the Betung Kerihun, Gunung Mulu, and Ulu Temburong National Parks.

Freshwater and Peat-Swamp Forests

As elevation increases on the island, the ecological characteristics of the lowland rainforest give ground to altitudinal zonation, through which species of flora and fauna gradually change mainly because of decreased temperatures and changes in rainfall patterns. The result is a changing mix of tropical, deciduous, needle-leaf trees to grasslands, alpine meadows, stunted trees, mosses, and lichen. Kayan Mentarang National Park in Indonesia and Mount Kinabalu in Malaysia (the tallest mountain on the island) are remarkable examples of this geophysical process.

The biological communities of lowland rainforests are similar to the swamplands and Sundaland heath forests of Borneo; differences in species stem mainly from the prevalence of standing water and certain other abiotic conditions. In the peat-swamp forests, dead leaves and wood never fully decompose in the soggy ground, thereby creating a spongy layer of rank, acidic matting called peat, which can be 66 feet (20 meters) thick and 5,000 to 10,000 years old. The forests now maintain an unusually high number of species of flowering plants and ferns. Trees can take root in peat, but the tallest trees grow around the margin. Maludam National Park in Sarawak is the largest peat forest of its kind on the island.

Sometimes, draining peat swamps for agriculture goes hand in hand with setting forest fires that ignite the peat in which fires can simmer for long periods. These fires can create harmful smoke and haze, which can spread throughout southeast Asia. One particularly notorious episode during the dry years of 1997–98 was exacerbated by the tropical atmospheric effects of El Niño.

The destruction of peat swamps amounts to a global warming double-whammy. Peat serves as a carbon sink, a reservoir for greenhouse gases that otherwise would be accumulating in Earth's atmosphere and adding to climate change. Research by the United Nations Collaborative Initiative on Reducing Emissions from Deforestation and Forest Degradation (UN–REDD) shows that peat is more effective than forests at sequestering carbon. Eliminating swaths of peat swamp reduces this carbon sequestration service, whether it is the result of burning the swampland or draining it, thus immediately degrading this function of the peat. By the same token, when masses of peat are consumed by fire, they release stored carbon to the atmosphere at a very high rate. Borneo, with one of the world's most massive peat areas, has thus become a UN–REDD priority area.

Sundaland Heath Forests

Acidic, sandy soils constitute the waterlogged terrain of the Sundaland heath forests, where a lack of nitrogen hinders a more complex variety of plant growth. Yet certain plants—such as sundews, nepenthes, and bladderwort—have adapted to the environment by becoming carnivores, seducing insects and even small frogs as entrees. Crocodiles, gibbons, and orangutans (the "men of the forest") inhabit this thoroughly watery region.

Sunda Shelf Mangroves

Along the coastal lowlands of Borneo—particularly where freshwater rivers, the longest being the Rajang of Sarawak, open into the seas—a remarkable ecoregion began to develop some 50 million years ago, centering on a family of trees called mangroves. The Sunda forest of mangrove trees and shrubs, or mangal, consists of an almost-unmatched assortment of endemic marine and terrestrial plants that have adjusted over time to differing degrees of the brackish water that washes the coastal plain.

The most prominent types of mangroves are Avicennia, Rhizophoras, and Nypa palms, which can be found in sizable stands in Bako National Park in Malaysia and Tanjung Puting and Gunung Palung National Parks in Indonesia. Most of the mangrove forests of Borneo, however, have been cleared for aquacultural, agricultural, or urban development.

Environmental Pressures

Extensive habitat loss and poaching, offset by new discoveries and evolutionary processes, continually alter Borneo's level of biodiversity. However, there has been a rapid decrease in the populations of large animals such as the Bornean clouded leopard (Neofelis diardi) and the island's quite rare Borneo Sumatran rhinoceros (Dicerorhinus sumatrensis harrissoni)—a subspecies of the Sumatran rhinoceros and considered a "living fossil" related to the extinct, ice-age woolly rhinoceros—boasting the unique distinction of bearing patches of reddish-brown hair. These decreases are a clear indication that human culture impairs the inimitable species-generating power of the Cool Equatorial Rainforest biome.

Even though an international consortium of green-culture enthusiasts agreed in 2007 to establish a Heart of Borneo, where about a third of the island's less accessible forested region will be preserved from industrial-scale logging, enforcement is typically not up to the task of facing down corporate interests. At the same time, plans for increases in oil, gas, and coal production; the further building of dams; expansion of palm oil plantations; and, as rumor has it, a new Indonesian capital city may finally put to rest the image of Borneo as an idyllic wilderness on the fringes of the known world.

Ken Whalen

Further Reading

Guhardja, Edi, Mansur Fatawi, Maman Sutisna, and Tokunori Mori, eds. *Rainforest Ecosystems of East Kalimantan: El Niño, Drought, Fire and Human Impacts.* New York: Springer, 2000.

Leinbach, Thomas R. and Richard Ulack. *Southeast Asia: Diversity and Development.* Upper Saddle River, NJ: Prentice Hall, 1999.

Weightman, Barbara A. *Dragons and Tigers: A Geography of South, East and Southeast Asia,* 3rd ed. Hoboken, NJ: John Wiley & Sons, 2011.

World Wide Fund for Nature (WWF). "Heart of Borneo Forests." WWF, 2012. http://wwf.panda.org/what_we_do/where_we_work/borneo_forests.

Brahmaputra River

Category: Inland Aquatic Biomes.
Geographic Location: South Asia.
Summary: The Brahmaputra River flows through Tibet, India, and Bangladesh; the flood-prone river is home to endangered species and is the focus of international wrangling over electricity generation and irrigation.

The Brahmaputra River originates on the Tibetan Plateau on the northern slope of the Himalayan mountains, where it is primarily fed by melting glaciers and snow from the high Himalayas. The river flows along the plateau, then breaks down through gorges in the Himalayas to the southern slope, leaving Tibet (China) for India, where it continues its journey toward Bangladesh. The waters of the Brahmaputra there become the Jamuna River, finally blending with the Padma before it meets the Ganges; this flow ends its roughly 1,800-mile (2,900-kilometer) journey into the Bay of Bengal.

The Brahmaputra River is a vital resource for plant and animal habitat, human irrigation and navigation, but annual floods can cause havoc for every type of community living near its flood-prone banks. As a river that runs through three countries, the Brahmaputra has also become a source of political tension as areas upstream plan to use the river in ways that may diminish its flow to dependent zones downstream.

The Brahmaputra is an example of a braided river, having many small channels that are separated by temporary and shifting islands that are essentially sediment bars. Such river types occur in systems that have either high slopes or have high loads of sediment carried by their waters. The Brahmaputra River meets both these conditions as it transports sediments from the Himalayas to the valleys and plains below. In some stretches, the river forks into separate major channels that flow more than 60 miles (100 kilometers) before rejoining. The course of the Brahmaputra changes with these shifting channels and sediment bars, which are influenced by annual floods.

Flooding can result in drastic changes of overall river width, spreading the Brahmaputra out to as much as 12 miles (19 kilometers) wide. Monsoon rains and annual river flooding serve to maintain a system of floodplain lakes; these become stocked with a variety of fish during flooding. However, increases in local flood control and irrigation—contributing to increased weed infestation and eutrophication—have in many cases reduced the productivity of such naturally occurring fish lakes. The annual flooding also creates hardships for river-adjacent communities, destroying infrastructure, homes, livestock, and lives.

Part of the 1,800-mile Brahmaputra River as it flows through India. Sediment bars and shifting, temporary islands are characteristic of the river, which is sometimes as wide as 12 miles across during flooding. (Thinkstock)

River Wildlife

Catfish, carp, featherbacks, and shad are among the 166 species of fish found in the Brahmaputra River. Giant freshwater prawn also thrive in the river and have long provided an important source of income for fishermen; however, aquatic farming of this resource has become more common than wild harvest. Other aquatic animals found in the river include the Gangetic turtle, soft-leathered turtle, the river dolphin, and the Indian gharial—all of which are endangered. The black soft-shell turtle has become extinct in the wild here. Until 2000, they lived in the wetlands of the Brahmaputra, where they fed on freshwater fish. Today, they only exist in a man-made pond sanctuary near Chittagong, Bangladesh, with a population of 150 to 300 turtles that are completely dependent on their human caretakers for survival.

The river dolphin, also known as the Ganges or Gangetic dolphin, is an endangered species numbering approximately 1,200 to 1,800 in the wild; it lives in the Ganges and Brahmaputra River systems of Nepal, India, and Bangladesh. These rivers run through one of the world's most densely populated areas, with pressures for river resources to provide water for irrigation and hydroelectricity. River habitat for the dolphins is threatened from more than 50 dams and other irrigation projects, as well as increased siltation and water pollution. The dolphins are also killed by being inadvertently caught in gill nets and other fishing gear.

The Indian gharial, a type of crocodile, is a critically endangered species that was previously thought to be locally extinct from the Brahmaputra River system. However, new fieldwork conducted from 2004 to 2007 found that the gharial population is fragmented, not extirpated, in the region. A limited number of individual gharials were found living in the upper Brahmaputra in isolated areas. The total wild population of Indian gharials is estimated at less than 235 individuals, down from thousands that once thrived throughout its historic range.

Threats and Conflict

With a river that spans three countries, the potential for conflict and tensions is high over issues of upstream river usage and its downstream impact. The Brahmaputra originates in the Tibet autonomous province of China, where the Chinese have built a series of dams to generate electricity and to divert river water for irrigation projects. An unexpected flood in 2000 that destroyed an island village in northeast India was blamed on a burst earthen wall dam on a Brahmaputra tributary in China, an example of the type of cross-purposes that can lead to political crisis.

For its part, China claims that its upstream regulation of the river's flow can reduce downstream floods, mitigate downstream droughts, and reduce the country's carbon footprint through increased electricity generation from a nonfossil fuel source. India also has similar pressures for building dams and diverting water for energy creation and irrigation, which then raises additional uncertainty for Bangladesh, the nation at the mouth of the Brahmaputra-Ganges system.

The political leaders of the region will have to work together more coherently in order to combat the effects of climate change. It is projected that global warming will increase the incidence and severity of both drought and flooding in areas touched by the Brahmaputra River. Faster glacial melt, shorter cold seasons and earlier snowmelt, more intense monsoons, lower overall sediment delivery, and coastal erosion are among the index of factors that must be considered and acted upon. There are also some likely positive effects to be anticipated from global warming, namely higher spring precipitation and temperatures in the Himalayan upland reaches of the river will permit extension of habitat those conditions favor—although higher levels of water in glacial lakes carries the potential of increased flooding events in those areas of the biome.

Shaunna Barnhart

Further Reading

Biswas, S. P. and Sanchita Boruah. "Fisheries Ecology of the Northeastern Himalayas With Special Reference to the Brahmaputra River." *Ecological Engineering* 16, no. 1 (2000).

Gray, Denis D. "Water Wars? Thirsty, Energy-Short China Stirs Fear." *Guardan*, April 16, 2011. http://www.guardian.co.uk/world/feedarticle/9599967.

Jianchu, Xu, Arun Shrestha, Rameshananda Vaidya, Mats Eriksson, and Kenneth Hewitt. *The Melting Himalayas.* Kathmandu, Nepal: International Centre for Integrated Mountain Development, 2007.

Saikia, B. P., B. J. Saud, M. Kakati Saikia, and P. K. Saikia. "Present Distribution Status and Conservation Threats of Indian Gharial in Assam, India." *International Journal of Biodiversity and Conservation* 2, no. 12 (2010).

Brahmaputra Valley Semi-Evergreen Forests

Category: Forest Biomes.
Geography Location: South Asia.
Summary: The Brahmaputra Valley's semi-evergreen forests are rich in species, but following exploitation by anthropogenic activities such as agriculture, fire clearing, livestock grazing, and tea plantation, the habitats are threatened by climate change.

The Brahmaputra Valley semi-evergreen forests are one of the 200 global ecoregions in India, supporting significant biological diversity even after many years of forest degradation. They lie in the foothills of the eastern Himalayas along the alluvial plains of the upper Brahmaputra River. The region can be characterized as tropical moist broadleaf forests covering an area of 21,892 square miles (56,700 square kilometers). It is situated in the eastern state of Assam, stretching to the neighboring states of Arunachal Pradesh, Nagaland, and southern reaches of the country of Bhutan. The plains of the valley are highly fertile, as the mighty Brahmaputra River (itself a main tributary of the Ganges) supports the formation of rich soils by bringing sediments from the Himalaya mountains, its point of origin. Two other rivers, Manas and Subansiri, add water to these plains. The vegetation experiences 59 to 118 inches (150 to 300 centimeters) of rainfall from June to September; local variances are generated in part by elevation.

The forests of the Brahmaputra Valley are inextricably associated with the welfare of millions of people and the rich biota here. The forests not only provide goods, but also support ecological services such as soil retention, carbon sequestration, climate regulation, and habitat for a wide range of flora and fauna. Under pressure from growing population, migration, and economic development, this biome has sustained fragmentation of its ecosystems.

Floral Biodiversity

The forest types here include such flowering evergreen plants as the genus Syzygium in the myrtle family, the Cinnamonium genus of the laurel family, the jackfruit or breadfruit family Artocarpus, and the family Magnoliaceae. Deciduous vegetation here includes such regionally common trees as the hardwood Terminailia genus, favoring riverine areas but also known to thrive in drier zones, and the typical Asian rainforest species Tetrameles nudiflora. Bamboo species such as spiny bamboo (Bambusa arundinaria) and giant bamboo (Dendrocalamus hamiltonii) also grow on the alluvial plains. The original forest types, however, occur mainly in patchwork; much has been lost because of the slash-and-burn agricultural practices employed here. Many areas of previous forest cover have been converted to grasslands by the high level of human influence. The grasslands along the Brahmaputra River often have soft and muddy soils; these wet grasslands provide habitat for such animals as the hispid hare (Caprolagus hispidus) and pygmy hog (Porcula salvania).

Faunal Biodiversity

The ecoregion has a significant percentage of unique animal species, and is targeted by the Convention on Biological Diversity to reduce the loss of biodiversity by encouraging conservation practices. The area supports various threatened large mammal species such as the tiger (Panthera tigris), Asiatic elephant (Elephas maximus), greater one-horned rhinoceros (Rhinoceros unicornis), gaur or Indian bison (Bos gaurus), clouded leopard (Pardofelis nebulosa), sloth bear (Melursus ursinus), and Asiatic black bear (Ursus thibetanus).

A hallmark monkey species here is the golden langur (Trachypithecus geei), restricted to the patch of semi-evergreen and tropical forest on the northern bank of the Brahmaputra River between where the Sankosh and Manas Rivers descend from the mountains. The Brahmaputra River here is host to the South Ganges river dolphin (Platanista gangetica gangetica), a freshwater type that is endangered; it is known as the only type of cetacean to swim on its side.

The biome is a favorite place for 370 species of birds, two of which are near-endemic (found exclusively in this biome): the marsh babbler (Pellomeum palustre) and the manipur bush quail (Perdicula manipurensis). The International Council for Bird Preservation has designated the Assam Plains an Endemic Bird Area. The overlapping Brahmaputra Valley semi-evergreen forests contain three restricted-range bird species. The endangered Bengal florican (Houbaropsis bengalensis) is also found here, strictly limited to protected areas. It has been reported that any damage or conversion of grasslands supporting the species could lead to its extinction.

Protected Areas and Threats

The ecoregion has 12 protected areas, the largest of which are Manas, Dibru–Saikowa, Kaziranga, and Mehao. Mehao is distributed between two ecoregions and falls into International Union for Conservation of Nature (IUCN) Category IV. The Kaziranga National Park is the largest protected area for the Indian rhinoceros, but seeks additional protection support to strengthen the security of the species.

The Brahmaputra Valley semi-evergreen forests are one of the eastern Himalayan hot spots of biodiversity. The Himalayan region has shown consistent warming trends during the past 100 years, with some research showing a higher-than-global-average increase rate here. It has further been recorded that the higher elevations of the Tibetan plateau are sustaining higher-than-average temperature increases. These warming trends bring with them glacier loss, earlier snowmelt patterns, alterations in water flow and sediment deposition, and thus the likelihood of multiplying impact on habitats downstream and at all elevations.

The Brahmaputra Valley ecoregion has already suffered considerable degradation of its forests due to land-use and land-cover changes, which in turn feed back into local and regional climate change impacts. The forests have a strong and complex role in regulating weather patterns; disruption of forest carbon cycle roles can add to temperature rise, shift rainfall patterns, and undermine the integrity of various habitats. The ultimate result is stress on the biota, and potential loss of species diversity.

Human communities also face myriad challenges in maintaining sustainable cultivation practices because of the modern crisis of urban development. The secondary forest lands of the ecoregion need climate-change mitigation practices such as community-based forest management and joint forest management. The measures required to respond to the changing climate include enhanced farming blending both contemporary techniques and traditional knowledge; conservation ethics applied at the local community and broad landscape levels; and input, support and enforcement from regional, national, and international governments, scientists, and nongovernmental institutions.

MONIKA VASHISTHA

Further Reading

Chatterjee, S., et al. *Background Paper on Biodiversity Significance of North East India for the Study on Natural Resources, Water and Environment Nexus for Development and Growth in North Eastern India*. New Delhi, India: World Wildlife Fund, 2006.

Chettri, N., et al. *Biodiversity in the Eastern Himalayas: Status, Trends and Vulnerability to Climate Change.* Kathmandu, Nepal: International Centre for Integrated Mountain Development, 2010.

Franco, Aldina M. A., et al. "Impacts of Climate Warming and Habitat Loss on Extinctions at Species' Low-Latitude Range Boundaries." *Global Change Biology* 12, no. 8 (2006).

Myers, Norman. "Threatened Biotas: 'Hotspots' in Tropical Forests." *The Environmentalist* 8, no. 3 (1988).

Brigalow Tropical Savanna

Category: Grassland, Tundra, and Human Biomes.
Geographic Location: Australia.
Summary: This biodiverse woodland region, running through a long stretch of Australia, is home to several endangered species that face further dangers in the future.

The Brigalow Tropical Savanna biome consists of the Northern and Southern Brigalow Belts, expanses of acacia-wooded grassland separating the tropical rainforest of the northeastern Australian coast from the semi-desert region of Queensland. The Southern Belt extends just past the Queensland/New South Wales border, where the eastern Australian temperate forests begin.

At 132,000 square miles (343,000 square kilometers) the Brigalow tropical savanna covers a wide area that includes sand and clay deposits, basalt and alluvium, the coal-producing Bowen Basin, the Great Artesian Basin, the sandstone gorges of the Carnarvon Range (part of the Great

Dividing Range); and the agriculturally rich Darling Downs. The biome is drained by the Fitzroy River system, running eastward toward the coast; the Belyando and Burdekin Rivers, near the tropic; and the Maranoa, Warrego, and Condamine rivers, draining into the Murray-Darling basin in the southwest.

Vegetation

Brigalow (Acacia harpophylla) is an endemic (unique to this biome) tree here, growing up to 82 feet (25 meters) tall in clay soils. It forms extensive open-forest communities with gidgee (Acacia cambagei); these forests are also home to smaller populations of A. tephrina, A. georginae, and A. argyrodendron. Overstory species in the forests include eucalyptus such as the coolabah (Eucalyptus microtheca) and Dawson River blackbutt (Eucalyptus cambageana), as well as the belah (Casuarina cristata). In the western regions, where rainfall is heavier, gidgee outnumbers brigalow. In the more humid valleys, brigalow mixes with C. cristata. Though gidgee has a limited capacity to resprout after fire damage, brigalow resprouts freely from the butt, roots, and living stems if necessary.

The brigalow once covered most of this area, but has typically been cleared by European settlers to make way for agricultural land and a more diverse woodland landscape. Much of the once-extant wildlife has become extinct, endangered, or threatened as a result of the destruction of habitat. The grasslands continue to be threatened today from introduced pasture grasses like buffelgrass (Pennisetum ciliare) and other invasive weeds. The clearing of brigalow and poplar box (Eucalyptus populnea) is a current threat, while protected areas are often limited to small park areas.

Fauna

Animals in the area include the wingless dung beetle (Onthophagus apterus) and numerous wallabies. The bridled nail-tail wallaby (Onychogalea fraenata), an endangered species, lives in the Taunton and Idalia National Parks, and is of scientific interest because of its unusually hardy immune system, which is better able to fight parasites and viruses than that of other marsupials. Another endangered mammal, the burrowing northern hairy-nosed wombat (Lasiorhinus krefftii), lives in Epping Forest National Park. Although its range at the start

A creek junction in the Southern Brigalow Belt in North Queensland, Australia, in 2009. While the Brigalow tropical savanna may be partially protected from the effects of global warming because of its location between a rainforest-bordered ocean and arid land, the area's endemic species are threatened by agricultural activities such as grazing and land clearing, and by flooding from the construction of large dams. (Flickr/Ian Sutton)

of the 20th century comprised most of New South Wales, Victoria, and Queensland, today the total population is thought to be less than 200. One of the threats to the wombat, apart from human encroachment, has been the introduction to its habitat of African buffelgrass, which outcompeted the grasses on which the wombat feeds.

The Fitzroy River turtle (Rheodytes leukops) is an endemic reptile, found only in the Fitzroy River watershed, and the sole survivor of the Rheodytes genus. Able to absorb oxygen from the water while submerged, through special subcutaneous sacs, it is an omnivore with a taste for everything from freshwater sponges to crustaceans. The turtle is also known for its acute sense of smell. Human activity, mainly agriculture and mining, has had a major effect on the turtle. The clear, clean water it was used to has become turbid and polluted; portions of the Dawson River that form part of its habitat have registered a water alkalinity range deleterious to the turtle's skin; and high levels of salinity, which are frequent by-products of forest clearing, have also impacted the species. It is currently unclear whether the species is able to propagate successfully in the wild.

A serious threat to the Fitzroy River turtle is the proposed Nathan Dam, which will be Queensland's fourth-largest dam. It will dam the Dawson River to irrigate 74,130 acres (30,000 hectares) of agricultural crops, primarily cotton. Pesticide and fertilizer in the runoff, joined by high levels of sediment, will jeopardize not only the turtles, but migratory birds and many species of fish, some of them endemic.

Global warming is not projected to markedly alter the Brigalow tropical savanna, due in part to its protected position between the rainforest-bordered ocean and the arid lands at its opposite extent. Endemic species here are not likely to be unduly stressed by climate change as it is currently understood. More dramatic threats are seen, however, as coming from land clearing, agricultural practices that favor monoculture and invasive species, displacement of native species as a secondary effect of grazing, flooding from dam construction, and habitat fragmentation.

Bill Kte'pi

Further Reading

Burrows, W. H., J. C. Scanlan, and M. T. Rutherford. *Native Pastures in Queensland: Their Resources and Management.* Brisbane, Australia: Queensland Government Press, 1988.

Grice, T. C. and S. M. Slatter, eds. *Fire in the Management of Northern Australian Pastoral Lands.* St. Lucia: Tropical Grassland Society of Australia, 1990.

Huntley, Brian J. and Brian Harrison Walker, eds. *Ecology of Tropical Savannas.* New York: Springer-Verlag, 1982.

Moll, Don and Edward O. Moll. *The Ecology, Exploitation, and Conservation of River Turtles.* New York: Oxford University Press, 2004.

Brisbane River

Category: Inland Aquatic Biomes.
Geographic Location: Australia.
Summary: Australia's Brisbane River features a variety of freshwater and estuarine habitats, which are threatened by habitat destruction and the impact of both upland and urban area dynamics.

The Brisbane River is the longest river in the northeastern Australian state of Queensland, flowing 214 miles (344 kilometers) from its source near Mount Stanley in the Great Dividing Range into Moreton Bay and the South Pacific Ocean. The lowest 50 miles (80 kilometers) form a flood-prone tidal stretch of the river. Draining a catchment of about 5,200 square miles (13,500 square kilometers), the Brisbane River ecosystem contains both freshwater and saltwater habitats. It flows through the port city of Brisbane, a fast-growing metropolitan area with the largest concentration of the catchment's one million inhabitants. Dredging and urban growth are the increasing environmental threats in downstream segments; upland concerns include clearing for agriculture, grazing, and logging. Both zones are impacted by the effects of global warming.

Vegetation Types

Overall, more than 85 percent of the Brisbane River watershed is cleared land. The furthest upland reaches of the Brisbane River watershed contain the least-disturbed native bush and forests. Alongside wild forest stands, replete with such hallmark varieties as blackbutt (Eucalyptus pilularis), forest red gum (Eucalyptus tereticornis), and tallowwood (Eucalyptus microcorys), are found such managed features as plantation forests of hoop pine (Araucaria Cunninghamii), a fine, straight-grained species cultivated for paneling and plywood use.

Middle reaches of the river system are more heavily impacted by agricultural land use. Two dams—Somerset and Wivenhoe—have also altered habitats in this area of the biome, with the trade-off being some improvement in flood control and mitigation. Farm runoff increases turbidity in tributaries and the mainstem, also challenging the nutrient balance of the river, augmenting the spread of certain bacteria and phytoplankton. Hoop pine plantation forests are still found here, but have given way to protected forest reserves in some quarters, such as the Benarkin State Forest.

In its estuarine stretch, the river winds through the heart of densely built-up Brisbane and into Moreton Bay. Here is found the Boondall Wetlands Reserve, of which a 2,500-acre (1,000-hectare) portion was recognized in 1993 as a Ramsar Wetland of International Importance. Mangroves, tea tree swamp (melaleuca wetland), salt marsh, grasslands, and open forest compose this transition area, supporting a full range of ecosystem services and a great diversity of fauna. Some characteristic plants here are the grey mangrove (Avicennia marina), swamp lily (Crinum pedunculatum), native jute (Corchorus cunninghamii), and angle-stemmed myrtle (Austromyrtus gonociada). The water hyacinth, historically introduced to the river from South America in the 19th century, has periodically grown out of control across the Brisbane River's surface.

Distinctive Fauna

In the mangroves and seagrass around Moreton Bay, the dugong (Dugong dugon) is a much watched-for marine grazer. Signature fish species found here and in the tidal zone of the Brisbane River include the Queensland lungfish (Neoceratodus forsteri), considered rare, and the aggressively predatory bull shark (Carcharhinus leucas), considered all too prevalent in terms of human interaction; swimming is discouraged in parts of the river where some fatalities have been ascribed to the bull in the past.

One of Australia's major populations of loggerhead turtle (Caretta caretta) is based in and around Moreton Bay. Its numbers are joined by the hawksbill (Eretmochelys imbricata) and green turtle (Chelonia mydas).

Notable extinct species of the river include the unique freshwater Brisbane River cod (genus Maccullochella), which suffered from overfishing and water quality degradation. A related subspecies of cod from the nearby Mary River (Maccullochella peelii mariensis) has been introduced in order to restock the Brisbane River. These fish are related to the largest freshwater fish of Australia, the Murray cod (Maccullochella peelii peelii).

The eel-tailed catfish (Tandanus tandanus) actually thrives in the turbid waters that deter many other species. It is also a broad feeder, consuming mollusks, shrimp, insects, and snails. The blue catfish (Arius graeffei) prefers more pristine environments; the empire gudgeon (Hypseleotris compressa) is often found in calm waters such as areas above dams.

A unique mammal in the mangrove area here is the grey-headed flying fox (Pteropus poliocephalus). Also known as fruit bats, they have thickly colonized the estuary zone of the Brisbane River. Osprey (Pandion haliaetus) nest here as well, and are commonly seen arcing over Moreton Bay, intent on their hunting missions, targeting fish that swim near the surface. Migratory wading birds such as the eastern curlew (Numenius madagascariensis) and grey-tailed tattler (Tringa brevipes) thrive here on stopover during their flights from Arctic or sub-Arctic breeding regions. The population here swells to about 50,000 during the nonbreeding season.

Human Impact

Water quality concerns in the city of Brisbane began to surface by the early 20th century as human

A grey-headed flying fox (Pteropus poliocephalus), or fruit bat. Their wingspan can reach 3.3 feet (1 meter), and they can live for as long as 15 years in the wild. (Flickr/Greg Schechter)

settlement increased both around the city and in communities much further upstream. Pollutants include pesticides, excess presence of hydrocarbons, bacteria, and nutrients. The river's high turbidity levels are also the result of human encroachment, the clearing of native vegetation, and the introduction of exotic species. The relationship between dredging the river for navigation and the growing turbidity of its waters led to the decision to halt dredging upstream of the city. Numerous land reclamation projects have also been instituted; the protection and replanting of native vegetation is an important component of water quality restoration initiatives because of the natural filtering role of many such species.

Global warming is projected to increase average rainfall, although its ultimate effect on the seasonality of precipitation in northeastern Australia is not clear. Sea-level rise, on the other hand, will clearly impact the lower reaches of the Brisbane River, threatening many upstream habitats with saltwater incursion while at the same time challenging the mangrove and other estuarine habitats with erosion and inundation. In the uppermost reaches of the watershed, higher average temperatures could lead to elevation-shifting among plant and animal communities as they vie to maintain their optimum climate regimes.

Marcella Bush Trevino

Further Reading

Davie, Peter, and Errol Stock, eds. *The Brisbane River: A Source-Book for the Future*. Moorooka: Australian Littoral Society and Queensland Museum, 1990.

Gregory, Helen. *The Brisbane River Story: Meanders Through Time*. Brisbane: Australian Marine Conservation Society, 1996.

Nock, Catherine J., Martin S. Elphinstone, Stuart J. Rowland, and Peter R. Baverstock. "Phylogenetics and Revised Taxonomy of the Australian Freshwater Cod Genus, *Maccullochella* (Percichthyidae)." *Marine and Freshwater Research* 61, no. 9 (2010).

Steele, John Gladstone. *The Brisbane River*. Adelaide, Australia: Rigby, 1975.

British Columbia Mainland Coastal Forests

Category: Forest Biomes.
Geographic Location: North America.
Summary: Extensive snowpack, glaciers, and rainforests grading to drier, rain-shadow forests characterize this spectacularly scenic and relatively intact ecosystem.

The British Columbia mainland coastal forests extend from the rugged North Pacific shoreline of western Canada inland for some 100 miles (160 kilometers). The inside passage to Alaska, a waterway between the mainland and Vancouver Island, extends alongside much of this biome. The landmass is buttressed by ancient granite, crowned with extensive glaciers and alpine meadows, over what is primarily a forested, continuous north-south mountain range the Coast Mountains, wet on the ocean side and drier on its eastern side, in the lee of the Pacific Ocean's moisture. From Puget Sound in the state of Washington on the south to the Yukon on the north, a distance of nearly 1,000 miles (1,600 kilometers), few areas present greater opportunity for the possible intact survival of an ecosystem in the face of climate change.

Extensive fjords make up the long coastline, which is similar to the coast of Norway, except that these fjords are largely sheltered from ocean storms by extensive offshore islands. Also different from Norway is the survival of much of the forest—as much as 40 percent is considered fully intact habitat—from logging, due primarily to the extremely steep and rugged nature of the mountains on which these forests grow. Rich biological resources, with such iconic features as cedar trees and salmon, have long supported Native American peoples here. Current threats also revolve around these same two resources, in the form of overfishing, fish farming, and clear-cut logging. Salmon and cedar trees are related because the salmon contribute significantly to forest fertility. Totem poles and large oceangoing dugout cedar canoes are the most familiar symbols of this culture.

The British Columbia mainland coastal forests encompass about 47,000 square miles (122,000 square kilometers). Elevation ranges from sea level to the tree line on the shoulders of Mount Waddington, the highest peak in the Coast Mountains at 13,186 feet (4,019 meters). The climate of the biome is temperate rainforest, with mild winters and cool summers. Rainfall of 58 to 138 inches (1.5 to 3.5 meters) annually makes this one of the most moist ecosystems outside of the tropics.

Plant Types

Vegetation regimes in the biome are generally classified according to three elevation zones: coastal plain forest, mountain forest, and alpine tundra. These can be further subdivided by moisture and temperature into such habitats as western slope forests, riparian areas, alpine meadow, and tree line scrub.

Western hemlock (Tsuga heteropylla), western red cedar (Thuja plicata), and amabilis fir (Abies amabilis) are typical arboreal denizens of the coastal plain forest. Below them are found ferns, flowering berry shrubs, a multitude of mosses and fungi. Riparian areas and other clearing areas often harbor a scattering of deciduous trees such as maples and cottonwoods. These low-lying habitats give way at higher elevations to a mix of mountain hemlock (Tsuga mertensiana), Alaska yellow cedar (Chamaecyparis nootkatensis), and amabilis fir. At the highest extent of the biome are found alpine tundra meadows that feature varieties of lupine (Lupinus spp.), sedge (Carex spp.), and colonies of moss and lichens.

Fauna of the Forest

Most stunning among the animal species here is the rare Kermode bear (Ursus americanus kermodei), a black bear subspecies known to the indigenous people here as the spirit bear. There are thought to be not more than 1,000 individuals in the wild. Other large mammals include the grizzly bear (Ursus arctos), moose (Alces alces), grey wolf (Canis lupus), and the woodland caribou (Rangifer tarandus caribou). Smaller denizens range from the red fox (Vulpes vulpes) and black-tailed deer (Odocoileus hemionus) to the American mink (Mustela vison), marten (Martes americana), and northern river otter (Lontra canadensis).

An iconic species of this well-watered biome is the coho salmon (Oncorhynchus kisutch). Its core status is earned, in part, by virtue of its ability to robustly fertilize almost any area of the forest through the distribution of its internal nutrients when excreted by the predators that have hunted it in downstream environments and brought its meat up and across the forest.

Avian species here range from the trumpeter swan (Cygnus buccinator), a very cold-tolerant waterfowl that favors estuary areas for its tidal feeding cycles; sandhill crane (Grus canadensis), at home in the peat bogs and wetlands of the coastal plain; to the northern saw-whet owl (Aegolius acadicus), an endangered species most comfortable in old-growth forest and especially on the many offshore islands that are an extension of the biome; and the bald eagle (Haliaeetus leucocephalus), which tends to overwinter in the rainforest.

Threats and Preserves

Significant levels of urban development encroach on the southern reaches of the British Columbia mainland coastal forests. Mining and logging have been and still are a threat to the ecology of the area. Logging tends to occur in low-altitude ranges. The region's intertidal and estuary zones are also affected by pollution from shipping and recreation craft.

A network of parkland and associated reserves help maintain the forest in its least-fragmented state; among the largest such areas are the Kitlope and Garibaldi Provincial Parks and the Hakai Recreation Park. Together, these three comprise some 2,400 square miles (6,200 square kilometers).

The British Columbia mainland coastal forests, as a mainly rainforest biome and thus a rich carbon sink, has the benefit of making positive contributions to the global warming challenge. On the other hand, it is predicted that the coastal areas of British Columbia will experience more frequent and intense rainstorms, as well as greater sea surges during storms. Overall precipitation, however, could potentially decrease and/or alter in seasonal patterns—this being the climate-change factor that could most dramatically stress this biome in its relatively stable current state.

ERIC BURR

Further Reading

Burr, Eric. *Ski Trails and Wildlife: Toward Snow Country Restoration.* Victoria, Canada: Trafford Press, 2008.

Cannings, Richard and Sidney Cannings. *British Columbia: A Natural History.* Toronto: Greystone Books, 2004.

Lyons, C. P. and Bill Merilees. *Trees, Shrubs, and Flowers to Know in Washington and British Columbia.* Vancouver: Lone Pine Publishing, 1995.

McAllister, Ian. *Following the Last Wild Wolves.* Vancouver: D&M Publishers, 2011.

Young, Cameron. *The Forests of British Columbia.* North Vancouver: Whitecap Books, 1985.

Brooks-British Range Tundra

Category: Grassland, Tundra, and Human Biomes.
Geographic Location: North America.

Summary: The Brooks-British Range tundra is the dividing line between interior Alaska and the Arctic. The region is full of magnificent, majestic, and spectacular mountain ranges, rivers, lakes, meadows, and wildlife.

The Brooks-British Range tundra, located in Alaska and Canada, is the northernmost mountain system of the North American Cordillera. The Brooks Range is considered to be the northern extension of the Rocky Mountains. The Brooks-British Range tundra extends about 60 miles (100 kilometers) from the Chukchi Sea eastward across northern Alaska into northern Yukon, and extends a shorter distance into the Northwest Territories. In Alaska, the Brooks Range separates interior Alaska from the Arctic.

The Brooks-British Range ecoregion is a combination of three large areas, including the Western Brooks; the Eastern Brooks Range/British Range; and the lower area of Anaktuvuk Pass, which divides these mountainous ranges. The Western Range has less rugged mountains compared with the Eastern Brooks Range/British Range. This mountainous belt includes the DeLong, Richardson, and British mountain subranges. The Canadian portion corresponds with the British-Richardson Mountain ecoregion. This area of Alaska has several small, predominantly Native Alaskan villages with a total population of less than 500, including Anaktuvuk (population 249) and Arctic Village (152), as well as the communities of Coldfoot (13), Bettles (43), and Wiseman (21). The Canadian side has no permanent settlements.

Climate and Geography

As a biome, the Brooks-British Range tundra spans 61,600 square miles (159,543 square kilometers). The climate of the region is continental sub-Arctic, with temperature variations corresponding to elevation. This type of climate has long, extremely cold winters and short, cool summers. Because the region is so large and includes various low-lying and mountainous regions, average temperatures vary. The average annual temperature in the Brooks-British mountains is 14 degrees F (minus 10 degrees C). The average winter temperature is minus 13 degrees F (minus 25 degrees C), and the average summer temperature is 44 degrees F (7 degrees C). Downdrafts from the mountain passes coupled with cold temperatures can result in severe wind-chill conditions.

Annual precipitation is 7 to 15 inches (180 to 390 millimeters), with the higher elevations and southern-facing slopes receiving more precipitation. The Brooks Range climate is comparable to an Arctic coastal plain, with increased precipitation occurring on the eastern side and at higher elevations.

The Brooks-British Range tundra has been shaped by the folding and faulting of sedimentary rocks, which created several groups of steep, rugged slopes with deeply dissected mountains and ranges from the Cretaceous Period. The elevation extends from 1,600 feet (488 meters) in low valleys up to Mount Isto or Mount Chamberlin, which reach approximately 9,000 feet (2,743 meters). The overall terrain lies from 2,624 feet (800 meters) in the Baird Mountains to the west to 7,800 feet (2,378 meters) in the central and eastern reaches. Small glaciers are scattered throughout the region at around 6,000 feet (1,829 meters); these are remnants of the Quaternary glaciations. Most of the Canadian region is unglaciated. Continuous deep permafrost dominates the region.

Above the permafrost, the dominant soils in the valleys are made up of glacial till and seasonally flooded sandy loams and the slopes of loose sediment and fragmented bedrock. This tundra is sparsely covered with vegetative material due to poor soil, windy conditions, cold temperatures, and the limited growing season. The Brooks-British Range is a drainage dividing line or continental divide, with the northern areas draining to the Arctic Ocean and the southern areas draining to the North Pacific.

Flora

The Brooks-British Range tundra is characterized by low biotic diversity and simple vegetation structures that can survive in its cold and mountainous terrain. Most vegetation grows in the valleys and on the lower slopes. The vegetation consists of prostrate and hemiprostrate dwarf shrubs such as skeletonleaf willow, white Arctic mountain heather, and bog blueberry, as well as erect dwarf shrubs such as Richardson's willow, grayleaf willow, and Lapland rosebay. Other common types of vegetation in the valleys, scree slopes, and alpine areas include forbs such as large-flowered wintergreen, arctic lupine, and alpine bistort; small awned sedge; purple mountain saxifrage; eight-petal mountain-avens; horsetails; lichens such as Arctic kidney, alpine snow, and globe ball; and mosses including splendid feather, turgid aulacomnium, and dicranum. All vegetation is vulnerable to frost at any time.

The southern slopes harbor black spruce and quaking aspens; this area is considered to be the northernmost limit for these trees. Trees in the subalpine areas include subalpine fir, lodgepole pine, stunted white spruce, and dwarf birch. Other than balsam poplar, very few trees grow on the northern drainage side of the Brooks-British Range tundra.

Fauna

The tundra supports several mammalian species. Porcupines and central and western caribou herds seasonally migrate through the river valleys. Common mammals include moose, mountain goats, and Dall sheep. Predators include brown and black bears, red and Arctic foxes, wolves, wolverines, lynxes, and martens. Smaller mammals include common and dusky shrews, Alaska and hoary marmots, northern red-backed and singing voles, ground squirrels, lemmings, pika, and snowshoe and tundra hares.

Avian species include tundra swans, long-tailed jaegers, rock and willow ptarmigans, northern shrikes, American golden plovers, grebes (horned and red-necked), American pipits, sparrows (savannah and white-crowned), redpolls (common and longspur), and horned larks. Birds of prey here are golden eagles, owls (great horned and snowy), marsh hawks, and gyrfalcons. Other common birds are black-billed magpies, common ravens, gray-cheeked thrushes, and sparrows (American tree, white-crowned, and fox).

The only amphibian is the wood frog. There are no reptiles.

Land Use and Conservation Status

Land use consists mostly of big-game/sport hunting and subsistence use. The area is considered to be relatively stable and intact. The Dalton and Dempster Highways run through the area, as does the Trans-Alaska Pipeline, which parallels the Dalton. Beyond the road systems, disturbance to the area is minimal. Threats to the area are small mining operations and increased numbers of recreational-park users damaging the fragile vegetation.

The Brooks-British Range tundra has several large protected areas. In Canada, protected areas include Ivvavik National Park and Vuntut National Parks. In Alaska, the Noatak basin was first proclaimed a national monument in 1978 and later a national preserve in 1980. The Noatak National Preserve covers 6.5 million acres (2.6 million hectares) and is the largest intact wilderness and watershed in North America. On the border of Noatak National Preserve are Kobuk Valley National Park and Gates of the Arctic National Park. Kobuk Valley National Park is 1.7 million acres (675,749 hectares) and was also established in 1980. Gates of the Arctic National Park is the second-largest national park, at 8.4 million acres (3.4 million hectares); like the Noatak basin, it was proclaimed a national monument in 1978 and a national park in 1980. This park is located entirely in the Arctic Circle.

All the U.S. areas were established under the Alaska National Interest Lands Conservation Act of 1980. Of these three national parks, Noatak is the only one that allows sports hunting.

There are also several federally designated wild and scenic rivers in the Brooks Range.

The two Canadian parks cover only part of the Brooks-British Range tundra's extent in Canada. Ivvavik National Park covers 2.5 million acres (1.0 million hectares) and was established in 1984. It lies on the shores of the Beaufort Sea and is adjacent to the Arctic National Wildlife Refuge. Ivvavik means nursery, and the park was created to protect the caribou herds' calving areas as well as a native land-claims settlement. Established in 1995, Vuntut National Park covers 1.0 million acres (434,500 hectares) and shares parts of its borders with Ivvavik National Park.

It is widely thought that global warming will increase the risk and spread of wildfires in the Brooks-British Range tundra, due to somewhat higher average summer temperatures, increased lightning activity, and drier surface conditions. Broader spread of fire will release more carbon into the atmosphere, a positive feedback effect of the greenhouse gas function. On the ground, fires will alter habitats and thereby put pressure on species to adapt or move. The tundra biome may actually increase in size as forests disrupted by wildfire convert to grasslands.

ANDREW HUND

Further Reading

Lange, Otto L., Sabine C. Hahn, Angelika Meyer, and John D. Tenhunen. "Upland Tundra in the Foothills of the Brooks Range, Alaska, U.S.A.: Lichen Long-Term Photosynthetic CO_2 Uptake and Net Carbon Gain." *Arctic and Alpine Research* 30, no. 3 (1998).

U.S. Forest Service. "Wildland Fire and Climate Change." June 27, 2011. http://www.fs.fed.us/ccrc/topics/wildland-fire.shtml.

Walker, Marilyn D., Donald A. Walker, and Nancy A. Auerbach. "Plant Communities of a Tussock Tundra Landscape in the Brooks Range Foothills, Alaska." *Journal of Vegetation Science* 5, no. 6 (1994).

Buru Rainforest

Category: Forest Biomes.
Geographic Location: Southeast Asia.
Summary: An island tropical rainforest home to unique Asian and Australian fauna, the Buru appears to face low levels of conservation threats.

North of Indonesia's Banda Sea and south of the Seram Sea lies Buru Island, a mountainous oval-shaped formation with an area of roughly 3,670 square miles (9,500 square kilometers). The Buru Rainforest, designated part of the Wallacean biogeographic zone, is home to a unique mix of Asian and Australian fauna. The relative isolation of the island, as well as the presence of two large protected areas, lends a stable outlook to its conservation prospects.

The two most significant bodies of water within the island are the Apo River and Lake Rana. Apo, which at 50 miles (80 kilometers) is Buru's longest river, winds northeast toward Kayeli Bay. Rana is a freshwater lake located almost dead center of the island at an elevation of 2,516 feet (767 meters). The island is mostly mountainous terrain, with the highest elevation being 7,969 feet (2,430 meters) at the top of Mount Kapalamadan. Flat areas, where the island's inhabitants are concentrated, exist mostly on the coast and in the valley around the Apo River.

Buru Island falls under a tropical wet climate regime, influenced largely by equatorial monsoons that bring the most rain from October to April. Due to the irregular terrain, precipitation amounts and patterns vary from the coast to higher elevations. In general, rain falls more than 90 days per year, totaling 60–100 inches (152–254 centimeters), while temperatures average about 80 degrees F (about 27 degrees C). These conditions result in very high humidity (70–90 percent), particularly within the rainforest.

Canopy of Dipterocarps

The Buru Rainforest is dominated by lowland evergreen and semi-evergreen trees, most of which are hardwood tropical trees in the dipterocarp family. Dipterocarps, which make up the emergent and canopy layers of the rainforest, have smooth, straight trunks that rise hundreds of feet toward the top of the forest and sunlight. Branches bearing small leaves suited to abundant sunshine appear only near the canopy, while broad trunk buttresses help to stabilize the tree in the shallow soil. Grooves and drip spouts at the leaf tips help keep the leaves dry and prevent mold growth in the constant rain and humidity.

A peculiar characteristic of dipterocarps is masting, a reproductive adaptation that involves simultaneous flowering of almost all the trees once every two to seven years. This phenomenon is believed to be triggered by patterns in temperature and relative drought. The result is a mass production of fruits and seeds that fall to the ground, ensuring a good supply of seedlings for species propagation. These seedlings can survive for many years in the deep shade of the forest floor, and possess a capacity for rapid growth once a gap in the canopy opens. Common genera of dipterocarps in the Buru Rainforest include Anisoptera, Hopea, Shorea, and Vatica.

Thick lianas, bromeliads, orchids, and other epiphytes complete the canopy layer. These plants grow on the branches of larger trees to get sunlight and water; they draw nutrients from the air and their host plants. Specialized leaves, like those found in pitcher plants, allow these varieties to take advantage of the humid conditions in the canopy to capture and keep abundant moisture.

Small-Sized Fauna

The majority of Buru's smaller animals, such as birds, tree frogs, bats, and snakes, live in this food-rich layer and never venture into the lower strata of the rainforest. Common adaptations include bright colors in distinctive patterns, loud vocalizations, and fruit-heavy diets.

Out of the 178 bird species that have been identified in Buru, 29 are endemic (found uniquely in a biome) or near-endemic to the area. Among these,

The wild pig Buru babirusa (Babyrousa babyrussa) is one of four mammals that have been found to be endemic to Buru Island. (Wikimedia/Masteraah)

six species are considered endangered or vulnerable: black-lored parrot (Tanygnathus gramineus), Moluccan scrubfowl (Megapodius wallacei), blue-fronted lorikeet (Charmosyna toxopei), Buru cuckoo-shrike (Coracina fortis), streaky-breasted jungle-flycatcher (Rhinomyias addita), and rufous-throated white-eye (Madanga ruficollis).

Three endemic bat species exist on the island: the Moluccan flying fox (Pteropus chrysoproctus), the Seram fruit bat (Pteropus ocularis), and the lesser tube-nosed bat (Nyctimene minutus). The latter two species are listed as threatened or vulnerable.

Beneath the canopy is the understory or lower canopy, a shady, cool environment with little air movement and high humidity. Plants in this layer include saplings, large-leaved shrubs, mosses, ferns, and epiphytes that thrive in low sunlight.

The lowermost rainforest strata is the forest floor, a completely shaded environment that is home to termites, fungi, worms, ant colonies, and other small animals that thrive on decomposing organic matter. Despite the large amounts of leaf litter that fall to the ground, and the constant activities of the decomposers, the soil in the rainforest is very low in nutrients and poor in quality. Any nutrients that are released are immediately absorbed by the roots of surrounding trees.

The cool, damp forest floor is where the largest animals are found. One of the four mammals so far identified to be endemic to Buru Island is the wild pig Buru babirusa (Babyrousa babyrussa), which the island's residents hunt for its meat.

Human Activity

Agriculture, logging, and residential activities have cleared away the old-growth forests in the coastal plains and northern parts of the island. However, the presence of many endemic species in the Buru Rainforest biome has afforded interest from scientists and conservationists alike. Two protected zones, Gunung Kapalat Mada and Waeapo, contain 550 square miles (1,424 square kilometers) of stable rainforests and offer a refuge for Buru's plants and animals—both those already discovered, and others believed yet to be identified. In addition, the relative isolation of the island contributes to a positive overall outlook for the preservation of this unique environmental treasure.

Still, human threats loom. Industrial-scale logging has been a factor, as has forest-clearing for agricultural uses near the settled areas. Both activities carry habitat-fragmentation risks. Global warming is projected to increase precipitation across much of the Indonesian archipelago; the impact could increase soil erosion and nutrient runoff on Buru when combined with logging and agriculture clearing. The national government has sponsored a tree-planting campaign launched in 2010; the goal is one million trembesi trees (Samanea saman) planted by 2020. However, the tree is not native to Indonesia, and side effects of the introduction of an invasive species are unpredictable.

Maricar Macalincag

Further Reading

Corlett, Richard and Richard Primack. "Dipterocarps: Trees That Dominate the Asian Rain Forest." *Arnoldia* 63, no. 3 (2005).

Newman, Arnold. *Tropical Rainforest.* New York: Checkmark Books, 2002.

Poulsen, Michael and Frank Lambert. "Altitudinal Distribution and Habitat Preferences of Forest Birds on Halmahera and Buru, Indonesia: Implications for Conservation of Moluccan Avifaunas." *Ibis* 142, no. 4 (2000).

Wikramanayake, Eric, Eric Dinerstein, and Colby J. Loucks, et al. *Terrestrial Ecoregions of the Indo-Pacific: A Conservation Assessment.* Washington, DC: Island Press, 2002.

Caatinga Dry Woodland

Category: Forest Ecosystems.
Geographic Location: South America.
Summary: The Caatinga dry woodland consists of dry forests and xerophytic woodlands in northeastern Brazil. The biome is fragile and under threat from human development, but is likely to expand due to global warming.

The term *Caatinga* comes from Tupi, the language of the indigenous people who originally lived in the region, and means *white forest,* denoting a rather open, deciduous vegetation formation. Today, Caatinga refers to the seasonal dry tropical forest biome in northeastern Brazil, covering an area of approximately 328,187 square miles (850,000 square kilometers), which corresponds to 10 percent of the Brazilian territory, and spans eight Brazilian states.

After European colonization, forest cover was reduced by timber harvest and by deforestation to open areas for cattle ranching. In consequence, today the biome consists mainly of dry shrub vegetation, mostly deciduous species in both the woody and herbaceous layers, and fragments of tall forests. Altogether, the biome includes a broad range of vegetation types, from cactus shrubland to tall forest. Some similarities exist to the Cerrado biome (central Brazilian savannas) and the Chaco in northern Argentina, western Paraguay, and southeastern Bolivia, both of which are likewise seasonal dry tropical forest biomes.

Climate in the Caatinga region is semiarid, with total annual rainfall of less than 30 inches (75 centimeters) in the larger part, and most precipitation is concentrated in the Southern Hemisphere summer months. Erratic rainfall and long drought periods, sometimes lasting several years, are common. Mean annual temperature lies around 73 to 81 degrees F (23 to 27 degrees C), with little variation around the year. Evapotranspiration is high, at 59 to 79 inches (150 to 200 centimeters) per year.

In terms of geomorphology, the Caatinga region is contained in a depression at an altitude of around 984 to 1,640 feet (300 to 500 meters), characterized by sandstone from the Cretaceous Period on Precambrian crystalline rock. Soils are heterogeneous, however—in general fertile, but with high rock and low organic-matter content. Altogether, it is clearly lack of rainfall, not soil fertility, that limits vegetation growth.

Today, we know that in the past 220,000 years, there have been several phases of rainforest expansion and retraction in what today is

Cacti growing below a rock formation in the Caatinga biome in Brazil. This biome can vary from cactus shrubland like this to tall forests, and covers about 10 percent of Brazil's land. (Thinkstock)

the Caatinga biome, as a consequence of shifting climate conditions. The biome has not been stable over the past millennia. Relicts of rainforests are the so-called *brejos de altitude* (highland swamps), which can be found at elevations above 1,640 feet (500 meters) and which, due to their geographic situation, receive orographic, or mountain-driven, rainfall of more than 47 inches (120 centimeters).

Vegetation and Fauna

The biome exhibits a wide range of vegetation physiognomies, which vary according to topography, type of soil, and rainfall. Biodiversity in the Caatinga biome is high for some species and groups. A compilation in 2005 in the journal *Conservation Biology* mentions 932 vascular plant, 510 bird, 240 fish, 167 reptile and amphibian, and 148 mammalian species. Not all parts of the biome or all species groups have been equally well studied, however.

The woody (tree and shrub) species of the Caatinga are characterized, in general terms, by the high importance of flowering species from the bean family (*Fabaceae*), the spurge family (*Euphorbiaceae*), and the cactus family (*Cactaceae*). Example of trees in the remnants of the original white forests are different Ipê or trumpet tree species (genus *Tabebuia*, family *Bignoniaceae*), with their beautiful flowers. Many of the woody plants are of short stature, with small leaves that they throw off during the dry season to minimize water loss, a strongly branched structure, and spines as defense against herbivory.

During the dry season, much of the vegetation seems to be dry and dead, conferring a somewhat hostile expression to the environment. As soon as there is some precipitation, however, plants rapidly develop leaves and flowers. Grasses and forbs produce their biomass only in the short rainy season. The herb layer is characterized by a high proportion of annual species that complete their entire life cycle in the short season with better moisture availability and survive dry periods in the form of seeds.

Among its animals, the Caatinga has high endemism (species found exclusively in this biome) in insects and fishes, but less in birds, mammals, and terrestrial invertebrates. Recently, the Caatinga woodpecker (*Celeus obrieni*) has been rediscovered; the species had been considered to be extinct for more than 80 years. The parrot Spix's macaw (*Cyaopsitta spixii*) is thought to be extinct in the wild as a consequence of habitat loss and illegal capture. An example of an endemic mammal is the Brazilian three-banded armadillo (*Tolypeutes tricinctus*), which had been considered extinct as well.

Amphibians here include the vulnerable frog species *Scinax fuscomarginatus, Adelophyrne baturitensis,* and *Odontophyrnus carvalhoi.* There are dozens of reptile species considered vulnerable or rare, such as the lizard *Leposoma baturitensis*, the snake *Atractus ronnie,* and the turtle *Mesoclemmys perplexa.*

Conservation and Human Land Use

The Caatinga is a fragile biome with respect to biodiversity and its conservation, but also with respect to land use and the situation of the human population. Large parts of the Caatinga biome are characterized by scarcity of water and are vulnerable to desertification, with very low aptitude for agricultural use. Due to the environmental conditions, the people of the Caatinga are among the poorest in Brazil. Large infrastructure projects—specifically, irrigation—are planned to improve conditions for planting fruits and crops, but intensification of land use may lead to further degradation of the biome; thus, big irrigation projects are subject to controversy and debate.

Currently, despite some considerable advances in conservation in the past few years, coverage of conservation units is still low, and not all major ecosystem types are included in protected areas. Ways need to be found to reconcile biodiversity conservation and reduction of rural poverty in the Caatinga biome.

Global warming is projected to create conditions that will lead to the expansion of the Caatinga, westward and southward, but also eastward, over the next 80 to 100 years—at the expense of rainforests there. Central parts of today's Caatinga likely may turn to desert. The fact that the Caatinga up to now has been very poorly considered in biodiversity related research can be considered a severe obstacle for addressing of environmental problems.

Fabíola Ferreira Oliveira
Gerhard Ernst Overbeck

Further Reading

Cook, Kerry. H., and Edward K. Vizy. "Effects of Twenty-First-Century Climate Change on the Amazon Rain Forest." *Journal of Climate* 21 (2008).

da Costa, Rafael Carvalho, Francisco Soares de Araújo, and Luiz Wilson Lima-Verde. "Flora and Life-Form Spectrum in an Area of Deciduous Thorn Woodland (Caatinga) in Northeastern Brazil." *Journal of Arid Environments* 68, no. 1 (2007).

de Oliveira, Guilherme and José Alexandra Felizola Diniz-Filho. "Spatial Patterns of Terrestrial Vertebrates Richness in Brazilian Semiarid, Northeastern Brazil: Selecting Hypotheses and Revealing Constraints." *Journal of Arid Environments* 74, no. 1 (2010).

Leal, Inara R., José Maria Cardoso da Silva, Marcelo Tabarelli, and Thomas E. Larcher Jr. "Changing the Course of Biodiversity Conservation in the Caatinga of Northeastern Brazil." *Conservation Biology* 19, no. 3 (2005).

Loebmann, Daniel and Célio Fernando Baptista Haddad. "Amphibians and Reptiles From a Highly Diverse Area of the Caatinga Domain: Composition and Conservation Implications." *Biota Neotropica* 10 (2010).

Werneck, Fernanda P. "The Diversification of Eastern South American Open Vegetation Biomes: Historical Biogeography and Perspectives." *Quaternary Science Reviews* 30, no. 13–14 (2011).

Cabayugan River

Category: Inland Aquatic Biomes.
Geographic Location: Asia.
Summary: The Cabayugan River in the Philippines is part of a critical mountain-to-sea ecosystem and threatened by soil erosion, illegal logging, and a decline in water quality, which the government has promised to mitigate.

The Cabayugan River flows down the slopes of Mount Bloomfield, on the island of Palawan in the Philippines, before disappearing beneath Mount Saint Paul to become an underground river for its last 5 miles (8 kilometers) before emptying into the Philippine Sea. The river has become popular with tourists because of its ease of navigability and its promotion in 2011 by the Swiss-based New-7Wonders Foundation. As a tourist destination, it is often called the Underground River; the Saint Paul Underground River; or the Puerto Princesa River. In the case of the latter, it was so named because of its association with Puerto Princesa Subterranean River National Park, located in the Saint Paul Mountain range 30 miles (48 kilometers) north of Puerto Princesa.

Until 2007, the Cabayugan was the longest known underground river in the world. The caves through which it flows include major stalactite and stalagmite formations and several large chambers; the lower portion of the river is tidal. The ecosystem here was named a World Heritage Site by the United Nations Educational, Scientific and Cultural Organization (UNESCO) in 1999, and the river contains important habitats for biodiversity conservation.

Vegetation and Wildlife

The lowland forest of the Cabayugan River includes large trees such as the dao (*Dracontomelon dao*), dita (*Alstonia scholaris*), and apitong (*Dipterocarpus gracilis*). Terrestrial plants include the almaciga (*Agathis philippinensis*) and rattan (*Calamus*); shoreline species here include Indian beech (*Pongamia pinnata*) and bitaog (*Calophyllum inophyllum*).

Numerous animals live in the vicinity of the river, among them at least 165 species of birds—including all 15 of the island's endemic (found only here) bird species—such as the blue-naped parrot (*Tanygnathus lucionensis*) and Palawan hornbill (*Anthracoceros marchei*). The nearby forest canopy is home to the island's only primate, the long-tailed macaque (*Macaca fascicularis*), as well as the bearded pig (*Sus barbatus*) and the Palawan stink badger (*Mydaus marchei*).

Reptiles include eight endemic species and 11 others, including the monitor lizard (*Varanus salvator*), green-crested lizard (*Bronchocoela cristatella*), and reticulated python (*Phython reticulatus*). The river itself is occasionally home to the Philippine woodland frog (*Rana acanthi*), the island's most common amphibian, and the endemic Philippine flat-headed frog (*Barbourula busuangensis*). In some of the tributaries and the aboveground stretch, the Philippine crocodile

Tourists preparing to enter the underground portion of the Cabayugan River, which passes under Mount Saint Paul and flows for five miles before it meets the Philippine Sea. The river is home to over 200 species of freshwater fish, while the cave, which features a number of large chambers, provides habitat for nine species of bat. (Flickr/Butch Dalisay)

(*Crocodylus mindorensis*) may be found. This crocodile is one of the smallest crocodile species, with males growing less than 10 feet (3 meters) long. It feeds principally on aquatic invertebrates and builds nests where the smaller female guards seven to 20 incubating eggs.

The underground cave through which the river passes is home to nine species of bat, two species of swiftlet, and the whip spider (*Stygophyranus*). The dominant species of freshwater crab in the river is a *Parathelphusa*, not described until 2004 but composing as much as 90 percent of the crab population. The remaining 10 percent are members of more than 30 species endemic to the Philippines. Two *Parathelphusa* species are endemic to the river: *P. cabayugan* and *P. manguao*. The former, which constitutes most of the river's crab population, has a flat carapace dorsal surface with deep cervical grooves, whereas *P. manguao* has a convex carapace dorsal surface with shallow cervical grooves. Unlike the more typical *P. palawenensis,* both species have a subdistal spine. These endemic crabs are found in headwaters of moderate to low turbidity, farther from areas of anthropogenic activity.

It is common in the Philippines for fish inhabiting the freshwater regions to return to the sea to spawn, and those of the Cabayugan River are no exception. More than 200 species of freshwater fish use the river on their way to the sea. Some of the country's endemic fish species, however, including *Puntius lindug, P. baoulan,* and *P. tumba,* have been pushed toward extinction by the introduction of carp (*Cyprinus carpio*) and goby (*Glossogobius giuris*). Both were introduced as food species for recreational and commercial fishing, and they have competed with the native endemics to the detriment of the latter.

Ecosystem Threats

The Cabayugan River in recent years has shown signs of murkiness and reduced water quality, partly as a result of erosion, which was principally caused by forest clearings. In many cases, the clearings were done illegally. Although the Philippines government has criminalized swidden agriculture, also known as slash-and-burn agriculture—in which agricultural fields are created by burning woodlands—the practice continues in the country, especially among the native Palawan people.

The loss of the forest root systems destabilizes the soil and destroys ecosystems. It has a destabilizing effect that ripples through the entire community. Destabilized soil near the riverbank—at the point where the Cabayugan goes underground—contributes to the buildup of silt in the river system that in turn reduces the amount of dissolved oxygen. This turns the waters muddy and destroys some of the river vegetation and spawning grounds. It also changes the salinity, acidity, and chemical balance of the river by introducing new solids that leach out of the eroding soil. The government, motivated by the river's importance to the tourism industry, has promised to work to repair erosion damage and prevent the illegal logging that has contributed to it.

Mining and agriculture contribute some level of threat to the river as well. Although erosion has received more attention because of the visible change to the quality of drinking water in the area, the contaminants introduced by even small-scale commercial agriculture are much more serious. Fertilizers, pesticides, livestock feces, and soaps enter the river waters via runoff carrying the potential to affect the whole food web of the ecosystem.

Bill Kte'pi

Further Reading

Freitag, Hendrik. "Composition and Longitudinal Patterns of Aquatic Insect Emergence in Small Rivers of Palawan Island, the Philippines." *International Review of Hydrobiology* 89, no. 4 (2004).

Freitag, Hendrik and Darren C. J. Yeo. "Two New Species of *Parathelphusa* H. Milne Edwards, 1853, From the Philippines." *Raffles Bulletin of Zoology* 52, no. 1 (2004).

Goldoftas, Barbara. *The Green Tiger: The Cost of Ecological Decline in the Philippines.* New York: Oxford University Press, 2006.

California Central Valley Grasslands

Category: Grassland, Tundra, and Human Biomes.
Geographic Location: North America.
Summary: Once the most widespread habitat in California, the Central Valley grasslands constitutes a unique prairie. A highly altered habitat, it still harbors diverse native plants and animals.

Few ecosystems in the world would be as unrecognizable to a 19th-century naturalist as California's Central Valley grasslands. Once a North American Serengeti, where perennial grasslands teemed with herds of ungulates and expansive marshes were covered in millions of ducks and geese, today there are cities, rice paddies, orchards, and cotton fields. Despite this, there are some isolated pockets of what once was so naturally abundant.

Geography and Climate

California's Central Valley spans an area 60 or more miles (97 kilometers) wide by 400 miles (644 kilometers) long. The area has a mediterranean climate, with hot, dry summers and cool, rainy winters. It is boxed in on all sides by mountains, with the Pacific Coast ranges to the west, the southern Cascades to the north, the Sierra Nevada to the east, and the Transverse Range to the south.

The valley exists in the rain shadow of the Coast Ranges, resulting in relatively low precipitation, though there is a strong latitudinal gradient in annual rainfall from north to south. The presence of mountains on all sides results in the development of strong inversion layers over the valley, causing both air-quality problems and the development of tule fog, a thick ground-covering phenomenon, in the winter. This famously thick fog is named after the tules, or marsh bulrush plants, that dominated the once-extensive wetlands in the center of the valley. Unfortunately, some 94 percent of these wetlands have been drained and/or converted to agricultural uses, including Tulare Lake, once the largest freshwater lake west of the Mississippi River.

Biota

The grasslands of the Central Valley were originally dominated by perennial bunchgrasses, specifically a few species of needlegrass. Between these bunchgrasses grew fields of annual wildflowers and other spring-flowering forbs, leading naturalist John Muir to describe it as a huge "central garden ... all one sheet of plant gold," humming with birds, the air rich with the bouquet of wildflowers. These grasslands were rooted in sediment as much as 10,000 feet (3,048 meters) deep, the results of thousands of millennia of erosion from the surrounding mountains and volcanic deposition.

This rich, diverse grassland was bordered by areas of foothill woodland on the perimeter, cut through with riparian forests (bordered by bodies of water such as a river), and the marshes. It is thought that the true Central Valley grasslands ecosystem covered some 13 million acres (5.3 million hectares) before the 18th century, in addition to another 9.5 million acres (3.8 million hectares) of botanically related savanna defined by the presence of scattered valley, blue, and interior live oaks.

Within the Central Valley grasslands today are a unique grasslands feature, a wetland hybrid known as vernal pools. These vernal pools result from several geologic conditions, including clay hardpans, volcanic rock below the topsoil, and the accumulation of alkaline minerals, but they are ecologically unified by the presence of seasonally ponding water during the wet season. This water accumulates in depressions but does not drain to major watercourses or local aquifers due to the geological barriers.

As the water in these pools seasonally evaporates, a series of plants specially adapted to saturated soils germinates, grows, and flowers, resulting in a series of concentric rings of wildflowers. Many of these plants, as well as invertebrates such as fairy shrimp, are restricted to these habitats. The species inhabiting these vernal-pool habitats are as isolated from one another as species occupying different mountain ranges; thus, each vernal pool represents an evolutionarily distinct lineage.

The Central Valley grasslands were the home of many species of wildlife. Huge herds of tule elk and pronghorn were present, in addition to black-tailed deer, which continue to be important grazers and browsers in the present-day habitat. The apex predator in this ecosystem was the grizzly bear, which once ranged over the foothills, river floodplains, and grasslands in great numbers. It was extirpated (locally extinct) by 1922, however, and the contemporary ecosystem is not conducive to its reintroduction. Mountain lions and coyotes still maintain their role in the Central Valley food web, though coyotes' more-adaptable nature makes them far more ubiquitous in human-modified habitats.

Effects of Human Activity

Before the California Central Valley was settled by humans, the strongly seasonal nature of precipitation in the valley and the resulting pulses in vegetation growth led to moderate grazing during the wet season, followed by migration of grazing ungulates to the surrounding, wetter foothill and montane habitats during the dry season. Native ungulates grazed the Central Valley grasslands seasonally. When the spring rains faded, they migrated into the surrounding woodland and montane habitats. In addition, Native Americans managed this region through the application of prescribed burning.

The native grasslands, however, were simply no match for the changes brought by European settlers. Europeans brought year-round, intensive grazing to the Central Valley. Native perennial bunchgrasses did not evolve with this grazing regime; thus, they succumbed to competition from Eurasian annual grasses, which were introduced in the coats and digestive tracts of imported livestock. Needlegrasses have been replaced by foxtails, filaree, wild oats, and other Eurasian grasses that have almost completely displaced native grasses, to the point that modern Californians have no clue that the golden hillsides that characterize the state in the dry season are the entirely anthropogenic result of human agriculture.

Global warming trends will likely be felt here in part by an increase in precipitation. Researchers believe this could lead to a near-explosion in the prevalence of the forb component of local vegetation, resulting in fundamental changes in soil composition, water retention capacity, and biological productivity.

It cannot be overstated that the true California Central Valley grassland ecosystem no longer truly exists, but it is still possible to find remains of its former glory. Perhaps the best extant example does not occur in the Central Valley, but in an adjoining valley of the Coast Ranges known as Carrizo Plain, most of which is protected by the United States as a national monument.

Carrizo Plain has persisted because it is a near-desert habitat unsuitable for most agriculture. It is a critically important habitat. Not only have many plant species managed to maintain their competitive edge against invasive species, but several imperiled animal species have held on as well. Carrizo Plain is the home of the federally endangered San Joaquin kit fox and the giant kangaroo rat, both of which are wholly reliant on the persistence of California Central Valley grasslands habitat. Carrizo Plain, with its roaming pronghorns, snoring spadefoot toads, and other species found nowhere else in California (or the world), is an unimpeachably important resource for international biodiversity conservation.

There are various conservation partners on board to help preserve the grasslands from further change, among them the Bureau of Land Management, Nature Conservancy, California Oak Foundation, and the California Department of Fish and Game. The Nature Conservancy alone identified 385 priority sites in the northern half of the valley.

MICHAEL DIXON

Further Reading

Barbour, Michael, Bruce Pavlick, Frank Drysdale, and Susan Lindstrom. *California's Changing Landscapes.* Sacramento, CA: California Native Plant Society, 1992.

Barbour, Michael, Todd Keeler-Wolf, and Allan A. Shoenherr. *Terrestrial Vegetation of California.* Berkeley: University of California Press, 2007.

Cunningham, Laura. *A State of Change: Forgotten Landscapes of California.* Berkeley, CA: Heyday, 2010.

Schoenherr, Allan A. *A Natural History of California.* Berkeley: University of California Press, 1992.

Stromberg, Mark R., Jeffrey Corbin, and Carla D'Antonio. *California Grasslands: Ecology and Management.* Berkeley: University of California Press, 2007.

California Chaparral and Woodlands

Category: Forest Biomes.
Geographic Location: North America.
Summary: California's chaparral and woodlands are interrelated ecoystems that form the heart of the mediterranean climate–influenced Pacific coast of North America.

The cold waters of the Pacific Ocean, high pressure from descending tropical air masses, mountain uplift from colliding tectonic plates, and numerous other forces combine on North America's west coast to create the temperate, rainy winters and hot dry summers that define California's mediterranean climate. This regime has favored the development of the biodiverse oak-dominated woodlands and fire-dependent chaparral (scrublands) that compose the classic rural California landscape, and also provided an attractive environment for human settlement for millennia.

Although they are structurally and botanically quite different, California's woodlands and chaparral are largely influenced by the same broad-scale climatic variables. In fact, they can often be found within a few dozen feet of each other. Perhaps nowhere in North America is the effect of slope aspect more profound than in lower and middle elevations of the Coast Ranges, Sierra Nevada, Peninsular, and Transverse Ranges of California. The south and west faces of these mountains receive much more solar radiation during the hottest parts of the day at the hottest time of the year, leading to higher drought stress on these often-chaparral-covered slopes. The opposing north and east slopes are comparatively cool and moist, and often contain woodlands. Adding to this contrast is the presence of riparian (water area) corridors between the mountains, rich in cottonwoods, willows, and even coast redwoods.

Woodlands

There are many geographic varieties of California woodlands, most of which are dominated by one or more species of oak. Among these, the most widely distributed is the habitat known as foothill woodland, which forms a continuous ring around the Great Central Valley and covers approximately 10 million acres (4 million hectares). Dominant trees in foothill woodland can include the evergreen interior live oak and the deciduous blue oak and valley oak, depending on the microclimate. The foothill pine, often interspersed with oaks in this habitat, has a distinctively bushy or branched appearance compared with most conifers, resulting from the open nature of the habitat in which it grows.

Much of the foothill woodland could more correctly be classified as savanna. On the flatlands of the peninsular ranges of Southern California are similar open woodlands of Engelmann oak; on the Pacific slope of the Coast Ranges, the woodlands are dominated by coast live oak. The climate of these coast-live-oak woodlands is influenced by the moderating influence of the ocean; thus, they may form denser, more forest-like stands. In areas influenced by frequent fog, they are often draped in beards of lace lichen, lending a Tolkienesque aesthetic to these forests.

Chaparral

Interspersed with these woodlands are expanses of scrubland or chaparral. This dry, or xeric, environment is home to shrubs that have evolved to be drought-tolerant in a variety of ways. Many have waxy or hair-covered leaves to reduce water loss. This adaptation allows them to continue photosynthesizing in the height of summer, unlike drought-deciduous shrubs in neighboring ecosystems. Some, such as many species of manzanita, have leaves with an up-and-down orientation to reduce direct sun exposure. On the more mesic, or wetter, slopes, short-statured scrub oak dominates the chaparral.

Chaparral is strongly fire-dependent; shrubs here have evolved to survive in this fire-prone environment in different ways. Chamise, one of the most common chaparral shrubs, quickly resprouts after a fire from a root ball known as a burl. It also exudes volatile chemicals in the height of summer, suggesting that it actively encourages fire. It is thought that fires cause a depletion of nitrogen in the soil, which may also favor shrubs such as Southern California lilacs (*Ceonothus* spp.), which can fix atmospheric nitrogen for their own growth. Unlike chamise, *Ceonothus* cannot resprout if the aboveground plant is killed by fire, but it survives in fire-stimulated seed banks.

Characteristic Animals

Because these habitats often occur quite close to each other, they share many species of animal wildlife. Mule deer are common in both California woodlands and chaparral; in fact, they can become so locally abundant that their browsing can have strongly negative effects on oak regeneration. These deer are the primary prey of mountain lions, the apex predator of this ecosystem. The range of this ecologically flexible cat extends from Canada to South America. Black bears also are common there, though they are far more omnivorous than the strictly carnivorous mountain lion. The habitat once was home to the much larger grizzly bear, but California's last wild grizzly was killed in 1922, and now is present only on the state flag.

California chaparral and woodland habitats are home to a diverse range of smaller mammals, including two species of woodrats or packrats, so named because they have a habit of collecting objects such as sticks, pine cones, bottle caps, and house keys. They stack these objects in their dens in piles known as middens. In some drier areas, middens have been preserved as subfossils and used to research climates thousands of years in the past.

Characteristic birds of the woodlands and chaparral include the raucous and wily scrub jay and the ever-busy acorn woodpecker. In the damp, dark areas underneath fallen logs, one can find the Ensatina salamander, a morphologically diverse group of amphibians. This is the textbook example of ring species, whose members are diverging or have diverged into different species around a central barrier to gene flow—in this case, the Great Central Valley. Flipping over a few more logs is likely to reveal some of the many snakes that make their home in this environment, including the western rattlesnake and its nonvenomous, visually striking predator, the California kingsnake. The chaparral also is home to the diminutive ring-necked snake, which has an olive body, yellow–red neck, and bright orange belly.

The California chaparral and woodlands are home to several types of snake, including the common kingsnake. (Thinkstock)

Human Activity

These habitats have been used by humans for thousands of years. The oak tree's fruit, the acorn, was a particularly important food source for Native Americans in this region. There are many easily accessible grinding rock sites in the foothill woodlands, where people such as the Miwok and Maidu, among others, ground their acorns into meal for countless generations.

Native Americans manipulated the landscape to suit their needs through the use of burning; these ecosystems have faced far more pervasive changes since European settlement. As in many ecosystems, nonnative species are a significant concern, with Eurasian brome grasses and star thistle crowding out native grasses and forbs in the open woodlands. Cattle browse and trample regenerating vegetation. Wild pigs are particularly voracious consumers of acorns, causing sharp declines in young blue oak trees in much of the Coast Ranges.

Agricultural land use has caused habitat loss; California's rightfully famous wine country is largely in the foothill woodland ecosystem. The chaparral of southern California has been particularly hard hit by development, and much of the acreage that has not been converted to cities has had its natural fire cycle disrupted because of fire suppression to protect adjacent subdivisions.

Fortunately, these threats are recognized, and active conservation initiatives are in place to preserve California woodlands and chaparral for future generations. These initiatives include the U.S. Fish and Wildlife Service's California Foothill Legacy Area. It is a far-reaching project, begun in 2011, that seeks to protect working rangelands and woodland by buying easements on the property, allowing the land to continue to be used for things such as grazing while protecting it from residential or commercial development.

Global warming effects on the oak savanna and chaparral of California are difficult to predict, as the biome has evolved in a complex manner that already accommodates extreme weather events, high winds, and great local divergence in temperature and moisture. These built-in factors mean that even if more extreme rainfall and drought events, or the arrival of Santa Ana winds later in the season, become more typical, it is not clear that this biome will be more than marginally disrupted. Fragmentation of habitat is a more prevalent and immediate threat; efforts to retain intact areas of habitat will have to take into account the likelihood of more extreme fluctuations in climate patterns.

MICHAEL DIXON

Further Reading

Barbour, Michael, Bruce Pavlick, Frank Drysdale, and Susan Lindstrom. *California's Changing Landscapes.* Sacramento, CA: California Native Plant Society, 1992.

Johnston, Verna R. and Carla J. Simmons. *California Forests and Woodlands: A Natural History.* Berkeley: University of California Press, 1996.

Pavlick, Bruce M., Pamela C. Muick, Sharon Johnson, and Marjorie Popper. *Oaks of California.* Los Olivos, CA: Cachuma Press, 1991.

Schoenherr, Allan A. *A Natural History of California.* Berkeley, CA: University of California Press, 1992.

U.S. Fish and Wildlife Service. "California Foothills Legacy Area Proposal." 2012. http://www.fws.gov/cno/refuges/planning/cfla.cfm.

California Coastal Forests, Northern

Category: Forest Biomes.
Geographic Location: North America.
Summary: The coastal redwood forests of northern California form a unique and cherished forest ecosystem that has been the focus of extensive conservation efforts to protect and preserve some of the largest trees on the planet.

The coastal redwood forest is the dominant ecosystem in coastal northern California. Commonly called redwood (*Sequoia sempervirens*), the tree belongs to the redwood family *Sequoioideae*, a relative of the giant sequoia (*Sequoiadendron giganteum*), found inland in California on the western slopes of the Sierra Nevada mountain range, and the dawn redwood (*Metasequoia glyptostroboides*), which is native to China. Coast redwoods are the tallest trees on Earth and the second-longest-lived. They can grow to more than 350 feet (107 meters) tall and live more than 2,000 years.

Redwoods are second only to giant sequoias in terms of their individual biomass volume, and the northern redwood forests are among the most productive on the planet, with productivity rates that rival those of tropical forests. These forests grow within a limited range in northern California, have several unique structural characteristics, and have been the focus of conservation efforts for more than 100 years.

Though coast redwoods and their close relatives were distributed across much of the mid-latitude Northern Hemisphere during the Cretaceous and early Tertiary Periods (144 million to more than 30 million years ago), they now are native only in a north-south stretch of approximately 466 miles (750 kilometers) from the southwestern tip of Oregon to Monterey County, California. They occur only within a narrow fog belt that stretches east-west from near the ocean up to 40 miles (65 kilometers) inland. In fact, the presence of fog during the otherwise dry summer is a defining characteristic of coast redwood forests.

Not only does the fog reduce water loss by the trees during dry seasons, but the redwoods can obtain water, and sometimes nutrients, from the fog as its molecules are stripped from the air by the redwood needles (leaves) and drip to the ground. The reduction in water loss and additional water supply are important because the overall climate of northern California is mediterranean. Almost all the rainfall occurs during the cool, wet winter, and there is little to no rain during the warmer, dry summer.

Vegetation

Redwood forests typically are found in alluvial areas near streams, and on hills and slopes with sandstone bedrock; they do not grow well in soils with very high iron or magnesium concentrations (ultramafic, or igneous-type soils). Toward the drier, southern end of their range, they do not grow well on exposed hilltops, which are dominated instead by chaparral or mixed evergreen forests consisting of Douglas fir, tanoak, bay laurel, and Pacific madrone. These other trees occasionally occur together with redwoods and as small mixed evergreen stands within redwood forests throughout the range. Nor do redwoods grow immediately next to the coast; the spray of sea salt inhibits their growth. Beach, coastal grasslands, and coastal chaparral ecosystems dominate the land areas close to the ocean, or they can contain western hemlock, coast live oak, Sitka spruce, and other tree species instead of redwood.

The tops of the tall, large redwood trees to the north are a canopy ecosystem unto themselves. Canopy soil that forms from decomposing litter in the nooks between large branching trunks can be more than 1 foot (0.3 meter) deep, allowing ferns and trees such as Sitka spruce and western hemlock to grow high up in the tree canopy.

Animals

The dominant redwood forest stands within this biome harbor a diverse range of animals, including the black bear, Roosevelt elk, and bobcat among the larger mammals, and the gray fox, big brown bat (*Eptesicus fuscus*), and Douglas squirrel or chickaree (*Tamiasciurus douglasii*) among the smaller species.

Along and within tidal pools are found such characteristic species as the purple shore crab (*Hemigrapsus nudus*), California mussel (*Mytilus californianus*), and the giant green anemone (*Anthopleura xanthogrammica*). Shoreline birds range from killdeer (*Charadrius vociferus*) to sanderling (*Calidris alba*) and willet (*Catoptrophorus semipalmatus*).

Birds within the forest proper include varied thrush (*Ixoreus naevius*), Steller's jay (*Cyanocitta stelleri*), chestnut-backed chickadee (*Parus rufescens*), and winter wren (*Troglodytes hiemalis*). Chief among avian predators here is the bald eagle (*Haliaeetus leucocephalus*).

More than 300 fungal species have been found in coast redwood forests, supporting many animals that prowl the well-shaded zones around the redwood trunks; another often-seen denizen of the forest floor is the banana slug (*Ariolimax columbianus*). The canopy above is home to diminutive species such as the California slender salamander (*Dicamptodon tenebrous*) and the Pacific tree frog (*Pseudacris regilla*).

Logging and Preservation

Old-growth coast redwood forests have unique structural characteristics that provide habitat for endangered species such as the northern spotted owl (*Strix occidentalis caurina*) and marbled murrelet (*Brachyramphus marmoratus*), as well as salmon spawning in their streams. The redwoods are highly cherished for their cultural and natural value, in part because they occur naturally within

Redwood Conservation

Redwood conservation efforts began as early as 1900, when California's oldest land trust, the Sempervirens Club, formed to protect the coast redwood forests of the Santa Cruz Mountains. In 1902, Big Basin Redwoods State Park (RSP), California's oldest state park, was established by members of the Sempervirens Club and others. In 1918, the Save the Redwoods League (SRL) was founded. Since then, the SRL has worked to purchase and preserve redwood forests throughout northern California. To date, the efforts have funded the purchase of more than 189,000 acres (76,486 hectares) of redwood forest and the development of 63 parks and preserves.

Over the years, some 27 redwood state parks have been established in California. Toward the northern end of the redwood range, Prairie Creek RSP, Del Norte RSP, and Jedidiah Smith RSP were purchased and established by the SRL in the 1920s, helping to create the California State Park System. In the same area, Redwood National Park was created federally in 1968 and then expanded in 1978. Together, these four parks are managed as one unit called Redwood National and State Parks (RNSP), covering more than 208 square miles (540 square kilometers), or 130,000 acres (52,600 hectares). Of this realm, about 39,000 acres (15,800 hectares) is old-growth forest. The RNSP contains 40–45 percent of the remaining old-growth coast redwood forests and has been designated as a World Heritage Site and an International Biosphere Reserve by the United Nations Educational, Scientific and Cultural Organization (UNESCO).

a very limited range of the world. For these reasons—and their rapid decline due to logging in the early 20th century—redwood forests have been the focus of intensive conservation efforts.

At the turn of the 20th century, there were more than two million acres (809,400 hectares) of old-growth redwood forest. Much of this was cut down for timber, and less than 5 percent of the old-growth redwood forest present before European settlement remains today. Of the remaining old-growth forest, about 80 percent is protected in a park or preserve, or by a land preservation agreement. The majority of the rest of present-day redwood forests are secondary forests that are operated by logging companies for timber production or have regrown after logging.

Damon C. Bradbury

Further Reading

Bourne, Joel. "Redwoods: The Super Trees." *National Geographic* 215, no. 10 (2009).

Evarts, J. and M. Popper, eds. *Coast Redwood: A Natural and Cultural History*. Los Olivos, CA: Cachuma Press, 2001.

Noss, R. F., ed. *The Redwood Forest: History, Ecology, and Conservation of the Coastal California Redwoods*. Covelo, CA: Island Press, 2000.

Camargue Wetlands

Category: Marine and Oceanic Biomes.
Geographic Location: Western Europe.
Summary: This ecosystem has lost one-third of its productive area in the past 50 years and is increasingly threatened by human activity.

The Rhône River spills into the northern Mediterranean Sea in a vast delta plain of sloping ridgelines, stranded riverbeds, salty lagoons, and coastal wetlands harboring significant biological diversity. This natural wealth in tangible forms of salt-tolerant herbs, rice fields, river and shore fisheries, birds, mineral salts, and livestock is associated with France's preeminent Mediterranean estuary. Flamingos, horses, rice, seasonal "mistral" winds, gypsies, bullfighting, and French faux cowboys are among the exotic subjects finding sanctuary in this remarkable wetland at the twin mouths of the Rhône, south and west of the city of Arles in the Languedoc region.

Extending behind 20 miles (30 kilometers) of dunes into the sea, the two diverging Rhône streams form the Grande Camargue. This sheltering core emerges as a mosaic of beaches, lagoons and marshes. Where rivers once flowed in slow loops, *étangs,* as these oxbows are called, evaporate under the Languedoc's simmering sun; the largest of these, Étang de Vaccares, forms the heart of this Biosphere Reserve. On the Rhône's west bank in the Petite Camargue lies Aigues-Morte, once a thriving medieval port, now stranded inland due to historically accumulating sediment that expanded the delta seaward. Since the Middle Ages, salt for curing food was dried on the flats.

The Rhône River mouth nourishes the Camargue's abundant diversity in an extensive estuary whose alpine-derived sediments are ensnared by shrubs, grasses, and halophytes. These maritime grasslands hold great biological wealth due to biotic nutrients trapped by a saltwater wedge, nourishing fish and birds attracted to the salt marshes and irrigated rice paddies. The expanse overflows into an area extending over 400 square miles (1,035 square kilometers) where the salt lagoons fed by the sea and dried by the summer sun persist beside the embanked rivers. Part of the Rhône's freshwater originates in glacial melt and snow, but is now diverted into 77 square miles (20,000 hectares) of cultivated rice in the Camargue's upland marshes.

Diversity

The unusual collection of estuarine plant communities here offers nesting or feeding for 35 bird species and 200,000 wintering ducks—making the Camargue a critical stopover point in their annual migrations. A fusion of fresh, brackish, and saltwater tolerant grasses, succulents, and diatoms thrive in an extended growing season, making for an uncommon amalgam of vegetative and bird species, some classified as threatened. Important bird rookeries were reserved dating from 1927, when nature preserves were created in the Rhône delta that later formed the essential wetlands protected in a 1962 agreement. The Camargue was designated a Ramsar site as a Wetland of International Importance on December 1, 1986.

The freshwater flow of the Rhône entering the lagoons works to stem the saltwater encroachment from the Mediterranean—but the sea is perhaps more relentless. In the 19th century the Digue à la Mer—a levee along 20 miles (30 kilometers) of the sea—was constructed to protect the Camargue's saline and hypersaline lagoons from the less salty Mediterranean seawater. Dams upstream of Arles along the Rhône's main stem have deprived the salt marshes of sediment in the last century, while providing hydroelectric power for the wine, chemical, salt, and tourist industries of the region. Less than a fifth of the sand, mud, and gravel historically provided by this northwestern Mediterranean river ever reaches the Camargue's sinuous waterways.

The loss of one-third of the productive marine wetlands over five decades in the Camargue has diminished the capacity of these salt marshes to absorb pollution while maintaining necessary productivity for livestock, fish, and birds. The Camargue's lagoons still give rise to a diversity of salt-tolerant succulents as well as brine shrimp that are a source of food for shore fisheries in the Gulf of Lyon, and also provide sustenance for the legendary flamingos and other wading birds. In 1997 an estimated 10,000 to 18,000 breeding pairs of flamingos nested here. Two threatened species have been documented in the Camargue: the European roller, which in 1997 was reduced to 50 to 100 breeding pairs, and the black-tailed godwit.

Shifting Influences

Despite the fact that ducks such as the teal (*Anas crecca*) help account for the spread of aquatic vegetation in the marshes, duck hunting has altered the hydrological character by introducing permanent freshwater ponds amid native salt marshes. Additionally horses and rabbits are agents of successional vegetation change in these wetlands, particularly by reducing the extent of rushes and sedges through grazing. Extensive rice cultivation—although helping to attract heron nesting—is responsible for pesticide contamination of the salt lagoons into which these paddies drain.

As a landscape shaped by the needs of salt industries, rice paddy farming, and livestock raising, these are contested grounds. Other forces also apply pressure, namely the region's chemical, steel, and petroleum refining interests and a preeminent port. Rice farmers allied with the native, evaporative, salt works have teamed up to defend regional biotic diversity. However, with declines in the nesting fowl of the Camargue, changes have developed due to both the hydrological alteration of the region's ecology and the pollution of its eel, wading bird, and microbial communities.

Clues to the shifting conditions of wildlife and fisheries reveal a complex disequilibrium creating the Camargue, strung out as it is between the Mediterranean-fed saline lagoons and diminishing sweet flows of the Rhône's watershed. Recent studies call for artificial buffers to retain freshwater runoff from rice fields to render persistent pesticide residues innocuous. This would assure some restoration of the natural cleansing capacity of the surrounding salt marshes to sustain rare plants and animals from the gradual, yet relentless, embayment from rising seas.

Joseph V. Siry

Further Reading

Arnaud-Fassetta, Gilles. "The Upper Rhône Delta Sedimentary Record in the Arles-Piton Core: Analysis of the Delta-Plain Subenvironments, Avulsion Frequency, Aggradation Rate, and Origin of Sediment Yield." *Geografiska Annaler: Series A, Physical Geography* 86, no. 4 (2004).

Comoretto, L., et al. "Runoff of Pesticides From Rice Fields in the Ile de Camargue (Rhône River Delta, France): Field Study and Modeling." *Environmental Pollution* 151, no. 3 (2008).

Golterman, H. L. "The Labyrinth of Nutrient Cycles and Buffers in Wetlands: Results Based on Research in the Camargue (Southern France)." *Hydrobiologia* 315, no. 1 (1995).

Roche, Helene, et al. "Rice Fields Regulate Organochlorine Pesticides and PCBs in Lagoons of the Nature Reserve of Camargue." *Chemosphere* 75, no. 4 (2009).

Tourenq, Christophe, et al. "Spatial Relationships Between Tree-Nesting Heron Colonies and Rice Fields in the Camargue, France." *The Auk* 121, no. 1 (2004).

Cameroon Savanna Plantation Forests

Category: Forest Biomes.
Geographic Location: Africa.
Summary: More than half of the African country of Cameroon is forest, and under serious threat of deforestation from commercial logging and the palm-oil trade.

More than half of the central-west African country of Cameroon is forested. In the south is an equatorial rainforest. The savanna of northern Cameroon covers about 7.4 million acres (three million hectares) and includes three national parks—Benoue, Boumba Ndjida, and Faro—among its protected areas. One of the last large savannas in central Africa, it is home to 300 species of birds, African wild dogs, rhinoceros, buffalo, giraffes, elephants, and many other animals. Intensive agriculture, cattle grazing, and unsustainable methods of fishing and hunting had already put the ecosystem at risk despite the best efforts of environmental groups. The recent adoption of plantation forestry in Cameroon will create yet another source of risk.

Plantation forests are monocultures: forests that are managed to produce a single type of tree as a crop. Monoculture forestry is a method that produces greater yields and more effective commercial growth than diverse or mixed forests. Because the trees are typically planted and harvested at the same time, they lack the diversity of tree sizes of naturally occurring monocultures.

Such forests also deprive the ecosystem of the niches that dead trees and openings can provide, and when the trees are harvested by clear-cutting, the habitats of the forest are deeply affected. Mechanical harvesting compacts soil, destroying much of the understory ecosystem. Opponents of monoculture forestry also claim that the forests are ideal breeding grounds for pests and disease because they no longer include some of the pests' natural predators. Theoretically, however, and depending on the particular monoculture, a plantation forest can benefit from biodiversity by limiting logging activities to that area.

Palm Oil

Palm oil is the cash crop driving the growth of Cameroon's plantation forests. The oil is extracted from the pulp of the fruit of the oil palm tree (*Elaeis guieensis*). Palm oil is a highly saturated vegetable fat. It has long been a common cooking ingredient in Africa, as well as southeast Asia, and has grown in popularity worldwide for its low price and commercial food industry applications.

The deforestation of oil palms by commercial interests has been a serious problem since the 19th century, when the king of Dahomey outlawed the practice. Industrial uses for palm oil have included machinery lubrication; soap products such as the Palmolive brand; the manufacture of napalm; and, most recently, the production of biodiesel. Although methods of producing biodiesel vary by manufacturer, the method of turning palm oil into biodiesel is fairly simple. The profit equation is currently driven by low production, labor, and land costs, along with the increased demand for biofuels.

Environmental groups such as Greenpeace and the World Wildlife Fund have pointed out that when palm oil is produced by clearing existing forests for oil-palm plantations, the environmental consequences of deforestation outweigh—or at the very least diminish—the environmental gains of using biofuel instead of fossil fuels. This consideration is apart from issues of social costs, which should be raised when the developed world adopts a technology dependent on cheap labor from developing nations. Another organization, Global Forest Watch, works with local organizations in Cameroon to collect and distribute information on forest development in order to hold public officials and forestry companies accountable and to foster better management of forest resources. Global Forest Watch Cameroon has published reports providing future plans on forest development and profiles of logging companies, and has explored the potential of employing satellite imagery to detect illegal logging.

Oil-palm plantations recently have spread to Cameroon when the government of Indonesia—a major producer—issued a two-year moratorium on new oil-palm plantations in that country. Agribusinesses such as Sime Darby (the largest palm-oil producer in the world) altered its expansion plans, buying up land in Africa so it could continue opening new plantations. In 2011, Sime Darby purchased 741,000 acres (300,000 hectares) in Cameroon.

New York–based Herakles Farms leased an additional 148,260 acres (60,000 hectares) from the Cameroonian government under a 99-year lease. This oil-palm plantation will replace small farms and local forests, and will be placed adjacent to several of Cameroon's nature reserves. As a result, the creation of oil-palm plantations in Cameroon will degrade forests and reshape agricultural land, jeopardizing Cameroon's food sovereignty to plant crops of oil palms. Animal migration routes will be disrupted, along with the lifestyles of the native primates, butterflies, and roaming mammals.

Workers skimming refined palm oil from a barrel during palm oil processing in central Africa. Low labor and production costs for palm oil biofuels are attracting large oil-palm plantations to Cameroon. (USAID/Ken Wiegand)

Sustainability Questions

Research in Cameroon, Ghana, and elsewhere on the reaction and sustainability of ant and termite colonies, as well as beetle populations, in forest areas disrupted by plantation clearance has indicated that there may be methods to mitigate some of the negative effects. For example, the more deliberate, manual clearing of forest areas prior to monoculture planting has been shown to be about half as disruptive in some cases as complete clearance effected by mechanized methods. Nevertheless, the evidence supports the thesis that plantation forests diminish species diversity in a given biome.

Situated in sub-Saharan Africa, the Cameroon savanna is projected to sustain annually reduced rainfall as an effect of climate change. Some scientists and land-use experts predict that this will lead to faster groundwater depletion, as irrigation for food and commercial croplands increases in order to offset the projected loss by 2050 of some 2 inches (50 millimeters) of yearly rainfall due to climate change. Conserving soil structure—which helps retain nutrients and support various habitat niches—will become more of a challenge in lower-moisture scenarios, whether a forest is in pristine condition, is converted to small-scale farming, or is cleared for plantation forestry.

Bill Kte'pi

Further Reading

Runge, Jurgen, ed. *Landscape Evolution, Neotectonics and Quaternary Environmental Change in Southern Cameroon.* Boca Raton, FL: CRC Press, 2012.

Topa, Giuseppe, Carol Megevand, and Alain Karsenty. *The Rainforests of Cameroon.* New York: World Bank Publications, 2009.

Watt, A., N. E. Stork, and B. Bolton. "The Diversity and Abundance of Ants in Relation to Forest Disturbance and Plantation Establishment in Cameroon." *Journal of Applied Ecology* 39, no. 1 (2002).

Yengoh, Genesis T., Sara Brogaard, and Lennart Olsson. "Crop Water Requirements in Cameroon's Savanna Zones Under Climate Change Scenarios and Adaptation Needs." In Peeyush Sharma and Vikas Abrol, eds., *Crop Production Technologies.* New York: InTech, 2012.

Campos Rupestres Montane Savanna

Category: Grassland, Tundra, and Human Biomes.
Geographic Location: South America.
Summary: The species-rich grasslands associated with rock outcrops in mountain ranges in eastern Brazil have a huge number of endemic plant species.

The *Campos Rupestres*—from the Portuguese, meaning "rocky grasslands"—are scattered in the mountain ranges of eastern Brazil and inserted into different biomes. Considered grassland/shrubland vegetation at an altitude of more than 2,625 feet (800 meters) above sea level in southeastern and northeastern Brazil, campo rupestre occurs principally in the Brazilian states of Minas Gerais and Bahia, on the ancient Precambrian rocks of the Serra do Espinhaço (Espinhaço Mountains), formed by the Chapada Diamantina (Diamantina Plateau) and the Cadeia de Espinhaço (Espinhaco Mountain Range). The term is sometimes used in a wider sense for grassland vegetation upon rock outcrops or on very shallow soils in other regions of Brazil (such as central and southern Brazil), but this definition is not the one adopted here.

Due to floristic and geomorphological differences, campos rupestres are distinct from *campos de altitude* (montane grasslands or altitude grasslands), which occur principally on mountaintops above 4,921 feet (1,500 meters) in the Serra do Mar mountain range. These campos de altitude are part of the Atlantic Forest biome, whereas the campos rupestres form a transition between the Cerrado (Brazilian savanna), Caatinga, and Atlantic Forest biomes. This transition is reflected in the species composition of campos rupestres vegetation. The terms *campo rupestre* and *campos de altitude* are not uniform among botanists or ecologists in Brazil, however. Recently, the researchers Ruy José Vàlka Alves and Jiří Kolbek synthesized the use of different terms and differentiated campos rupestres from other vegetation types in the region that have similar overall appearances.

Climatic conditions in the distribution range of campo rupestre are continental, with dry winters (July is the driest month), classified as Cwa in Köppen's system. Annual rainfall throughout the region with campo rupestre varies from less than 39 inches (1,000 millimeters) to more than 59 inches (1,500 millimeters). Campo rupestre occurs at sites with rock outcrops, mainly quartzite rocks and derived soils, but also on sites with arenite rocks or iron ore. Environmental conditions at these sites are rather extreme: Soil is shallow and often characterized by high concentrations of heavy metals, low organic-matter content and water availability coupled with high solar radiation.

The site conditions are reflected by morphological and physiological adaptations in the plant community: The leaves of many species are strongly reduced, succulent, leathery, or covered by leaf hairs. Many plants manage to retain heavy metals in their tissues, and the presence of ecotypes specific to the region, such as dwarf growth, is common in many species. Though a grassland formation, campo rupestre can contain a large number of shrubs or treelets, especially at rock outcrops. Physiognomically, some patches of campo rupestre may resemble savannas or even shrublands rather than grasslands. Usually, different formations are found in a mosaiclike fashion.

Flora

The region's flora are characterized, in general terms, by the high importance of species from the sunflower family (*Asteraceae*), the bromelia or pineapple family (*Bromeliaceae*), cactus (*Cactaceea*), sedge (*Cyperaceae*), pipewort (*Eriocaulaceae*), iris (*Iridaceae*), mint (*Lamiaceae*), legume (*Leguminosae*), myrtle (*Myrtaceae*), orchid (*Orchidaceae*), and grasses (*Poaceae*), among others.

Although campo rupestre grasslands are very species-rich, even by Brazilian standards, their high rates of endemism (species that are found nowhere else on Earth) are even more astonishing. Campos rupestres are considered to have the highest levels of plant endemism in Brazil, which may be a consequence of high speciation rates due to patchy distribution and isolation of vegetation. Examples of flowering plant families with large numbers of endemic species are *Cactaceae, Eriocaulaceae, Velloziaceae,* and *Xyridaceae.* The ecosystem as a whole has been recognized as a center of endemism for plants.

Fauna

Together, more than 200 bird species have been cited for campos rupestres ecosystems. This includes many hummingbirds, with hallmark species such as the hyacinth visorbearer (*Augastes scutatus*) and hooded visorbearer (*Augastes lumachella*). Additional signature avian species are the lesser nothura (*Nothura minor*) and the crowned solitary eagle (*Harpyhaliaetus coronatus*).

The campos rupestre region is habitat for a range of mammals, including jaguar (Panthera onca), tiger cat (Leopardus tigrinus), maned wolf (Chrysocyon brachyurusi), bush dog (Speothos venaticus) and different bat species. Many of these species are endangered. The campo rupestre fauna also includes a number of reptiles, such as the lizards Tropidurus erythrocepalus and T. mucujensis. Characteristic amphibians include the frogs Proceratophrys cururu and Physalaemus deimaticus.

Conservation

The region where campo rupestre grasslands can be found has been exploited principally for mining diamonds and other ornamental stones since the region was first colonized. Today, mining is mostly for iron ore. These activities, together with other ventures (collection of ornamental plants, tourism, and urban development), constitute a threat to the species-rich and unique ecosystem, which is still poorly known, especially in the international literature. Currently, only a small part of the campos rupestres is included in conservation units, which further adds to the vulnerability of this ecosystem.

Fabíola Ferreira Oliveira
Gerhard Ernst Overbeck

Further Reading

Alves, Ruy José Vàlka and Jiří Kolbek. "Can Campo Rupestre Vegetation be Floristically Delimited Based on Vascular Plant Genera?" *Plant Ecology* 207, no. 1 (2010).

Alves, Ruy José Vàlka and Jiří Kolbek. "Plant Species Endemism in Savanna Vegetation on Table Mountains (Campo Rupestre) in Brazil." *Plant Ecology* 113, no. 2 (1994).

Ribeiro, Katia Torres and Leandro Freitas. "Potential Impacts of Changes to Brazilian Forest Code in Campos Rupestres and Campos de Altitude." *Biota Neotropica* 10, no. 4 (2010).

Vasconcelos, Marcelo Ferreira and Marcos Rodrigues. "Patterns of Geographic Distribution and Conservation of the Open-Habitat Avifauna of Southeastern Brazilian Mountaintops (Campos Rupestres and Campos de Altitude)." *Papéis Avulsos de Zoologia* 50, no. 1 (2010).

Canary Islands Dry Woodlands and Forests

Category: Forest Biomes.
Geographic Location: Northwestern Africa.
Summary: These singular island communities, with highly diverse vegetation, are influenced by the proximity of the Sahara Desert and trade winds.

The Canary Islands is a group of seven major volcanic islands—La Gomera, El Hierro, La Palma, and Tenerife in the west; and Gran Canaria, Lanzarote, and Fuerteventura in the east—located in the Atlantic Ocean close to the northwestern coast of Africa. An archipelago, the island chain belongs to Spain and comprises 2,875 square miles (7,447 square kilometers). Due to the islands' geographic location in the tropics, vegetation communities here are determined by three main biophysical driving forces: the influence of trade winds, the proximity to the Sahara Desert, and the local topography.

Trade winds are most evident in the western islands, where the ocean moisture carried by these northeastern air currents is stopped by the main mountain barriers—such as the Teide peak at 12,198 feet (3,718 meters)—producing the "sea of clouds" effect, which produces high humidity and rainfall at altitudes above 2,625 feet (800 meters). By contrast, the southern parts of these islands and the eastern end of the archipelago undergo a rain-shadow effect characterized by drier conditions.

The African influence occurs mainly in the eastern islands, where the hot winds that blow from the Sahara Desert produce the Calima effect, which increases average temperatures and water scarcity.

The general hilly topography of the Canary Islands, derived from their volcanic origin and subsequent erosive processes, determines the existence of microclimate variations on a local scale. Based on these factors, several vegetation communities can be described according to their altitudinal range and their location in the archipelago.

Vegetation

Basal areas in all the islands are dominated by xerophytic (low-moisture) shrub communities composed of succulent plants that have adapted to water scarcity; among these are balsam spurge (*Euphorbia balsamifera*) and leafless spurge (*Euphorbia aphylla*). Near the coastline, these communities become enriched with coastal-environment plants such as Canary sea fennel (*Astydamia latifolia*) and dwarf sea lavender (*Limonium pectinatum*). Also in these areas, two species of singular dragon trees appear: common drago (*Dracaena drago*), distributed in the western islands, and drago (*Dracaena tamaranae*), limited to Gran Canaria. Also, from 656 to 1,312 feet (200 to 400 meters) are Canary palm (*Phoenix canariensis*) grove communities located in gullies, particularly on La Gomera and Gran Canaria.

First forest-land covers occur between the upper limit of shrubland and 2,297 to 2,625 feet (700 to 800 meters). These thermophilous (heat-loving) evergreen communities are dominated by

Canary juniper (*Juniperus turbinata canariensis*) on the western part of the archipelago and by wild olive (*Olea cerasiformis*) on the eastern islands. Other trees are mocan (*Visnea mocanera*), rarely deciduous and widely scattered, and Atlas mastic (*Pistacia atlanticus*), sometimes as a shrub.

In midmountain areas above these thermophilous formations and up to 4,921 feet (1,500 meters), the transition to more humid conditions occurs, resulting in two typical evergreen hardwood forest communities: laurisilva and the fayal-brezal forest. The first, very similar to a tropical rainforest due to its dense tree canopy and its abundant herbs and ferns understory, is dominated by a group of characteristic laurel-leaved trees such as Canary laurel (*Laurus novocanariensis*), barbusano (*Apollonias barbujana*), til (*Ocotea foetens*), and Indian bay (*Persea indica*). Laurisilva occurs predominantly on the north slopes of the western islands, where the influence of wet winds is greater, and in certain southern areas with humid microclimate conditions. Fayal-brezal forest, dominated by Canary heather (*Erica arborea canariensis*) and wax myrtle (*Myrica faya*), occurs in areas where moisture is lower.

From 3,937 feet (1,200 meters) to the timberline at 6,562 feet (2,000 meters), the upper forest communities are represented by Canary pine forest (*Pinus canariensis*), especially on Gran Canaria, Tenerife, and La Palma. This open forest community is characterized by its resilience in the face of wildfires due to Canary pine's high resprouting capacity.

Finally, summit areas are occupied by shrubs, mainly of the *Leguminosae* family. At that altitudinal range, the number of endemic herbaceous plants highlights the already rich group of Canary endemic (found nowhere else) plants.

A male Gallotia lizard on the island of Tenerife in Spain's Canary Islands. With four subspecies, Gallotia lizards are thriving in the islands. (Wikimedia/Jörg Hempel)

Native Animals

While thousands of plant species are native to the Canary Islands, animal diversity isn't as varied. Currently the islands are home to some 1,200 animals species, including the *camello Canario* (*Camelus dromedarius*), a dromedary (one hump, long limbs, and short hair) that is called a camel (two humps, short legs). It was brought to the islands from Africa in 1405, adapted to the local climate and conditions, and is now found throughout the archipelago, especially in the south of Gran Canaria and Tenerife and across the islands of Fuerteventura and Lanzarote.

More in keeping with an island's expected animal life are the offshore cetaceans, such as bottlenose dolphin and long-finned pilot whale. The La Palma giant lizard, thought to have disappeared from the islands for nearly 500 years, was discovered again in 2007. The species population is again on the decline due to more human development on the islands. Also declining in numbers is the El Hierro giant lizard, which grows to 7.9 inches (20 centimeters). Now a population of between 300 and 400, El Hierro giant lizards are declining due to seagull attacks and a scarcity of plant food. Increased human activity could further stress the creature's numbers. Gallotia, or western Canaries lizards, on the other hand, seem to be a very resilient genus, with representatives of at least four sub-species found all across the archipelago. These wall lizards thrive on a wide variety of fruits, and have colonized habitats from the sere volcanic slopes of Mount Teide, to riverine areas of the woodlands, to the more harsh environs near the coastlines.

One animal with growing numbers is the Barbary ground squirrel, introduced to the island of Fuerteventura in the 1960s from North Africa. Arriving as pets, some escaped to the wild and have been reproducing without check, as they have no natural predators.

Human Activity and Conservation

Human influence is a key factor in landscape configuration. In that sense, deforestation has taken place throughout the archipelago, mainly involving fayal-brezal forest and laurisilva, which have suffered the most outstanding regression. On Gran Canaria, scarcely 1 percent of the original laurisilva remains, and on Tenerife, this amount is just 10 percent. The declines are linked to the historical conversion of forests to croplands, especially for growing bananas and sugarcane, and also associated with timber extraction, firewood and charcoal appropriation, and more recently to urban sprawl and the growth of the tourism industry since the 1970s.

Further driving forest loss is fire. In 2007, wildfires ravaged an area equivalent to almost 5 percent of the archipelago. Similarly, in 2012, wildfires burned across more than 11 square miles (28 square kilometers) on La Gomera, devastating parts of Garajonay National Park, an internationally recognized ecological treasure.

Some environmental efforts have been created to stop deforestation and improve biodiversity conservation. Forest management and reforestation linked to soil conservation and water-flow management has been utilized since the 1950s, mainly with Canary pine and other pine species. Approximately one-third of total pine forest land is planted. Protected areas also have played key roles in biodiversity conservation. About 40 percent of the Canary Islands archipelago is protected in various categories. The islands have four national parks, two of which have been declared as World Heritage Sites and two as MAB Biosphere Reserves by the United Nations Educational, Scientific and Cultural Organization (UNESCO). There are also 11 natural parks and 26 reserves.

In terms of global climate change, the Canary Islands are likely to see warmer overall sea surface temperatures and warmer nighttime air temperatures over land, increasing humidity but diminishing precipitation, and a shift in wind direction due to more frequent high-pressure system incidence. The projected impacts through 2025, according to government-sponsored research, could include more extreme weather events, higher incidence of tropical disease, and great pressure on biodiversity.

Francisco Javier Gómez Vargas

Further Reading

Cox, S. C., S. Carranza, and R. P. Brown. "Divergence Times and Colonization of the Canary Islands by Gallotia Lizards." *Molecular Phylogenetics and Evolution* 56, no. 2 (2010).

Kunkle, Gunther, ed. *Biogeography and Ecology in the Canary Islands.* The Hague, Netherlands: Junk Publishers, 1976.

Perez, Miguel Angel. *Native Flora of the Canary Islands.* Leon, Spain: Editorial Everest, 2000.

Candaba Swamp

Category: Inland Aquatic Biomes.
Geographic Location: Asia.
Summary: A main wetlands site in the Philippines, the Candaba Swamp is a fertile and biodiverse area noted for being a primary nesting habitat for numerous migratory birds.

An ecologically important wetland ecosystem spanning the provinces of Pampanga and Bulacan in the Philippines, Candaba Swamp is known as a primary nesting ground for a host of migratory birds. Spanning more than 150 square miles (388 square kilometers) of arable land, the Candaba Swamp is a rich mosaic of freshwater ponds, swamps, and marshes on a sprawling alluvial floodplain surrounded by seasonally flooded grassland, nipa palm savanna, mangrove forests, and farms known for their production of watermelon and muskmelon. The site of communal fishing grounds that provide critical habitat for fish, including mudfish and catfish, the Candaba also is an ecosystem facing a trio of threats: illegal hunting; the introduction of exotic fish species; and siltation, or the deposition by moving water of fine sand, organic matter, and the like as sediment.

At 36 feet (11 meters) above sea level, Candaba represents one of the lowest points of interior Luzon Island, the largest island of the Philippines archipelago, and acts as a natural flood basin for capturing the wet-season overflow of five small rivers—Maasim, San Miguel, Garlang, Bulu, and Penaranda—that empty into the larger Pampanga River. Just 37 miles (60 kilometers) north of metro Manila, Candaba Swamp is bounded by the towns of San Luis and Arayat in the province of Pampanga; Baliuag, San Ildefonso, and San Miguel in the province of Bulacan; and Cabiao in the province of Nueva Ecija. The swamp is the area closest to the nation's capital that offers very rich biodiversity.

Critical to Bird Species

A main stopover along the East Asia–Australasian Migratory Flyway, which includes Siberia, China, Japan, Hong Kong, South Korea, Singapore, the Philippines, Indonesia, Papua New Guinea, Australia, and New Zealand, Candaba Swamp is filled with a wide array of migratory bird species wintering from October to April. During the dry season of December to May, the swamp itself is mostly dried out. In the January-February time frame, it is largely drained and converted to melon plantations and rice fields. During this period, large numbers of egrets descend on the swamp's shallow pools to feed on small fish and snails. When the wet season is at its peak, from June to November, the swamp is flooded and used for fish-pond aquaculture.

Located 5 miles (8 kilometers) north of the town of Baliuag, in Bulacan Province next to the Pampanga River, is a 1-square-mile (3-square-kilometer) shallow water reservoir that is the primary refuge for waterfowl, featuring small islands and floating mats of aquatic plants. Candaba is a critical winter refuge that supports rare migrants such as purple swamp hen (*Porphyrio porphyrio*), Chinese pond heron (*Ardeola bacchus*), black-crowned night heron (*Nycticorax nycticorax*), Von Schrenck's bittern (*Ixobrychus eurhythmus*), great bittern (*Botaurus stellaris*), coot (family *Rallidae*), and Eurasian spoonbill (*Platalea leucorodia*). Among threatened bird species that have been observed are the spot-billed pelican (*Pelecanus philippensis*) and the streaked reed warbler (*Acrocephalus sorghophilus*).

The large populations of migratory birds arriving from as far as Australia and Siberia make Candaba a popular bird-watching site for locals, tourists, and scientists. In 1982, approximately 100,000 ducks (family *Anatidae*) were observed here in one day—a record in the Philippines. Candaba Swamp supports more *Anatidae* than any other site in the Philippines. The two most abundant species in the swamp are garganey (*Anas querquedula*) and the Philippine mallard (*Anas luzonica*), though the latter is listed as a vulnerable species by the International Union for Conservation of Nature (IUCN).

There are significant numbers of other ducks, including the northern pintail (*Anas acuta*), wandering whistling duck (*Dendrocygna arcuata*), Eurasian widgeon (*Anas penelope*), northern shoveler (*Anas clypeata*), common pochard (*Aythya ferina*), and mottled duck (*Anas fulviga*). In the late 1970s, the first observations in the Philippines of gadwall (*Anas strepera*) and the endangered Baer's pochard (*Aythya baeri*) were made in Candaba.

Effects of Human Communities

Candaba Swamp is being looked at as a potential source of potable water for the communities of Pampanga, Bulacan, metro Manila, and neighboring provinces. This decision follows a 2011 water summit that projected a water crisis within 15 years due to aggressive extraction of the region's groundwater. The creation of a 2,471-acre (1,000-hectare) water reservoir adjacent to the 79,074-acre (32,000-hectare) swamp could also support year-round fishing. Proponents of turning the swamp into a lake say that in addition to solving the flooding problem, it would create a nautical highway connecting the eastern provinces of Nueva Ecija and Bulacan to the western provinces of Pampanga and Bataan.

In recent times, the swamp has been drained earlier in the year—in December or January as opposed to March or April—and has therefore supported fewer wintering birds. More control of water levels to support agriculture will continue to threaten the biodiversity for which Candaba is famous. The Department of Environment and Natural Resources has declared Candaba a bird sanctuary to protect the area, but although there is a permanent ban on hunting, the killing of waterfowl for recreation or food remains a persistent problem. Additional threats to this important ecosystem include the introduction of nonnative fish species and siltation.

In 1971, The Convention on Wetlands of International Importance, known as the Ramsar Convention, recommended that Candaba be protected, especially as a waterfowl habitat. The convention's mission is to achieve sustainable development for a variety of wetlands, including swamps such as the Candaba, using "wise use" of the resources for the benefit of humankind.

REYNARD LOKI

Further Reading

BirdLife International. *Philippine Wetlands.* Cambridge, UK: BirdLife International, 2011.

Kennedy, Robert S., et al. *A Guide to the Birds of the Philippines.* New York: Oxford University Press, 2000.

Society for the Conservation of Philippine Wetlands (PSDN). *Conserving Arayat Watershed for the Protection of Candaba Wetlands.* Pasig City, Philippines: PSDN, 2011.

Cardamom Mountains Rainforests

Category: Forest Biomes.
Geographic Location: Asia.
Summary: The Cardamom Mountains Rainforests contains very rich habitats, largely untouched but now in need of protective intervention by governmental and environmental organizations.

The Cardamom Mountains rainforests in western Cambodia are among the most species-rich and intact natural habitats in southeast Asia, and among the least explored. The rainforests form a biome covering the entire Gulf of Thailand coastal region of Cambodia, and extending inland as much as 200 miles (320 kilometers) across rugged terrain. The northwestern edge of the Cardamom region abuts the Thai border and the easternmost part stretches to within 60 miles (100 kilometers) of the Cambodian capital of Phnom Penh. The region's area is 2.5 million acres (1.0 million hectares), and the highest point in the range (and in Cambodia) is Mount Aural, at 5,948 feet (1,813 meters). Five main rivers run through the Cardamom, creating dozens of waterfalls.

About 25,000 people live in this remote region, some of whom are ethnic tribal minorities. There are two wildlife sanctuaries in the Cardamom Mountains that were decreed by King Norodom Sihanouk in 1993 solely on the basis of aerial photographs. Khmer Rouge guerrillas retreated to the Cardamom Mountains after losing power in 1979, and for the next 20 years, few would enter the area for fear of the Khmer Rouge and the land mines they left behind. As a result, the region remained largely untouched and undeveloped.

Plant Life

Despite the political upheaval and the years of war that ravaged the countryside and the rainforests, Cambodia is home to a wide range of plant life. The area's fauna and flora are not uniformly distributed, providing unique plant and wildlife species throughout the Cardamom Mountains.

Plant life is largely divided by elevational zones; it has been poorly studied overall.

The lowland evergreen forest of the Cardamom range is abundantly furnished with vegetation, including tall trees such as ficus (*Moraceae* spp.) and various species of bamboo, rattan, and liana. At altitudes of 2,300 to 2,600 feet (700 to 800 meters), shorter trees such as lithocarpus (*Fagaceae* spp.) and syzygium (*Myrtaceae* spp.) are common. The forest floor is covered with leaf litter from numerous plant species including small orchids and terrestrial mistletoe.

Two to three million pileated gibbons like this one once lived in Thailand, but they are now classified as a Vulnerable Species. (Thinkstock)

One of the most abundant canopy species in the wet evergreen forests is the *Hopea pierrei,* a small tree of limited distribution outside this area. Another is the *Anisoptera costata,* an emergent tree of up to 213 feet (65 meters).

A carnivorous pitcher plant, *Nepenthes holdenii,* named for the British photographer Jeremy Holden of Fauna and Flora International (FFI) who discovered the species in 2010, grows on steep ridges in peat-type soil, fully exposed to sunlight. It has large leaves with a wide opening at the top. The leaves, formed into a deep pitcher-like structure, catch and digest insects that enable the plant to nourish itself rather than rely on nutrient-poor soil. They also hold water, making the plant more drought-resistant; its roots are also fairly fire-resistant.

Endangered Animal Species

The forests support endangered species such as the pileated gibbon, tiger, Siamese crocodile, and Asian elephant. Distinctive mammals include the sun bear, clouded leopard, gaur, banteng, and kouprey. Some species here are otherwise found only on Thailand's southern peninsula, such as the flying lemur, moustached hawk cuckoo, buffy fish-owl, silver oriole, and greater mouse deer. There are more than 450 species of birds here, including several that are endemic (found exclusively in this biome).

The kouprey (*Bos sauveli*), also known as the Cambodian forest ox, is one of the most elusive mammals in existence. It was not known to Western science until 1937, and since then has been seen only a few times. The animal looks like a gray forest ox with frayed horns and a long dewlap hanging around its neck. The kouprey inhabits low, rolling hills with patches of dry forests, grazing in the open areas during the day and entering the forests for shelter from predators and sunlight.

The kouprey's range is primarily in northern and eastern Thailand, southern Laos, and western Vietnam. Sihanouk designated it the national animal of Cambodia in the 1960s, partly due to its mystique. The kouprey has always been elusive; indeed, the destruction of its habitat due to slash-and-burn agriculture, logging, warfare, hunting, and disease has caused many experts to believe that it has become extinct—it has not reliably been seen since 1988.

Pileated gibbons are found in the tropical forests here. The gibbon uses its long arms to throw itself from tree to tree, covering gaps of 33 feet (10 meters) or more. An estimated 2 million to 3 million pileated gibbons lived in Thailand prior to aggressive deforestation in recent decades. While the International Union for Conservation of Nature classified the pileated gibbon as Endangered in 1994, preservation programs have improved the creature's status to Vulnerable. In the medium-term, pileated gibbons in the wild are still at risk for extinction due to both habitat degradation and reduced hunting grounds for food.

Ecological Threats

Today, the Cardamom Mountains region is the largest wilderness in mainland southeast Asia, preserving a huge number of endangered species not found elsewhere in the world. Overall, the

The Siamese Crocodile

The Siamese crocodile, also known as the soft-belly crocodile, is among the most endangered of the world's crocodilian species. These animals have disappeared from many of the countries in their dispersal range, including Brunei, Malaysia, and Vietnam, but are found in the Cardamom Mountains rainforests. Siamese crocodiles have been bred extensively in captivity, but captive crocodiles are of very mixed geographic and genetic origin, and have frequently been hybridized with Cuban or saltwater crocodiles for commercial purposes; these tend to be more aggressive toward humans.

The skin of the Siamese crocodile is extremely valuable—which, ironically, may be a boon for conservation, as it means there are economic as well as environmental imperatives for keeping the species alive. The animal is still under significant threat from habitat destruction and poaching.

rainforests of Cambodia are relatively intact, but the areas in southeast Thailand are under pressure from illegal logging, animal poaching, and widespread hunting. About one-third of the ecoregion has been designated as protected areas, the largest of which is Phnom Sankos Wildlife Sanctuary. Other sizeable parks include Phnom Aural, Preah Monivongm, and Botum-Sakor, although the level of active protection in all parks in the mountains is uneven.

The Cardamom Mountains are sparsely populated by humans, most of whom are very poor. In 1992, FFI initiated a study with the Cambodian government to identify the needs of the local population and to instruct the people in basic agricultural skills. The goal was to help the communities become self-sufficient and to prevent escalating social problems and environmental destruction related to hunger. Since 2000, FFI has led a program that integrates community-based environmental conservation and sustainable development in the greater Cardamom ecosystem. The program is now linked with national and nongovernmental organizations to conserve natural resources and indigenous peoples.

Global climate change is projected to generate increases in temperature, rainfall, and extreme weather events in the Cardamom Mountains rainforests. The biome as it is provides carbon storage of an estimated 230 million tons; it also provides protection against floods while serving as a vital natural water reservoir for drier areas to the east. With growing economic and political stability in Cambodia, pressure has increased from logging, mining, and agribusiness interests to develop parts of the rainforest. Most such activity would degrade this ecosystem's capacity to support species diversity and deliver its recognized ecological services.

RHAMA PARTHASARATHY

Further Reading

Boulos, Nick. "Cambodia's Cardamom Mountains, Full of Secrets." *Washington Post*, August 10, 2012.

Daltry, Jenny and Frank Momberg, eds. *Cardamom Mountains: Biodiversity Survey 2000.* Cambridge, UK: Fauna & Flora International, 2000.

McPherson, Stewart, Alastair Robinson, and Andreas Fleischmann. *Pitcher Plants of the Old World.* Volume One. Dorset, UK: Redfern Natural History Productions, 2009.

Caribbean Sea

Category: Marine and Oceanic Biomes.
Geographic Location: North America.
Summary: The Caribbean Sea region is one of the world's biodiversity hot spots; its widely varying habitats are vulnerable to the negative impacts of human impact and climate change.

An enclosed basin of the western Atlantic Ocean, the western and southern sides of the Caribbean Sea form the boundaries of Central and South America, while it encompasses the Antilles Islands chain to the east and northeast. In the northwest, the Yucatan Channel links the Caribbean to the Gulf of Mexico. Stretching for approximately 8,389 miles (13,501 kilometers), the coastal areas around the Caribbean Sea include 26 countries as well as 19 territories of the United States, France, Britain, and the Netherlands. The coast is made up of coral reefs, mangroves, expanses of seagrasses, sand beaches, and rocky shores.

Millions of years ago, the Greater Caribbean was separated from the Pacific Ocean by the closing of the Isthmus of Panama. Separated by water ridges and sills, the floor of the Caribbean Sea today is comprised of the Basins of Grenada, Venezuela, Colombia, Yucatan, and the Cayman Trough. The depth is recorded at more than 7,874 feet (2,400 meters) through three-quarters of the Caribbean, but it reaches 21,326 feet (6,500 meters) in the Cayman Trough. The area of the Caribbean Sea is approximately 1.05 million square miles (2.72 million square kilometers).

Biodiversity

The Caribbean has been identified as one of the world's five hot spots for marine biodiversity. While many species remain unknown, a team of experts led by marine biologist Patricia Miloslavich in 2010 identified 12,046 species of marine life in the Caribbean.

Among the species found in the Caribbean's ecosystem is the tiny bee hummingbird, among the smallest known birds in the world, which is found only in Cuba and on the Isle of Youth. Weighing only 0.004 pound (1.8 grams), the bee hummingbird is often taken for an insect. In 2008, the smallest snake in the world was discovered on the Caribbean island of Barbados by Penn State biology professor Blair Hedges. Less than four inches (102 millimeters) long, this snake, named *Leptotyphlops carae*, looks like a strand of spaghetti. Blair determined that the snake's genetic makeup was new to science, with a unique color pattern and scales. At the other end of the size spectrum, the Caribbean Sea is home to at least two dozen species of marine mammals, from spotted dolphin (*Stenella attenuata*) to the humpback whale (*Megaptera novaeangliae*).

A total of 6,550 plant species are known in the Caribbean region. At least seven species of seagrass are present in the Caribbean Sea, with turtle grass (*Thalassia testudinum*) and manatee grass (*Syringodium filiforme*) the most widely distributed. While many animals tend to remain in a single habitat type for most of their activity cycles, a number split their time between seagrass beds, for instance, and coral reefs. Among fish, the bluestriped grunt (*Haemulon sciurus*) is known as a denizen of the reef, where it shelters during the day. However, the bluestripe migrates by night to seagrass beds for feeding. This behavior has been documented in both juveniles and adults; it is seen as evidence for habitat diversity in the support of maintaining strong populations of native fauna in protected areas. The interaction of mangrove-based species and reef-based species is another area of similar research.

The Atlantic hawksbill sea turtle (*Eretmochelys imbricata*), an endangered omnivore, depends on isolated beaches for nesting, sometimes migrating to the Caribbean from as far north as Labrador, where it feeds during summer. While enjoying some protection, the hawksbill continues to be a target of tortoiseshell poachers who trade in this illicit product.

Resident birds of the Caribbean Sea range from the common pelican (*Pelecanus occidentalis*) and magnificent frigatebird (*Fregata magnificens*) to snowy plover (*Charadrius alexandrinus*) and clapper rail (*Rallus longirostris*). Migratory species rely on many of the islands and wetlands surrounding the sea to make stopovers on their North American routes. These include the peregrine falcon and least tern.

Reefs and Atolls

The Caribbean region is home to some 65 to 75 species of hard coral and at least 500 to 700 reef-associated fish species. The center of reef biodiversity in the Caribbean region is the area around the Belize Barrier Reef, which hosts 40 to 50 species of hard coral and 80 percent of reef-associated species found in the biome. In addition to providing a home to a vast number of creatures and their ecosystems as well as offering a wealth of economic resources for local inhabitants and tourists, the reefs of the Caribbean offer crucial protection from the destructive hurricanes that frequent the region in the summer months.

More than one-fourth of all fish species in the world are found in coral reef ecosystems. Across the planet, these reefs are under threat from climate change and direct human activity. That behavior generally involves such practices as overfishing, fishing techniques that employ explosives and cyanide, and coastal development. As climate change causes temperatures to rise in oceans and seas, associated coral bleaching is a challenge to coral growth and health. In addition, increased carbon dioxide concentrations are causing seawater acidity levels to rise, a related threat to coral sustainability.

Researchers at the Global Coral Reef Monitoring Network estimate that 20 percent of coral reefs are threatened or already harshly impacted. Another 35 percent of coral reefs could be lost in the first few decades of the 21st century. Only 45 percent of all reefs are considered to be at low risk. In the Caribbean Sea, about 38 percent of coral reefs have been identified as lost or critically threatened; another 24 percent are considered moderately threatened.

The region around the Caribbean Sea has two true barrier reefs. The largest of these is the Belize Barrier Reef, which stretches for some 137 miles (220 kilometers) from the Yucatan in southern Mexico to the Gulf of Honduras. In the southwest Caribbean, the second major barrier reef is located north of Providencia Island, Colombia. In the Caribbean, the most dominant reefs are those identified as fringing reefs, which means that they are adjacent to mainland areas or to continental islands. Fringing reefs also encircle the majority of the smaller Caribbean islands, including Aruba, Bonaire, Antigua, and the Caymans.

There are only about 10 to 20 true atoll reefs in the region of the Caribbean; most of these are in the offshore area that extends from the Yucatan to Nicaragua. Unlike most of the atoll islands of the Pacific, these atolls do not appear to be volcanic. Along the coast of southern Belize lies Glover's Reef, the best-developed atoll in the region. The reefs of the Caribbean region also are distinct from those in the Indo-Pacific region in other ways, including the fact that Caribbean reefs generally do not have prominent algal ridges on reef crests. In the Indo-Pacific region, reef surfaces tend to be made up of hard corals and microalgae, while those in the Caribbean instead produce an abundant mixture of sponges and octocorals. Because of the differences present here, biodiversity is enriched and reef topography is more complex here.

In 2000, the Seaflower Biosphere Reserve was established, sponsored by the government of Colombia. This reef-based sanctuary includes about one-tenth of the Caribbean Sea. In 2005, a core area of the reserve was designated as a Marine Protected Area.

Human Impact

Environmental damage and species disruption in the Caribbean Sea have come in the form of habitat destruction resulting from the development of farm runoff, aquaculture, petrochemical exploitation, and other threats. When the habitat of an endemic species is encroached upon, by definition there is nowhere else for the species to go. At least 38 species are known to have become extinct recently in the Caribbean. Many others, such as two species of giant shrews and the Cuban crocodile, are considered highly threatened. Additionally, 48 bird species, 18 mammal species, and 143 amphibian species that are endemic to the Caribbean are now threatened. On the positive side, nearly 18,640 square miles (30,000 kilometers) have been classified as protected areas by various governments within the Caribbean region.

Since the late 20th century, the reefs of the Caribbean area have become increasingly more vulnerable to overdevelopment along the coast and to overuse from recreational purposes. As the quality of the waters of the Caribbean Sea have been overtaxed by human uses, algae have begun to smother coral in some locales. To deal with reef degradation, Caribbean nations have identified Marine Protected Areas that are organized around protection of coral reefs and related habitats. However, environmentalists warn that much more needs to be done to adequately protect the ecosystems here. They are calling for better management of protected areas, improved monitoring of protected species and sites, stricter enforcement of environmental laws, and significant expansion of existing protected areas.

Climate change impacts in the Caribbean Sea are already particularly evident in sea-level rise, seawater temperature increase, heavier rains, and more violent hurricane events. Such effects are expected to multiply in the decades ahead. Scientific predictions include a doubling of hurricane damage for every 39-inch (1-meter) rise in sea level, for example. Beach erosion, intertidal zone habitat degradation, coral reef stress, and associated undercutting of biological diversity are among the unfortunate, but likely, outcomes.

ELIZABETH RHOLETTER PURDY

Further Reading

Beets, J., L. Muehlstein, K. Haught, and H. Schmitges. "Habitat Connectivity in Coastal Environments: Patterns and Movements of Caribbean Coral Reef Fishes With Emphasis on Bluestriped Grunt, *Haemulon sciurus*." *Gulf and Caribbean Research* 14, no. 2 (2003).

McKeown, Alice. "Coral Reefs Under Threat." *World Watch* 23, no. 1 (2010).

Miloslavich, Patricia, et al., eds. *Caribbean Marine Biodiversity: The Known and Unknown.* Lancaster, PA: DEStech Publications, 2005.

Norton, Robert. L. *An Inventory of Sea Birds of the Caribbean.* Gainesville: University Press of Florida, 2009.

Schrope, Mark. "Conservation: Providential Outcome." *Nature* 451, no. 1 (2008).

Spalding, Mark. *A Guide to the Coral Reefs of the Caribbean.* Berkeley: University of California Press, 2004.

Caroni Swamp

Category: Marine and Oceanic Biomes.
Geographic Location: Trinidad.

Summary: The Caroni Swamp is an estuary mangrove supporting a wide variety of wildlife; the Caroni Bird Sanctuary here protects the endangered scarlet ibis. Pollution is its greatest threat.

The Caroni Swamp is an estuary mangrove swamp situated on the west coast of Trinidad, southeast of the capital, Port of Spain. The area is bordered by the Caroni River to the north, Madame Espagnol River to the south, Princess Margaret Highway on the east, and the Gulf of Paria on the west. The Caroni Swamp contains approximately 60 percent of the growing mangroves in Trinidad, encompassed within areas of marshland and tidal pools. Climate here is tropical: hot with high relative humidity. The average high temperature is 84 degrees F (29 degrees C) and the low 74 degrees F (23 degrees C). Average rainfall is 62 inches (157 centimeters). The area does have a seasonal climate, with the dry season extending from January to April and the wet from June to December.

The Caroni Swamp can be divided into three regions: mangrove, herbaceous swamp, and the numerous channels and lagoons found between them. The area's vegetation consists of halophytic (salt-loving) trees and plants growing in brackish water (containing a higher saline content than freshwater but not as salty as seawater). Four natural estuaries are associated with the Caroni Swamp: the Caroni, Blue, Guaymare, and Madame Espagnol Rivers. These rivers offer a mix of freshwater and saltwater and, along with the dry and wet seasons, result in water and soil concentrations with fluctuating salinity. The soil here is often waterlogged and low in oxygen.

The Mangroves

Of prime importance to this ecosystem are the mangroves: Seven mangrove species can be found here. Red mangrove (*Rhizophora mangle*) is found growing closest to the water's edge, while the black mangrove (*Avicennia germinans*) is located farther inland, and the white mangrove (*Laguncularia racemosa*) is sited still further inland. In open areas where there are no mangroves, coarse grasses grow. Such zones, called

The Reeds by locals, consist of grasses that are major feeding grounds for the scarlet ibis and other swamp birds.

All plant life in the biome is subject to water submersion from the tide fluctuations. The mangroves are able to extract the freshwater they need while preventing the seawater from harming them. They accomplish this through reverse osmosis, which enables the roots to absorb water and nutrients while filtering the salt. Any excess salt is stored in the plants' leaves. The mangroves also serve to support the growth of various other plants; a variety of vines, orchids, and climbing ferns can be seen growing on and around these trees, along with some lichens that live entirely upon the mangrove trees.

Mangroves have many uses. They are a source of fuel, pilings, boat timber, medicine, and charcoal, and their bark is used to make tannin for tanning leather. In addition, they provide a habitat for oysters, conchs, mussels, and crabs that can be found living in the mud surrounding their broad roots. Mangroves protect the coastlines as well, as their complex root system prevents erosion and protects the inland areas from the worst effects of severe tropical storms, hurricanes, and tidal waves.

Diverse Ecosystem

The Caroni Swamp is a rich ecosystem supporting numerous animals, birds, and fish. The mangroves provide a habitat for the many barnacles, protozoa, and worms that attach themselves to the roots of the trees. These organisms, along with those living in the mud, provide food for the fish and shrimp that, in turn, feed the wading birds and crocodiles. Many coral reef fish species use the mangroves as nurseries to raise their young. Tree boa, spectacled caiman, fiddler crab, and more than 180 species of birds including snowy egrets, blue herons, pelicans, ospreys, jacanas, ducks, gallinules, sandpipers, flycatchers, and grackles call the area home.

With its brilliant red feathers, the endangered scarlet ibis is the national bird of Trinidad. (Thinkstock)

Additional denizens include 32 species of bats, red howler monkeys, white-fronted capuchin monkeys, and West Indian manatees. Tree rats, water rats, crab-eating raccoons, silky anteater, iguana, alligator, and capybara also can be found in the Caroni Swamp.

More than 80 varieties of fish call the swamp home, including tarpon, grouper, moonfish, cavalla, snapper, shrimp, mullet, salmon, tilapia, and the four-eyed anableps (*Anableps anableps*), which can see both above and below the water. Approximately 450 acres of the swamp have been allocated to the Caroni Bird Sanctuary for the protection of the country's national bird, the endangered scarlet ibis.

An ecological investigation of the Caroni Swamp began in 1965. Pollution has been found to be an ongoing problem threatening the ecosystem. Each day, the swamp receives water contaminated with sewage, wastewater from surrounding industry, and pollutants from agricultural runoff. Petrochemical plant oil spills, petroleum residue, and pollutants from residential development also are a significant problem. A further threat is the unsustainable harvesting of the oyster population, as the mangrove's roots are severed to harvest the oysters.

Sandy Costanza

Further Reading

Food and Agriculture Organization of the United Nations (FAO). *Global Forest Resources Assessment 2005—Thematic Study on Mangroves—Trinidad and Tobago.* Rome: FAO, 2005.

Verweij, Marieke C., Ivan Nagelkerken, Suzanne L. J. Wartenbergh, Ido R. Pen, and Gerard van der Velde. "Caribbean Mangroves and Seagrass Beds as Daytime Feeding Habitats for Juvenile French Grunts, *Haemulon Flavolineatum*." *Marine Biology* 149, no. 6 (2006).

Warne, K. "Forests of the Tide." *National Geographic* 211, no. 2 (2007).

Carpathian Montane Conifer Forests

Category: Forest Biomes.
Geographic Location: Europe.
Summary: The montane-zone flora of the Carpathian Mountains covering central and eastern Europe consists almost exclusively of coniferous trees that emerge from the lowland and midaltitude beech and oak woodlands.

The wide eastward bow of the Carpathian Mountains extends from the western border of the Czech Republic to the Danube River in Romania. The 870-mile (1,400-kilometer) range is typically divided into three sections: the Southern Carpathians (in Romania), the Eastern Carpathians (Poland, Romania, Slovakia, and Ukraine), and the Western Carpathians (Czech Republic, Hungary, Slovakia, and Poland). Vegetation biodiversity corresponds to three elevation zones: submontane, montane, and alpine. Although humans have affected Carpathian forests since antiquity, 19th- and 20th-century deforestation has significantly altered the floral landscape.

Before human settlement, beech woods dominated nearly the entire montane zone, located from 1,969 to 3,937 feet (600 to 1,200 meters). Silver fir (*Abies alba*) and Norway spruce (*Picea abies*), both conifer species, have replaced European beech (*Fagus sylvatica*), especially in cooler sites—high altitude and deep valleys. Norway spruce became a preferred species of silviculturists, and since the 19th century, diverse old-growth forests have been replaced by spruce, particularly thriving in the montane-alpine transition.

While beech flourishes on limestone, conifers excel in granite-rich soils. The highest ridges of the Carpathians—Retezat, Tatras, Făgăraş, and Parîng—are composed mainly of granite, providing the right conditions for the conifers at the higher-elevation montane and alpine zones. Among the timberline conifers, stone pine (*Pinus cembra*) and mountain pine (*Pinus mugo*) are abundant.

The floral and faunal diversity of the Carpathian montane conifer forests is low relative to that of its other European counterparts. Endemic (exclusive to this biome) species in the Carpathians are fewer in number than in the Alps, the mountains of the Balkan Peninsula, or the Pyrenees, but this region does provide natural habitat for a bevy of rare or threatened species. Both the stone pine and the mountain pine are among the endangered plants of the conifer forests.

Home to brown bears, wolves, lynx, deer, and boars, the Carpathians also provide refuge for golden eagles to nest, and the only free-ranging European bison population roams within its lowlands. Additionally, eight scarce species of indigenous Romanian amphibians can be found, and almost 200 endemic bird varieties reside in the montane conifer forests, including rare species of mountain eagle (*Aquila chrysaetos*), serpent eagle (*Circaetus gallicus*), mountain cock (*Tetrao urogallus*), black stork (*Ciconia nigra*), and the migratory falcon (*Falco peregrines*).

The interior mountain area has been settled for centuries by Walachian shepherds from the Balkans, who burned parts of the forest along the mountain ridges to create numerous glades and meadows. These areas remain a distinct feature of the Carpathian landscape. Grazing cattle, sheep, and horses are common in the southern and eastern areas (Romania and Ukraine), but are rapidly declining in the western reaches. Agriculture is restricted to the valleys and foothills and is not an economic factor.

Conservation

Before the 18th century, the montane conifer woodlands flourished; however, subsequent Ottoman Empire market demands prompted deforestation of much of the Romanian Carpathian forests. Late 20th-century estimates of Romania's montane conifers reveal 90 percent depletion of the pre-18th-century coverage. Starting in the 1930s, efforts were made to establish protection areas for native forests but only a fraction (approximately 16 percent) is currently under regulation, and the extent of protection management varies among the various nations of the region.

Modern challenges to endemic species include illegal and legal logging, air and water pollution, alpine tourism, and road construction. Hungary's present timber needs, for example, are negatively impacting Carpathian conifer survival; meanwhile, Romania exports timber to supply Hungarian markets and enhance its own economic prosperity.

Climate change is projected to affect the altitudinal zonation of the conifer forests of the Carpathian Mountains. Rising temperatures and prolonged heavy rains have already aggravated erosion and landslides. The timberline, on the other hand, has begun to advance to somewhat higher elevations, in effect adding to the upper reaches of the biome. However, this can only go so far toward reducing habitat stress at lower elevations, as plant communities and the animals that depend on them must adapt or migrate: pressures that often result in a loss of species diversity—especially in combination with the habitat fragmentation via human activities.

There are bright spots. The future shows promise for an anthropogenic correction geared toward long-term preservation of endemic Carpathian conifers. The past three decades have ushered in aggressive measures to salvage this unique ecoregion. In 1979, the United Nations Educational, Scientific and Cultural Organization (UNESCO) listed Romania's Retezat and the Tatras ranges on the border of Poland and Slovakia as biosphere reserves, to support substantial local ecological protection. Additionally, in 2001, the United Nations Environment Program (UNEP) began supervising a cooperative effort between the Carpathian countries to negotiate a framework for protection and sustainable development of resources. The Carpathian Convention, the product of UNEP's collaboration with Carpathian nations, is in particular designed to foster preservation, conservation, and restoration of the endemic Carpathian Mountain biota.

MATTHEW ALEXANDER

Further Reading

Administratia Parcului National Retezat. "Biodiversity—The Flora and Plant Communities." http://retezat.ro/index.php/english/about-the-park/biodiversity.html.

Baker, Richard St. Barbe. *Green Glory: The Forests of the World*. New York: A. A. Wyn, 1949.

Dolezal, J.and M. Srutek. "Altitudinal Changes in Composition and Structure of Mountain-Temperate Vegetation: A Case Study From the Western Carpathians." *Plant Ecology* 158, no. 1 (2002).

Mankovska, B., B. Godzik, O. Badea, Y. Shparyk, and P. Moravcik. "Chemical and Morphological Characteristics of Key Tree Species of the Carpathian Mountains." *Environmental Pollution* 130, no. 1 (2004).

Röhrig, E. and B. Ulrich. *Temperate Deciduous Forests*. Amsterdam, Netherlands, Elsevier, 1991.

United Nations Educational, Scientific and Cultural Organization. "Biosphere Reserve Information: Romania, Retezat." 2005. http://www.unesco.org/mabdb/br/brdir/directory/biores.asp?code=ROM+02&mode=all.

Carpentaria Tropical Savanna

Category: Grassland, Tundra, and Human Biomes.
Geographic Location: Australia.
Summary: This semi-arid savanna has proven resilient over the millennia, but is threatened by ranching and global warming.

The Carpentaria Tropical Savanna biome stretches inland from the Gulf of Carpentaria on the north-central coastal region of Australia. It comprises portions of both the Northern Territory and the state of Queensland. The savanna lies in a band of up to 125 to 280 miles (200 to 450 kilometers) that encompasses the shoreline of the gulf up to rolling hill areas, extending east to west some 745 miles (1,200 kilometers). Tropical and subtropical grasslands, savanna proper, and shrublands are all contained within the biome.

Mangroves and grasslands along the Gulf transition to open woodlands of eucalyptus trees with an understory of grasses in the genera *Spinifex* and *Triodia*. Along streams, the river red gum (*Eucalyptus camaldulensis*) is prominent; small stands of rainforest are also seen, especially in the rocky uplands. Rainfall here is mainly a summer phenomenon; the average range is 16–48 inches (40–120 centimeters) annually, and cyclones account for a not insubstantial part of that. The driest areas inland may receive only 8 inches (20 centimeters) of rain. Temperatures at the high end reach 97–102 degrees F (36–39 degrees C).

The grasslands here, among the largest expanses on the continent of Australia, have long supported grazing and agriculture. While a boon to the settler economy, grazing is now a considerable threat to habitats in the Carpentaria, as invasive grasses have upset the balance of some local biota. Likewise, various nonnative animal species of donkey, cat, goat, and pig have become feral, engendering new inputs while stressing native grasses.

Morning Glory Tour

Along the northern coastline, cool Pacific Ocean winds breach inward over the land until they collide with a thick blanket of warm, humid air. Upon this U-shaped inlet, a rare type of cloud forms, dubbed the Morning Glory. This only happens in spring, typically on a 90-degree F (32-degree C) day. The last rain was not enough to satiate the cracked clay ground; the soils are hardened with dry roots that have been baking in the sun for the past seven months. Relentless, the sun evaporates the moisture from the ground, giving rise to the horizontal tubelike Morning Glory cloud.

The great wavelike cloud rolls from Cape York Peninsula, entering over the Gulf of Carpentaria and spans across the coastal shores where white and Aboriginal fishermen mingle. The indigenous inhabitants own or lease nearly one-third of the land within this region. In the wetlands along the gulf, dugongs share the shallows with migratory wading birds. Terrestrial birds here include the ubiquitous king quail (*Coturnix chinensis colletti*) and long-tailed finch (*Poephila acuticauda*). A rarely seen, endangered avian species that haunts the biome is the red goshawk (*Erythrotriorchis radiatus*).

The Morning Glory cloud now rolls above patches of wetlands and seasonal ponds shaded by mangroves. Edging southwestward, the cloud soars over yellow plains of hummock grass, open eucalyptus forests, and acacia trees. The plains are interspersed by plateaus and giant rock towers. They are horizontally stratified, exposing the years in which sediments have been washed along the floodplain, deposited by streams and rivers, and compressed with yet another year of weather and new sand.

Soils vary in color from yellow and red to shades of brown. Laterite soils are embedded with heavy metals, providing the vegetation above with meager nutrition. The mining pits found in the area give testament to the metals hidden deep within the soils that have yielded zinc, lead, and silver.

The land appears dry and desolate, with the exception of a few cattle. In taking a closer look though, exotic and native organisms are interacting with one another on a daily basis. A cane toad (*Bufo marinus*) hides under the brush of a prickly acacia tree (*Acacia nilotica*). Both the toad and the tree are exotics once brought to the continent for practical intentions. They have adapted to the region, competing with resources of other native animals; various native reptiles and amphibians have declined with the rise of cane toad populations. Efforts to control the spread of acacia trees have been ongoing since the 1980s, but with limited success.

Empires of termites rise and fall within dirt mounds, some of which are taller than people. Over 300 species of termites are found in Australia, a mix of both native and exotic species. Under the ground, rodents and mammals have created hideouts. The endangered northern quoll (*Dasyurus hallucatus*) sleeps away the afternoon keeping cool in its den while snakes bask in the sun. Meanwhile, the kangaroo and wallaby can be found grazing on grass.

Finally, the Morning Glory cloud approaches eagerly waiting hang gliders in an open, rolling field. They take off and drift upward toward the horizontal tunnel cloud. As the afternoon presses on, the hang gliders ride out the cloud with great skill until it dissipates over the horizon.

Threats and Conservation

The expansion of cattle ranching is a growing concern in maintaining a balance with both fire and the native species. Exotic feed grasses have overtaken the habitat of many Australian plants. And when ranchers occupy a land with too many cattle, the fire regime changes, with more frequent but less intense fires. It is unclear what long-term effect this has on habitat and regrowth.

Strategies for fire management in this region incorporate both modern technology through predictive modeling and traditional indigenous knowledge. Some scientists, for example, have calculated that if ranch owners clear pastures later in the dry season, they can reduce the amount of carbon released into the atmosphere through fire. Aboriginals, for their part, are cautious to invite any changes to a system that has worked well for centuries.

Climate change seems projected to intensify droughts, which already visit their severe effects upon the region once every 20 years or so. The shallow soils of this biome, although in part regenerated by fire, are not helped by the trampling of cattle nor the disruption of invasive species. Additional stress in the form of longer, hotter droughts coupled with potentially stronger cyclones and the erosion from higher storm surges will not favor currently viable habitats. The overall flat nature of the land here is another characteristic that makes the biome more vulnerable, as such a surface is not favorable for capturing rainfall, making inland erosion a danger and doing less to provide for habitat niches than a more topologically diverse landscape could.

Across the savanna region here, there is evidence that small mammals are already under high stress; global warming effects are not thought likely to reverse this trend. On the other hand, studies of the tropical savanna bird species here indicate that, while the habitat range for some species might decrease, territorial opportunities for many of the migratory species may increase with global warming.

Reserve areas are patchy at best in the Carpentaria tropical savanna. The Boodjamulla National Park, near the inland extreme of the savanna, and blessed with waterfalls and limestone and sandstone gorges, consists of about 1,500 square miles (3,880 square kilometers). Chillagoe-Mungana Caves National Park protects another 14 square miles (36 square kilometers), characterized by such plants as kurrajong or bottle tree (genus *Brachychiton*) and the coral tree (*Erythrina variegata*).

Michelle Cisz

Further Reading

Booth, Carol and Barry Traill. *Conservation of Australia's Outback Wilderness.* Maleny, Australia: The Pew Environment Group and The Nature Conservancy, 2008.

Greening Australia. *Native Vegetation Management—A Needs Analysis of Regional Service Delivery in Northern Territory—Savanna.* Canberra, Australia: Natural Heritage Trust, 2003.

Grice, Tony and Shane Campbell. *Weeds in the Tropical Savannas.* Brisbane, Australia: Queensland Department of Natural Resources and Mines, 2001.

Cascades Forests, Eastern

Category: Forest Biomes.
Geographic Location: Northwestern United States.
Summary: The biodiversity of the eastern slopes of the Cascade Mountains is endangered by climate change and the ongoing impact of various land-use projects. The area features high turnover in species along environmental gradients, commonly referred to as high beta diversity.

Underlain by Cenozoic volcanic rock and associated deposits, the Eastern Cascades Forests biome is defined by the eastern slopes of the Cascade Mountains in Oregon and Washington that owe their existence to the subduction of the Juan de Fuca Plate under the North American Plate some four million years ago. Numerous dormant and active volcanoes are part of an extensive volcanic arc that rims the larger Pacific region. Cinder cones and fields of pumice are evidence of past volcanic activity in eastern Oregon; the 1980 eruption of Mt. St. Helens in southwest Washington is a more recent reminder. To the north is Mt. Rainier, the tallest peak in the region at 14,409 feet (4,393 meters), and the most extensively glaciated mountain in the lower 48 states. Pleistocene glaciations created numerous alpine lakes; however, most glaciers are restricted to small alpine areas.

This forest of fir trees common to the eastern Cascades forests, including subalpine fir and white fir, was killed by a fire in the Wenatchee Mountains in Washington State around 2005. (Wikimedia/Walter Siegmund)

The largest river in the Pacific Northwest (and fourth-largest in the continental United States), the Columbia, flows out of the Rockies of British Columbia, heads northwest and then south into Washington, and marks the boundary between the states of Oregon and Washington before reaching its terminus at the Pacific Ocean. Fourteen hydroelectric dams block the mainstem and are part of ongoing controversy over management of wild salmon runs, many of which are threatened with extinction.

The southern extent of these forests is marked just across the Oregon border in northern California by the Klamath River, which was once the third-largest producer of salmon in the continental United States. Numerous instream diversions, hydroelectric dams, logging, and livestock grazing have jeopardized wild salmon runs on the Klamath; however, an agreement is currently in place to remove the four mainstem dams by 2020, restoring flow and upstream access for fish. A moist temperate climate currently exists west of the Cascades; a drier, warmer continental climate is to the east.

Ecoregional and Vegetation Types

Vegetation types in these forests is adapted to the prevailing dry climate and to periodic fire of varied severity. Close proximity to the southern Cascades (especially their western moist slopes), as well as the Klamath-Siskiyou, Sierra-Nevada, and Snake-Columbia shrub steppe ecoregions allows for mixing of plant communities and numerous "ecotones" (zones of overlap). Natural vegetation is a complex and largely fire-generated mosaic of shrublands, grasslands, and coniferous forests; seven different forest zones and numerous plant associations have been recognized.

Dry forests include ponderosa (*Pinus ponderosa*) and mixed conifer forests dominated by Douglas fir (*Pseudotsuga menziesii*), grand (*Abies grandis*) and white (*A. concolor*) fir, western larch (*Larix occidentalis*), lodgepole pine (*P. contorta*), sugar pine (*P. lambertiana*), incense cedar (*Calocedrus decurrens*), and western juniper (*Juniperus occidentalis*). Topographic moisture gradients such as sheltered valleys or north-facing slopes provide microclimatic conditions for mesic (moderately moisture-dependent) plants to coexist in this dry region. High-elevation areas include Shasta red fir (*A. magnifica var. shastensis*) and subalpine fir (*A. lasiocarpa*).

Role of Fire

Fire is a predominant disturbance factor in this region and its historic versus contemporary influences have been widely debated by land managers and researchers. Many have assumed that fires in dry mixed conifer forests were historically low-severity (burning primarily along the ground, removing shrubs and small trees), creating open parklike conditions that have been altered by livestock grazing, high-grade logging, and fire suppression. This scenario has created forests that today are more dense and likely to burn severely, which is uncharacteristic for the area.

However, recent studies using surveys from the late 1800s and historic stand reconstructions challenge prevailing views, providing evidence that historic fire regimes actually varied greatly in extent, frequency, and severity, and that parklike conditions were rare. It remains to be determined whether contemporary fires will become more severe due to climate change. Studies show a recent uptick in fire extent; however, climate projections are uncertain for future precipitation patterns. Nonetheless, fire prevention through suppression and forest thinning has dominated land management concerns.

Biodiversity Importance

The Eastern Cascades Forests ecoregion is considered bioregionally outstanding by the World Wildlife Fund, with intermediate levels of taxa richness and endemism. Beta-diversity is particularly high due to numerous ecotones along the Cascade Crest (e.g., portions of the Okanogan-Wenatchee National Forest) and overlap of plant communities from surrounding ecoregions. Biodiversity of the region was ranked as Endangered due to high numbers of taxa with conservation status (state or federal listed), and to high rates of conversion due primarily to logging, livestock grazing, invasive species, and hydroelectric development. Logging has significantly depleted old-growth forests, particularly large sugar pines, and most such habitat zones, outside inventoried roadless areas, are highly fragmented.

Wildlife in the region is diverse; the mountains, forests, and clean water attract many species. Marmot, black bear, mountain goat, several species of deer, golden mantle ground squirrel, chipmunk, and many species of birds are native to the area. The list of threatened, endangered, and candidate species within the biome is long, and includes the bald eagle, common loon, northern spotted owl, and pileated woodpecker, among others.

Many animal populations have been adversely affected by the encroachment of human development at the same time that similar animal and fish populations are thriving in some areas. The Skagit River, flowing toward Puget Sound, is a major waterway for rainbow, cutthroat, and Dolly Varden trout (*Salvelinus malma malma*) and several types of salmon. The Upper Snake River region is home to 14 species of fish. Reaches of the Columbia River below the Snake River is home to as many as 35 species of fish and high concentrations of freshwater mollusks.

Climate Change

Downscaled global circulation models for the central Oregon portion of the Eastern Cascades Forests ecoregion predict an increase in annual average temperature of 5 to 9 degrees F (2.8 to 5 degrees C) by the late 21st century; summer projections show the greatest degree of warming. Projections for annual average precipitation vary from 7 percent less to 22 percent more by late century. Future winters are likely to be somewhat wetter, but all other seasons have variable projections for precipitation trends.

Increasing temperatures suggest snowpack levels will continue to decline despite wetter winters; this could affect seasonal salmon migrations. Vegetation model results indicate climatic conditions may favor mixed conifer forests dominated by pine, at the expense of mixed conifer dominated by fir and subalpine species. The extent of wildfire is projected to increase by late century; however, there is high uncertainty due to varied precipitation projections.

Conservation

The Northwest Forest Plan (of the U.S. Department of the Interior) continues to provide the foundation for ecosystem management and biodiversity conservation on federal lands in this and related Pacific Northwest ecoregions. The plan's reserve network, combined with inventoried roadless areas on federal lands, provide protection for the bulk of the region's remaining late-successional forests.

There are gaps in the reserve network, however. Climate change will require building onto the existing reserves so they are robust, expanding the network along elevation and latitudinal-longitudinal gradients to accommodate climate-forced shifts in plant and wildlife communities, and restoring aquatic floodplains to prepare for potential flooding if precipitation increases. Such measures can help to recover wild salmon runs and to reduce land-use stressors (e.g., livestock grazing) that contribute to a shift from native communities toward more invasive ones.

Restoration of native grasslands, shrublands, and riparian areas is essential for climate adaptation. More research, monitoring, and adaptive management is needed on fire to determine what truly constitutes "uncharacteristic" events and whether contemporary fire management is likely to be restorative or create novel ecosystems that worsen climate-related impacts. Development of renewable energy sources (e.g., biofuels) is a growing concern in this and surrounding regions. Partnerships with conservation researchers, land managers, and energy developers are essential for reducing habitat fragmentation problems.

Dominick A. DellaSala

Further Reading

Baker, W. L. "Implications of Spatially Extensive Historical Data From Surveys for Restoring Dry Forests of Oregon's Eastern Cascades." *Ecosphere* 3, no. 1. (2012).

Geos Institute, "Planning for Climate Change in Central Oregon." 2010. http://www.geosinstitute.org/completed-climatewise-projects/planning-for-climate-change-in-central-oregon.html.

Ricketts, T. H., E. Dinerstein, D. M. Olson, C. J. Loucks, W. Eichbaum, D. DellaSalla, K. Kavanagh, P. Hedao, P. Hurley, K. Carney, R. Abell, and S. Walters. *Terrestrial Ecoregions of North America: A Conservation Assessment.* Washington, DC, Island Press, 1999.

Caspian Lowland Desert

Category: Desert Biomes.
Geographic Location: Asia.
Summary: A salt-rich lowland desert ecosystem located along the northern and northeastern margins of the Caspian Sea, this biome sustains salt-tolerant scrub vegetation and both indigenous and migratory fauna.

The Caspian lowlands resulted from the collision of tectonic plates that pinched shut a portion of an ancient ocean in the relatively recent geologic past. The salt-rich depression that formed was key to inducing the conditions that eventually created this unique lowland desert ecosystem. The salt deposits define, in part, the character of the ecosystem and offer various economic benefits.

Situated in an endorheic basin—one with no outflows to the ocean—the Caspian Sea is actually a vast lake in central Asia, but is situated in a climate zone that lends to the brackish nature of its waters. This, in turn, has created a high-saline desert that extends around much of the eastern and northern shores of the sea, through parts of the bordering countries of Iran, Turkmenistan, Kazakhstan, and Russia.

Climate

The lowland desert continues to receive various salts from the weathering of local mountain ranges and the adjacent Caspian Sea during its regressive stages. In the latter case, the salt contributions occur along the margins of the desert because it is in contact with the brackish water of the Caspian Sea as it recedes and evaporates. Contrarily, episodic expansions of the Caspian Sea correspond to freshening events and limited exports of water into the adjacent western portions of the lowland desert. This spurs higher but temporal levels of biological activity.

Although the modern climate of the Caspian lowland desert is temperate and dry, geologic data indicates that the climate of this region was once cool, moist, and supportive of a more diverse ecosystem. Nowadays, the nearby mountains strip moisture from the air and hinder significant precipitation to the region. This orographic effect exacerbates the effects caused by the lack of precipitation in this region, which is typically less than 15 inches (38 centimeters) annually. Notably, during periods of extended drought, the hard lowland desert soils prevent water infiltration into the ground following precipitation events. This process promotes runoff and the creation of surface pools in low-lying areas.

Vegetation and Animals

Ponded water eventually evaporates and concentrates salt deposits, affecting the soil and water quality of the region, and ultimately the diversity and sustainability of plant and animal groups in the lowland desert. As a result, salt-tolerant (halophytic) and drought-tolerant (xeric) vegetation including shrubs such as tetyr and sagebrush, sedge (*Carex* spp.), and grass genera such as *Stipa* and *Panicum* dominate this ecosystem. Among the predominant halophyte types are saltworts (*Salicornia* spp.), orache or saltbush (*Atriplex* spp.), and goosefoot (*Chenopodiaceae* spp.). Tulips are also represented here.

Despite the harsh conditions, the lowland desert ecosystem supports an assortment of fauna, particularly small to moderate-size mammals that include various cat groups, antelope, jackal, rodents, and other animals that typically occupy dry temperate wastelands. The grass and low vegetation particularly allow grazing by wild and domesticated herbivores; the region is also home to reptiles, whether vegetarian or omnivorous.

Multitudes of avian flocks winter here, mainly in close proximity to the Caspian Sea; these include massed migratory formations of coot (*Fulica atra*), Sandwich tern (*Thalasseus sandvicensis*), and flamingo (*Phoenicopterus roseus*). Among the predatory birds recorded here are the eastern imperial eagle (*Aquila heliaca*) and white-tailed eagle (*Haliaeetus albicilla*).

Human Interaction

This ecosystem is viewed as a viable economic resource because the region supports livestock grazing, salt mining, and petroleum exploitation. The indigenous inhabitants view each of these activities as a necessity for sustainability, due to the constraints imposed by the harshness of the environment, which inhibits extensive development. However, each activity also imposes stress on this delicate ecosystem. Sheep and cattle commonly overgraze in ecosystems where precipitation is limited, because food supplies are of poor quality or scarce. This often causes grazers to consume and destroy even the root systems of the plants they forage. Consequently, natural food sources diminish with time, inducing further stress on both the ecosystem and those that use it.

Similarly, salt mining may alter the landscape for long periods because the limited precipitation impedes the process of plant reestablishment and succession even when it is supported by restorative efforts. Likewise, tailings from underground mining of locally occurring salt domes enter soil, surface water, and aquifer systems if the wastes are not properly managed. Further, the highly soluble salts migrate extensively and have wide-ranging effects on surrounding freshwater systems.

The Caspian lowland desert is susceptible to damage from accidental petroleum spillage. Some petroleum additives are soluble and miscible in water, and will severely damage potable water supplies. Further, the drilling and support activity tends to disrupt fragile vegetation networks in the desert soils. The costs of recovering spilled petroleum and restoring affected regions within this ecosystem could surpass the economic benefits derived from petroleum exploration.

The Caspian lowland desert is expected to expand, as global warming affects this region. However, this is predicted to come at the cost of lowered biodiversity, as relatively rapid changes in average temperature, seasonality of precipitation, and extreme weather events combine with man-made pollution to fragment the existing habitat web.

Careful planning and exhaustive cost-risk-benefit analyses of the effects of tapping the region's natural resources—and counterbalancing the negative effects of climate change—will ultimately affect the short- and long-term quality and sustainability of this ecosystem.

John A. Anton

Further Reading

Abdelly, C., M. Ozturk, M. Ashraf, and C. Grignon, eds. *Biosaline Agriculture and High Salinity Tolerance.* Basel, Switzerland: Birkhauser, 2008.

Dumont, H. J. "The Caspian Lake: History, Biota, Structure, and Function." *Limnology and Oceanography* 43, no. 1 (1998).

Firouz, E.. *The Complete Fauna of Iran.* London: I. B. Tauris & Co., (2005).

Zehzad, B., B. H. Kiabi, and H. Madjnoonian. "The Natural Areas and Landscape of Iran: An Overview." *Zoology in the Middle East* 26, no. 7–10 (2002).

Caspian Sea

Category: Inland Aquatic Biomes.
Geographic Location: Asia.
Summary: Due to large oil reservoirs in the Caspian Sea, the world's largest enclosed body of water is facing geopolitical and environmental issues from the five countries that border its shores.

The Caspian Sea is the largest enclosed body of water on Earth with a surface area of 143,244 square miles (371,000 square kilometers), a volume of 18,761 cubic miles (78,200 cubic kilometers), composing at least 40 percent of the total inland waters of the world. The land-locked body of water, considered a sea by some and a lake by others, has no outflows. Kazakhstan is located on the sea's north and northeast coasts, Russia is to the northwest, Azerbaijan is on the west, Turkmenistan to the east, and Iran is on the southern shore. Oil and gas reserves and pipelines, along with the ancient and modern trade routes, define the region's geopolitical and economic climate.

The Caspian Sea is a relic of the ancient Thetis Ocean, which served as a link between the Atlantic and Pacific Oceans. Due to tectonic uplift and a fall in the worldwide sea level, the Caspian became landlocked some 5.5 million years ago. Its water is a mix of fresh and saline, depending on the location. The body of water has freshwater in its northern portions and is brackish elsewhere, with salinity of about one-third that of the global oceanic average.

More than 130 rivers empty into the Caspian. The Volga River drains 20 percent of Europe's land area and enters from the northwest, its delta channels providing three-fourths of the Caspian's freshwater inflow. Among others, the Ural and Kura Rivers empty into the sea from the north and the west, respectively.

To the west and south, the Caspian Sea is framed by the Caucasus and Alborz Mountains; in the latter range is Mount Damavand, at 18,600 feet (5,670 meters) the tallest peak in Eurasia west of the Hindu Kush range in central Asia. By contrast, the lands to the north and east of the sea tend to be flat steppe. Indeed, the semi-arid desert of the Caspian Depression is as much as 430 feet (130 meters) below sea level.

Depths in the Caspian Sea vary dramatically, too. The northern part is a shallow shelf with an average depth of just 16 to 20 feet (2 to 6 meters). Most of the Caspian islands lie in this region, with a total area of approximately 772 square miles (2,000 square kilometers). The islands are small and mostly without human settlements. The island of Ogurja Ada, at 23 miles (37 kilometers) long,

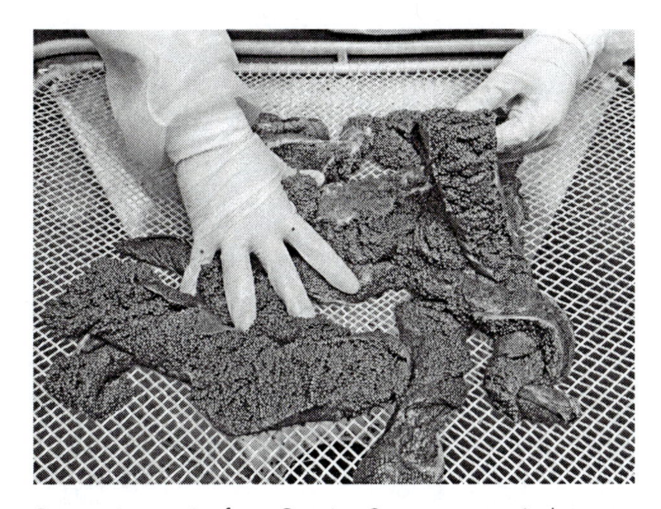

Processing caviar from Caspian Sea sturgeon. As late as 1980, the Caspian Sea held over 75 percent of the world's sturgeon, but overfishing, caviar harvesting, and pollution have decreased their numbers. (Thinkstock)

is the largest. The middle part of the Caspian has an average depth of 623 feet (190 meters), and the deepest parts of the southern basin plunge more than 3,281 feet (1,000 meters).

The northern portion freezes in the winter; if the weather turns very cold, the water's surface freezes in the Caspian's southern area. Cold continental desert climate rules the northern region, except for a more temperate and moist zone around the Volga delta, while a warm temperate climate embraces the southernmost extent.

Biota

Due to its enclosed nature, the aquatic plants and animals in the Caspian reach a very high level of endemism (species found exclusively here), as much as 80 percent. Despite the challenges of its variable salinity conditions, the Caspian Sea is home to 87 species of microphytes and many species of green algae.

Wetlands at the inflow points of its rivers give the Caspian various grassland and forest habitats as well. The Volga delta harbors such characteristic wetlands vegetation as waterwheel plant (*Aldrovanda vesiculosa*) and Caspian lotus (*Nelumbo caspica*). In the floodplain of the Samur River, which enters the Caspian from the Caucasus, a liana forest is found.

The most emblematic aquatic species here are the various types of sturgeon; the Caspian Sea was home to more than three-quarters of all sturgeon in the world as recently as 1980. Among the key species are Russian sturgeon (*Acipenser gueldenstaedtii*), Persian sturgeon (*A. persicus*), star sturgeon (*A. stellatus*), and beluga or European sturgeon (*Huso huso*). Being anadromous, these fish use the major rivers as spawning grounds—a fact that has acted to deplete their populations here in part due to dam construction.

Other Caspian fish, important both commercially and to the ecosystem, include the kutum or Caspian whitefish, salmon, marine shad, Caspian roach (*Rutilus caspicus*), and white-eyed bream (*Abramis sapa*).

Tyuleniy Archipelago in the northeastern Caspian Sea is an important area for waterbirds such as swans, coots, and egrets. Caspian gull (*Larus cachinnans*) and Caspian tern (*Hydroprogne caspia*) are among the endemic avian species here.

Among mammals unique to this biome, the Caspian seal (*Pusa caspica*) is one of just two inland seals in the world, the other being endemic to Lake Baikal in Siberia. The Caspian seal seems to be comfortable in all areas of the Caspian Sea, as well as some river stretches. It also varies in its birthing regimen, perhaps preferring ice floes in the northern waters but also seen giving birth on sandbars further south. These seals feed on crustaceans as well as fish of the *Clupeonella* genus.

Environmental Concerns

Since the Caspian Sea has a closed basin, it has no natural outflow except evaporation. Therefore, its sea level has been inconsistent over the centuries. Scientists have noted that the sea level changes in the Caspian are linked to atmospheric conditions in the North Atlantic. The cycles of the North Atlantic Oscillation, a climatic phenomenon of atmospheric pressure at sea level, causes variations in the interior depressions of Eurasia, resulting in the fluctuating rainfalls in the Volga River valley. This climate regime eventually affects the level of the Caspian Sea and increases its significance in global climate change studies.

The rivers of the Caspian carry large quantities of industrial pollutants from their catchment basins, endangering the biological environment of the Caspian. Overfishing is another threat, particularly to the sturgeon, whose eggs are processed into quite valuable caviar. Caviar harvesting targets reproductive females of this slow-to-mature fish family, putting its sustainability even more at risk. Conservationists have suggested bans on sturgeon fishing in the Caspian until the population fully recovers.

Because of the construction of water channels and canals like the Volga-Don during the 20th century, many species were introduced to the Caspian from the Black and Mediterranean Seas. Some of these could not survive properly, while others were hostile to some native species. Mineral extraction, underwater oil drilling, and gas pipelines, coupled with transport activities, are additional environmental threats to the Caspian's ecosystem.

Geopolitical Issues

Various islands near the coast of the Caspian Sea have strong geostrategic significance. Nargin, the largest island in Baku Bay, served as a base for the former Soviet Union. The region also is important because of its oil resources, production of which dates as far back as the 10th century. Some of the world's first offshore wells and machine-drilled wells were made in Bibi-Heybat Bay in Baku, present-day Azerbaijan. The greater Caspian-Caucasus area once provided more than two-thirds of the Soviet Union's entire oil production. The fields are still producing; in 2006, Azerbaijan began transporting oil to Turkey through the Baku-Tbilisi-Ceyhan pipeline.

The fate of fossil-fuel projects here is subject to agreements among the five littoral states of the Caspian about the use of the sea for oil excavation, transportation—and the impacts on commercial fishing. Various issues concerning how to divide the resources of the Caspian Sea remain to be settled between Russia, Iran, Azerbaijan, Kazakhstan, and Turkmenistan. Not only overland pipelines but also the use of riverways and canals hold weight in negotiations. The Volga River, for instance, connects the Caspian Sea through the canals with the Black Sea, Baltic Sea, Northern Dvina, and White Sea. It is the only way at present that the landlocked states of Azerbaijan, Kazakhstan, and Turkmenistan can gain access to international waters.

MUHAMMAD AURANG ZEB MUGHAL

Further Reading

Anker, Morten, Pavel K. Baev, Bjørn Brunstad, Indra Øverland, and Stena Torjesen. *The Caspian Sea Region Towards 2025: Caspia Inc., National Giants or Trade and Transit?* Delft, Netherlands: Eburon Academic Publishers, 2010.

Barannik, Valeriy, Olena Borysova, and Felix Stolberg. "The Caspian Sea Region: Environmental Change." *AMBIO: A Journal of the Human Environment* 33, no. 1 (2004).

Gelb, Bernard A. and Terry Rayno Twyman. *The Caspian Sea Region and Energy Resources.* New York: Novinka Books, 2004.

Ivanov, V. P. *Biological Resources of Caspian Sea.* Astrakhan, Russia: KaspNIRKH, 2000.

Kostianoy, Andrey and Aleksey Kosarev, eds. *The Caspian Sea Environment.* Berlin, Germany: Springer, 2005.

Kroonenberga, S. B., G. V. Rusakovb, and A. A. Svitochc. "The Wandering of the Volga Delta: A Response to Rapid Caspian Sea-Level Change." *Sedimentary Geology* 107, no. 3–4 (1997).

Salmanov, M. A. *Ecology and Biological Productivity of the Caspian Sea.* Baku, Azerbaijan: Institute of Zoology, 1999.

Caucasus Mixed Forests

Category: Forest Biomes.
Geographic Location: Eurasia.
Summary: The forests of the Caucasus are biologically diverse; they are one of the planet's endangered hot spots, a refuge for many unique species.

Considered to be part of the geographic boundary between Europe, Asia, and the Middle East, the Caucasus Mountains cover about 205,660 square miles (532,658 square kilometers) between the Black and Caspian Seas. The nations of Georgia, Armenia, Azerbaijan, Russia, Turkey, and Iran all share portions of this mountainous region. The highest peak, Mount Elbrus, reaches 18,500 feet (5,640 meters). The region is glaciated in places, has expansive lowlands, woodlands to the north, deserts—and river deltas—in the east, temperate forests at middle elevations, and temperate rainforests in the humid western portion.

Climate and Species

The Caucasus functions as a biological crossroads for converging biogeographical provinces from central and northern Europe, central Asia, the Middle East, and north Africa. For a temperate area, the Caucasus has extraordinary numbers of endemic species (those found nowhere else on Earth) and is considered to be globally significant in this regard.

A pine tree in the fog in the Caucasus region, which supports 21 threatened species of wildlife. An international plan calls for increasing protected areas from 13 percent of the overall land to 17 percent. (Thinkstock)

Climate is quite varied here, with nine of the world's 11 major climate types. Precipitation ranges from less than 6 inches (150 millimeters) in the east to more than 157 inches (4,000 millimeters) along coastlines in the west of the range. More than 6,400 plants have been identified in the region, 25 percent of which are native to the area. Rare in Europe, temperate rainforests (or humidity-dependent forests) occur on the eastern portion of the Black Sea catchment (as colchic forests) and the southern coastal area of the Caspian Sea (as the hyrcanian type). Climatic stability has resulted in these forests being among the oldest in western Eurasia.

Colchic forests in particular are small in spatial extent, covering about 7.4 million acres (three million hectares) of the Caucasus and supporting lowland hardwood forests, foggy gorges consisting of mixed broadleaf trees, sweet-chestnut, beech, dark coniferous, and oak woodlands. Notably, areas with limestone scree (accumulation of broken rock fragments found on mountain cliffs or valley shoulders) support extraordinary levels of unique plants and up to 80 percent of plant species are endemic. Relict species include rhododendrons, oaks, Persian ironwood, Caucasian salamanders, Caucasian vipers, and many others.

Hyrcanian or Caspian Sea rainforests, by contrast, cover nearly 4.9 million acres (2 million hectares) and are found primarily along the southeastern reaches of the biome. Oak and mixed broadleaf forests, ironwood, and ravine forests with high levels of endemic plants characterize them. Hyrcanian forests also support relict species, including Caucasian zelkova, wing nut, date plum, and mountain cranberry.

Endemic animal taxa of the Caucasus Mixed Forest biome include birds (at least one of the 378 species), mammals (18 of 131 species), reptiles (20 of 86 species), amphibians (three of 17 species), and freshwater fishes (12 of 127 species). Leopards (critically endangered), bears, wolves, bison, golden eagles (endangered), and rare Caucasian black grouse can be found.

A total of 21 wildlife species are on the International Union for Conservation of Nature (IUCN) Red List as globally threatened. Because of its overall conservation significance, Conservation International has ranked the region among the 34 most diverse and endangered hot spots (the richest and most threatened reservoirs of plant and animal life on Earth), and the World Wildlife Fund considers it to be a Global 200 ecoregion (area of global biological importance). The Caucasus also has high diversity of ethnic groups, languages, and religions.

Human Impact

Lowland areas in the Caucasus mostly have been transformed by thousands of years of human settlement. Low economic status and high demand for natural resources, particularly firewood, as well as illegal logging and wildlife poaching, have more recently transformed some of the region's densest forests; demand for natural resources has also degraded wetlands, woodlands, and deserts here. Logging and grazing have triggered the spread of invasive weeds and landslides in places.

By the end of the 21st century, climate change could include an increase of as much as 6.3 degrees F (3.5 degrees C) in average annual temperature range, while annual precipitation could drop by 10 percent throughout the Caucasus. The combination of unabated land-use and climate chaos are likely to trigger unprecedented declines of the region's world-class biological diversity, particularly in the temperate rainforests, where many species have adapted to stable climates.

The Caucasus Ecoregional Conservation Plan was prepared by more than 150 experts from all six Caucasian countries. The recent strategic plan adopted by the parties to the Convention on Biological Diversity for the Aichi Biodiversity Targets (2011–20) calls for 17 percent of the countries' land bases to be placed in "ecologically representative" reserves; it has resulted in increased attention to protected areas and transboundary cooperation. This is exemplified by several developments:

- A total 2.5 million acres (1 million hectares) have been added to the protected areas network across the region since 2004, yielding about 13 percent protected areas throughout;
- Protection in Azerbaijan has been doubled since 2000;
- A total of 230 new natural monuments have been established in Armenia since 2009;
- Nearly 12 percent of the region is under protection in Russia, including 93,900 acres (38,000 hectares) of new protected areas since 2010; and
- Increased cross-border conservation areas have been initiated by Armenia and Georgia and by Georgia and Turkey.

In addition, national action plans are underway, engaging local stakeholders in support of protected areas through sustainable financing, and the Caucasus Nature Fund was recently established for long-term cofinancing of protected areas in the southern Caucasus. Other conservation financing mechanisms being tried include payment for ecosystem services (PES) coming from protected areas.

Like many other regions, the Caucasus Mixed Forests biome is at an important cultural-ecological crossroads in terms of whether expanding interest in conservation can keep pace with ongoing demand for natural resources, rising population, and climate disruptions.

Dominick A. DellaSala

Further Reading

DellaSala, Dominick, ed. *Temperate and Boreal Rainforests of the World: Ecology and Conservation.* Washington, DC: Island Press, 2011.

Ministry of Nature Protection. Biodiversity Strategy and Action Plan. Yerevan: Government of Armenia, 2000.

Nakhutsrishvili, George. *The Vegetation of Georgia (South Caucasus).* New York: Springer, 2012.

Celtic Broadleaf Forests

Category: Forest Biomes.
Geographic Location: Western Europe.
Summary: The lush Celtic broadleaf forests of the British Isles have become a fragmented biome, but still provide habitat for a surprising range of species.

Celtic broadleaf forests are located throughout Ireland and Britain, and as far north as Newfoundland. Characterized by their mixed deciduous-dominated canopy, year-round rainfall, and humid, temperate climate, they have a long history of deforestation, with the most intense activity occurring over the past 80 years. Recently, conservation measures have been enacted to protect this expansive ecoregion.

Celtic broadleaf forests have a mixed population of deciduous (broadleaf) and some coniferous trees. These forests are classified as part of the Palearctic ecozone, a vast temperate area that extends across all of Europe and the northern reaches of Asia and Africa. The Celtic Broadleaf Forests biome is both a temperate broadleaf and mixed-forest biome.

The combined size of the Celtic broadleaf ecoregion is approximately 81,000 square miles (210,000 square kilometers). The growing season ranges from 200 to 330 days because of the consistent rainfall and warm temperatures, an average range of 36–64 degrees F (2–18 degrees C); the ambient climate is in large part generated by the warming, humid influence of the Gulf Stream and its North Atlantic Current extension. Frost is a short-term annual occurrence, and most temperature extremes are limited due to the influence of the seas that immediately surround the land mass of this biome.

Celtic broadleaf forests are influenced by a variety of ecological parameters. In addition to native species, the temperate weather of this ecoregion supports plant communities of the mediterranean, alpine, and Arctic-alpine climate regimes.

Water also plays an important role in these forests; in addition to rainfall, surrounding water bodies affect the forests, encompassing adjacent marine, estuarine, and freshwater environments. Water helps distinguish Celtic broadleaf forests from the English lowland type. Although these forests are similar in flora, fauna, and location, English lowland forests are defined by their characteristically drier landscape.

Soils and Vegetation

The soils of the Celtic Broadleaf Forests biome support a variety of plant life. Based on the topography, soils composition can range from sandy (arenosols) to clayey (cambisols, gleysols) to well-drained (podzols). The plant population throughout the forest depends on soil moisture and its ability to retain water. As suggested by the name, broadleaf trees are the dominant species in these forests, and mixed oak-ash canopies are especially characteristic of this ecoregion. Additional common species include the English oak, downy birch, weeping birch, English holly, European mountain ash, honeysuckle, and a variety of ferns. All these species require well-drained soils.

In waterlogged, poorly drained soils, peat bogs have formed over thousands of years. Peat is formed from the slow, incomplete decomposition of plants in an acidic, low-oxygen, cold environment; for this reason, dead plants and animals found in peat bogs are very well preserved. In fact, bog people—bodies of people who date back to the Iron Age (ca. 1200 B.C.E.)—have been found remarkably well preserved in this soil. Archaeologists and historians have greatly benefited from the opportunity to study ancient cultures of western Europe thanks to the natural preservation of these bodies.

Animal Life

Celtic broadleaf forests also support abundant and relatively diverse animal life. More than 200 species of birds are found in these forests, including the common redshank (*Tringa totanus*), the Eurasian blackbird (*Turdus merula*), and the common black-headed gull (*Larus ridibundus*). Mammals include the western roe deer (*Capreolus capreolus*) and European badger (*Meles meles*). European grass snake (*Natrix natrix*) is a characteristic reptile; a representative amphibian is the European frog (*Rana temporaria*).

A wide variety of invasive and exotic plant species have been introduced to the Celtic broadleaf forests, threatening the existing diversity. Some exotics, such as lime trees, were introduced hundreds of years ago in Ireland to enable citizens to show their status in society. Animal invasives include the mallard duck (*Anas platyrhynchos*), European starling (*Sturnus vulgaris*), house sparrow (*Passer domesticus*), red fox (*Vulpes vulpes*), and European hare (*Lepus europaeus*). In part because of these introductions, the forests contain a remarkable number of endangered species, especially plants.

Effects of Agriculture

Due to plentiful rainfall and humid climate, Celtic broadleaf forests are ideal agricultural lands. The majority of the land in this ecoregion has been converted to agricultural fields for silage, hay, and arable crops; pastures for sheep or cattle; or land to be harvested for peat, a powerful and popular fuel source. About 1 percent of the land area in Ireland is untouched Celtic broadleaf forest, whereas approximately 9 percent of the land has been converted for the timber industry over the past 80 years.

As these environmental industries grew in Ireland, the populations also expanded; the converted forests now sustain populations of 10 to 250 people per 0.4 square mile (one square kilometer). As farming continued, however, it became apparent that the lushness of the Celtic broadleaf ecosystem was not an indicator of its resilience. Expansive deforestation more than 6,000 years ago by Neolithic farmers resulted in significant loss of soil from County Clare, for example, where bare limestone is still visible for miles today.

Conservation and Protection

As of 1998, the Irish government started offering subsidies for sustainable farming and management schemes that limit new deforestation. Additionally, 30 percent of all new reforestation projects plant broadleaf species. One of the largest protected areas of Celtic broadleaf forest is in Killarney National Park in Killarney, Ireland (39 square miles, or 102 square kilometers). The park was established as a Biosphere Reserve in 1981 by the United Nations Educational, Scientific and Cultural Organization; because of this, Killarney National Park is recognized as an area of international conservation and protection.

CATHERINE G. FONTANA

Further Reading

Fossitt, J. A. *A Guide to Habitats in Ireland.* Kilkenny, Ireland: Heritage Council, 2000.

Jeanrenaud, Sally. *Communities and Forest Management in Western Europe.* Gland, Switzerland: World Conservation Union, 2001.

Olson, David M. and Eric Dinerstein. "The Global 200: A Representation Approach to Conserving the Earth's Most Biologically Valuable Ecoregions." *Conservation Biology* 12, no. 3 (1998).

Central American Atlantic Moist Forests

Category: Forest Biomes.
Geographic Location: Central America.
Summary: This large Central American biome is characterized by permanent heavy rainfall and tall trees with dense canopy, a tropical jungle of important species richness.

The Central American Atlantic Moist Forests biome encompasses all the Caribbean lowland forests of Central America from Guatemala to Panama. The region is characterized by the heavy rainfall and dense canopy that creates an extensive structural habitat for a wide range of species. Set at elevations mainly below 1,640 feet (500 meters), this ecoregion represents a unique combination of North and South American biotas, featured within expanses of jungle, swamp area, and unique estuary systems and coastal lagoons.

The unique geographical position and contours of the Central America region generate conditions of high humidity and precipitation for the Caribbean slopes. The tropical and subtropical portions of the biome vary in terms of precipitation, falling within a range of 100–200 inches (250–500 centimeters) of rain per year across the region.

Plant Life

Vegetation distribution shows a slight difference between the northern and southern portions of the biome, mainly related to changes in palm abundance and dominance in the understory, but sharing most of the main plant groups and species. Some of the most common palms here are those in genera *Welfia*, *Socratea*, *Iriatea*, and *Raphia*. Among other tree types, genera such as *Pentaclethra*, *Carapa*, *Dipteryx*, *Apeiba*, *Bursera*, *Carapa*, *Cedrela*, *Cordia*, *Ficus*, *Terminalia*, *Swietenia*, and *Lecythis* dominate the canopy.

Pachira aquatica is a common riverside tree and often serves as the perch for one of the many species of kingfisher which live along riparian corridors. There are swamped and flooded areas with characteristic palm vegetation formations that give way at the coastline to mangroves and floating vegetation.

Animal Types

The volume of endemic (found only here) species is not as high as in some neighboring biomes, but unique combinations here characterize Central America as the bridge between two continental ecoregions. There is remarkable variety in birds here, however, with species such as Honduran emerald (*Amazilia luciae*), yellow-billed cotinga (*Carpodectes antoniae*), snowy cotinga (*Carpodectes nitidus*), streak-crowned antvireo (*Dysithamnus striaticeps*), grey-headed piprites (*Piprites griseiceps*) and the great green macaw (*Ara ambiguus*).

For Central America, this biome represents a vital refuge for several large mammal species, as much of the Pacific slopes are highly disturbed and historically altered by human activities. Most of the heavier forest cover of Central American countries today is located in the Caribbean lowlands; these represent some of the most viable habitats. Species such as puma (*Puma concolor*), jaguar (*Panthera onca*), tapir (*Tapirus bairdii*), brocket deer (*Mazama* spp.), Central American spider monkey and brown-headed spider monkey (*Ateles geoffroyi* and *A. fusciceps*), the endemic Roatan agouti (*Dasyprocta ruatanica*), and the highly persecuted peccaries (*Tayassu peccary, Pecari tajacu*) find in this biome their most readily available habitats in all of Central America.

The jungle, swamps, estuaries, and coastal lagoons of the Central American Atlantic Moist Forests are populated by an immense variety of bird species, including the great green macaw, seen here. (Wikimedia/Dick Daniels)

Among the amphibians and reptiles, characteristic species include the threatened O'Donnell's salamander (*Bolitoglossa odonnelli*), Moravia de Chirripo salamander (*Bolitoglossa alvaradoi*), and worm salamander (*Oedipina* spp.); rare frogs such as the blue-sided tree frog (*Agalychnis annae*), tree frogs (*Duellmanohyla* spp., *Pristimantis* spp.), rainfrogs (*Craugaster* spp.), marsupial frog (*Gastrotheca cornuta*), and Hernandez tree frog (*Pristimantis hernandezii*); the Honduran dwarf spiny-tailed iguana (*Ctenosaura palearis*); and Central American river turtle (*Dermatemys mawii*).

Human Interaction

The Central American Atlantic Moist Forests biome is threatened by increasing human intervention. Historically, banana plantation development degraded large portions of the lowland floodplains, and much of the remaining forests are retained in areas with some physical relief that evaded plantation agriculture. Logging, from selective to clearance, is still a threat with many tree species at risk in the region. Hunting, pollution, and expansion of the agricultural boundaries seriously threaten to inflict habitat fragmentation on the biome.

Some important protected areas are present in several countries as response to these threats has become better organized. Representative areas include Pico Bonito, Rio Platano, and Bosawas in Honduras; Wawashan, Cero Silva, and Indio Maíz in Nicaragua; Tortuguero, La Selva, Maquenque, and Barra del Colorado in Costa Rica; and Chagres in Panama.

Global warming is predicted to lead to higher temperatures and somewhat lower precipitation levels in this region. Hurricane activity, however, is expected to increase in frequency and intensity, combined with higher storm surges in combination with sea-level rise. The outfall from these trends will be greater erosion, disruption of lower-altitude and riverbank habitat, altered timing of seed germination and bird migration, increased threat to trees from insects and invasive mammalian species, and related changes that will exert pressure on the sustainability of this biome's species diversity and species range.

José F. González-Maya
Jan Schipper

Further Reading

Emmons, L. *Neotropical Rainforest Mammals.* Chicago: University of Chicago Press, 1990.

Guyer, C. "The Herpetofauna of La Selva, Costa Rica." In A. H. Gentry, ed. *Four Neotropical Rainforests.* New Haven, CT: Yale University Press, 1990.

Reid, F. *Field Guide to the Mammals of Central America and Southeast Mexico.* New York: Oxford University Press, 2009.

Sánchez-Azofeifa, G. A., C. Quesada-Mateo, P. Gonzalez-Quesada, S. Dayanandan, and K. S. Bawa. "Protected Areas and Conservation of Biodiversity in the Tropics." *Conservation Biology* 13, no. 2 (1999).

Central American Montane Forests

Category: Forest Biomes.
Geographic Location: Central America.
Summary: Home to many unique habitats, this cloud-forest biome represents an evolutionary crossroads of montane species from North and South America.

Montane forests occur in isolated patches in a mosaic along the mountaintop region from southern Mexico to northern Nicaragua. Predominantly cloud forests—marked by persistent seasonal or frequent low-level clouds—this biome is unique in representing part of a temperate dispersal corridor (together with Talamancan montane forests) between North and South America, where many elements from the north and south have mixed. The region has very high endemism (species found nowhere else on Earth) among plants, invertebrates, and vertebrates.

The biome consists of at least 40 small habitat islands occurring in a chain along the Sierra Madre del Sur in Mexico, the Sierra de las Minas in Guatemala, and isolated regions of Honduras, El Salvador, and Nicaragua, most of it poorly represented by the national protected-areas systems of these countries. Montane elements typically occur at elevations above 4,921 feet (1,500 meters) and extend to more than 13,123 feet (4,000 meters) along the highest peaks. Above 9,843 feet (3,000 meters), forests become scrubby, and temperature extremes can be dramatic, resulting in unique alpine-like elements such as grasslands.

From 4,921 to 9,843 feet (1,500 to 3,000 meters), cloud forests are dominated by conifers and oaks, forming the pine-oak forest belt. Within these formations are bromeliad associations and bamboo stands. The origins of these mountains are diverse, ranging from Paleozoic sedimentary and metamorphic rocks to much more recent volcanic (some still active) formations dating back to the Pliocene Epoch.

Vegetation

The vegetation typical of this biome is dominated by components from the north with a mix of elements from the south that worked their way north following the closure of the Central American land bridge. The dominant natural cover type is pine-oak forests, and much of the land has been converted to human-dominated agricultural and urban matrices. This ecosystem represents the southernmost limit of native pine forests.

Among the unique features of the vegetation is high species endemism, reaching as high as 70 percent in some areas. The isolated mountain peaks along the central spine of northern Central America are not only isolated from one another, but also represent unique climatic gradients moving up from lowland rainforests to high, sometimes-snow-covered peaks. The pine-oak forests typically are a mix of conifers (*Abies, Cupressus, Juniperus,* and *Taxus*) and broadleaf evergreens such as oak (*Quercus*). This biome represents the southern limit of the conifer's native range, whereas oaks continue southward to the Talamancan montane forests and beyond.

The climate is temperate and precipitation is high, which is usual for tropical montane systems, and is mostly due here to the narrow extent of land that Mesoamerica represents. Weather patterns create unique microclimates, and the mountain systems give way to rain shadows and other effects that influence both flora and fauna. Precipitation typically is 79 to 157 inches (2,000 to 4,000 millimeters) per year, much of it occurring as rainfall but some as cloud drip. These cloud forests form an extremely high diversity of epiphytic species such as orchids, bromeliads, and many others that depend on—and can live only on—atmospheric moisture. Above 9,843 feet (3,000 meters), frost is typical, and plant communities become limited to frost-tolerant species.

Avian Life

Bird diversity and endemism are particularly high in this biome. Among the flagship species is the majestic horned guan (*Oreophasis derbianus*), which is increasingly threatened due to loss of its cloud-forest habitats. Another charismatic species is the resplendent quetzal (*Pharomacrus mocinno*), prized by bird-watchers and one featured in the art and culture of many local pre-Hispanic and current indigenous peoples.

Other species in the biome include the endangered ocellated quail (*Cyrtonyx ocellatus*); pink-headed warbler (*Ergaticus versicolor*); azure-rumped tanager (*Tangara cabanisi*); and the extinct Atitlan grebe (*Podilymbus gigas*), which disappeared in 1986. The biome has 50 Important Bird Areas, established to identify conservation priorities.

Other Animals

Like birds, mammals are characterized by high diversity, but rates of endemism are lower than for birds or plants. One of the most interesting mammal species of the region is the very small volcano rabbit (*Romerolagus diazi*), native to Mexico and endangered. Also present are threatened species such as the Central American spider monkey (*Ateles geoffroyi*), the Mexican long-nosed bat (*Leptonycteris nivalis*), and the Baird's tapir (*Tapirus bairdii*). The biome represents one of the last habitats for these species across their range.

The area's amphibians and reptiles are unique, with diversity decreasing with elevation. One of the most remarkable characteristics is the high number of salamander and frog species. At least 28 threatened species occur across the biome, and the actual number probably is larger, considering the low documentation efforts in most of the countries. At least 13 species are critically endangered, including nine frogs and four salamanders. The decline of amphibians in this biome is associated with the threat from the chytrid fungus infection, which affects this group mostly at higher elevations. Climate change also seems to be an important issue for amphibian species in this biome, with altered temperature and moisture patterns leading to the necessity for some species to transit to new localities to find their comfort zones.

Effects of Human Activity

Due to their biological richness and their isolation, many of these mountain peaks have been declared protected areas, but little environmental enforcement is extended to them. Typically, however, the cloud forests are not only habitats. They have become isolated as much of the lowlands that previously spanned the mountains has been converted to agriculture and other human-dominated environments. Little connectivity now exists between fragments, so species have little dispersal ability. A secondary effect is that many species in this area, especially birds, rely on elevational migration routes that have become fragmented, if not completely altered.

The largest remaining forest block in the biome is Guatemala's Sierra de las Minas, which covers more than 4.9 million acres (2 million hectares). Although protected, it is also the only remaining native habitat of sufficient size to maintain all native species and ecological processes. Although many of the other montane areas within this biome are protected, most are remnants of their former distribution and today are isolated fragments of forest in a sea of agriculture. Typically, lower elevations are more highly converted, with intactness increasing with elevation—exactly the opposite trend required for biodiversity. Even though this biome is fairly intact in general, the areas with the highest biodiversity are the most at risk.

José-F. González-Maya
Jan Schipper

Further Reading

Halffter, G. "Biogeography of the Montane Entomofauna of Mexico and Central America." *Annual Review of Entomology* 32, no. 1 (1987).

Land, H. C. *Birds of Guatemala.* Wynnewood, PA: Livingston, 1970.

Reid, F. *Field Guide to the Mammals of Central America and Southeast Mexico.* New York: Oxford University Press, 2009.

Rich, P. V. and T. H. Rich. "The Central American Dispersal Route: Biotic History and Paleogeography." In D. H. Janzen, ed. *Costa Rican Natural History.* Chicago: University of Chicago Press, 1983.

Sutton, Susan Y. "Nicaragua." In David G. Campbell and H. David Hammond, eds. *Floristic Inventory of Tropical Countries.* Bronx, NY: New York Botanical Garden, 1988.

Cerrado Grasslands

Category: Grassland, Tundra, and Human Biomes.
Geographic Location: South America.
Summary: The Cerrado grasslands, a large neotropical biome with tropical and subtropical grassland, savanna, and shrubland, is threatened by agricultural expansion, water projects, and other development.

The Cerrado biome is a diagonal strip in the center of South America that is roughly oriented from northeast to southwest, occupying large areas in central Brazil (nearly 25 percent of the country), eastern Bolivia, and northeastern Paraguay. The biome is mostly coincident with the Brazilian Precambrian shield, one of South America's largest and oldest geological areas. These rocks of this area, mainly granite and gneiss, have been affected by several erosion cycles since Paleozoic times, and now form a landscape of plateaus and undulating peneplains. Some have low hills, composed of less ancient geological materials such as sandstone, limestone, and shale.

Because the region has been mostly unaffected by subsequent marine transgressions, the soils have developed over long periods under warm, semi-humid, tropical-climate conditions, with strongly seasonal rainfall. There is a general predominance of old, weathered, and impoverished red and yellowish soils. The Cerrado biome is highly adapted to soils that are poor in nutrients, and to the strong seasonal rainfall, which can reach 31 to 67 inches (800 to 1,700 millimeters) annually. The rain occurs primarily in the warmer months, November to April, with an opposing significant water deficit that extends over four to five months. This prolonged dry season favors the high incidence of fires, both naturally caused and human-induced. Consequently, many researchers consider the Cerrado to be a pyrophytic or pyrogenic ecosystem that has at least partially evolved by adapting to periodic fire disturbance.

Biota

The vegetation of the Cerrado is comprised of several dynamic savanna physiognomies and structures, varying in their respective extension and landscape importance on the basis of the incidence and intensity of fires, as well as human management modalities. This physiognomies include almost pure grassland savanna (*campo limpo*); open scrub and shrubby savanna (*campo sujo*); arboreal savanna (*campo cerrado*); and patches of low, partly open woodland (*cerradão*). Both the herbaceous and woody Cerrado floristic components show a set of morphological and anatomical specializations that various scientists have interpreted as being adaptations to fire, the poverty and toxicity of soils, and seasonal drought. Among these adaptations, the most significant is the low-lying, contorted trees with thick, corky, fire-resistant barks; the frequent sclerophyllous leaves with thick cuticles and or silicified tissues; and the presence of several swollen woody underground vegetal structures called xylopodia.

Animals of the biome include the maned wolf (*Chrysocyon brachyurus*), giant anteater (*Myrmecophaga tridactyla*), and the pampas cat (*Oncefelis colocolo*). Among the bird species are the rhea (*Rhea americana*); red-legged seriema (*Cariama cristata*); and the Spix's macaw (*Cyanopsitta spixii*), which is critically endangered due to habitat loss and capture for the pet trade.

Environmental Threats

Fire and livestock pressure on vegetation are very important factors determining the density and frequency of the Cerrado woody component. Overgrazing and excessive frequency or intensity of the fire regime both favor the progressive reduction or disappearance of woody plants and, conversely, grass predominance, in transforming the savanna woodlands and arboreal savanna (*cerradão, campo cerrado*) to savanna grasslands and thickets (*campo limpo, campo sujo*).

According to several researchers, the natural potential vegetation of the Cerrado is the woodland and arboreal savanna physiognomies. The grasslands and thicket–grasslands are successional degraded phases; they can evolve through ecological succession to woody savanna if the excessive fire or livestock pressure ceases or diminishes. These ideas are consistent with field observations that show a recovery of woodland and arboreal savanna in the interior of existing Cerrado protected areas in Brazil, as well as the general dominance of woody savanna aspects in the better-preserved Cerrado areas of Bolivia and Paraguay.

Effects of Human Activity

Most of the Cerrado biome is not suitable for agriculture, because of strong soil and climatic constraints. Therefore, the main natural land potential use is cattle ranching on natural pastures and browsing in the woodland savanna, where fire is used annually to encourage the growth of tender grass after burning. Extensive range management has been the dominant economic activity since the 17th century when cattle were introduced by Portuguese and Spanish settlers. In Bolivia and Paraguay, the traditional use of the Cerrado has continued, leading to a predominance of semi-natural landscapes with still-remarkable levels of ecosystem conservation. Today, both countries have the best naturally preserved areas of the Cerrado biome.

In Brazil, by contrast, different social characteristics and greater economic capacity to transform the environment have resulted in a dramatic conversion of much of the Cerrado to large areas of mechanized agriculture. With the extensive addition of lime and fertilizer after deforestation, the Brazilian Cerrado has largely been replaced by cultivated soybeans, cotton, rice, and oil palms, among other crops. Today, this biome has practically disappeared in Brazil outside national parks, where some more or less representative areas are preserved.

Studies have shown that climate change is likely to affect the Cerrado by breaking up tree habitats, leading to lower diversity among the woody species here, and limiting a successful rebound in the aftermath. The current, relatively high level of diversity is seen transitioning to a moderate level of diversity, which will be found in more isolated distribution. Such research points to a conservation strategy that would strengthen the core areas of the Cerrado—including its central expanse in Brazil—in order to preserve the maximum diversity of tree species and thereby support the health of the habitats they underpin.

Gonzalo Navarro

Further Reading

De Siqueira, Martinez Ferreira and Andrew Townsend Peterson. "Consequences of Global Climate Change for Geographic Distributions of Cerrado Tree Species." *Biota Neotropica* 3, no. 2 (2003).

Killeen, T. J. A. and T. S. Schulenberg, eds. *A Biological Assessment of the Parque Nacional Noel Kempff Mercado, Santa Cruz, Bolivia.* Washington, DC: Conservation International, 1998.

Navarro, Gonzalo. *Clasificación de la Vegetación de Bolivia.* Santa Cruz de la Sierra, Bolivia: Editorial Centro de Ecología Simón I. Patiño, 2011.

Oliveira-Filho, Ary T. and James A. Ratter. *The Cerrados of Brazil: Ecology and Natural History of a Neotropical Savanna*. New York: Columbia University Press, 2002.

Chad, Lake

Category: Inland Aquatic Biomes.
Geographic Location: Central Africa.
Summary: Lake Chad supports an abundance of life, but is shrinking at an alarming rate; conservation efforts are widespread but not entirely effective.

Lake Chad is a unique inland sea that receives water from four rivers and has no water outlet—yet has shrunk to less than 10 percent of its original size within 40 years. This lake and its vast resources support more than 25 million people, and its location in the southern Sahara Desert provides a perfect sanctuary for animals, making it a biome with rich biological diversity of global importance and an international meeting point for important animals, especially migrant birds. Crude oil deposits within its basin are set to affect the economies of the concerned countries, perhaps for the better. The lake, however, has become a point of contention for communal conflicts over rights to water use as it shrinks, threatening the livelihoods of pastoralists, fishers, and farmers in the area.

The name *Chad* is a local term meaning "a large expanse of water." Lake Chad, in contrast to several other brackish or high-saline inland seas, such as the Caspian and Dead Seas, is still nearly entirely a freshwater lake. The lake is believed to be a remnant of an ancient sea that had drainage into the Atlantic Ocean through the Benue River in present-day Nigeria. Today Lake Chad is shared by four countries: Niger in the northwest, Chad in the east, Cameroon in the south, and Nigeria in its southwest portion.

The main influent rivers of Lake Chad are the Chari, which flows through Central African Republic, Chad, and Cameroon, and also collects the waters of the Logone River; the Komadugu-Yobe from Nigeria; and the El-Beid from Cameroon. The Chari River accounts for more than 90 percent of the water supplied to the lake, and it flows year-round, unlike the rest, which seasonally flow for only a few months each year.

This satellite photograph of Lake Chad from October 21, 2001, shows evidence of the lake's decline over the past 40 years, as its previous surface area included the rippled area near the top of the lake. (NASA)

Biodiversity

The lake and the wetlands of its basin combine to support biodiversity that is of global importance, including 176 fish species and 354 resident and 155 migrant bird species. Birds here include ostriches, bustards, pelicans, cranes, herons, ducks, marabous, storks, and egrets, which gather in the dried floodplains in populations as high as 450,000. Lake Chad is the wintering ground for several European bird species, too, such as white storks, ducks, waders, white-faced whistling ducks, and little egrets.

A stronghold for wildlife, Lake Chad provides habitat for animals in search of water and grasslands. These include several antelope species—dama gazelle, addax, korrigum, red-fronted gazelle, sitatunga, and kob—black-crowned cranes, giraffes, monitor lizards, elephants, lions, hippopotamuses, crocodiles, and two otter species that thrive in Lake Chad's surface waters.

Conservation Efforts

The animals of Lake Chad have suffered from excessive hunting by the local population and habitat degradation by the activities of contending armies and rebel groups. Worse, the western black rhinoceros and scimitar oryx have gone extinct in the area. Since 2000, the Lake Chad Basin Commission (LCBC), a cooperative body of the four national governments here, has declared the whole of the lake a transboundary Ramsar Wetlands Site of International Importance.

The basin area features nine national parks, three biosphere reserves, one World Heritage Site, and one faunal reserve. The preserved areas include Lake Chad National Park (Nigeria), Zina-Waza National Park (Cameroon), Andre-Felix National Park (Central African Republic), and Fitri Biosphere Reserve (Chad). The management of these protected areas, however, has generally been evaluated as being very poor.

Human Activity

According to the LCBC, well over 20 million people depend on Lake Chad and its flora and fauna for their daily survival. The basin is rich in cultural diversity, drawing people from ethnic groups and tribes such as the Kanuri, Mobber, Buduma, Haoussa, Kanembu, Kotoko, Shewa Arabs, Haddad, Kouri, Fulani, and Manga. In Chad alone, more than 130 languages are spoken by people living in the Lake Chad basin. They are mainly pastoralists (camel and cattle rearers); fishers; and farmers growing cereals, legumes, vegetables, and forage crops.

Resources collected within and around the lake also contribute to the livelihood of the people. These resources include atroun, a sodium carbonate complex that is rich in minerals; and dihe (*Spirulina*), an alga with high nutritional content that grows naturally in the northeastern pools of the lake at the end of the rainy season. Dihe is a local dietary product; its nutritional value was confirmed by the United Nations World Food Conference in 1974, when it was declared an outstanding foodstuff for the future. Dihe has wide usefulness in food industries and in human and animal diets. Since 1981, the international market for dihe took off, and the product now sells for up to $250 per pound (0.4 kilogram) in developed countries.

Crude oil was discovered within the lake's basin and has recently been found in Cameroon. Furthermore, the receding areas of the lake have provided another opportunity of farming on the land via a farming system called *fadama*.

Atroun

More than 7,716 tons (7,000 metric tons) of a mineral-rich substance called atroun is produced in Chad; it is an important source of income for the country, especially in the Wadi area. Atroun replaces salt in traditional dishes; has ethnomedicinal significance for both animals and humans; and is used for purposes such as fixing indigo dyes on clothes, tanning skins, and making soap. The highest grade of atroun sells for hard cash regionally and is used mainly in pharmaceutical preparations.

Environmental Threats

Research has shown that since 1963, the lake has shrunk to as little as 5 percent of its original size, from 9,653 square miles (25,000 square kilometers) to 521 square miles (1,350 square kilometers). The shrinking size of the lake is due to the effects of excessive grazing; deforestation; and unsustainable irrigation projects built by Niger, Nigeria, Cameroon, and Chad that have diverted water from the lake and its feeder rivers for irrigation and hydroelectric power. As a result, fish harvests have been greatly reduced, and several friendly communities along the shore are now in conflict about boundaries and access to water and pasture. Half of the remaining lake is invaded by an alien tree, *Prosopis* spp., that was introduced to stop desertification in the Sahel region. Furthermore, pastoralists moving their animals to greener pastures come into conflict with farmers.

LCBC plans to construct canals to supply water from the Ubangi River (one of the tributaries of the Congo River) into Lake Chad, but the project will cost millions of dollars and its launch has been hampered by financial, political, and organizational constraints.

TEMITOPE ISRAEL BOROKINI

Further Reading

Batello, Caterina, Marzio Marzot, and Adamou Harouna Toure. *The Future Is an Ancient Lake.* Rome: Food and Agriculture Organization, 2004.

Lake Chad Basin Commission. *Survey of the Water Resources of the Chad Basin for Development Purposes—Surface Water Resources in the Lake Chad Basin.* Rome: United Nations Development Programme (UNDP) and FAO, 1972.

Lemoalle, Jacques. *Lake Chad: A Changing Environment—Dying and Dead Seas, Climatic Versus Anthropic Causes.* New York: Springer-Verlag, 2004.

Chad, Lake, Flooded Savanna

Category: Inland Aquatic Biomes.
Geographic Location: Africa.
Summary: This savanna is a strategically located home of migrating birds, currently threatened by drought, underground seepage, and anthropogenic influences.

Africa's fourth-largest lake and the pride of west-central African hydrology is now a shadow of what it used to be. Lake Chad is located in the west of Chad, bordering on northeastern Nigeria. The main source of Lake Chad's water is the Chari River, fed by its tributary the Logone, with a small amount coming from the Yobe River in Nigeria and Niger. The lake is freshwater in spite of its high level of evaporation.

Lake Chad's landscape is characterized by many small islands, including the Bogomerom Archipelago; reed beds; mud banks; and a belt of swampland. The shorelines are largely composed of marshes. Lake Chad is very shallow, with a depth of only 34 feet (10.5 meters) in its deepest area, which makes it highly vulnerable to small changes in average rainfall, leading to seasonal fluctuations in average depth of about three feet (one meter) every year.

The lake's current water surface of 965 square miles (2,500 square kilometers) is miniscule compared to its 9,653 square miles (25,000 square kilometers) in the 1960s. Severe droughts in the 1970s as well as underground seepage have adversely affected the volume and depth of this lake. These factors, in addition to increased pressure from the approximately 30 million humans nearby and persistent desertification, have led to large-scale destabilization of the ecosystem. They have also made the lake extremely shallow—from an average depth of 12 feet (39 meters) in the 1960s to 1.5 to 5 feet (0.5 to 1.5 meters) today—and have in recent years led to its division into northern and southern pools.

Biodiversity

The lake's rich flora include more than 44 species of algae and yaéré grassland of *Echinochloa pyramidalis, Vetiveria nigritana, Oryza longistaminata,* and *Hyparrhenia rufa.* The grassland savanna area is a habitat for numerous species of wildlife, including a fast-diminishing number of hippopotamuses and crocodiles; large species of migrating birds, such as wintering ducks; wading birds such as the ruff (*Philomachus pugnax*); and other waterfowl and shore birds. The decreasing water levels in Lake Chad have affected bird life across the savanna, particularly reducing the availability of nesting sites for the endangered West African subspecies of black-crowned crane and decreasing the supply of adequate wintering grounds for birds such as the ruff. The savanna surrounds, a semi-arid Sahel formation, have not been favorable for the habitation of mammal species.

The decrease in crocodile and hippopotamus has adversely affected fish populations. Crocodiles prey heavily on catfish, which consume the eggs and fry of tilapia and other cichlid fish. Without crocodiles to control catfish populations, the catfish have dramatically reduced tilapia stocks here.

Environmental Threats

The severe droughts that have plagued the Lake Chad savanna area have contributed to the shrinking of the lake and have affected food security and the livelihoods of the people who have made the basin their home. Low rainfall and underground seepage of water out of the lake have also affected the agricultural productivity of the savanna and greatly undermined people's ability to provide for their basic needs. Water-resource harnessing, in the form of damming the rivers that flow into Lake Chad, has adversely affected the timing and extent of seasonal flooding—undercutting the extent of nutrient-rich sediment deposition around the savanna—as well as disrupting fish migration patterns. Furthermore, each of these factors as well as evaporation have led to increasing salinization of the lake, which negatively impacts agricultural activities and habitat sustainability around the basin.

Because the lake and savanna are vital to four countries—Cameroon, Chad, Niger, and Nigeria—there have been governance difficulties. These include the challenge of formulation of water-use and land-use policies that would benefit all of the nations as well as take into consideration the peculiarities of each one. The basin's location within some of the poorest countries of the world has not helped resolve these issues, as some of these governments have very weak legislatures and have not manifested concerted efforts to protect the area.

Conservation Efforts

Considering the savanna's ecological and agricultural importance, the Lake Chad Basin Commission (LCBC) has proffered some means for ameliorating the challenges. Suggestions have been made concerning augmenting Lake Chad with water from the Zaire River basin, for example. One such plan proposes moving 3.5 trillion cubic feet (100 billion cubic meters) of water annually from the Zaire River in a navigable canal 1,491 miles (2,400 kilometers) in length. Dams would be built in the receiver and donor areas for electricity generation.

There was an earlier proposal to divert the Ubanghi River into Lake Chad, which would revitalize the dying lake and provide livelihoods in fishing and enhanced agriculture to tens of millions of central Africans and Sahelians. Interbasin water-transfer schemes were proposed in the 1980s and 1990s. In 1994, the LCBC proposed a similar project, and at a March 2008 summit, the heads of state of the LCBC member countries committed to the diversion project. A World Bank-funded feasibility study is underway.

Peter Elias

Further Reading

Bene, C., A. Neiland, T. Jolley, et al. "Inland Fisheries, Poverty, and Rural Livelihoods in the Lake Chad Basin." *Journal of Asian and African Studies* 38, no. 1 (2003).

Coe, Michael T. and Jonathan A. Foley. "Human and Natural Impacts on the Water Resources of the Lake Chad Basin." *Journal of Geophysical Research* 106, no. 4 (2001).

Kolawole, A. "Environmental Change and the South Chad Irrigation Project, Nigeria." *Journal of Arid Environments* 13, no. 2 (1987).

Sarch, M. T. and C. Birkett. "Fishing and Farming at Lake Chad: Responses to Lake-Level Fluctuations." *Geographic Journal* 166, no. 2 (2000).

Chagos Archipelago

Category: Marine and Oceanic Biomes.
Geographic Location: Indian Ocean.
Summary: The Chagos Archipelago, located in the middle of the Indian Ocean, is among

the world's largest marine reserves. A group of pristine coral atolls, its reefs provide a breeding and repopulation ground for land, shallow-water, and deep-sea species.

U.S. Navy personnel cleaning a beach with British Royal Navy sailors and civilians on Diego Garcia on April 16, 2010. That year, one of the world's largest nature reserves was established when the United Kingdom declared the archipelago and its marine ecosystem a no-take zone. (U.S. Navy)

The Chagos Archipelago is an area of seven atolls consisting of some 60 islands in the Indian Ocean, located approximately 310 miles (500 kilometers) south of the Maldives. The atolls are composed of several types of coralline rock structures sitting atop volcanoes in a subterranean mountain ridge, and they include the Great Chagos Bank, the world's largest atoll. The land areas of the archipelago are part of a tropical and subtropical moist broadleaf forest biome, consistent with the region's Maldives and Lakshadweep Archipelagos.

In the early 1970s, the United Kingdom (UK) evicted the native Chagossians from several of the archipelago's islands to accommodate the construction of a military base that would jointly serve UK–U.S. military interests. Currently, military personnel inhabit only Diego Garcia, the largest island. The depopulation of the islands decreased the human impact in the area and the surrounding waters, leaving it all relatively pristine so that biodiversity has flourished in a somewhat uninterrupted way. In 2010, the islands and surrounding waters were declared a no-take zone by the United Kingdom, making this marine ecosystem among the world's largest nature reserves.

Coral Reefs

There are 220 species of coral composing the reefs of the Chagos; coral cover is consistently dense and healthy throughout the reefs. In a symbiotic relationship, each polyp houses a class of photosynthetic algae, or zooxanthellae, that provides nutrients for the polyp as well as giving the coral its color. Abnormally high sea surface temperatures in 1998 caused widespread coral bleaching, a phenomenon in which the polyp expel these algae, causing color loss. The overall health of the region allowed the coral to recover, and today the surface area increases yearly.

Because the algae are photosynthetic, they require clear water to maximize sun exposure. Though most of their energy is supplied by the zooxanthellae in the form of glycerol, corals can prey on zooplankton and, if polyps are larger, sometimes even small fish. Notable coral species in the Chagos include the endemic (native only here) shallow-water brain coral and the deep-water staghorn corals. Human activity impacts the health and vitality of the coral, which also are vulnerable to the predatory starfish "crown of thorns." Outbreaks of these starfish can kill a reef.

The reefs provide a home for more than 1,000 species of reef-dwelling and pelagic fish. The reef fish that inhabit the shallower areas of the reef close to shore include the endemic Chagos clownfish, wrasse, and grouper, and such pelagic fish as manta rays, sharks, and tuna. Migratory fish, including skipjack and yellowfin tuna, pass through the Chagos for roughly two months every year, and are then subject to catch by commercial fisheries. Illegal fishing, however, also occurs.

Sharks, top predators in the reef's ecosystems, have decreased in number due to poaching and accidental netting by tuna fisheries. The grey reef shark lives in shallow water near the drop-off zone of the reefs and is an important predator in the shallow-water ecosystem. They feed primarily on bony fish and cephalopods living among the reef. Nurse sharks, also hunted by humans, have similar prey, as do octopus, urchin, and rays.

Flora and Fauna

Winds and passing birds seeded the flora native to the islands, and humans introduced foreign plants; these factors have considerably changed the landscape here over the centuries. Notably, humans introduced coconut palms around 1800 to support the coconut oil industry, clearing natural habitats such as *Typha* (commonly known as bulrush or cattail) swamp and peat habitats. Few other plant species are able to grow in the abandoned coconut groves, but bird species such as noddies, fairy terns, black rats, and coconut crabs have established habitats there.

Native plant life consists of 41 flowering plants, four ferns, and several species of mosses, liverworts, fungi, and cyanobacteria. There also are forests of *Pisonia* and *Barrington asiatica*, the fish poison tree, and other hardwood trees that are favored by such birds as boobies and frigates for nesting. Open, lightly vegetated areas serve as breeding areas for ground-nesting birds, including terns, noddies, and gulls.

In addition to land-nesting birds, there are large numbers of seabirds that breed in areas free of such predators as rats and cats, usually on the islands that were not used for coconut plantations. The red-footed booby is a remarkable case of a species whose population in the Chagos recovered in the 1970s after a 100-year absence. Also present are species of *Puffinus*, *Sternidae*, and sooty terns that nest in groups of great density to discourage predators.

The shores of the Chagos islands are home to hawksbill and green turtles, both species on the International Union for Conservation of Nature Red List. Though they were originally exploited for their shells, their populations are recovering because of strict nature reserves that allow them to breed and forage unmolested by humans.

Yasmin M. Tayag

Further Reading

Koldewey, H. J., D. Curnick, S. Harding, L. Harrison, and M. Gollock. "Potential Benefits to Fisheries and Biodiversity of the Chagos Archipelago/British Indian Ocean Territory as a No-Take Marine Reserve." *Marine Pollution Bulletin* 60, no. 11 (2010).

Sheppard, C. R. C., et al. "Reefs and Islands of the Chagos Archipelago, Indian Ocean: Why it is the World's Largest No-Take Marine Protected Area." *Aquatic Conservation: Marine and Freshwater Ecoystems* 22, no. 2 (2012).

Speight, Martin R. and Peter A. Henderson. *Marine Ecology: Concepts and Applications.* Hoboken, NJ: Wiley-Blackwell, 2010.

Cham Islands Coral Reefs

Category: Marine and Oceanic Biomes.
Geographic Location: South China Sea.
Summary: This biodiverse system of coral reefs is an ecologically rich area that has been adversely affected by excessive coral harvesting, aggressive commercial fishing, and exploitation from tourism.

The Cham Islands are a group of eight islands, nine nautical miles (17 kilometers) off the coast of Vietnam in the South China Sea. The islands are: Hon Lao, Hon Dai, Hon Mo, Hon Kho Me, Hon Kho Con, Hon La, Hon Tai, and Hon Ong. Sandy beaches and forested hills are the rule here; most of the population is engaged in fishing. The islands, known locally as Cu Lao Cham, are part of Cu Lao Cham Marine Park, a world Biosphere Reserve recognized by the United Nations Educational, Scientific and Cultural Organization

(UNESCO), in large part because of its coral reefs and diverse marine life.

The islands were first settled about 3,000 years ago and are the only Vietnamese islands included in the old Silk Road network of international trading. Today, ecotourism is important to the islands' economy; tourists explore the coral reefs either by diving or by riding in glass-bottom boats. The nearest city is Hoi An, on the mainland coast. In 2011, due to proximity to Cu Lao Cham Marine Park, Hoi An was designated Vietnam's first ecological city, and United Nations funds were released to help fund a project that will include mitigating climate-change-related environmental problems and the redevelopment of the city's infrastructure.

The Cham Islands are somewhat cooler and more humid than the mainland; the humidity remains around 85 percent year-round. Typhoons and other heavy storms are common in September and October. The islands experience tidal activity on a semi-diurnal cycle: two high tides and two low tides a day, on most days of the month, with two periods of flood tide every month. These floods can reach dangerous levels from September to December.

Biodiversity

A total of 947 marine species have been identified in the biome encompassing the waters around the islands. Surrounding the islands are about 410 acres (166 hectares) of coral reef and 1,200 acres (486 hectares) of seaweed beds. The reefs include 135 species of coral, along with 84 species of mollusks and four species of tiger shrimp. Six of the coral species are endemic (found nowhere else on Earth) to Vietnam. The biosphere has at least 202 fish species in 85 genera and 36 families.

A major component of the marine ecosystem of the coral reef community is the butterflyfish, of which 29 species have been recorded in the area, the most common of which are: *Chaetodon lunula, C. octofasciatus, C. auriga, C. melannotus, C. speculm, C. trifascialis, C. trifasciatus,* and *Heniochus chrysostomus.* Butterflyfish feed on coral and are useful as an indicator species because of their sensitivity to coral degradation.

The blackback butterflyfish (Chaetodon melannotus), shown here, is frequently seen on the Cham Islands coral reefs off the coast of Vietnam. (Thinkstock)

New species are discovered in the ecosystem on a regular basis. During a sample collection from 1989 to 1994 comprising 813 specimens, 326 fish species were collected from around the coral reefs, in 117 genera and 44 families; 69 of them had never been recorded before. In addition to butterflyfish, damselfish (*Pomacentridae*), wrasses (*Labridae*), parrotfish (*Scaridae*), snappers (*Lutjanidae*), rock cods (*Serranidae*), and surgeonfish (*Acanthuridae*) are abundant. The reef-edge grooves are frequented by large species of these fish, as well as *Epinephalae, Letherinae,* and *Cheilinae.*

The reef flats and clusters are home to smaller fish, including butterflyfish; surgeonfish; and species of *Abudefduf, Thalassoma, Halichoeres, Siganus,* and *Caesio.* Damselfish seem to be the most abundant,with 53 species of 12 genera found among the reefs, all of them fairly small and feeding among the reef flats and clusters. Given the abundance of species, it is likely that there are scores more yet to be discovered.

Environmental Threats

The area continues to be important to the commercial fishing industry, particularly for lobster, grouper, snapper, crab, shrimp, and clam. One of the major threats to the coral reef ecosystem is dynamite fishing, which involves detonating explosives to kill large numbers of fish. This practice causes significant damage to organisms irrelevant to the catch. In the case of butterflyfish, the population density has suffered, with numerical recovery very slow.

Destructive fishing methods are not limited to dynamite fishing. Other such techniques include electric fishing (in which clusters of marine life are electrocuted) and the use of sluice traps, trawling nets, and gill nets. Destructive fishing is still the norm in the islands, as in most of

Vietnam. The various methods of catching fish are employed to maximize the catch while mitigating the economic risk of the fishers.

Another victim of exploitation is the local population of sea cucumber. When processed as a dried animal product, sea cucumber is valuable in markets in China, Japan, Hong Kong, Singapore, and Taiwan. The coral trade is also seriously threatened, affected by increased tourism and the popularity of coral exports. Since 2000, the harvesting and cleaning of live corals sold as coral skeletons has advanced. At least 33,069 pounds (15,000 kilograms) of coral skeletons were exported in 2000, but no one is managing or maintaining the informal industry, or maintaining statistics.

The condition of the coral reefs is suffering. Only 1.4 percent of the reef areas in all of Vietnam's waters are considered to be in excellent condition; less than 33 percent are in good condition. Approximately half of the coral reefs are in fair condition and 37 percent are considered to be in poor condition. While Cu Lao Cham's coral reefs are in generally better shape than those located closer to major population centers, the intensity of commercial fishing operations and storms, exacerbated by climate change, are two major threats to the ecosystem. Global warming means that seawater temperature rises, a threat indicator for coral, as it tends to drive off the beneficial symbiotic organisms that help keep coral healthy and prevent bleaching events.

In 1997, Typhoon Linda destroyed a major area of coral-reef habitat in the Con Dao islands, considered the best-protected coral reefs in Vietnam. This weather episode was followed months later by significant bleaching events in the same reefs. A similarly strong storm in the Cu Lao Cham Islands produced long-lasting reef damage.

Conservation Efforts

As part of the Biosphere Reserve, the area is the subject of managed integrated conservation and development efforts, with emphasis on land/seascape diversity, habitats, species, and genetic resources. One of the challenges of the reserve is to conserve biodiversity, species abundance, and the health of the coral reefs without impinging on the economy of the islands, which currently is almost entirely dependent on fisheries. Tourism is believed to be the best remedy for the reefs, helping to convert the islands' economy to one more dependent on visitors seeking to witness the area's biodiversity.

Bill Kte'pi

Further Reading

Latypov, Y. Y. "Changes of Community Near Ku Lao Cham Islands (South China Sea) After Sangshen Typhoon." *American Journal of Climate Change* 1, no. 41–47 (2012).

Morton, Brian, ed. *The Marine Biology of the South China Sea*. New York: Hong Kong University Press, 1998.

South China Sea Project. *Reversing Environmental Degradation Trends in the South China Sea and Gulf of Thailand.* Bangkok, Thailand: United Nations Environmental Programme, 2006.

Champlain, Lake

Category: Inland Aquatic Biomes.
Geographic Location: North America.
Summary: Lake Champlain, the sixth-largest freshwater body in North America, is home to a variety of water and land species, but human activities have compromised the ecosystem with toxic runoff and non-native species.

Often referred to as the sixth Great Lake, Lake Champlain is a large freshwater lake bordering Canada and the United States. The lake and its basin are home to a wide variety of aquatic and terrestrial inhabitants, resulting in the creation of a complex community in which many species of plants and animals are dependent on one another to survive. Lake Champlain is an important resource of food and shelter for plants and animals as well as humans, 600,000 of whom rely on the lake for its natural and recreational resources.

More than 12,000 years ago, glacial ice covered the area where Lake Champlain exists today. As the glaciers melted, the Atlantic Ocean advanced, creating the Champlain Sea. As the glaciers retreated and the Earth's surface rebounded, the sea became isolated from the ocean and became the freshwater lake it is today. Named for Samuel de Champlain, whose explorations brought him to the area in 1609, the lake has a surface area of about 435 square miles (1,127 square kilometers) and an average depth of 64 feet (20 meters), though at its deepest point, it reaches 400 feet (122 meters). Its borders lie within the states of Vermont and New York as well as the province of Quebec in Canada.

The lake is divided into five major parts: South Lake, which is very thin and narrow in appearance, like a river; Main Lake, containing approximately 80 percent of the lake's water and including the deepest and widest points of the lake; Mallets Bay, the most isolated section due to causeways, resulting in restricted circulation of its water; the Inland Sea, often referred to as the Northeast Arm, and much smaller than Main Lake; and Missisquoi Bay, the most northeastern point of the lake, very shallow with a maximum depth of 14 feet (4 meters).

Biodiversity

Lake Champlain's numerous inhabitants include a large variety of plants and algae, as well as several species of birds and fish, creating a very intricate food web. While the lake does contain blue-green algae, which can create toxic blooms, the largest algae community is comprised of unicellular algae called diatoms. This algae captures solar energy and acts as food for the zooplankton living in the lake. Diatoms feed the fish living in the lake, which currently contains more than 100 native underwater plant species. These plants may appear to be weeds but actually provide food and shelter for numerous kinds of fish, snails, and eels.

More than 70 fish species are native to the lake, including minnows, dace, perch, bass, catfish, pike, walleye, and sturgeon. Lake Champlain once was home to Atlantic salmon and trout, but both have likely disappeared, due in large part to the construction of dams.

Throughout the year, the lake provides food and shelter to more than 300 species of birds, as part of the migration routes of the Atlantic Flyway. Warblers, hawks, ducks, snow geese, gulls, and double-crested cormorants are among these itinerant inhabitants.

Threats and Conservation

Lake Champlain helps sustain the area's human population as well, with 35 percent of the local community using the lake as a source of drinking water. It also serves as a food source for local fishers and it provides for the livelihood of others who make their living on tourism and recreational pursuits. The United Nations has designated Lake Champlain's watershed an International Biosphere Reserve due to its natural resources.

Currently, the lake faces several environmental issues that threaten to disrupt its ecosystem, among them phosphorus, a naturally occurring nutrient that is necessary for plant growth. Agricultural activities and industrial discharges near the lake have created a surplus of phosphorus, leading to toxic algae blooms and increased plant growth that harms fish and humans. Pollution from agricultural and industrial activities has led to the introduction of other toxins into the lake, such as mercury, that poison the fish and the predators and people who eat them. Human activities, including recreational boating, and nonnative or exotic species that have found their way into the water via connecting canals and rivers, have added further damage. Lake Champlain currently has 46 exotic species, such as zebra mussels and the underwater plant Eurasian watermilfoil. These invasive species threaten the native plants and animals.

Still another threat, one that may exacerbate all the others, is climate change. Scientists report that the average annual air temperature in the Lake Champlain region has already seen a rise of 2.1 degrees F (1.2 degrees C) since 1976. Ice cover has diminished, meaning that more water evaporates from the lake surface during winter. Lack of ice is also a hardship for various plant and animal species. With more precipitation coming in the form of rain rather than snow, springtime groundwater recharge is challenged, putting stress on the wetland areas around the lake. Warmer water temperatures tend to favor the invasive species over some of the vital native types such as salmon and northern pike. Overall, each of these trends and impact factors is expected to intensify in the years ahead.

The Lake Champlain Basin Program—a partnership of Vermont, New York, Quebec, and other governments—is working to eliminate the environmental threats by creating plans of action to reduce pollution and phosphorus runoff, and by engaging the public to help restore and preserve the lake for future generations. Another organization, The Lake Champlain Land Trust, is working to preserve significant islands, shoreline areas, and natural communities in the Lake Champlain region.

ELIZABETH STELLRECHT

Further Reading

Howland, William, et al. *Lake Champlain—Experience and Lessons Learned Brief.* Grand Isle, VT: Lake Champlain Basin Program, 2006.

Marsden, J. Ellen and Michael Hauser. "Exotic Species in Lake Champlain." *Journal of Great Lakes Research* 35, no. 2 (2009).

Stickney, Michaela, et al. "Lake Champlain Basin Program: Working Together Today for Tomorrow." *Lakes & Reservoirs: Research and Management* 6, no. 3 (2001).

Winslow, Mike. *Lake Champlain: A Natural History.* Burlington, VT: Lake Champlain Committee, 2008.

Chatham Island Temperate Forests

Category: Forest Biomes.
Geographic Location: Pacific Ocean.
Summary: Located off the coast of New Zealand, the ecosystem of Chatham Island features high species diversity, which had deteriorated under invasive pressures but is slowly beginning to recover.

The Chatham Islands lie some 500 miles (800 kilometers) east of the South Island of New Zealand. They consist of two large inhabited islands, Chatham and Pitt, and 40 rock stacks and islets. Various ecosystems are associated with the archipelago, including forests, dunes, cliffs, and peats. The importance of the forested ecosystems is in part due to their high rate of endemism, meaning species that are found nowhere else on Earth.

The climate of the archipelago is temperate, and is often filled with cloud cover, rain, and windy conditions. Rarely does weather give way to extremes; annual sunlight is approximately half that of mainland New Zealand. The climate has produced adaptations in many of the plants, with 47 of the 388 indigenous flowering varieties endemic to the archipelago. The interaction of such species, and with the local birds, insects, and other animals, makes the Chatham Islands temperate forests a unique and valuable ecosystem.

The impact of humans here is noteworthy, as deforestation and invasive species have altered the balance, potentially destroying it from within. Fortunately, conservation efforts have been put in place over the past 40 years, and species that might have otherwise disappeared are now recovering.

Biodiversity

The high rate of endemism in plants here is primarily due to the absence of several of the dominating forest species associated with New Zealand, such as conifers and beeches, allowing locally evolved species instead to influence the landscape. Five main forest habitats exist that are unique to the archipelago: akeake, mixed broadleaf, kopi-broadleaf, tarahianu, and akeake/karamu swamp. The windy climate has resulted in the tarahinau tree's developing wind-resistant needle leaves, while numerous types of tree fern dominate the forest floor, providing shelter for tree seedlings to germinate.

Many of the endemic species exhibit gigantism traits, a feature that is common among some island populations of animals, and in the Chatham Islands among some plants as well. The forest tree koromiko and the forest shrub karamu, which matures into a tree, each are here represented by the largest species of their genus; the tree daisy akeake, too, is one of the largest of its kind.

Similar rates of endemism exist for insect and bird populations in the forests here, but there are no native mammals or amphibians, and there is only one indigenous lizard.

The island of Rangatira is famous for its bird populations, and the forest on the island is also important to many insects. These species are often vital to forest ecosystems: pollinating, decomposing, and playing a key role in the food web. Rangatira is home to the Rangatira spider, giant click beetle, and numerous other native insect species. Approximately two-thirds of the flowering plants in the temperate forests here are pollinated by insects, which is higher on the Chatham Islands than in many other lands, due to the relative lack here of indigenous mammals and birds.

Threats and Conservation

Although the Chatham Island temperate forests originally evolved without the influence of humans, their current survival is dependent on them. The islands experienced a similar settlement trend to that of the New Zealand mainland, with the indigenous populations, the Moriori, arriving on the archipelago from about 800 to 1000 B.C.E., European discovery in 1791, and Maori settlement shortly afterward. Before the Moriori, more than 90 percent of the archipelago's landscape was a mosaic of forest, scrub, and swamp, but the majority of this land cover has since been transformed for agricultural purposes. Although agriculture puts pressure on many of the forest floral species by fragmenting and restricting their ranges, this land cover change also has affected birds. More than half of the identified bird species inhabiting the archipelago have become extinct since the first human settlers arrived.

Another major threat to these forests is the introduced Argentine ant (*Linepithema humile*). This species outcompetes other insects for food and has even been shown to outcompete the kiwi (family *Apterygidae*) for food on mainland New Zealand. The Argentine ant could potentially destroy the forest ecosystem from within, winning out over a large number of other insects that have evolved to occupy small niches left from the absence of one dominating species.

The Black Robin

The forest-floor dwelling black robin (*Petroica traversi*) is often cited as a successful conservation story of the Chatham Islands. Local extinction of this species occurred on the main islands of the archipelago after pressures from deforestation and the introduction of exotic mammalian predators. Rats and cats were introduced to the islands; their presence has altered the balance of the forest ecosystem. Many native bird species had adapted to the local conditions by evolving as flightless species; they survived because there were no natural predators. However, such flightless birds could not escape from the introduced predators.

The black robin became restricted to Little Mangere Island, and in 1976, only seven individuals remained, with one fertile female (nicknamed Old Blue by researchers). This situation sparked conservation efforts that included relocating the individuals to government-managed Mangere Island, where extensive reforestation and predator eradication had been undertaken. Sustainable populations of the black robin now exist on both Mangere and Rangatira.

An effort is underway to remove the Argentine ant from the archipelago; initiatives have been directed toward creating awareness among people traveling to the Chatham Islands from mainland New Zealand—including searches of picnic baskets, baggage, and other items. The potential for one species of ant to alter the balance of this ecosystem is huge, and the mitigation strategies implemented on the islands are imperative for the forests' survival.

Heavier rainfall events, higher sea level, and more extreme storm surges are among the impacts projected from global climate change here. Average temperature rise is anticipated to be slower than the global average, and thus less of a major factor on habitats in the Chatham Islands. Stronger westerly winds are projected; along with greater annual rainfall, this will mean western shorelines of the islands will be at substantial risk for wetlands degradation and inland-slopes habitat erosion.

According to a report released in 2009 by the United Nations Environment Programme World Conservation Monitoring Centre (UNEP-WCMC), there is no current protected-area coverage for the Chatham Island temperate forests. This compares to an average of 12.3 percent for other island ecosystems. Many of the conservation efforts on the archipelago are overseen by the New Zealand government, with the two government-owned islands Mangere and Rangatira, identified as conservation hot spots. Both islands are undergoing reforestation.

PAUL HOLLOWAY

Further Reading

Dawson, J. W. "New Zealand Botany with a Difference—The Chatham Islands." *Tuatara* 31, no. 7 (1991).

Miskelly, C. *Chatham Islands: Heritage and Conservation.* Christchurch, New Zealand: Canterbury University Press, 2009.

Mullan, Brett, Jim Salinger, Craig Thompson, Doug Ramsay, and Michelle Wild. *Chatham Islands Climate Change.* Wellington: New Zealand Ministry for the Environment, 2005.

Wardle, P. *Vegetation of New Zealand,* New York: Cambridge University Press, 1991.

Chesapeake Bay

Category: Marine and Oceanic Biomes.
Geographic Location: North America.
Summary: The Chesapeake Bay is the largest estuary in the United States, of great importance to wildlife and fish, and the subject of numerous programs to improve and reverse past environmental degradation.

The Chesapeake Bay is the largest of the 130 estuaries occurring in the conterminous United States and the third-largest in the world. Bordering the Atlantic Ocean, it extends from its mouth near Norfolk, Virginia, north to Havre de Grace, Maryland. The bay is 180 miles (290 kilometers) long and 3 to 35 miles (5 to 56 kilometers) wide. The main bay averages less than 30 feet (9 meters) deep, and the entire bay, including tributaries, averages about 20 feet (6 meters). Several deep holes indicating the remnants of the ancient Susquehanna River exceed 171 feet (52 meters). Substrate is comprised primarily of gray silt and clay in the bay's center, with fine and medium sands near the shore.

More than 11,000 miles (17,000 kilometers) of shoreline, including mainland and islands, are associated with the bay. The bay's shorelines are a mixture of developed and conserved lands. The Baltimore, Maryland, metropolitan area, with 2.7 million people, is the most populous city directly on the bay's shore. Two major roadways cross the bay: the 4-mile (7-kilometer) Chesapeake Bay Bridge near Annapolis to Kent Island, Maryland; and the 17-mile (28-kilometer) Chesapeake Bay Bridge Tunnel from Virginia Beach to Cape Charles, Virginia. Over 90 million tons of cargo from international trade pass annually through the Chesapeake Bay waterways. Substantial revenues are generated from trade and commercial fishing. In addition, tourism and recreation contribute billions of dollars to the regional economy.

Hydrology

The Chesapeake Bay started to form its present shape over 3,000 years ago, following repeated glacial melting that began about 18,000 years ago in the ancient Susquehanna River watershed. Evidence suggests that a bolide (meteor or comet) struck the region 35 million years ago, creating a 55-mile-wide (89-kilometer-wide) crater. These conditions eventually led to the formation and placement of the Chesapeake Bay and its watershed, an area that stretches across 64,000 square miles (165,000 square kilometers) of land.

The bay is fed by 19 large rivers and over 150 smaller rivers, which in turn are fed by more than 100,000 streams in the watershed. The Susquehanna River provides the largest—nearly half—of freshwater inflow. Rivers on the western shore generally are larger than those on the eastern shore. The Chesapeake Bay watershed encompasses portions of Delaware, Maryland, New York, Pennsylvania, Virginia, and West Virginia. This land area is primarily forested (57 percent) although much reduced from historic levels due to agriculture and development. An estimated 17 million people live in the watershed, and that number continues to climb.

Tides rise and fall twice daily, with tidal range varying from 0.1 inch to 3 feet (2.5 millimeters to 1 meter). Salinity varies based on location, season, and depth, averaging 15 parts per thousand and varying from 0 to 36 parts per thousand. Salinity is generally greater near the bottom, due to the lighter freshwater emptying into the bay from the streams and stratifying above the heavier ocean-derived saltwater. The mouth of the Chesapeake Bay is saltier and generally deceases farther north due to the greater influence of streams. The bay occurs in the humid subtropical climate zone, which exhibits hot, humid summers and mild to cool winters. Ice often builds up during the winter along the Delmarva Peninsula, located on the eastern shore, due to the prevailing northwest winds. During extreme cold periods, ice may even accumulate on the western shore.

Biodiversity

The Chesapeake Bay estuary supports more than 3,600 species of plants and animals. Typical habitat features of the bay include sand beaches, intertidal flats, shallow subtidal zones, wetlands, oyster reefs, and subtidal deepwater. Sand beaches are most common in the southern part of the bay but overall are less abundant than on the Atlantic Coast proper.

Beach "fleas" (genus *Emerita*), ghost crabs (*Ocypode quadrata*), and horseshoe crabs (*Limulus polyphemus*) are common inhabitants of sandy beaches. Horseshoe crabs, with their characteristic domed bodies and spiked tails, are primitive crustaceans dating back 360 million years. These crabs lay their eggs in the sand, particularly in the southern and middle portions of the bay, during high spring tides. Although not as abundant as in the Delaware Bay, horseshoe crabs are still extremely common here. Each crab lays 10,000 to 100,000 eggs, attracting shorebirds that consume many of the eggs. Humans harvest the crabs for use in the biomedical industry and as bait for commercial fishing.

Intertidal flats composed of mud and sand are valuable areas for a variety of invertebrates, including fiddler crabs (*Uca* spp.), marsh crabs (*Sesarma reticulatum*), and snails. Shallow subtidal zones are extremely productive areas, with an abundance of submerged aquatic vegetation (SAV) and a suite of fish, invertebrates, and waterbirds. SAV is a common and integral component of Chesapeake Bay; the 26 species occurring in the bay contribute to clean, productive water. Areas of higher salinity are dominated by widgeon grass (*Ruppia maritima*) and eelgrass (*Zostera marina*),

A young boy examining snails living on rocks along the Chesapeake Bay in Maryland in 2009. That year, the Chesapeake Bay was named a U.S. national treasure. (U.S. Fish and Wildlife Service/Ryan Hagerty)

which harbor a less diverse assemblage of species than freshwater areas of the bay. SAV is important for filtering pollutants and nutrients, including nitrogen and phosphorous, from the water and for providing food for fish, crabs, and waterfowl. If pollution levels become too high, SAV declines. At its peak recorded distribution, the bay had more than 927 square miles (2,400 square kilometers) of SAV, but in recent years, this figure has declined to 124 to 156 square miles (320 to 405 square kilometers).

Chesapeake Bay is well known for native American or Eastern oyster (*Crassostrea virginica*) beds. These dense oyster reefs attract diverse assemblages of micro- and macro-organisms that live in and among the oyster beds. Eastern oysters are ecologically important inhabitants of the bay and are central to the region's commercial fishery. Their abundance has declined since the mid-1800s, but at their peak, they filtered the entire volume of water in the bay every three to four days, removing algae and particulates from the water.

Oyster beds also provide habitat for numerous other species. Today, the oysters are at less than 5 percent of their historic densities, and harvests are less than 2 percent. The decline is attributed to increased efficiency of harvest techniques, degradation of water quality and habitat, and increased prevalence of diseases and parasites. Historic harvest records are rare, but in 1980, 3.1 million bushels (109,000 cubic meters) were harvested, compared with only 370,000 bushels (13,000 cubic meters) in 2001.

The forested, shrub, and emergent wetlands, including both tidal and nontidal areas, span the gradient from freshwater to saltwater, providing the plant material that breaks down into detritus, forming one of the primary foundations of the bay's complex food web. Indeed, consumption of detritus by organisms is about 10 times greater than herbivorous grazing in the bay. The eastern shores have larger expanses of marshes, providing habitat for migratory waterfowl during the spring and fall migrations, as well as during the winter. About 40 percent of the waterfowl in the Atlantic Flyway overwinter on Chesapeake Bay. Avian types range from ducks, terns, and gulls to apex predators including osprey (*Pandion haliaetus*) and bald eagles (*Halieaeetus leucocephalus*).

Large fish, including bluefish (*Pomatomus saltatrix*) and red drum (*Sciaenops ocellatus*), are among the most impressive inhabitants of the subtidal deepwater zones, but phytoplankton and zooplankton are dominant, integral contributors to the bay's productivity. Several anadromous fish species, including striped bass (*Morone saxatilis*), American shad (*Alosa sapidissima*), and American eel (*Anguilla rostrata*), live in the deepwater zones and enter the streams feeding Chesapeake Bay to spawn.

Blue Crabs

Blue crabs (*Callinectes sapidus*) are common and ecologically important to the food web of the shallow subtidal zones in the Chesapeake Bay. More than 100 million blue crabs live here, representing almost one-third of the United States' total commercial blue crab harvest, which is a highly sought human delicacy. Harvests in the bay have declined to less than 25,000 tons (22,700 metric tons) but had averaged more than 35,000 tons (32,000 metric tons) annually during the 1970s to 2005, but crabbing is still a $50 million-per-year industry. Populations have declined due to habitat loss, and harvests have declined both from the lower populations and from restrictions placed on harvesting.

Human Impact and Conservation

Chesapeake Bay has a long human history; paleo-Indians first inhabited the Chesapeake Bay watershed about 13,000 years ago, following the retreat of the glaciers. Oyster shells were used extensively by these early inhabitants as tools and jewelry, but the extent to which they were consumed for food is unclear. Certainly, mastodons (*Mammut* spp.), mammoths (*Mammuthus* spp.), and other large mammals were pursued for their food value. As the climate warmed in the region, the paleo-Indians were replaced by the archaic Indians about

9,000 years ago; they also primarily subsisted on hunting but apparently had a stronger connection to the aquatic resources, including fish and shellfish, and the migratory birds provided by the Chesapeake Bay. Woodland Indians moved into the Chesapeake Bay region some 3,000 years ago, remaining until after European settlers arrived; evidence of many villages and stronger relation to bay resources survives.

The Chesapeake Bay has been negatively affected and will continue to be so, both by direct and indirect human impact According to U.S. Fish and Wildlife reports, there is a long list of threatened and endangered species of flora and fauna in the bay area. Climate change is predicted to lower salinity due to increased flows from major rivers such as the Susquehanna, which could potentially affect important bay species that are intolerant of low salinity conditions, such as oysters and crabs. Moreover, sea-level rise from melting glaciers, due to global warming trends, will submerge some salt marshes. Models predict greater future increases in mean temperatures, and lower dissolved oxygen levels, which may affect numerous organisms.

The bay's productivity is limited by degradation of water quality caused by excessive chemical, nutrient, and sediment input. Sediment accretion rates in the bay have increased one and a half to seven times over the pre-European settlement rate. Europeans started settling the region in the early 17th and 18th centuries. Between 1830 and 1890, 80 percent of the land was cleared of timber, and much of it was plowed and put into agriculture. Clearing of the land led to increased flow of freshwater into the bay, resulting in greater sediment deposition. With the advent of commercial fertilizers since the mid-1800s—most notably those containing nitrogen and phosphate compounds—fertilizer discharge has increased, as have sewage discharge and industrial waste. The 1940s brought about common agricultural practices implementing a variety of pesticides. Starting in the 1960s, efforts have been made to reduce the effects of contaminated runoff. Water quality continues to fluctuate, as do population numbers of various species in the region.

The Chesapeake Bay in 2009 was declared a national treasure, and President Barack Obama issued an executive order to promote and restore its health, heritage, and natural resources. The U.S. Environmental Protection Agency has mandated goals in the form of Total Maximum Daily Load for the amount of pollutants that can enter the bay. These goals and related solutions to reduce sediment, nitrogen, and phosphorous runoff are implemented for each state in the watershed. In addition to federal and state agencies, numerous local and regional watershed groups, such as the Chesapeake Bay Foundation, are working to improve water quality in the bay.

JAMES T. ANDERSON

Further Reading

Baird, Daniel and Robert E. Ulanowicz. "The Seasonal Dynamics of the Chesapeake Bay Ecosystem." *Ecological Monographs* 59, no. 4 (1989).

Committee on Nonnative Oysters in the Chesapeake Bay. *Nonnative Oysters in the Chesapeake Bay.* Washington, DC: National Academies Press, 2004.

Cooper, S. R. and G. S. Brush. "Long-Term History of Chesapeake Bay Anoxia." *Science* 254, no. 5034 (1991).

Lippson, A. J. and R. L. Lippson. *Life in the Chesapeake Bay.* 2nd ed. Baltimore, MD: Johns Hopkins University Press, 1997.

National Oceanic and Atmospheric Administration. "Chesapeake Bay Office." http://chesapeakebay.noaa.gov.

United States Geologic Survey. "The Chesapeake Bay Bolide: Modern Consequences of an Ancient Cataclysm." http://woodshole.er.usgs.gov/epubs/bolide.

Chihuahuan Desert

Category: Desert Biomes.
Geographic Location: North America.
Summary: The vast Chihuahuan Desert, covering a large portion of the southwestern United States

and northern Mexico, hosts wildlife adapted to this isolated, harsh, and extreme ecoregion in diverse and unique ways.

The Chihuahuan Desert is the largest desert in North America, stretching across more than 170,000 square miles (440,000 square kilometers). This is an area larger than the state of California. The northernmost extent of the desert is in central New Mexico, while its southern extreme is in the southeastern corner of the Mexican state of Zacatecas. The desert was formed approximately 8,000 years ago as a result of its location in the mid-latitudes, combined with the rain-shadow effect of two major mountain ranges: the Sierra Madre Occidental to the west and the Sierra Madre Oriental to the east. Elevations in the Chihuahuan Desert range from 2,000 feet (600 meters) to 8,750 feet (2,700 meters), the majority of this occurring between 3,600 feet (1,100 meters) and 4,900 feet (1,500 meters).

Much of the Chihuahuan Desert is characterized by having basin-and-range geomorphology including a central highland that accounts for the relatively low mean annual temperature of 65 degrees F (18.6 degrees C). Abrupt changes in elevation often reveal unique habitats known as Sky Islands that are home to endemic species (native only to a particular biome), and also support one of the most critical migratory flyways in North America.

Precipitation in the Chihuahuan Desert varies greatly, depending on elevation and geographic location, but on average, the desert has annual precipitation of less than 9 inches (24 centimeters), most of which falls during the months of July and August. The desert is composed mainly of marine sedimentary rock, resulting from the region being underwater as part of the Cretaceous Seaway, which filled the entire region approximately 90 million years ago. The rich gypsum deposits that are widespread throughout this desert have resulted in the subterranean formation of expansive cave systems such as Carlsbad Caverns and Lechuguilla Cave in southern New Mexico. The largest body of water in the Chihuahuan Desert is the Rio Grande (called Rio Bravo in Mexico), which dissects the desert into northern and southern halves. Other important sources of water are seeps and groundwater springs: groundwater is the source of most surface water in this desert. Seeps and springs support complex ecosystems with a large number of endemic plants and animals. Of special interest is the valley at Cuatro Cienegas in Coahuila, Mexico, which supports more than 500 freshwater pools and an enormous and varied amount of wildlife.

The ecosystems of the Chihuahuan Desert are very diverse and range from xeric habitat to riparian zones along major rivers. Other types include vast grasslands, savannas, and alpine habitat. The vast majority of the Chihuahuan Desert is xeric habitat categorized as desert scrub. Desert-scrub habitat tends to occur at lower elevations and tends to have the least precipitation in the desert. Grasslands and savannas occur at low to midrange elevations and tend to have slightly more precipitation than desert scrub but still remain quite arid. Grasslands have been extremely modified over the past 200 years as overgrazing and water diversion have led to desertification and drastic changes in the flora and fauna of the region. The basin-and-range topography along with changes in climate over the past 10,000 years has resulted in isolated areas that support, collectively, up to 1,000 endemic species.

Flora and Fauna

The organisms of the Chihuahuan Desert are often unique, including the indicator species *Agave lechuguilla,* a small agave species found only in the calcareous soils of this desert. The region is home to an estimated 3,500 plant species, of which 25 to 30 percent are endemic. Among the most common are creosote bush (*Larrea tridentata*), various species of mesquite (*Prosopis* spp.), tarbush, acacia, an abundance of grasses, and hundreds of species of cacti in the *Coryphantha* and *Opuntia* genera. As many as one-fifth of the world's cacti species can be found in this desert.

Many of these plant species have evolved mechanisms to tolerate long periods of drought and intense heat. Deep root systems allow species of mesquite to tap into groundwater reservoirs, while extensive networks of shallow roots allow species like creosote bush to maximize water capture during the short, intense rainfall periods. Some species have even evolved ways to improve their photosynthetic efficiency in response to the extreme environmental conditions of the desert.

Animal species diversity is higher in the Chihuahuan Desert than in any other North American desert. Vertebrates include approximately 120 species of mammals, 300 species of birds, more than 100 species of fish, and more than 150 species of amphibians and reptiles. The diversity of habitat types in the desert provides homes for several endemic species of vertebrates and countless species of insects. The species diversity of bees, which are important pollinators here, is higher than in any other North American desert.

Many of the species have adapted to the extreme desert environment by modifying their behavior (such as nocturnal activity) and evolving physiological adaptations that maximize water retention and allow organisms to function for long periods without food or water. Large mammals that can still be found in the Chihuahuan Desert, though in small populations, include: brown bear, pronghorn antelope, American bison, wolves, peccaries, and even jaguars and mountain lions. Common birds of this desert include several species each of hummingbirds, owls including the great horned owl (*Bubo virginianus*), hawks, vireo, migrating sparrows, and quail. Also commonly found here are roadrunners and other ground birds.

Human Impact

Native Americans have occupied the Chihuahuan Desert throughout its entire history. The Mogollon people occupied much of the desert for more than 1,000 years and eventually were displaced by Pueblo and Apache cultures. The Tarahumara people of the highlands of Chihuahua, Mexico, still occupy much of the harshest mountainous terrain of the Chihuahuan Desert and have adapted to the area with unique hunting and agricultural practices. Urban, agrarian, and industrial development over the past 200 years, especially along the United States-Mexico border, has led to major threats to the ecological balance in the desert: overgrazing (and the related invasion of nonnative species); depletion of water sources due to irrigation and pumping of groundwater; mining; and off-road vehicle use in some fragile areas.

Conservation and protective measures have been put in place in the United States in the regions that fall under National Park status, and in Mexico with the creation of the National Commission for the Knowledge and Use of Biodiversity (CANABIO).

Such agencies will have to contend with changes wrought by global warming, such as wider opportunities for invasive species as both droughts and floods become more severe. It is expected that salt cedar (*Tamarix ramosissima*) and Russian olive (*Elaeagnus angustifolia*), for example, will spread through riparian areas by crowding out native species and reducing diversity as groundwater supply declines.

A lengthening of the frost-free season and higher minimum temperatures will expand the desert to the north and east, while higher elevations are likely to be colonized by arid savanna vegetation; some sky island communities will be severely stressed and may disappear as the warming climate alters their fundamental temperature, humidity, and snowpack regimes.

Israel Del Toro

Further Reading

National Park Service. "Inventory and Monitoring Program." May 2009. http://science.nature.nps.gov/im/units/chdn.

Skroch, Matt. "Sky Islands of North America: A Globally Unique and Threatened Inland Archipelago." *Terrain.org, 2008.* http://www.terrain.org/articles/21/skroch.htm.

Sowell, John. *Desert Ecology: An Introduction to Life in the Arid Southwest.* Salt Lake City: University of Utah Press, 2001.

Tweit, Susan J. Barren. *Wild, and Worthless: Living in the Chihuahuan Desert.* Tucson: University of Arizona Press, 2003.

Chilean Matorral Forests

Category: Forest Biomes.
Geographic Location: South America.
Summary: The Chilean Matorral ecosystem is rich with wildlife, much of it endemic, but human activity and low protection status present a continuing challenge.

Outside of the Mediterranean Sea area itself, mediterranean-type ecosystems worldwide are found in only a few areas, including the southwestern coast of Australia, the Cape of Good Hope area in South Africa, the coastal chaparral zone of California, and one part of South America: the Chilean Matorral. The term *matorral* refers to the Spanish word *mata,* for scrub vegetation. The Chilean Matorral covers an area of roughly 57,300 square miles (148,000 square kilometers), predominately a patchwork of small parks, agricultural areas, private lands, towns, and cities. Chile's capital, Santiago, is located here, as are other urban centers.

The Matorral is a narrow stretch of land in central Chile, extending south from one of the driest deserts in the world, the Atacama, to the mixed deciduous-evergreen temperate zone known as the Valdivian forests. The Matorral is about 350 miles (563 kilometers) long and 62 miles (100 kilometers) wide. Here, the summers are hot and dry and prone to drought conditions; winters are wet and mild. The mean annual temperature is 54 degrees F (12.2 degrees C). The native plant and animal communities in the Chilean Matorral biome are species-rich with a very high proportion of endemism (found only in this ecosystem), particularly among plants.

A chilla (Lycalopex griseus) in the Parque Nacional La Campana in Chile in 2009. The chilla is one of the increasingly threatened small mammals unique to the Chilean Matorral biome. (Flickr/Andrea Ugarte)

Flora and Fauna

As a plant community, Matorral refers to a zone of sclerophyll shrubs and trees (i.e., evergreen "hard-leaved" woody vegetation with small, waxy leaves that prevent water loss in the dry summer); cacti, bromeliads, and palms; and diverse understories of herbs, vines, and grasses. Most of the present scrubland was created by human activity and is a successional remnant of the native sclerophyllous forest. It now exists as a mosaic of shrubs and trees within a matrix of naturalized herbaceous plants.

This land hosts animals that are specially adapted to their unique habitat and cannot be found anywhere else on Earth—making them extremely rare and dependent upon protected areas for their survival. These include many small mammals such as the chilla, a fox-like animal; the yaca (mouse opossum); and the kodkod, the smallest wildcat in the Americas. Several lizard species are also endemic to the Chilean Matorral, as are 15 known species of birds. Among the latter are the Chilean mockingbird; three varieties of tapaculo; two species of parrots; the giant hummingbird; and some carnivorous species such as the aplomado falcon, cinereous harrier, and the short-eared owl.

Human Impact

The ecoregion's core, the Central Valley, constitutes Chile's most intensively inhabited area. It is very fertile and is the agricultural heartland with booming wine, vegetable, and fruit industries. In the more southerly parts fruit, crops, pasture, and fire-prone pine and eucalyptus plantations are widespread. Because of the high agricultural value, the Central Valley of Chile has been highly modified since the arrival of Europeans. Although early settlers introduced fires around 14,000 years ago, it is only since Spanish colonization that fires have become frequent. The Matorral is poorly fire-adapted; human-induced fires cause major and long-lasting damage. Seeds of native sclerophyllous species do not survive even low-intensity fires. Moreover, the capacity of regeneration of these sclerophyllous species is very low—even after cessation of livestock grazing—because of the constant soil disruption and shoot consumption pressures from introduced rabbits and hares.

Logging and mining, with the ensuing roadways and pollution that are created, have contributed to habitat loss here. The increasing density of the road network goes hand-in-hand with invasions by exotic species and with deforestation, both being positively correlated with distance to primary roads. The current rate of introductions of invasive plant species is unprecedented in regional history. Intentionally and unintentionally, nonnative species have spread fast and uncontrolled, further promoted by the secondary plant and animal invaders that come with the introduction of controlled populations of livestock. The consequences can be severe. Introductions not only modify patterns of abundance and distribution of native species, they also cause local extinctions and, especially in the case of plantations of exotic trees, can significantly modify soil, microclimate and fire characteristics, thus irreversibly altering the environmental physiology of the ecosystem.

Despite its highly unique biodiversity, the Chilean Matorral is perhaps the least protected of the world's five major mediterranean regions. Moreover, it is the least protected region in Chile at large. The World Wildlife Fund has assessed this ecoregion's conservation status to be Critical/Endangered. There remains a significant proportion of currently unprotected natural and semi-natural land with conservation potential, while human pressure increases continually.

More intensive and efficient protection and conservation action is urgently required; some conservation schemes on private land and neighborhood initiatives have recently emerged. In 2010, the new Altos de Cantillana Nature Sanctuary, located approximately 25 miles (40 kilometers) from Santiago, was established by the Chilean government, protecting 6,778 acres (2,743 hectares) of Matorral. Although there is relatively little protected public land in Chile, the concept of private land trusts has recently gained momentum. These small parcels are considered complementary to the National Public System of Protected Areas—Sistema Nacional de Áreas Silvestres Protegidas (SNASPE), which unifies conservation efforts within and between the country's national parks, reserves, and monuments.

Global warming scenarios generally point to warmer and drier climate in mediterranean biomes such as the Chilean Matorral forests. The pressure this will exert on plant and animal species to relocate will only be exacerbated by increased likelihood of fire; together these vectors point to accelerated habitat fragmentation.

Stephan M. Funk

Further Reading

Cox, Robin L. and Emma C. Underwood. "The Importance of Conserving Biodiversity Outside of Protected Areas in Mediterranean Ecosystems." *PLoS One* 6, no. 1 (2011).

Figueroa, Javier A., S. A. Castro, P. A. Marquet, and F. M. Jaksic. "Exotic Plant Invasions to the Mediterranean Region of Chile: Causes, History and Impacts." *Revista Chilena de Historia Natural* 77, no. 3 (2004).

Funk, Stephan M. and John E. Fa. "Ecoregion Prioritization Suggests an Armoury Not a Silver Bullet for Conservation Planning." *PLoS One,* 5, no. 1 (2010).

Gomez-Gonzalez, S., C. Torres-Diaz, G. Valencia, P. Torres-Morales, L. A. Cavieres, and J. G. Pausas. "Anthropogenic Fires Increase Alien and Native Annual Species in the Chilean Coastal Matorral." *Diversity and Distributions* 17, no. 1 (2011).

Chile Intertidal Zones

Category: Marine and Oceanic Biomes.
Geographic Location: South America.

Summary: Chile has one of the world's longest and most diverse intertidal zones, nourished by the Humboldt Current that flows along its entire coast.

Chile is a long, narrow country that is defined by the Andes mountain range on the east and by the Pacific Ocean on the west. Chile is 2,672 miles (4,300 kilometers) long and an average of 109 miles (175 kilometers) wide. This long strip of coastal land runs parallel to the Andes and extends from midcontinent to Cape Horn, the southernmost point of South America.

The coastal climate is strongly influenced by the Humboldt Current. This slow, cold Pacific current sustains one of the world's richest fisheries along the Chilean coast. Many forms of wildlife are also found in abundance along the adjacent land areas. The Humboldt Current begins north of the Antarctic and runs the full length of the Chilean coast, bringing nutrient-rich waters that nourish the intertidal zones. Intertidal zones along shorelines are the areas between the high tide and the low tide and can include many types of habitat: steep rocky cliffs, sandy beaches, wetlands, and marshes.

The organisms that live here have become specially adapted to survive the significant changes in temperature, salinity, moisture, acidity, dissolved oxygen, and food supply that occur on a daily basis due to the movement of the tides. Different plants and animals that have excelled at dealing with different sets of conditions dominate zones at different tidal heights. This distribution pattern of different species along the shore in vertical bands is referred to as intertidal zonation.

Varied climatic conditions along Chile's long coast also contribute to the diversity of intertidal species: An extremely dry climate is typical of the north, but the coast exhibits a rainforest microclimate, as moisture from the sea is trapped by cliffs and valleys; the central region of the country is characterized by the hot, dry summers and mild, wet winters of a Mediterranean climate; and the southern region is cooler and wetter, classified as an oceanic temperate climate.

The complex relationship between the Humboldt Current and the intertidal zones supports a wealth of oceanic and land organisms. Near the coast are found some of the world's greatest populations of shrimp, anchovies, squid, sardines, salmon, and mackerel. Marine birds and mammals—such as penguins, pelicans, flamingos, bottle-nosed dolphins, whales, and sea lions—live off of the smaller of these fish, mollusks, and crustaceans. Intertidal species such as sea urchins, starfish, and coral also contribute to this diverse, yet fragile eco-community, as do seaweeds, kelp, and shoreline grasses.

Many of these organisms are adapted to extremes, as the local water may be fresh, brackish, or salty. They may be subject to varied wave action, and must be able to live exposed to air and sun, or submerged in water. Many types of shellfish thrive here, most notably the Chilean blue mussel, which is harvested by humans regularly. Wild fisheries and salmon fish farms have become very economically important to Chile in the past 30 years.

Human Impact and Conservation

Because there has been an increasing worldwide demand for fish over the past several decades, wild fisheries are being overexploited. Salmon farms consume many of these wild fish, and also have caused problems on some coastal areas due to pollution runoff, low oxygen zones, algal blooms, and the spread of viruses into the marine environment. Inland mining practices have also had a negative impact on shoreline species. Forestry, pulp, and paper industries are present in the south, releasing pollutants that harm the intertidal zones.

The Chilean coast, including the intertidal zone, is administered by the Ministry of National Defense through the Marine Subsecretary, which addresses issues related to national and state public goods in the intertidal zones—including harvesting of marine species. Under this Marine Subsecretary is another division that regulates such diverse activities as aquaculture and harbors. State ownership of the coast tends to occur in the extreme north or south of Chile, as the other coastal lands became private during the long colonial period.

Since the 1990s, there has been extensive Chilean research on intertidal zones, and two coastal marine research stations have been created: Estación Mehuín in southern Chile of the Universidad Austral de Chile, and Estación Costera de Investigaciones Marinas in Las Cruces in Central Chile of Pontificia Universidad Católica de Chile.

In 2001, Oceana was founded. It is the largest international organization focused solely on ocean conservation. The Oceana offices in North America, Central America, South America, and Europe work together on a limited number of strategic, directed campaigns to achieve measurable outcomes that will help return the oceans to former levels of abundance. (Oceana Chile is located in Santiago, Chile.) In its first decade, Oceana has achieved some policy victories for marine life and habitats. As oceans play a key role in regulating the Earth's temperatures, this organization is currently considering the impacts of global warming and is moving forward with public education, legislation advocacy, and ecosystem protection.

Magdalena Ariadne Kim Muir

Further Reading

Bamfield Marine Sciences Centre Public Education Program. "Oceanlink—All About the Ocean: Exploring the Intertidal." http://oceanlink.info/biodiversity/intertidal/intertidal.html.

Castilla, J. C. "Roles of Experimental Marine Ecology in Coastal Management and Conservation." *Journal of Experimental Marine Biology and Ecology* 250, no. 1–2 (2000).

Heileman, S., R. Guevara, F. Chavez, A. Bertrand and H. Soldi. "XVII-56 Humboldt Current: Large Marine Ecosystems #13." 2009. http://www.lme.noaa.gov/lmeweb/LME_Report/lme_13.pdf.

China Loess Plateau Mixed Forests, Central

Category: Forest Biomes.

Geographic Location: Asia.

Summary: This plateau, also known as the Plateau of Yellow Earth, is the cradle of the Chinese civilization. After more than 1,000 years of deforestation and soil erosion, it is sparsely vegetated but parts of it are being rehabilitated to its natural state.

The central loess plateau, extending across 250,000 square miles (650,000 square kilometers) of north central China along the Yellow River, is the largest loess deposit region in the world and home to some 100 million people. Tilted upward from southeast to northwest, its elevation ranges from 1,600 to 10,000 feet (488 to 3,048 meters) above sea level. The lush Qinling Mountains to the south and the Mongolia Plateau to the north set up its geographic limits, stretching across broad climatic gradients.

The yellowish silt deposits called loess were formed here up to 2.5 million years ago during repeated waves of glaciation and wind from northern desert regions. Loess tends to make fertile farm lands, but because of its vertical cleavage, loose structure, and high porosity, is one of the most erosion-prone soils of the planet. Affected by centuries of deforestation and overgrazing, and exacerbated by summer monsoon rainfall and subsequent flash floods, the China Loess Plateau biome has become one of the most eroded areas of the world. The Yellow River, its name given by the silt that colors it, picks up roughly 1.8 billion tons (1.6 billion metric tons) of runoff deposits every year. Periodic dust storms also move large quantities of the yellow silt, while causing visibility problems in urban areas.

Flora and Fauna

Natural vegetation of the loess plateau has experienced degradation during the history of human habitation (a period spanning several thousand years), leaving only about 5 percent of forest cover in the entire region. The existing forest patches are more commonly found in the southeast part of the plateau, where climate is more favorable and where agricultural activities are limited only by rocky soils and more mountainous topography.

These mixed forests are comprised of trees like aspen, oak, willow, birch, arborvitae, and pine.

The majority of the current plateau is characterized by a sparsely vegetated landscape of numerous dry gullies and fragmented flat-top ridges and table lands, with slopes encircled by terrace farms. Shrubs and grass are now more common natural vegetation. Despite the loss of an abundance of habitats that forests may have supported earlier in human history, some animals that still may be found in the region are: the rhesus macaque, musk deer, giant salamander, pheasants, black stork, and the golden eagle.

Human Impact and Conservation Efforts

The China Loess Plateau forests have gradually been lost to centuries of traditional, yet unsustainable, agricultural practices, such as planting on hillsides and free-range of livestock. Increasing population and farmland expansion, overexploitation of timber for buildings and fuels, and quite often damages incurred by wars and famines have also contributed to the decline of this vast ecoregion. Toward the drier part of the plateau, overgrazing has caused disruption of grassland soil structure, leading to severe desertification and frequent dust storms. The absence of a stable vegetation cover has caused fast erosion of life-supporting soil layers and may have modified the regional climate toward an even drier and more inclement one. Further, sediment loads carried by the Yellow River are quickly deposited in the plains east of the plateau, creating perched river channels 20 feet (6 meters) higher than surrounding areas, constantly posing flood hazards to inhabitants along the river.

Recovering a deeply deteriorated ecosystem is by no means an easy task. However, starting in 1978, national reforestation projects have made steady progress in many areas of northern China, including the Loess Plateau. In 1994, the Chinese government, with help from the World Bank, launched the Loess Plateau Watershed Rehabilitation Project, one of the largest ecological restoration projects in the world in terms of the level of governmental support and public participation.

Farmers in the region were educated about, and subsidized to implement new agricultural practices. Livestock grazing is now restricted, and seeding grasses are encouraged. Herders are subsidized by the government for constructing sheepfolds and nurturing high-productivity grasslands to compensate for the reduced access to natural forage. Farmers are subsidized to plant trees among their own crops. These projects are not only aimed at restoring ecosystems, but also used as opportunities to transform the economic structure of rural regions, and to improve the quality of people's lives.

Already, there has been a noticeable improvement in land quality. According to the government conservation plan, 30,000 square miles (78,000 square kilometers) of forests and 33,000 square miles (85,000 square kilometers) of grasslands will be restored to the plateau before 2030. Subsequent benefits will include increased ecosystem capacity to clean up air, retain water, and resist dust storms.

It is hoped that if desertification trends here are reversed, there will be some protection provided against future global warming pressures. The China Loess Plateau Mixed Forests biome, forming part of the heartland of China, has for centuries constantly experienced intensive interactions between human and nature. A sustainable future requires incessant efforts toward both recovery and protection of vegetation, soil, and water.

LIANG LIANG

Further Reading

National Development and Reform Commission, People's Republic of China. "China's National Climate Change Programme." 2007. http://www.ccchina.gov.cn/WebSite/CCChina/UpFile/File188.pdf

Okuda, Setsuo, et al. *Loess: Geomorphological Hazards and Processes.* Reiskirchen, Germany: Catena-Verlag, 1991.

The World Bank. "Restoring China's Loess Plateau." March 15, 2007. http://www.worldbank.org/en/news/2007/03/15/restoring-chinas-loess-plateau.

Youming, Wang. *Ecological Characteristics of Loess Plateau Forest Plantations.* Beijing: Chinese Forestry Publishing House, 1994.

Chittagong Hill Tracts Plantation Forests

Category: Forest Biomes.
Geographic Location: Bangladesh.
Summary: The forests of the Chittagong Hill Tracts support a variety of species that are threatened mainly by the activities of many groups of indigenous peoples whose hunting and agricultural methods are degrading the habitat.

Chittagong Hill Tracts, formerly Korpos Mohol, is an area in southeastern Bangladesh, inland from the Bay of Bengal. Its terrestrial borders run along India and Burma (Myanmar) and cover an area of 8,000 square miles (13,000-square kilometers), about one-tenth of the area of Bangladesh. The Chittagong Hills constitute the only significant elevated area in the country, creating the western shoulder of the north-south mountain ranges of Burma and eastern India. The Chittagong Hills rise steeply to narrow ridge lines, with altitudes from 2,000 to 3,000 feet (600 to 900 meters) above sea level. The highest elevation in Bangladesh is at Mowdok Mual, at 3,500 feet (1,050 meters) altitude, in the southeastern part of the hills.

Fertile valleys lie between the ranges, which generally run north-south. West of the Chittagong Hills is a broad plain, cut by the rivers that drain into the Bay of Bengal, then an area of low coastal hills, and finally a narrow, wet coastal plain located between the seaport cities of Chittagong in the north and Cox's Bazar in the south. The Chittagong Tracts extend into and overlap with the Mizoram-Manipur-Kachin Rainforests of Burma (Myanmar) to the east. This large rainforest ecoregion represents the semi-evergreen submontane that supports high biological diversity.

Chittagong Hill Tracts region has a tropical monsoon climate based around a seasonal change in prevailing wind direction. The mean annual rainfall is roughly 79 inches (200 centimeters) in the north and east, and up to 118 inches (300 centimeters) in the south and west. The mean annual temperature is 73 degrees F (23 degrees C). In the summer, air moves from the ocean toward the land, and in winter air moves from the land toward the ocean. The tropical climates here create the conditions that allow a significant amount of vegetative production.

Flora and Fauna

The forests of tropical and subtropical evergreens in the Chittagong Hill Tracts cover about 15 percent of the region. Bangladesh is known as a land of trees. The hills, rivers, and cliffs of the Chittagong Hill Tracts are covered with dense bamboo breaks, tall trees, and creeper jungles. The valleys are covered with thick forest. The vegetation is characterized by semi-evergreen (deciduous) to tropical evergreen, dominated by tall trees belonging to the two-winged-fruit tree (*Dipterocarpaceae*), spurge (*Euphorbiaceae*), laurel (*Lauraceae*), pea (*Leguminosae*), and madder (*Rubiaceae*) families.

Domesticated elephants walking along a wooded road in Bangladesh. Migrating wild elephants, along with other large mammals like buffalo and bears, have become rare in the Chittagong Hills. (World Bank/Thomas Sennett)

Teak (*Tectona grandis*) was introduced to the Hill Tracts in the last century; the quality of the wood grown here approaches that of Burma's (Myanmar's) and is much better than Indian teak.

Each season produces its special variety of flowers; among them, the prolific water hyacinth (*Eichhornia* spp.) flourishes. Its carpet of thick green leaves and blue flowers gives the impression that solid ground lies underneath. Other decorative plants that are also widely spread here are jasmine, water lily, rose, hibiscus, bougainvillea, magnolia, and an incredible diversity of wild orchids in the forested areas.

Plantation trees that grow in this area include coconut, mango, banana, jackfruit, teak, palm, and bamboo. Crops of cotton, tobacco, rice, tea, and oilseeds are also raised in the valleys between the hills. Unfortunately, the creation of these plantations and croplands contributes to the degradation of natural forests. Of recent significance is the recognition that many of the different groups of indigenous peoples in the area use various plants for medicinal purposes. These plants and their uses are only very recently being documented.

Many animal species in the Chittagong Hills have become critically endangered due to indiscriminate poaching and rapid depletion of forests; these include rare species of wildlife like yak (goyal), the royal Bengal tiger, panther (chitabagh), large and small civet cat, pangolin, gibbon (Ulluk), barking deer, samber, slow loris (Lajjabati Banor), and peacock. Some of these animals may already be on the verge of extinction in the Chittagong Hill Tracts forests. Meanwhile, many common animals such as deer, wild pigs, and wild birds are also fleeing the forests across the border as increasing population, illegal poaching, and degradation of forest reserves are causing their habitats to diminish.

Other animals that are now only rarely seen in the Chittagong Hills are elephants (migratory), wild buffalo, bear, and porcupine. Reptiles that can be found in this ecosystem include the mud turtle, river tortoise, tiki-tik gecko, crocodile, python, king cobra, and a variety of other poisonous snakes. There are more than 250 species of birds here, including little grebe, heron, egret, common moorhen, common coot, Asian openbill stork, white-winged wood duck, and other waterfowl.

William Forbes
Kaleshia Hamilton

Further Reading

Chowdhury, Khairul. "Politics of Identities and Resources in Chittagong Hill Tracts, Bangladesh: Ethnonationalism and/or Indigenous Identity." *Asian Journal of Social Science* 36, no. 1 (2008).

Gain, Philip, ed. *The Chittagong Hill Tracts: Life and Nature at Risk*. Dhaka, Bangladesh: Society for Environment and Human Development, 2000.

Gunter, Bernhard G., Atiq Rahman, and Ataur Rahman. *How Vulnerable are Bangladesh's Indigenous People to Climate Change?* Falls Church, VA: Bangladesh Development Research Center, 2008.

Christmas and Cocos Islands Tropical Forests

Category: Forest Biomes.
Geographic Location: Indian Ocean.
Summary: These isolated oceanic islands possess large tracts of tropical forests that provide habitat for endemic terrestrial species; they also serve as important seabird rookeries and stopover points for migratory birds.

Christmas Island and Cocos (Keeling) Islands are two Australian territories located southwest of the island of Java in Indonesia and northwest of Australia in the Indian Ocean. Christmas Island is about 560 miles (900 kilometers) northeast of the Cocos Islands. Both territories have equatorial oceanic climates, with distinct tropical wet and dry seasons; average temperatures range from about 73–84 degrees F (23–29 degrees C), and average annual precipitation may exceed 79 inches (200 centimeters).

Christmas Island and the Cocos Islands are both volcanic islands, the latter being more typical of atoll ecosystems in the Pacific. The islands possess large tracts of broadleaf forests that are comprised of a combination of Indo-Pacific and Melanesian tree species, and serve as breeding grounds for colonizing birds. They also function as habitat for endemic (exclusively found in this biome) plant, bird, bat, reptile, and crab taxa.

Christmas Island

Christmas Island is situated at the peak of an extinct volcano rising from the ocean floor. Its landscape is comprised of steep coastal cliffs surrounded by a series of terraces and an interior plateau with large tracts of primary monsoonal rainforest. The highest points on the island are approximately 1,150 feet (350 meters) above sea level. Rainwater falling here filters through the soil and limestone, supplying several freshwater streams on the island. The role of Christmas Island as an important seabird rookery is underscored by the buildup of phosphate from the deposition of large amounts of bird guano over the millennia. Although phosphate mining has threatened the island's unique forest community in the past, approximately two-thirds of Christmas Island's land area is now protected within Christmas Island National Park, managed by Parks Australia.

The island's plant communities are dependent on soil type, depth, moisture, and distance from the sea, with tropical broadleaf forest making up 75 percent of the island's total area. The inland plateau is somewhat sheltered from the ocean, allowing for the development of relatively deeper soils compared with the terraced coastal regions of the island. These deeper soils permit the establishment of rainforest vegetation that creates a thick evergreen canopy and an open understory that supports a rich epiphyte community. Common canopy species include Malayan banyan (*Ficus microcarpa*), Tahitian chestnut (*Inocarpus fagifer*), satinash (*Syzygium nervosum*), soapberry tree (*Tristiropsis acutangula*), and Jack-in-the-box tree (*Hernandia ovigera*). Two endemic species—the Christmas Island palm (*Arenga listeri*) and a species of screwpine (*Pandanus elatus*)—dominate the understory.

Christmas Island is home to 16 land-crab species, representing very high diversity among island populations of such crabs. In addition to the Christmas Island red crab (*Gecarcoidea natalis*), another species of note is the robber crab (*Birgus latro*), which is the largest extant land-crab species.

Endemism within the forests of the island is especially pronounced. The Christmas Island fruit bat (*Pteropus natalis*) as well as a variety of endemic bird species are found here. The forest serves as important breeding habitat for Abbott's booby (*Papasula abbotti*); the Christmas Island frigatebird (*Fregata andrewsi*); the Christmas Island white-eye (*Zosterops natalis*); the Christmas Island imperial pigeon (*Ducula whartoni*); and the Christmas Island hawk owl (*Ninox natalis*), which is one of the rarest owls on Earth.

Only about 1,400 people inhabit the island. A concern to residents is the effect global warming could have on this small ecoregion. Although Christmas Island is less vulnerable to the potential impacts of climate change and sea-level rise than the Cocos Islands, being almost completely surrounded by a sea cliff of about 50 feet (15 meters) above sea level, there are a number of low-lying areas (such as the Kampong and Settlement) that are potentially exposed to sea-level rise and the effects of storm surge associated with climate change.

Cocos Islands

Also known as the Keeling Islands, the Cocos Islands are among the most isolated islands in the Indian Ocean. They consist of two separate groups of coral islands and reefs that make up two atolls, each sitting atop a volcanic seamount. The atolls are relatively flat, with soils composed of guano, pumice, and coral, and with thick stands of pisonia (*Pisonia grandis*) and coconut palm (*Cocos nucifera*). Other species present include *Laportea aestuans*, *Canavalia cathartica*, and *Erythrina variegata*. Freshwater is available in lenses found only on the larger islets here. Vegetation of the southern atoll has been largely replaced by coconut plantations or other introduced plant species. However, the uninhabited North Keeling Island possesses large areas where the native forest community is thought to be largely intact.

Pulu Keeling National Park on North Keeling Island serves as a seabird rookery, supporting large

The Christmas Island red crab (Gecarcoidea natalis). *Sixteen land-crab species, including the largest extant land-crab species in the world, the robber crab* (Birgus latro), *inhabit the island. (Thinkstock)*

breeding colonies of red-footed booby *(Sula sula)*, great frigatebird (*Fregata minor*), lesser frigatebird (*Fregata ariel)*, common noddy (*Anous stolidus*), and white tern (*Gygis alba*). It is also home to the one existing population of the endemic Cocos buff-banded rail (*Rallus phillippensis andrewsi*).

Because of isolation, low total land area, frequent inundation by oceanic cyclones, and likely recent (geologically speaking—about 4,000 years ago) submergence, the Cocos Islands terrestrial fauna has low species richness, and endemism is not as evident as on Christmas Island. No land mammals are native to the Cocos Islands, but there are a number of land crabs, insects, birds, and four reptiles, including three gecko and one blind snake species, that are native here.

Land crabs are common in the forest, including the little nipper crab (*Geograpsus grayi*), the red land crab, the robber crab, and three species of hermit crabs. As on Christmas Island, these crabs quickly consume tree seeds, seedlings, and leaf litter on the forest floor and play a major role in regulating the plant community.

Human Impact

Human presence here has only been since the early 1800s. About 600 people inhabit this region—and only on the Southern Keeling Islands. Wildlife in the South Islands continues to be threatened by loss of forest due to coconut plantations. Although not inhabited by humans, there has been a great loss of bird habitat on the North Island, due to hunting and poaching parties that visit from the south.

One notable, human-induced habitat threat was the accidental introduction in the early 20th century of the yellow crazy ant (*Anoplolepis gracilipes*) to Christmas Island. With few predators, an ideal climate, and very abundant food sources, these ants became widely established within a few decades. By the latter 20th century, the thriving yellow crazy ant had built supercolonies on Christmas Island, and had devastated many native crab species, among other fauna; they also negatively altered the balance of rainforest species and canopy structure. Mitigation since the 1990s has come in the form of baiting with contagious poisons, sometimes by aerial spraying; these measures seem to have depleted the population of this tramp ant to manageable levels.

A larger concern may be that of global warming trends. The Cocos Islands are extremely vulnerable to the potential impacts of climate change and sea-level rise. These atolls have elevations between 3 and 12 feet (1 to 4 meters) above sea level, and any change in the mean sea level, combined with the effects of storm surge associated with large storms or cyclones, are likely to have dramatic consequences, especially for settlements on South Keeling. These islands are already becoming increasingly vulnerable to extreme weather events; among the most substantial impacts of climate change would be losses of coastal infrastructure and land resulting from inundation, storm surge, and shoreline and coral reef erosion.

Kerry Bohl Stricker

Further Reading

Australian Government Publications. "Climate Change Risk Assessment for the Australian Indian Ocean Territories: Cocos (Keeling) Islands and Christmas Island." January 2009. http://www.regional.gov.au/territories/publications/files/Final+Report+CKI+and+CC.pdf.

Du Puy, D. J. "Christmas Island." In A. S. George, A. E. Orchard, and H. J. Hewson, eds. *Flora of Australia, Volume 50, Oceanic Islands 2.* Canberra: Australian Government Publishing Service, 1993.
Green, Peter T., Dennis J. O'Dowd, and P. S. Lake. "Control of Seedling Recruitment by Land Crabs in Rainforest on a Remote Oceanic Island." *Ecology* 78, no. 8 (1997).
O'Dowd, Dennis J., Peter T. Green, and P. S. Lake. "Invasional 'Meltdown' on an Oceanic Island." *Ecology Letters* 6, no. 9 (2003).

Chukchi Sea

Category: Marine and Oceanic Biomes.
Geographic Location: Arctic.
Summary: This diverse aquatic landscape, named after the indigenous Chukchi people, connects Russia's Siberian territory to the state of Alaska; it is a focus of dramatic climate change.

The Chukchi Sea, located off the coast of northwestern Alaska, is approximately 232,000 square miles (600,000 square kilometers), with most depths not exceeding 164 feet (50 meters). North of the Bering Sea, east of the East Siberian Sea, and west of the Beaufort Sea, it connects the Arctic and Pacific Oceans. Its shallow seafloor and stratified water column are populated by a vast array of organisms due to the influx of nutrients, warmth, and freshwater from warmer Pacific waters. This water flows from the Bering Sea through the Bering Strait before breaking into three different currents under the ice of the Chukchi Sea.

As with other areas of the Arctic, the Chukchi basin is overlain by pack ice for most of the year. However, indicators of global climate change are greatly in evidence here: Sea ice is disappearing at a rate of roughly 4 percent per decade, and alarmingly, this trend may be accelerating; sea temperatures are rising; shorelines are dramatically eroding; and summer sea ice is receding to historic lows. In addition to a changing climate, overfishing and commercial interest in natural resources such as gas and oil threaten the region. Thus, while rich in nutrients, marine species, and minerals, the Chukchi is threatened and its future remains uncertain.

Hydrology and Sea Ice

The Chukchi Sea gets its name from the indigenous Chukchi people who inhabit the Chukchi Peninsula within the Russian Federation. Despite being 193 miles (500 kilometers) wide and 309 miles (800 kilometers) long, one of the defining physical characteristics of the sea is its shallow continental shelf. The water column is highly stratified, with melting sea ice and runoff providing for a fresher surface layer and denser, saltier water near the bottom. The Arctic and Pacific Oceans are bonded by the Chukchi, whose waters are an important source of nutrients, heat, and freshwater for the Arctic Ocean.

Nutrients here support Arctic ecosystems; heat influences ice-melt; freshwater can stratify the Arctic Ocean and help protect the surface ice from the warmer waters below. Freshwater throughflow is an important part of global water cycles, and the pathways of Pacific waters in the Arctic is thought to affect the entire Arctic system.

Melting sea ice constantly floating above the ocean currents is another source of nutrients, providing life to a vast array of species. Sea ice in the Arctic grows and recedes seasonally, with a peak around March and a trough typically arriving in late September that leaves the Chukchi at its sea-ice minimum. Chunks of ice, or floes, are prone to converge and form thick ridges, creating a unique landscape of extremely dense sea ice. Because of these ridge formations, sea ice in the Arctic tends to have a longer life cycle than its flatter, thinner counterpart in the Antarctic. Sea ice in the Arctic, which receives very little snow due to the region's being surrounded by land, is on average about 7 feet (2 meters) thick but can grow up to 16 feet (5 meters). Sea ice is formed by cold temperatures—a minimum of 29 degrees F (minus 2 degrees C)—acting on ocean saltwater, causing tiny ice platelets, or frazil ice, to eventually form pack ice. As it freezes, the crystals of the ice expel salt into the water, yielding crystals that consist of nearly pure freshwater.

Up until very recently, the average mean Arctic sea-ice extent (or the average annual maximum ice area) has been around 6 million square miles (15 million square kilometers), and the average minimum sea-ice extent has been around 3 million square miles (8 million square kilometers), which means that more than half of the ice melts every year. During the summer of 2012, however, Arctic sea-ice extent fell to just over 1.5 million square miles (4 million square kilometers), surpassing the previous low set in September 2007, as reported by the U.S. National Snow and Ice Data Center. The comprehensive outcome of these significant ecosystem changes will unlikely be positive for many of the organisms now sustained in the region.

Biodiversity

When sea ice starts to retreat in the spring, polynyas, which are bodies of open water in the middle of ice floes, can be formed. Polynyas are unique to the polar oceans and formed via ocean currents, wind, or upwelling. When the sunlight hits one of the approximately 20 polynyas in the Chukchi every year, a microalgae bloom occurs, creating the foundation of the marine food web. The combination of sea ice and the influx of nutrient-rich Pacific water provides for one of the most productive ecosystems in the world, with a multitude of seals, seabirds, walruses, polar bears, whales, fish, and invertebrates inhabiting its realm.

Beneath the sea ice, various species of fish such as the capelin, Arctic cod, and sculpin live in the open water, while anemones, mollusks, and numerous other benthic organisms are common sights on the seafloor. Meanwhile, rivers and inlets along the Chukchi provide more than adequate breeding, nesting, and feeding spots for dozens of bird species. Spectacled eiders, another endangered species, feed on mollusks in shallow polynyas before migrating south for breeding.

Endangered fin, bowhead, and humpback whales, as well as the formerly endangered gray whale, feed and migrate throughout the Chukchi Sea. Beluga whales numbering about 3,500 molt, feed, and calve along the Kasegulak Lagoon off the coast of Alaska. Many other marine mammals call the Chukchi home, with bearded, ribbon, spotted, and ringed seals as well as walrus living their lives both on and below the sea ice. Walruses, which are long-lived social animals, are considered to be a keystone species of the Arctic ecosystem, with Pacific populations spending their summers and falls in the Chukchi before migrating south to the Bering Strait for the cold, harsh winters.

Ringed seals are the most abundant, wide-ranging seals in the Northern Hemisphere, identified by the small head, plump body, and silver rings on their back, from which the seals get their name. They are the primary food source for the icon of the Arctic: the polar bear. The polar bear, now on the Endangered Species list, is a symbol of wildlife loss due to global warming.

Environmental Threats

With an estimated population of about 2,000, more than half of America's polar bears inhabit the Chukchi ecoregion. They depend on the sea ice for all aspects of life, from breeding and denning to hunting. In 2007, it was reported that surveys were showing that numerous polar bears were swimming far from land and risking drowning due to record loss of sea ice. Because of this increase of ice-melt due to climate change, the U.S. Geological Survey predicts a comprehensive loss of Alaska's polar bears by 2050.

Like most of the Arctic, the Chukchi is also environmentally threatened due to commercial interests in its natural reserves. Commercial fishing and whaling are putting pressure on the fragile ecosystem, and as of August 2012, Shell Oil was seeking extended time for drilling purposes. The area is projected to contain more than 30 billion barrels (4.8 billion cubic meters) of oil.

For now, the Chukchi remains a diverse, pristine, and balanced ecosystem, providing a habitat for over a dozen threatened or endangered species, among others. However, with global climate change causing sea-ice extent retreat, a rise in sea level, and erosion along coasts due to thawing permafrost, legislative advocacy is in the forefront of all discussions regarding this biome. More research needs to be conducted; potential impacts and mitigation measures need to be adequately addressed.

KEVIN BAKKER

Further Reading

Audubon Alaska. "Chukchi Sea." http://ak.audubon.org/chukchi-sea.

Moore, Sue E., Douglas P. DeMaster, and Paul K. Dayton. "Cetacean Habitat Selection in the Alaskan Arctic During Summer and Autumn." *Arctic* 53, no. 4 (2000).

Nagel, P. A. and John F. Turner. *Results of the Third Joint US-USSR Bering & Chukchi Seas Expedition (BERPAC): Summer 1988.* Washington, DC: U.S. Fish and Wildlife Service, 1992.

National Snow and Ice Data Center. "All About Sea Ice." http://nsidc.org/seaice/characteristics/difference.html.

Clipperton Island

Category: Marine and Oceanic Biomes.
Geographic Location: Pacific Ocean.
Summary: Uninhabited Clipperton Island is the only coral atoll in the eastern Pacific Ocean; its enormous seabird colonies are currently threatened by recently introduced black rats and possibly by commercial fishing.

Clipperton Island is an isolated, uninhabited atoll in the middle of the eastern tropical Pacific Ocean, approximately 800 miles (1,300 kilometers) off the west coast of Mexico. It is the only coral atoll in the entire eastern Pacific Ocean and one of the most isolated reefs anywhere in the world. Although French-owned, it has sometimes been claimed by Mexico. It was intermittently inhabited by guano miners from the 1800s until 1917, and was also occupied by the U.S. military briefly during World War II. Since then, there have been only brief visits by fishers, shipwreck survivors, ham-radio operators, and researchers, except in 2005, when a scientific expedition spent four months on the island investigating the island's biota and physical aspects.

The island is a narrow coral ring approximately 2.4 miles (4 kilometers) long and 1.9 miles (3 kilometers) wide, with a large, completely enclosed central lagoon. The lagoon is mostly brackish and often smells of hydrogen sulphide, although a lens of potable water forms on it after heavy rains. The lagoon has few fish, although it supports dense algal mats at times. Most of the island is low coral rubble with an average elevation of seven feet (two meters), except for Clipperton Rock, a 95-foot (29-meter) volcanic outcrop. The island receives as much as 200 inches (500 centimeters) of rain annually during the rainy season (June to November), and because it lies on a hurricane belt, during heavy storms waves sometimes break over the outer perimeter and into the lagoon.

Biodiversity

A total 115 species of shore fish have been identified at Clipperton, with about equal proportions of eastern and Indo-West Pacific affinities; nine endemic (found exclusively here) species or subspecies have been identified to date. The island itself is largely devoid of vegetation except for a few scattered clumps of introduced coconut palms, interspersed with spiny grass and low thickets of scrub vegetation. A few tiny islets that dot the lagoon are lushly covered with rushes and other weedy plants.

Two species of reptiles inhabit the island: an endemic skink (*Emoia cyanura*) and an introduced gecko (*Gehyra mutilata*). In addition, a large, vivid orange land crab (*Gecarcinus planatus*) is often present in huge swarms. These crabs feed on bird carcasses and excrement as well as on any vegetation that tries to establish on the island; they are the main reason why the island is nearly without plant life.

Visitors to Clipperton have mostly been impressed by the huge numbers of seabirds that blanket the island and fill the air. These species include: white terns (*Gygis alba*), masked booby (*Sula dactylatra*), sooty tern (*Sterna fuscata*), brown booby (*Sula leucogaster*), brown noddy

(*Anous stolidus*), black noddy (*Anous minutus*), and, greater frigate *(Fregata minor*). Ducks are also reportedly common in the lagoon, and migratory landbirds make Clipperton a part-time home during winter trips to the Southern Hemisphere.

Despite the island's being isolated and uninhabited, humans have dramatically altered the ecology of Clipperton Island at times. When the last guano miners departed the island in 1917, they left behind introduced palm trees and pigs. At that time, there were innumerable seabirds nesting on the island, countless numbers of land crabs, and almost no vegetation. However, when a scientific expedition next visited the island in 1958, the effects of feral pigs on the local ecology were clearly obvious. The pigs fed mainly on land crabs, ground-nesting birds, and bird eggs, and because the number of herbivorous land crabs was much reduced, the island was now covered with low, weedy vegetation. Only an estimated 500 brown boobies and 150 masked boobies still survived.

The entire pig population (apparently, 58 individuals) was then killed by expedition members, and when the next expedition visited the island 10 years later (in 1968), no pigs were sighted, an estimated 11.5 million land crabs infested the island, nearly all the vegetation was gone, the masked boobies had increased to more than 4,000, and the brown boobies to an estimated 25,000. By 2003, it was determined that there were more than 110,000 masked boobies at Clipperton, a figure that probably represented more than half the world population for that species.

*The Clipperton angelfish (*Holacanthus limbaughi*) is one of nine species of fish that are endemic to Clipperton Island. The fish are prized for aquariums, selling in Asia for as much as $10,000. (U.S. Fish and Wildlife Service)*

Human Impact

Now there are new threats to the Clipperton avifauna. Black rats were found on the island for the first time in 2000, after two fishing boats wrecked on the beach during a storm. In addition to feeding on smaller birds such as terns, rats preyed on land crabs. By 2005, the crab population had been reduced to the point that vegetation was once again spreading across the island and altering the ecosystem. Also, during the 2005 scientific expedition, almost half of the masked booby chicks on the island perished during an apparent food shortage. This occurred at the same time that a fleet of purse seine vessels were illegally fishing tuna around the island, catching entire schools of yellowfin tuna (*Thunnus albacares*) at one time. This could have important ramifications, because for successful feeding, boobies rely on tuna schools to drive prey to the surface, and without those schools, food shortage is inevitable.

Because Clipperton's isolation has not kept it completely immune from human impact, the island will require monitoring and protection if this important coral reef and bird colony is to persist. The effects of climate change due to global warming are now being studied here. A research team from Mexico in 2012 launched a small expedition; these scientists say Clipperton Island provides an interesting example of how a terrestrial ecosystem survives in the middle of the ocean, and is also a thermometer of sorts for how global warming affects the planet. The director of the team, Johnathan Bonfiglio, said their research aims "to create a new kind of discourse and presentation of climate change, using Clipperton Island as a prism through which this broad theme can be seen."

Robert Pitman

Further Reading

Charpy, Löic, ed. *Clipperton, Environnement et Biodiversité d'un Microcosme Océanique*. Marseille, France: Muséum National d'Histoire Naturelle, 2009.

Pitman, Robert L., Lisa T. Ballance, and Charly Bost. "Clipperton Island: Pig Sty, Rat Hole and Booby Prize." *Marine Ornithology* 33, no. 2 (2005).

Rueda, Manuel. "Mexico: Expedition Sails Towards the Remote and Uninhabited Clipperton Island." March 1, 2012. http://univisionnews.tumblr.com/post/18558688508/clipperton-island-expedition-mexico-france.

Sachet, Marie-H. "Flora and Vegetation of Clipperton Island." *Proceedings of the California Academy of Sciences* 31, no. 10 (1962).

Coiba Island Coral Reefs

Category: Marine and Oceanic Biomes.
Geographic Location: Central America.
Summary: Coiba Island, off the coast of Panama, is a biodiversity hot spot, unique in its tropical location, providing habitat and breeding grounds for thousands of terrestrial and marine species, and hosting as a stopover site on complex migratory routes.

The largest island in Central America was once the site of a notorious prison but is now home to the largest protected marine ecosystem in the tropical eastern Pacific region. Coiba Island (Isla Coiba) is located approximately 20 miles (32 kilometers) off the southern coast of Panama and hosts one of the richest assemblages of terrestrial and marine biodiversity in the entire eastern Pacific Ocean realm. Coiba National Park was created in 1992 and includes Coiba Island itself, 38 other smaller islands, and their respective surrounding coral reefs.

The island's location on the eastern edge of the Gulf of Chiriquí buffers the waters around Coiba from large temperature extremes caused by upwelling events (common elsewhere in the region), and keeps the water warm and clear year-round. This unique set of characteristics makes the coral reefs of Coiba Island an incubator for an abundance of varied marine life.

The climate of the tropical eastern Pacific is characterized as highly variable due to the convergence of warm oceanic currents and cooler coastal currents. The Northern Equatorial Counter-Current supplies warm water from the western Pacific. Closer to the coast, the Panama Current supplies water from the north, while the Columbia Current supplies water from the south. All these currents combine to supply the reefs of Coiba Island the larvae of fish, corals, and marine invertebrates that settle and grow on the reefs. In general, the tropical eastern Pacific is considered to be one of the most isolated marine regions on the planet, with high rates of local endemism (species not found elsewhere), indicating that the region is not well connected to other parts of the Pacific. However, because of the confluence of currents around Coiba Island, its reefs are host to some fish species from far across the Pacific. In fact, it is the only place in the tropical eastern Pacific region where trans-Pacific marine species can be found.

At least 24 species of hard corals have been identified on the reefs of Coiba National Park, two of which are thought to be endemic. Although the region regularly experiences extreme warming that is caused by El Niño that can damage and even kill corals, the reefs surrounding Coiba Island persist. Scientists believe this is due in part to the sheltering effects of the Gulf of Chiriquí. It is even thought that Coiba Island reefs may supply coral larvae to damaged reefs downcurrent. The island is situated at the northern end of a large underwater mountain range that extends southwest from Panama and reaches all the way to the Galapagos Islands more than 620 miles (1,000 kilometers) away. It is possible that these far-off islands are connected via oceanic currents and that reefs off the Galapagos are recharged with larvae from Coiba Island after bad El Niño years.

Because of the island's isolation and the brutal reputation of the prison, the island was sparsely developed, and at least 80 percent of its area remains covered by intact rainforest. The forests of Coiba National Park are host to many endemic plants, but all this richness pales in comparison to the treasures on the coral reefs surrounding Coiba Island. The reef on the eastern side of the island is the second-largest coral reef in the tropical eastern Pacific. Coiba National Park is one of the few places in the world where fish species from the Indo-Pacific and the eastern Pacific are found on the same reef.

The island's rich community of marine fauna includes the vast coral reefs, mollusks, echinoderms, and cetaceans (whales, dolphins, and porpoises). More than 760 species of fish, including 33 species of sharks and rays, have been identified. These include the whale shark, bull shark, white-tip reef, black-tip and scalloped hammerhead shark. Four species of endangered sea turtles nest on the beaches of Coiba Island. The island and its surrounds are home to many threatened and endangered species, including the endemic Coiba Island howler monkey and an endemic rodent, the Coiban agouti. At least 147 species of birds can be found living in the forests of Coiba, including one of the few remaining flocks of scarlet macaws in the world.

Human Impact

There is evidence among the islands and reefs that make up Coiba National Park that Central American natives, the Chibcha, established fishing camps on several of the islands around 4,000 years ago. Remnants of their now ancient fish traps can still be seen on the intertidal zones around the park. In the early 1500s, the first Spaniards visited Coiba, and by 1560, the Indians had been exterminated, or relocated to mine for the Spanish. Coiba Island remained uninhabited until 1919 when a prison was built and it served as a penal colony. At one time it held more than 3,000 prisoners, but was slowly emptied until it closed in 2004.

The national park that was created in 1992 was expanded to include the surrounding reefs, and a Special Zone of Marine Protection was created to protect more than 390,000 acres (160,000 hectares) of marine habitat. The new zone created one of the largest protected marine areas in Central America and placed new rules on fishing within the zone boundaries. These rules ban commercial fishing within the zone and place limits on the types of traditional fishing practices. In 2005, the entire Coiba National Park plus the Special Zone of Marine Protection were designated a United Nations Educational, Scientific, and Cultural Organization (UNESCO) World Heritage Site.

Despite these protections, the islands and reefs of Coiba National Park face threats to the fragile balance of life in this dynamic ecosystem. Illegal logging and poaching in the rainforest threaten the rare tropical hardwoods and endangered birds. Cattle that were brought over to support the prison now threaten areas of virgin rainforest. Illegal fishing goes on inside the park boundaries. In the face of these threats, the coral reefs of Coiba Island continue to teem with life unmatched elsewhere in the eastern Pacific. There are presently plans to build a research station on the island to monitor the impact of human activity, including the effects of climate change.

ROBERT D. ELLIS

Further Reading

Foulkes, Jenifer Austin. "Coiba, Panama: Mission Blue Expedition Launched." March 9, 2012. http://www.huffingtonpost.com/2012/03/19/coiba-panama-mission-blue_n_1354115.html?ref=green.

Lopez, E., P. Cladera, and G. San Martin. "Orbiniidae Polychaetes (Polychaeta: Scolecida) From Coiba Island, Eastern Pacific of Panama, With Description of a New Species." *Revista de Biologia Tropical* 54, no. 4 (2006).

Milton, K. and R. Mittermeier. "A Brief Survey of the Primates of Coiba Island, Panama." *Primates* 18, no. 1 (1977).

Robertson, D. Ross, et al. "Marine Research, Education, and Conservation in Panama." Balboa, Panama: Smithsonian Tropical Research Institute, 2008.

Colorado Plateau Shrublands

Category: Desert Biomes.
Geographic Location: North America.

Summary: This truly spectacular landscape of rugged red-rock sandstone plateaus and canyons was carved by the Colorado River and its tributaries. Although mainly semi-arid, this region supports a wide variety of plant and animal life.

The Colorado Plateau stretches across vast portions of southeastern Utah and western Colorado, and has extensions into northern Arizona and northwest New Mexico. This region is a semi-arid landscape of rocky plateau and canyon with some of the most spectacular scenery in the world. Open woodland of pine and juniper, as well as low shrubland, blanket the landscape. The plateau's long human presence ranges from archaic hunter-gatherers to ancestral Pueblo cliff dwellers, and from the Navajo Nation to Mormon settlers and today's outdoors enthusiasts.

Geography

The sedimentary rock here was carved by the Colorado River and its tributaries to form sheer-walled canyons, buttes, mesas, badlands, and plains. Several large rivers flow through this region, mostly extending from the Rocky Mountains and draining through the Grand Canyon. Isolated mountain ranges are scattered throughout, reaching elevations of 11,500 feet (3,500 meters) where in the past they have supported glaciers. The wide elevation range, from mountaintops to the bottom of the Grand Canyon, creates a broad diversity of biological zones, from alpine meadows too cold to support trees down through montane forests, foothill woodlands, and desert scrub on dry, dusty plains and rocky flats.

Precipitation amounts range from a low of 5 to eight inches (13 to 20 centimeters) per year in the 8 deserts and arid canyonlands, to almost 20 inches (50 centimeters) per year in the higher pinyon-juniper woodlands. The southern part of the ecoregion differs from the north in having a summer monsoonal precipitation pattern.

Vegetation

By virtue of its isolation, complex geology, specialized landforms, and range of elevation, the Colorado Plateau supports numerous species that are endemic, that is, found nowhere else in the world. More than 300 plant species are known to be endemic to this region. Two-needle pinyon pine and Utah juniper dominate among the tree types. These species occur as short trees and shrubs, either in dense clumps or in more open, rocky savannas. Pinyon-juniper woodlands and shrublands typically are found at elevations ranging from 4,000 to 8,000 feet (1,200 to 2,400 meters) above sea level. Severe climatic events, such as frosts and drought, are thought to limit the distribution of pinyon-juniper woodlands and scrub vegetation to relatively narrow elevation belts on mountainsides. Rocky Mountain juniper may codominate or replace Utah juniper at higher elevations.

At lower elevations, often where soils are very thin and drought-prone, dwarfed pine trees and short shrubs are common. Under the tree canopy may be low shrubs, grasses, or bare ground. Other common plants include greenleaf manzanita, big sagebrush, Gambel's oak, bitterbrush, and James's galleta grass.

Additional characteristic types of vegetation are found in the short shrublands, some dominated by blackbrush and/or Mormon tea. This scrub vegetation is abundant throughout the Colorado Plateau on rolling hills and gentle slopes at elevations of 1,800 to 5,000 feet (550 to 1,500 meters). Soils in these areas are shallow, alkaline, and gravelly or sandy, lying over sandstone or limestone bedrock. Other common plants include Torrey's jointfir, spiny hopsage, and (especially where soils are sandy) sand sagebrush. Where the bedrock is mostly limestone, it is common to see black sagebrush or Bigelow sage. Semi-arid grasses such as Indian ricegrass, purple threeawn, blue grama, and muttongrass are common.

In arid environments, water evaporation often leads to salt accumulation on the soil surface, limiting the plant species that can survive there. Throughout the Colorado Plateau, saline basins often include open shrubland composed of one or more saltbush species, such as four-wing saltbush, shadscale, cattle saltbush, or spinescale. Other shrubs may include Wyoming big sagebrush, rubber rabbitbrush, and winterfat.

Wildlife Fauna

Of known animal species that dwell on the Colorado Plateau, there are more than 100 types of mammal, 30 reptile species, 10 amphibians, 40 fish, and more than 250 bird species. A number of once-common native animal species on the Colorado Plateau have become increasingly rare, and some, particularly native freshwater fish, have become extinct.

Due to natural endemism and dynamic land-use history, the Colorado Plateau supports some 27 species listed under the U.S. Endangered Species Act and more than 200 other species considered by various scientific authorities to be vulnerable. Mammals that may be found in various parts of the Plateau are: coyote, gray fox, ringtail, long-tailed weasel, badger, mountain lion, bobcat, mule deer, bighorn sheep, marmot, antelope, desert woodrat, long-tailed vole, porcupine, and jack rabbit. Among the birds here are Mexican spotted owl, turkey vulture, golden eagle, red-tailed hawk, northern harrier, American kestrel, and blue grouse. Collared lizard, long-nosed leopard lizard, short-horned lizard, and sagebrush lizard number among the reptiles, along with tiger salamander, Great Plains toad, and Great Basin spadefoot toad. Representative fish include cutthroat trout, Apache trout, and mottled sculpin.

Human Impact

People have lived throughout the Colorado Plateau for perhaps 15,000 years, with the earliest known inhabitants being hunter-gatherer societies. Some researchers believe that the introduction of ancient human societies and their big-game hunting led to the extinction of various large animals across the region. These prehistoric animals included mammoths, mastodons, camels, protohorses, a stately deer called the stag-moose, as well as giant ground sloths and beavers the size of today's black bears. These large herbivores were prey for the also-large carnivores: saber-toothed cats, savage short-faced bears, cheetahs, maned lions, and dire wolves. It is still debated how these and scores of other large animal species underwent a mass extinction around 13,000 years ago—roughly the same time this region first became populated by humans. Regardless, these first people continued to exist here in small numbers.

Farming as a primary subsistence strategy began in the Colorado Plateau shrublands about 2,000 years ago. While most of the region is marginal for agriculture, indigenous communities continued to use mountains and intermountain valleys for hunting and gathering wild food plants. Among the best-known early societies was that of the ancestral Pueblo, who built cliff dwellings among overhanging sandstone throughout the region's canyons. Mesa Verde National Park protects some excellent examples; these dwellings provided both safety and comfort while maintaining access to nearby water sources.

These cliff dwellings built by the ancient Pueblo can be seen in Mesa Verde National Park in Colorado. Almost half of the Colorado Plateau region is managed by the National Park Service and federal agencies. (Thinkstock)

Only in the past 150 to 200 years has human existence on the plateau impacted the environment in a way that threatens this ecoregion. Colonization and settlement by peoples of European decent began in the 1850s, largely by followers of the Mormon Church who had established a new base at Salt Lake City, Utah. At that time, many indigenous communities, including the Navajo and Hopi nations, were resettled within tribal reservations. This period through the early 20th century saw the rapid increase in pressure on lands and waters throughout the Colorado Plateau.

Grazing by domestic cattle and sheep reached their peak of intensity around the turn of the 20th century, causing severe soil compaction and erosion, as well as trampling damage to vegetation. Timber harvesting and mining also occurred throughout surrounding mountain ranges, leading to larger-scale mining activities, especially for coal and uranium. The water resources of the Colorado River and its tributaries were harnessed through dams (large and small) and water diversions for intensive agriculture. Lake Powell here, named for 19th-century explorer John Wesley Powell, is among the largest of Western reservoirs. It resulted from the construction of the Glen Canyon Dam along the Colorado River at the Utah-Arizona border.

Much of what one sees today throughout the Colorado Plateau is the legacy of the past 150 years of intensified land use. On the whole, this region appears to be largely undeveloped, but past overgrazing, infrastructure development, and other land uses have caused biodiversity losses, decreased soil productivity, and the introduction of invasive plant and animal species. Among the many invasive plant species, annual grasses have become ubiquitous in many semi-desert shrublands. In many cases, the introduction of these fine fuels has enabled wildfires to occur at unprecedented frequencies and intensities.

Along rivers, invasive tree and shrub species have in many cases replaced the native cottonwood-willow vegetation. Overlaying these legacies are the effects of more recent development. In recent decades, an influx of new residents to small towns has put pressure on local infrastructure. This population growth has been accompanied by a surge in recreational tourism, including hikers, bikers, and climbers who find their ways to once-isolated corners of the region.

Today, tribal lands encompass more than one-third of the region. Nearly half of the regional landscape is under public ownership, with the U.S. Bureau of Land Management being the largest federal land manager, along with national forests and parks. Several world-renowned parks characterize the Colorado Plateau, including the Grand Canyon, Arches, Canyonlands, and Mesa Verde National Parks.

Global warming effects on the Colorado Plateau shrublands may be severe. The region generally has suffered drought on a roughly 20-year natural cycle (tied to the Pacific Decadal Oscillation, a sea-surface temperature cycle). Ahead, however, some scientists predict a period of megadrought during the second half of the 21st century. Already there seems to be a trend toward drier winters and wetter summers, which counters many of the built-in connections and structures of this biome. Habitat boundaries will likely shift within the plateau, as they tend to during any multiyear drought, animal migration patterns and timing will change, and biodiversity will be threatened to the degree and speed of the deepening of arid conditions.

Patrick J. Comer

Further Reading

Betancourt, Julio L. "Late Quaternary Biogeography of the Colorado Plateau." In Julio L. Betancourt, Thomas R. Van Devender, and Paul S. Martin, eds., *Packrat Middens: The Last 40,000 Years of Biotic Change.* Tucson: University of Arizona Press, 1990.

Blakey, Ron and Wayne Ranney. *Ancient Landscapes of the Colorado Plateau.* Grand Canyon, AZ: Grand Canyon Association, 2008.

Grahame, John D. and Thomas D. Sisk, eds. "Canyons, Cultures and Environmental Change: An Introduction to the Land-Use History of the Colorado Plateau." 2002. http://www.cpluhna.nau.edu.

Schwinning, Susan, Jayne Belnap, David R. Bowling, and James R. Ehleringer. "Sensitivity of the

Colorado Plateau to Change: Climate, Ecosystems, and Society." *Ecology and Society* 13, no. 2 (2008).
Waring, Gwendolyn. L. *A Natural History of the Intermountain West: Its Ecological and Evolutionary Story.* Salt Lake City: University of Utah Press, 2011.
Wilkinson, Charles F. *Fire on the Plateau: Conflict and Endurance in the American Southwest.* Washington, DC: Island Press, 1999.

Colorado River

Category: Inland Aquatic Biomes.
Geographic Location: North America.
Summary: The Colorado River is the main freshwater artery in the southwestern United States and northwestern Mexico.

Rising in the Rocky Mountains and passing through great expanses of arid land on its way to the Pacific Ocean, the Colorado River is the main freshwater artery in the southwestern United States and northwestern Mexico. It is a critical water resource for the people and wildlife in this part of the world. Most of the water is used for agricultural irrigation, but water diverted from the river is also used for drinking as well as for recreation and industry Reservoirs produced by dams on the river and its tributaries provide long-term water storage for communities and farms along the river. Management of this precious resource must include provisions for wildlife sustained by the river, as well as for the inhabitants of Mexico, who receive only a small portion of all initial runoff.

Geography

The Colorado River is approximately 1,450 miles (2,300 kilometers) in length, flowing mostly west and south, draining a large portion of the arid regions of the western slopes of the Rocky Mountain range. Most of the water that courses through the Colorado River canyons and its tributaries is from snowmelt. The watershed of the Colorado River is extremely large, encompassing approximately 200,000 square miles (50,000 square kilometers). This drainage area includes portions of seven of the United States—Wyoming, Colorado, Utah, New Mexico, Nevada, Arizona, and California—and two states of Mexico: Sonora and Baja California. For 17 miles (27 kilometers), the Colorado River makes up the boundary between the United States (in Arizona) and Mexico.

The Colorado River originates from the Continental Divide at La Poudre Pass in Rocky Mountain National Park, Colorado, approximately 25 miles (40 kilometers) north of Lake Granby. The elevation at this location is 10,000 feet (3,000 meters). As the river leaves Rocky Mountain National Park, it empties first into Shadow Mountain Lake and then into Lake Granby. These lakes are parts of the Colorado-Big Thompson Project, a large water storage and delivery system that diverts water from the Colorado River to provide water for agricultural and municipal uses for the northern Front Range and the plains of Colorado.

The Roaring Fork River empties into the Colorado at the city of Glenwood Springs; then the Colorado is joined by the Gunnison River at Grand Junction. Both the Roaring Fork and Gunnison Rivers are swift-flowing major streams that provide the Colorado massive amounts of water. After this junction, the Colorado flows through Ruby Canyon and crosses into Utah and into Westwater Canyon. The river continues on through Utah and Arizona, through more canyons; is joined by additional rivers, such as the Dolores and Green Rivers; and then flows into Lake Powell, which is formed by the Glen Canyon Dam in Utah.

In Arizona, at the southern end of Marble Canyon, the Little Colorado River empties into the Colorado, and the river enters the Grand Canyon and Grand Canyon National Park. The Grand Canyon is 217 miles (350 kilometers) in length, and the distances between the upper cliffs (South and North Ridges) vary from 4 to 20 miles (6 to 32 kilometers). The walls of the Grand Canyon are 4,000 to 6,000 feet (1,200 to 1,800 meters) high, dropping in successive escarpments of 500 to 1,600 feet (150 to 500 meters). The rocks are striated in columns of striking colors, layered in an abundance of fossils, and create one of the most exposed and explicit geologic formations on the planet.

In Nevada, the Hoover Dam, which was constructed during the Great Depression, forms Lake Mead, which serves as a popular recreation area as well as the major water supply for most of the Las Vegas metropolitan area. From Hoover Dam, the Colorado River continues south. Along the California-Arizona stretch of the river, four additional dams currently function to divert water for agricultural irrigation and municipal water uses, and to form reservoirs for recreational purposes.

Formerly, the river emptied into the Gulf of California between the Baja California peninsula and the mainland of Mexico, but it no longer reaches the Gulf of California on a regular basis. This is due to the large number of diversions of water from the river for agricultural irrigation, use of the river for urban water supplies, and significant evaporation losses from reservoirs produced by damming the river. More than 20 major dam projects have been completed on the Colorado River and its tributaries.

The lower course of the Colorado River forms the border between the Mexican state of Sonora on the mainland and the state of Baja California on the Baja California peninsula. The river at this location is either dry or a small stream for most of the year, due to the numerous diversions upstream, and especially due to the diversion of water for agricultural irrigation in the Imperial Valley of California. The All-American Canal is the major water supply route from the Colorado River to the Imperial Valley; the water is used to irrigate orchards and row crops in the valley. This canal is the largest irrigation canal in the world; it carries a volume of water that ranges from 15,000 to 26,000 cubic feet (425 to 735 cubic meters) per second. Before the middle of the 20th century, the Colorado River delta was a diverse estuarine ecosystem. At present, it is mostly dried up, but the river still continues to be an important ecological estuary.

Flora and Fauna

The Grand Canyon is one of the richer areas for plants and animals along the Colorado River. Above and below the walls of the Grand Canyon, there are 75 species of mammals, 50 species of reptiles and amphibians, 25 species of fish, and over 300 species of birds. On the canyon rims at elevations above 7,000 feet (2,000 meters), ponderosa pine is the dominant tree. Douglas fir, blue spruce, and Gambel oak are other common trees here. Below this elevation, pinyon pine and Utah juniper are the dominant trees. The trees are interspersed with drought-resistant shrubs like cliffrose, fernbush, and serviceberry. Warm, sunny areas along the rim may be home to desert plants such as yucca, sage, and various cacti.

Down in the canyon itself, it is like another world. The temperature within the inner canyon can be as much as 30 degrees F (18 degrees C) higher than temperatures on the rim. Summertime highs along the Colorado River can reach 120 degrees F (49 degrees C). Much of the inner canyon is considered desert, excluding the areas along the river and tributary streams which have rich riparian habitat. Much of the vegetation in the inner canyon is typical of that found in deserts to the south: cacti and drought-resistant shrubs. Riparian plants include thickets of willow and tamarisk.

Prominent mammal species in the Colorado canyons include several types of squirrels, mule deer, elk, and coyote, and bighorn sheep. Around 90 species of birds inhabit this region, including the red-tailed hawk, the golden eagle, the peregrine falcon, and the endangered California condor. The Colorado River Basin harbors 14 native species of fish. Four of these species are endemic (found nowhere else on Earth) and endangered: Colorado pike minnow, razorback sucker, bonytail chub, and humpback chub. The Upper Colorado River Endangered Fish Recovery Program is an effort by the U.S. Fish and Wildlife Service, in cooperation with the Arizona Game and Fish Department, the Colorado Division of Wildlife, and the Utah Department of Wildlife, to rebuild the populations of these endangered fish species.

The Colorado River is a critical water resource for the people who live in the arid southwestern United States and Mexico. The water from this river is essential for irrigation, drinking water, and other uses by people throughout the region.

Unfortunately, Mexico receives little of the enormous quantity of water that surges through parts of the Colorado canyons.

Allocation of the water in the Colorado River is determined by the Colorado River Compact, an agreement signed in 1922 by seven states in the Colorado River basin. Nearly 90 percent of all of the water diverted from the Colorado River is used for irrigation purposes in the United States. Several cities—such as Phoenix and Tucson (the Central Arizona Project), Los Angeles and San Bernardino (Colorado River Aqueduct), Las Vegas, and San Diego—have canals or aqueducts that run from the Colorado River to these major urban centers.

The large number of dams on the Colorado River has had negative environmental effects on the ecology and hydrology of the river. The dams have prevented much of the seasonal flooding normal to the river that would clean the river of debris. The lack of flooding has also caused erosion of sand bars in the Grand Canyon, which are essential for a variety of wildlife in that region of the river. The diversion of water to irrigate farmlands and to supply large cities has taken a huge toll on the quantity of water available to existing habitats. The Colorado River and its wildlife face the challenges of water loss due to overuse, drought, and future climate changes.

DANIEL M. PAVUK

Further Reading

Benke, Arthur C. and Colbert E. Cushing, eds. *Field Guide to Rivers of North America.* Boston: Elsevier/Academic Press, 2010.

Cushing, Colbert E., Kenneth W. Cummins, and G. Wayne Minshall, eds. *River and Stream Ecosystems of the World.* Berkeley: University of California Press, 2006.

Fradkin, Philip L. *A River No More: The Colorado River and the West.* Berkeley: University of California Press, 1996.

Waterman, Jonathan. "Colorado River Basin: Lifeline for an Arid Land." *National Geographic,* March 2010. http://www.natgeomaps.com/assets/downloads/ColoradoRiverBasin-side1-lowrez.pdf.

Colorado Rockies Forests

Category: Forest Biomes.
Geographic Location: North America.
Summary: This large ecosystem in the southern Rocky Mountains is characterized by a diversity of forest types, but is threatened by an unprecedented outbreak of tree-killing insects and human-driven factors.

The Rocky Mountains cover much of the central and southwestern portions of the state of Colorado. Bounded on the east by the high plains and on the west by the Colorado Plateau, the Colorado Rockies are easily distinguished from these relatively low, flat landforms. From either the eastern piedmont or the western plateau country, the Colorado Rockies rise over 8,000 feet (2,400 meters) in elevation, ultimately reaching the state's highest point atop Mount Elbert at 14,440 feet (4,390 meters).

In addition to their steep and often snow-capped mountainous character, the Colorado Rockies are distinguished by the vast forests that cover most of their slopes and valleys. These forests are among the most diverse of all forested landscapes in the entire Rocky Mountain system, and together they form part of a complex ecosystem that provides habitat, resources, and ecological functions for a multitude of other plant species, wildlife, and people. The Colorado Rockies Forests are currently undergoing massive changes due to the largest outbreak of tree-killing insects in Colorado's recorded history, as well as the expansion of the wildland-urban interface, logging, wildfire, and wildfire-suppression policies, and other natural and human-driven processes.

The state of Colorado includes 36,000 square miles (93,000 square kilometers) that are classified by the U.S. Forest Service as forest land. Using their definition, most of the forest land in Colorado is found within the Rocky Mountains, save the small patches found along waterways and in other suitable habitats. Within that expanse of forest land are an estimated 12.7 billion live trees spread across four major forest types that can be broadly characterized by elevation gradients.

Forest Composition and Wildlife

Beginning along the base of the mountains in the foothill zone below 6,000 feet (1,800 meters), the most common forest type is a scattered patchwork of dry-adapted pinyon pine (*Pinus edulis*), Rocky Mountain juniper (*Juniperus scopulorum*), and scrub oak (*Quercus gambelii*). Directly above is the montane zone at 6,000 to 9,300 feet (1,800 to 2,800 meters), where ponderosa pine (*Pinus ponderosa*) and Douglas fir (*Pseudotsuga menziesii*) predominate. Patches of Colorado blue spruce (*Picea pungens*), aspen (*Populus tremuloides*), and lodgepole pine (*Pinus contorta*) occur along this zone's upper limits.

Still higher, in the subalpine zone at 9,300 to 11,400 feet (2,800 to 3,475 meters), the composition of the forest shifts again. Here, dense stands of Engelmann spruce (*Picea engelmannii*) and subalpine fir (*Abies lasiocarpa*) are the most common species, with occasional stands of limber pine (*Pinus flexilis*), aspen, and lodgepole occupying recently disturbed areas. Toward the upper limit of the subalpine zone, the spruce-fir forests become increasingly deformed and stunted, forming low-lying clusters of gnarled trees known as *krummholz* (a German word meaning "twisted wood"). These determined trees gradually give way to the perennial grasses and sedges that characterize the alpine tundra zone above 11,400 feet (3,475 meters), marking the upper end of the mosaic of forest types that characterize the Colorado Rockies Forests.

A lone pine tree in the Colorado Rockies. The forests here, while facing numerous threats, are some of the most diverse in the Rocky Mountain system. (Thinkstock)

These forests contribute to the integrity, functionality, and health of the Colorado Rockies Forests biome in many ways. Habitats are provided for many animal organisms, a few of these being: elk, mule deer, black bear, wolverine, cougar, lynx, American marten, coyote, bighorn sheep, hundreds of bird species, and a variety of native and nonnative fish species. In terms of the region's hydrological cycle, the forests help maintain stream flows by absorbing precipitation and snowmelt and releasing these via groundwater channels at regular rates. This is particularly important in mountainous areas like the Colorado Rockies, where the majority of the region's precipitation occurs in the high country but is used downslope along the base of the mountains.

The forests also help maintain water quality, aquatic habitats, and even dams and reservoirs by controlling soil erosion, siltation processes, and slope stability. In addition to sustaining key hydrological functions like these, the forests cycle nutrients, provide habitat for numerous forms of wildlife, influence weather patterns at local and regional scales, and even affect global climate change through their carbon sequestration functions. To all these ecological services can be added the more strictly anthropocentric ones, which include aesthetic, recreational, and economic roles.

Human Impact

All these ecological functions—and the vast Colorado Rockies Forests ecosystem that they help maintain—face significant short- and long-term threats from both natural and human forces. The most immediately pressing issue is the ongoing infestation of mountain pine beetle (*Dendroctonus ponderosae*), which has reached epidemic proportions since foresters began tracking the outbreak in 1996. In 2010, active infestations were present in 1,370 square miles (3,550 square kilometers) of forest land, with the most severe outbreaks occurring in mature lodgepole and limber-pine forests.

The 2010 figure is actually a decrease from 2009, when 1,630 square miles (4,220 square kilometers) were found to be infested. While seemingly positive, this decrease underscores the ongoing severity of the outbreak, because the decline is due in part to the mortality of most of the susceptible trees in many areas, such as Colorado's western slope. In addition to mountain pine beetles, major outbreaks of spruce beetle (*Dendroctonus rufipennis*), western spruce budworm (*Choristoneura occidentalis*), western tent caterpillar (*Malacosoma californicum*), dwarf mistletoe (*Arceuthobium* var. species), Marssonina blight (*Marssonina populi*), and other insects and diseases currently threaten various forest types in the Colorado Rockies.

Beyond these immediate issues, the Colorado Rockies Forests face long-term challenges. Chief among these is climate change, which many authorities believe has already begun to influence regional weather patterns and alter the forests due to changing temperature and moisture conditions. Another important long-term challenge comes from the growing number of people living in the wildland-urban interface. As larger numbers of people live in close proximity to forest land, the risks of forest fragmentation, habitat degradation, increases in wildfires, and loss of wildlife all rise.

To these challenges can be added the millions of dead trees left behind by the recent insect and disease outbreaks. Because the rate of decay is slow in many areas of the semi-arid Colorado Rockies, these dead trees pose significant long-term risks for homeowners, infrastructure, and watershed health. These and other threats to the Colorado Rockies Forests underscore the need for sound forest management decisions on both the private and public level. Because about three-fourths of all of the Colorado Rockies Forests are managed by public agencies (the U.S. Forest Service alone administers nearly half the total), government leaders will bear a large role in shaping the future of the region's forests and the diverse Colorado Rockies Forests biome that they help maintain.

George Vrtis

Further Reading

Colorado State Forest Service (CSFS). *Continuing Challenges for Colorado's Forests: Recurring and Emerging Threats*. Fort Collins, CO: CSFS, 2011.

Kupfer, John A., George P. Malanson, and Scott B. Franklin. *Identifying the Biodiversity Research Needs Related to Forest Fragmentation*. Missoula, MT: National Commission on Science for Sustainable Forestry, 2004.

Lynch, Dennis L. and Kurt Mackes. *Wood Use in Colorado at the Turn of the Twenty-First Century*. Fort Collins, CO: U.S. Department of Agriculture, Forest Service, Rocky Mountain Research Station, 2001.

Thompson, Michael T., et al. *Colorado's Forest Resources, 2002–2006*. Fort Collins, CO: U.S. Department of Agriculture, Forest Service, Rocky Mountain Research Station, 2010.

Veblen, Thomas T. and Diane C. Lorenz. *The Colorado Front Range: A Century of Ecological Change*. Salt Lake City: University of Utah Press, 1991.

Columbia River

Category: Inland Aquatic Biomes.
Geographic Location: North America.
Summary: The largest river in the Pacific Northwest of North America, the Columbia River is known for historically abundant salmon runs, but is fragmented by hydroelectric dams.

The Columbia River basin encompasses 279,500 square miles (724,000 square kilometers) in Canada and the western United States. The major tributaries are the Snake, Yakima, and Willamette Rivers. The

Columbia River is 1,200 miles (2,000 kilometers) long, and is the fourth-largest river in the United States by volume, with an average discharge at the mouth of 264,860 cubic feet (7,500 cubic meters) per second, more than any other river draining to the Pacific Ocean from North America.

The source of the Columbia River is in the Rocky Mountains of northern Montana. The river proper begins in the Canadian province of British Columbia, flows south through the states of Washington and Oregon, and eventually empties into the Pacific Ocean. The river drains portions of British Columbia, Alberta, Montana, Wyoming, Idaho, Utah, Nevada, Oregon, and Washington, with some of the water originating as snowmelt. The geology of the basin is largely glaciated in the upper reaches, and unglaciated in the lower reaches. The basin includes parts of the uplift Rocky Mountains and the volcanic Cascade Mountains, punctuated by low valleys and plateaus.

The catastrophic Missoula floods, occurring near the end of the last ice age, shaped the geology of the Columbia River basin. Ice-dam failures of glacial Lake Missoula occurred at least 40 times, resulting in floods with discharges 10 times higher than the combined flow of all the rivers in the world. These floods created many prominent landforms of the region, including the scablands of eastern Washington, the Columbia River Gorge, and many smaller tributary canyons. Depositional areas from the floods also created rich sediments, including the Palouse region of eastern Washington and the Willamette River Valley of western Oregon, making them valuable agricultural areas.

Precipitation in the Columbia River basin is strongly seasonal, occurring mainly in the winter months. Due to elevation differences and the fact that much of the basin is quite arid—located in the rain shadow of the Cascade Mountains—precipitation and temperatures have a wide range within the drainage region.

Wildlife

The uplands in the region typically are dominated by coniferous forests, grasslands, and shrub steppe ecosystems. Much of the grasslands area has suffered degradation due to overgrazing and the introduction of varied croplands. The U.S. Geological Survey reports approximately 1.7 million acres (688,000 hectares) of irrigated crops in the greater Columbia Basin. The World Wildlife Fund classifies the basin in three freshwater ecoregions: Columbia Glaciated, Columbia Unglaciated, and Upper Snake.

The Columbia River system was once one of the largest salmon-producing rivers in the world. Salmon were present in the river up to 12,000 years ago; the abundant runs of salmon became the cornerstone of indigenous hunter-gatherer societies of the basin, and the region was an important area for commerce and culture. The availability of marine-derived nutrients were important to the people of the area and to the ecosystem itself, as postspawn carcasses of adult salmon annually reseeded the basin with ocean-sourced nutrients.

The trophic ecology in less affected areas is now typically dominated by a rich and productive array of invertebrates, although impounded, or dammed, areas have been dramatically altered. The fish diversity of the Columbia River is low, while endemism (species found nowhere else on Earth) is high. The sheer number of migratory salmon runs make up for the low diversity, and runs of Chinook salmon, chum salmon, coho salmon, sockeye salmon, and steelhead trout were enormous before 1850.

Large runs of migratory Pacific lamprey, cutthroat trout, white sturgeon, and eulachon also were present once, and their numbers have been dramatically reduced in recent years. It is estimated that recent runs of salmon are only 3 to 20 percent of original numbers. Extensive freshwater mussel beds were once abundant as well, harvested by Native Americans. The impounded reaches of the Columbia River basin have made juvenile and adult salmon and sturgeon migrations more difficult, while improving habitat for nonnative fish like walleye, catfish, and smallmouth bass.

Human Impact

Prehistorically, the Columbia River had salmon runs that made the area an important gathering and cultural center for many indigenous peoples of the region. The Lewis and Clark expedition of the early 1800s charted the area for the incoming American settlers. The basin became an important area for ranching, timber harvesting, and agriculture. The river provided transportation for people and goods, water for irrigation, and (in recent history) power production. The basin's high gradient results in high-velocity flows, which are ideal for producing hydroelectric power. As such, 14 major dams have been constructed on the mainstem Columbia and Snake Rivers, with many small dams built for hydropower and flood control on smaller tributaries.

Fish ladders are present on many of the dams, allowing passage of adult fish, but upstream migration is permanently blocked by the Grand Coulee Dam on the Upper Columbia River and by the Hells Canyon Dam on the Snake River. Free-flowing reaches of the Columbia River mainstem are rare, most notably the Hanford Reach in east-central Washington. Consequently, 13 stocks of salmon species are listed as endangered or threatened under the U.S. Endangered Species Act; other species and stocks are declining, and still others are extinct or extirpated. Quality of water in the Columbia River has also deteriorated over the past century due to agricultural runoff (such as nitrates), increased use of herbicides and pesticides, and logging practices. The alteration of river flows, fragmentation of the river, and other anthropogenic effects (such as hatcheries, habitat loss, and harvest) have directly threatened the historically abundant fish runs.

Fisheries restoration has been largely unsuccessful, although diverse and expensive restoration and mitigation actions are in place, including hatchery production, irrigation screening, and juvenile-dam-bypass systems. Management of the basin to conserve wild salmon is overseen by multiple entities, including federal (such as the Army Corps of Engineers); state; tribal agencies; and numerous power companies, notably the Bonneville Power Administration. These efforts are often confounded by the lack of a consistent management paradigm, further reducing prospects for successful recovery of salmon stocks.

Climate change is becoming an increasingly important component of water demand and supply forecasting in the biome. The potential consequences to water resources in the Columbia River basin associated with warmer temperatures and greater precipitation include a shift from snow to rain precipitation in winter. The resultant reduced snowpacks, higher winter streamflows, increased flood potential, and lower summer flows will exacerbate streambank erosion, nutrient deposition, and habitat disruption in many reaches of the basin.

Jeffrey C. Jolley

Further Reading

Allen, John E., Marjorie Burns, and Sam C. Sargent. *Cataclysms on the Columbia.* Portland, OR: Timber Press, 1986.

Chatters, James C. "A Paleoscience Approach to Estimating the Effects of Climatic Warming on Salmonid Fisheries of the Columbia River Basin." *Canadian Special Publication of Fisheries and Aquatic Sciences* 121, no. 1 (1995).

Meengs, Chad C. and Robert T. Lackey. "Estimating the Size of Historical Oregon Salmon Runs." *Reviews in Fisheries Science* 13, no. 1 (2005).

Stanford, Jack A., et al. "Columbia River Basin." In A. C. Benke and C. C. Cushing, eds., *Rivers of North America.* Burlington, MA: Elsevier Academic Press, 2005.

White, Richard. *The Organic Machine: The Remaking of the Columbia River*. New York: Hill and Wang, 1995.

Williams, Richard, N. *Return to the River: Restoring Salmon to the Columbia River.* San Diego, CA: Elsevier, 2005.

Congaree Swamp

Category: Inland Aquatic Biomes.
Geographic Location: North America.
Summary: The Congaree Swamp, actually known as a floodplain forest, supports some of the last remaining stands and largest examples of old-growth tree species in the United States.

The Congaree Swamp, located in south-central South Carolina, is the largest intact expanse of old-growth bottomland hardwood forest remaining in the southeastern United States. This small, now protected, area covers approximately 42 square miles (107 square kilometers). However, it is not scientifically classified as a swamp because it does not contain standing water throughout most of the year. In general, this floodplain habitat floods about 10 times a year. Waters from the Congaree and Wateree Rivers move through this floodplain, carrying nutrients and sediments that nourish and rejuvenate the ecoregion.

Habitat Biodiversity

The Congaree Swamp biome supports a diversity of wildlife. Preliminary surveys of the park list 30 species of mammals, 170 birds, 32 reptiles, 30 amphibians, and 49 fish species. The nationally threatened bald eagle is occasionally seen flying over the Congaree swamp, but is not known to be nesting there at present. The wetting and drying of the floodplain are beneficial for a number of amphibians and crayfish, which include swamp crayfish and chimney crayfish. Yellow-bellied sapsuckers and many woodpeckers can be found here, as can bobcats, deer, river otters, alligators, and feral pigs.

The value of the Congaree is also in its ability to provide clean water. Floodplains and wetlands store water and, with the assistance of vegetation, filter out pollutants. Thus, these systems are critical for providing clean drinking water, habitat support, and also water for recreational purposes.

Human Impact

The first humans documented as living in what is now known as Congaree National Park were those of the indigenous tribe whose name is that of the park and its river: Congaree.

A small band, it is thought that the Congaree peoples were related to or part of one of the isolated eastern extensions of the Sioux nation of the Great Plains. They also had close connections to the Catawba and Shawnee, although they spoke a different dialect. The Congaree were a relatively small group, numbering 800 at most. They claimed

Redwoods of the East

Congaree National Park is host to some of the largest trees in the United States, many of which were species of interest for the timber industry here in the 1900s. These old-growth specimens survived because the timber company that owned the land was not able to develop an efficient method to harvest them. The Congaree Swamp is thought to have the tallest specimens of sweet gum, cherry bark oak, American elm, swamp chestnut oak, overcup oak, common persimmon, and laurel oak. These particular honors have earned this area the moniker Redwoods of the East.

Two signature trees of the Congaree Swamp, both adapted to continuously flooded environments, are the bald cypress and water tupelo. Because these trees live in swampy soils, they have adopted wide bases that help support them. The bald cypress trees also have modified roots or "knees" that extend above the water like small towers. These knees are thought to help the tree obtain oxygen during floods as well as help maintain structural support.

Bald cypress "knees" in water filled with duckweed. The Congaree Swamp is the largest intact expanse of old-growth bottomland hardwood forest in the southeastern United States. (U.S. Fish and Wildlife Service)

the floodplain land where they had long subsisted on farming, hunting, and fishing. The arrival of the European settlers around 1700 and the introduction of smallpox devastated this population.

Throughout much of the 1700s and 1800s, attempts were made by the new settlers to adapt to the land and make it suitable for crops and livestock. However, since the floodplain was flat and flooded often, these attempts failed. Later, the land attracted various logging operations because of the abundant population of bald cypress. By 1905, the Santee River Lumber Company had purchased most of the land in the Congaree Swamp. The company soon found that logging would not be profitable because of the wet soils, which were difficult to navigate. However, when timber prices spiked in 1969, the lumber company again considered opportunities to harvest the area.

In 1976, to help protect the swamp from logging operations, the U.S. Congress, with pressure from the Sierra Club and local groups, established the Congaree Swamp National Monument. The swamp has since gained international attention. In 1983, it was designated an International Biosphere Reserve and a Globally Important Bird Area. In 2003, it was designated a National Park.

There are new ecological threats to the Congaree Swamp. Climate change is expected to affect temperature and precipitation patterns, accelerate rising sea levels, and increase the frequency and intensity of hurricanes and tropical storms. Although the subsequent impacts on coastal wetlands will vary, rising sea level and even small changes in the quality of tropical storms and hurricanes are expected to have substantial impacts on coastal wetlands. They will likely affect the community structure of ecosystems, exacerbate extinction rates, and undercut biodiversity, in part by disruption of essential processes such as nutrient cycling in the Congaree Swamp.

MARCUS W. GRISWOLD

Further Reading

Jones, R. H. *Location and Ecology of Champion Trees in Congaree Swamp National Monument.* Blacksburg: Virginia Polytechnic Institute and State University, 1996.

National Park Service, U. S. Department of the Interior. "Water Resources Management Plan—Congaree Swamp National Monument, 1996." http://nature.nps.gov/water/planning/wrmp.cfm.

Putz, F. E. and R. R. Sharitz. "Hurricane Damage to Old-Growth Forest in Congaree Swamp National Monument, South Carolina, U.S.A." *Canadian Journal of Forest Research* 21, no. 12 (1991).

Congolian Lowland Forests

Category: Forest Biomes.
Geographic Location: Central Africa.
Summary: About one-third of this vast forest is classified as remote or frontier; it is a stronghold of the western lowland gorilla.

The Congolian lowland forests of central Africa stretch across Cameroon, Gabon, Republic of Congo, northeastern Democratic Republic of Congo, and southern Central African Republic. Forest savanna lies to its north and south, the Atlantic Equatorial Coastal Forest biome lies to its west, and montane forest to its east. Elevations are in general higher than those of the other large global block of tropical forest (in the Amazon), here reaching 980 to 3,280 feet (300 to 1,000 meters).

Rainfall averages 63 to 78 inches (160 to 200 centimeters) annually across the Congolian lowlands. Much of the greater regional precipitation is generated by evaporation from these central rainforests. Geologic elements include the alluvial-derived soils of the Cuvette Centrale basin, within the arc of the Congo River, but also ancient pre-Cambrian bedrock in the northern sections of the biome. Oxisols, a fairly fertile soil type, also exist that are typical of highly leached tropical rainforest.

Plant Types

Of the roughly 10,000 higher plant species found here, researchers estimate that 30 percent are endemic, or evolved for this biome and found nowhere else on Earth. The limbali (*Gilbertiodendron dewevrei*) is a characteristic evergreen tree species here. Much like some of the other flowering hardwoods found in the Congolian lowland forests—such as muhimbi (*Cynometra alexandri*) and mnondo (*Julbernardia serettii*)—the limbali is considered a gregarious grower, often found in dense stands that permit almost no other tree species. These trees, all members of the *Leguminosae* (also called *Fabaceae*) family, are vital to the animals, insects, and humans here for various reasons, not least being their ample production of nutritious pods, seeds, and essential oils.

Characteristic Fauna

The vast expanse of remote forest harbors large mammals that include the iconic western lowland gorilla (*Gorilla gorilla gorilla*) and forest elephant (*Loxodonta africana cyclotis*). The western lowland gorilla is by far the most common gorilla species here, numbering up to 200,000. This primate feeds on as many as 100 different fruit species, and its preferred habitat ranges from sea level to 5,250 feet (1,600 meters) in elevation. Some of this gorilla's forest habitat is so remote that, in 2007, the Wildlife Conservation Society discovered 125,000 lowland gorillas, previously unrecorded, living in a remote northern region of the Republic of Congo.

Some savanna wildlife types have evolved forest species or subspecies in this ecosystem; among them are the forest elephant and Okapi giraffe. The northwestern reaches of the Congolian Lowland Forest biome have the highest mammal species richness, and among the highest rates of endemism, of any forest in Africa, including for primates. The countries of Gabon and Congo have 190 and 198 endemic mammal species, respectively. Bird, reptilian, and amphibian species endemism is also high.

Threats and Conservation

Human populations are relatively low in density here. They include long-standing pygmy tribes such as the Baka, along with Bantu farmers. More than 60 million people live in the wider Congo River basin, however, and their presence is felt. Critical threats to wildlife species such as elephants and bonobo include logging and poaching. The effect of the past 20 years of armed conflict in the eastern Congo has stressed the region's cash-based and subsistence economies, and has made local populations even more immediately dependent on natural resources. Enforcement of poaching laws has proven cumbersome in the large reserves; results tend to be frustratingly uneven.

Major reserves that include Congolian lowland forest in part or whole are Kahuzi-Biega, Okapi, and Salonga National Parks. The latter is also a United Nations Educational, Scientific and Cultural Organization (UNESCO) World Heritage Site, as are the Lope-Okanda Ecosystem and Relict Cultural Landscape in Gabon, as well as the relatively new, 9,650-square-mile (25,000-square-kilometer) Sangha Tri-National Protected Area a vast reserve located within Cameroon, Central African Republic, and the Republic of Congo.

Since Wildlife Conservation Society biologist Michael Fay's highly publicized Megatransect project documented vast expanses of relatively pristine forest in the region, the government of Gabon has been especially supportive of new protected areas. By the late 1990s and early 2000s, Gabon gazetted approximately 10 percent of its land area into nature reserves, including Minkebe Forest in the northeast, rivaling as a conservation leader such renowned green nations as Costa Rica.

As a valuable and powerful carbon sink, the lowlands forests of central Africa deserve protection as a hedge against the runaway greenhouse effect that drives global warming. Deforestation here, by releasing stored carbon-rich gases in the leaves, wood, and soils—and by undercutting these forests' ability to absorb carbon dioxide already in our atmosphere—can literally affect local and regional environments around the world.

William Forbes

Further Reading

Auzel, P. and D. Wilkie. "Wildlife Use in Northern Congo: Hunting in a Commercial Logging

Concession." In J. Robinson and E. Bennett, eds. *Hunting for Sustainability in Tropical Forests.* New York: Columbia University Press, 2000.

Bryant, D., D. Nielsen, and L. Tangley. *The Last Frontier Forests: Ecosystems and Economies on the Edge.* Washington, DC: World Resources Institute, 1997.

Minnemeyer, S. *An Analysis of Access into Central Africa's Rainforests.* Washington, DC: World Resources Institute, 2002.

Weber, W., L. J. T. White, A. Vedder, and L. Naughton, eds. *African Rainforest Ecology and Conservation.* New Haven, CT: Yale University Press, 2001.

Congolian Swamp Forests

Category: Forest Biomes.
Geographic Location: Central Africa.
Summary: A vast, intact swamp forest, one of the largest on the planet, contains many large mammals but is not particularly high in species diversity.

The Congolian Swamp Forests biome of central Africa stretches from the eastern Republic of Congo across the Democratic Republic of Congo to encompass 47,880 square miles (124,000 square kilometers), including the central Congo Basin and some of the Congo River's largest tributaries. The topography is a large alluvial plain called the Cuvette Congolaise, thought by some researchers to have lain under a lake at one time. It is generally agreed the swamp forests retreated to along the river's edge during colder, drier Pleistocene ice ages and expanded during warmer, wetter periods.

Annual precipitation in the present day is approximately 70 inches (180 centimeters). The region has limited seasons due to its equatorial location. Although seasons are not highly pronounced, habitats can be flooded to a depth up to three feet (one meter), yet dry out for short periods annually. Habitats include not only swamp forest but also open water, seasonally flooded forest, dryland forest, and seasonally inundated savannas.

Essential Vegetation

The eastern extent of the swamp forests reaches to the Stanley Falls area near Kisangani. Swamp forest vegetation includes *Guibourtia demeusei, Mitragyna* spp., *Symphonia globulifera, Entandrophragma palustre, Uapaca heudelotii, Sterculia subviolacea, Alstonia congensis*, and species of *Manilkara* and *Garcinia*.

More frequently flooded swamps can harbor large, nearly monoculture expanses of *Raphia* palm. Levee forests often host liana species along with *Gilbertiodendron dewevrei* and *Daniellia pynaertii*. Open areas harbor giant ground orchids (*Eulophia porphyroglossa*), while riverbanks can be lined with arrowroot (*Marantochloa* spp).

Swamp Forest Animals

Large mammals include the western lowland gorilla (*Gorilla gorilla gorilla*), chimpanzee (*Pan troglodytes*), and forest elephant (*Loxodonta africana cyclotis*). Larger numbers of forest buffalo (*Sycerus caffer nanus*) were once present, but have been overhunted. Elephants, gorillas, and the remaining forest buffalo use open grassland areas within the swamps for refuge and feeding.

The Congo River, up to nine miles (15 kilometers) wide in places, acts as a significant barrier to movement of wildlife, especially primates. Occurring only on the left bank, or facing downriver, are communities of the Angolan colobus monkey (*Colobus angolensis*), Wolf's guenon (*Cercopithecus wolfi*), bonobo (*Pan paniscus*), golden-bellied mangabey (*Cercocebus galeritus chrysogaster*), black mangabey (*Lophocebus atterimus aterrimus*), southern talapoin (*Miopithecus talapoin*) and dryad guenon (*Cercopithecus dryas*).

Their counterparts on the right bank, facing upriver, include the crowned guenon (*Cercopithecus pogonias*), chimpanzee (*Pan troglodytes*), agile mangabey (*Cercocebus agilis*), and gray-cheeked mangabey (*Lophocebus albigena*). A minor number of primate species are found on both sides of the Congo River; counted among them are Allen's swamp monkey (*Allenopithecus nigroviridis*).

One of the great and abiding myths of the region is said to be found at Lac Télé-Likouala-aux-Herbes in the Republic of Congo, where the giant, dinosaur-like animal called Mokele Mbembe is reputed to be master of its stomping grounds.

Human Impact and Conservation

Human populations are relatively low in the swampland, and tend to concentrate in villages along river settings. Threats to wildlife species such as elephants and bonobo include logging and poaching, especially closer to the Congo River, due to its ease of accessibility both to wildlife and markets. Salonga National Park in the Democratic Republic of Congo contains the most swamp forest area under official protection, while a large Ramsar Wetland of International Importance site covers 1,695 square miles (4,390 square kilometers) in the Republic of Congo. One leading characteristic of the Congolian swamp forest is its relative inaccessibility—yet new roads are increasing access by both loggers and poachers in the commercial timber and bushmeat trades. Even with increased poaching, fortunately, bonobo numbers are still relatively healthy.

Salonga National Park, the largest protected area of dense humid forest in Africa at 8.2 million acres (3.3 million hectares), is also listed as a United Nations Educational, Scientific and Cultural Organization (UNESCO) World Heritage Site in danger. Two major native groups live within the park: The Kitwalistes currently number between 3,000 and 4,000 in the northeast section; the Iyaelema in the south currently number about 2,340 inhabitants in eight villages.

The effect of the past 20 years of conflict in the eastern Congo region has stressed cash economies and made local populations more dependent on natural resources. Fishing makes up approximately 65 percent of incomes in some communities. Enforcement of poaching laws can be largely futile, as only 200 guards cover the vast Salonga park, for example. Once apprehended, transporting an offender to a nearby tribunal may require 125 miles (200 kilometers) of travel by foot or bicycle. The scale of conservation needed in the Congolian swamp forests requires a "landscape" approach that will incorporate the economy of local inhabitants.

Some of the Congolian swamp forest is protected because it is so difficult to access. This handmade wooden bridge facilitated travel for visitors walking through a swamp in the Okapi Fauna Reserve in the Democratic Republic of Congo in 2007. (USAID)

As much as two-thirds of the Congolian swamp forests are thought to be under threat of deforestation by the year 2040. Such a scenario would create havoc in efforts to slow or reduce global warming, as this biome is a major contributor to natural sequestration of carbon. Like moist forests and bog lands elsewhere, these swamp forests absorb carbon-heavy gas from the atmosphere—an ecological service that grows more precious with each passing hour as concentrations of greenhouse gases continue their implacable buildup.

William Forbes

Further Reading

Blake, S., E. Rogers, J. M. Fay, M. Ngangoué, and G. Ebéké. "Swamp Gorillas in Northern Congo." *African Journal of Ecology* 33, no. 3 (1995).
Mitsch, William J. and James G. Gosselink. *Wetlands.* 4th ed. Hoboken, NJ: Wiley and Sons, 2005.
Nishihara, Tomoaki. "Feeding Ecology of Western Lowland Gorilla—Nouabale-Ndoki National Park, Congo." *Primates* 36, no. 2 (1995).
Van Krunkelsven, E., I. Bila-Isia, and D. Draulans. "A Survey of Bonobos and Other Large Mammals in the Salonga National Park, Democratic Republic of Congo." *Oryx* 34, no. 3 (2000).

Congo River

Category: Inland Aquatic Biomes.
Geographic Location: Africa.
Summary: The most important river in equatorial Africa and the second-largest worldwide in flow volume, the Congo has been a key means of transportation and commerce for hundreds of years, and more recently holds promise as a major source of hydroelectric power.

The Congo River, also known as the Zaire River, is the longest river in the Democratic Republic of the Congo (DRC), with a length of 2,700 miles (4,344 kilometers). It is the second-longest river in Africa (after the Nile), the sixth-longest river in the world, and the second-largest (after the Amazon) river in the world in terms of flow volume; it is also the only major river of the world to cross the equator; the Congo crosses it twice. The Congo and its tributaries include about 9,000 miles (14,500 kilometers) of navigable waterway. The Congo Plume, created where the Congo empties into the Atlantic Ocean, is one of the largest carbon sinks in the world.

Geography and Climate

The Congo Basin is one of the four primary geographic regions of the DRC, covering about one-third of the country's area. In all, the Congo Basin covers more than 782,000 square miles (2 million square kilometers), and extends into six countries: Cameroon, the Central African Republic, the DRC, Equatorial Guinea, Gabon, and the Republic of Congo. The Congo Basin lies at about 2,000 to 2,600 feet (600 to 800 meters) above sea level, while the surrounding land is relatively even and flat and sits at about 1,600 feet (500 meters) above sea level.

The Congo River flows consistently throughout the year; because it is so long, and crosses so much north-south territory, when it is the dry season in one part of the river, it is the rainy season in another. The climate of this ecoregion is tropical and very wet, especially near the equator. Average annual rainfall is approximately 75 inches (190 centimeters) and average annual temperature is 75 degrees F (24 degrees C).

The geography of the Congo River is unusual in that it has waterfalls and rapids near its mouth; from Malebo Pool to the Atlantic, the Congo has 32 cataracts (known collectively as Livingstone Falls), and drops a total of about 980 feet (300 meters). The Congo forms an estuary at Matadi, a port city—and one of the largest harbors in central Africa—about 93 miles (150 kilometers) from the Atlantic coastline, and from this point onward is navigable by ocean-going ships. A canyon on the ocean floor continues the course of the river for about 500 miles (800 kilometers) into the ocean.

Wildlife

The Congo Basin is an extremely diverse ecological region and is the world's second-largest expanse of tropical forest; about one-sixth of the world's total remaining tropical forest is in this region. The basin is estimated to harbor well over 11,000 plant species. Plant types in the biome that are endemic, or found nowhere else on Earth, include many species in of the spiny, perennial herb family *Caesalpiniaceae,* as well as the family of okoumé trees (*Burseraceae* spp.). Meadow habitats feature barnyard grass (*Echinochloa* spp.), as well as papyrus (*Cyperus papyrus*) and other sedge types (*Cyperaceae* spp.).

About 400 mammal species, almost 1,000 bird species, about 80 species of amphibians, and almost 400 reptile species live in the Congo Basin. The region is home to most of Africa's remaining great apes and forest elephants. The critically endangered western lowland gorilla and eastern lowland gorilla, as well as the extinction-threatened Cross River gorilla and mountain gorilla are among the primate species most in focus of conservation efforts here. Among the threats to ward off are poaching for bushmeat, and habitat destruction from human conflict and commercial interests.

Researchers believe that up to 700 species of fish live in this river; as many as 80 percent are endemic. Among the most commonly found are the generally nocturnal, somewhat electrified elephantfish (*Mormyridae* spp.), and the cichlids (*Cichlidae* spp.), remarkable among fishes for their extended care of offspring. Also found in the Congo are African lungfish (*Protopteridae* spp.), which are able to survive dry seasons by burrowing into the mud, covering themselves in mucus and obtaining oxygen by direct breathing from the air.

Human Impact

The Congo has been a major source of transportation and commerce in equatorial Africa for centuries, and remains so today as roads and railroads remain fairly sparse in the region. Notably, European exploration and colonization of the region was initially inhibited by the presence of cataracts near the mouth of the river. Today, railroads have been constructed to bypass the major cataracts (replacing the human porters used in previous centuries), and with its tributaries, the Congo forms a system of thousands of miles of navigable waterways, with goods and people transported by gasoline-powered barges and ships as well as more traditional vessels.

The primary threats to the ecology of the Congo Basin stem from expanding human populations and increased exploitation of the region's substantial natural resources. Much of the region's forests are contracted to logging companies, and the World Wildlife Fund estimates that 70 percent of the forests in the Congo Basin may be gone by 2040 if logging continues at the current pace. Slash and burn agriculture is another source of deforestation, although as of 2012 damage was localized. As the population expands, so will the need for new agricultural lands. Both traditional and commercial fishing practices have rapidly expanded, particularly in the coastal regions.

Other abundant, and often exploited, natural resources in the Congo River Basin include petroleum, diamonds, gold, coltan, and ivory; the latter is driving the rapid depletion of the elephant and hippo populations through poaching. The spread of logging and other extraction industries into previously remote areas brings with them new workers and roads, and is a primary cause of the loss of biodiversity. Presently, the leading cause of wildlife loss here is the commercial bushmeat trade, which is expected to double in the next 25 years. Animals such as monkeys, bonobos, and antelope fall prey to this lucrative business.

During the 1970s and 1980s, the DRC built two hydroelectric dams (Inga I and Inga II) on the Inga Falls on the lower Congo River, about 30 miles (50 kilometers) upstream from the river's mouth. Although the dams are constructed in an excellent location due to the large drop and water flow over the falls, they have been much less successful than expected and currently operate far below their capacity. An internationally financed project is currently underway to rehabilitate these facilities, and to build the Grand Inga Hydroelectric Project.

This new dam would constitute the largest hydroelectric project in the world, with 52 generators potentially producing over twice as much electricity as the Three Gorges Dam project in China. However, some international organizations oppose the new project on the grounds that it will change the river's oceanic plume, require flooding large regions of farmland, and require the destruction of large swaths of the rainforest in order to run power transmission lines. Some organizations also point to ongoing problems with corruption in the DRC, and lack of compensation provided to those displaced by the construction of the first two dams.

Though hydroelectric power may cut down on the use of fossil fuel use, the forest loss and degradation within the basin would likely be responsible for some global climate change, as deforestation causes the release of more carbon to the atmosphere than any other land-use practice on the continent.

SARAH BOSLAUGH

Further Reading

Butcher, Tim. *Blood River: A Journey to Africa's Broken Heart*. London: Chatto & Windus, 2007.

International Rivers. "Grand Inga Hydroelectric Project: An Overview." http://www.internationalrivers.org/resources/grand-inga-hydroelectric-project-an-overview-3356.

Nelleman, C., I. Redmond, and J. Refisch, eds. "The Last Stand of the Gorilla: Environmental Crime and Conflict in the Congo Basin." United Nations Environment Programme, 2010. http://www.unep.org/pdf/GorillaStand_screen.pdf.

Tayler, Jeffrey. *Facing the Congo: A Modern-Day Journey into the Heart of Darkness*. St. Paul, MN: Ruminator Books, 2000.

Tideland near the mouth of the Connecticut River. The light areas are patches of land in which native plant communities are being restored by invasive plant control. (USDA Natural Resources Conservation Service)

Connecticut River

Category: Inland Aquatic Biomes.
Geographic Location: North America.
Summary: The Connecticut River watershed is a vital and diverse ecosystem designated as an American Heritage River; it is undergoing significant rehabilitation efforts.

The Connecticut River is approximately 410 miles (660 kilometers) in length, with a watershed that drains approximately 11,000 square miles (28,490 square kilometers). The Connecticut River is one of 14 American Heritage Rivers, and its watershed is New England's largest river ecosystem. The river is home to a wide variety of flora and fauna, including some that are federally endangered or threatened species. Human habitation began in the Paleolithic Age, with the area's rich farmland, timber, wildlife, and energy resources enabling population growth and development over the centuries. Human use has created numerous environmental problems, including deforestation, riverbank erosion, water pollution, alterations of natural water flow, loss of habitat, and species endangerment.

The headwaters begin at Fourth Connecticut Lake, along the border between Canada and the United States. The Connecticut River then flows through New Hampshire, Vermont, Massachusetts, and Connecticut before emptying into Long Island Sound, a lobe of the North Atlantic Ocean. The river's many tributaries include the Hall Stream and the Ammonoosuc, Sugar, Cold, and Ashuelot Rivers in New Hampshire; the Nulhegan, Passumsic, White, and West Rivers in Vermont; the Millers, Deerfield, Chicopee, and Westfield Rivers in Massachusetts; and the Farmington, Salmon, and Eightmile Rivers in Connecticut. The river is considered to be tidal for as much as 60 miles (100 kilometers) inland from the mouth.

The watershed ecosystem is comprised largely of forest cover. Other components include cropland; pasture land; water and wetlands; shrubs; and urban, rural residential, and industrial areas. The hydrologic system includes surface and ground water, bogs and swamps, and currents and tides with flows that range from gentle, meandering waters to churning falls, cataracts, and rapids.

Ancient and Modern Biota

Archaeologists have found evidence of the presence of dinosaurs and very ancient fish in watershed locations such as Rocky Hill, Connecticut, and Barton Cove in Gill, Massachusetts. The ecosystem currently supports a wide variety of animals, including more than 50 species of mammals, approximately 250 species of birds, more than 20 species of reptiles, more than 20 species of amphibians, more than 140 species of fish, and approximately 1,500 species of invertebrates. Shore and waterbirds live, rest, and feed along the watershed; migratory fish such as American shad, salmon, and blueback herring travel its lengths to live and spawn.

The Connecticut River biome also supports thousands of different types of plants. The presence of invasive species, such as the water chestnut, has disrupted the area's natural balance, leading to organized eradication efforts. Federal-level endangered or threatened animal species residing here include the American bald eagle, peregrine falcon, piping plover, shortnose sturgeon, dwarf wedge mussel, yellow lamp mussel, and puritan tiger beetle. Endangered or threatened plants include the small whorled pogonia orchid, Jesup's milk-vetch, and northeastern bulrush.

Human Settlement

Humans have inhabited the Connecticut River ecosystem since the first Paleo-Indians of approximately 12,000 years ago. The area's rich soils fostered agricultural settlements beginning with Algonquian corn cultivators. European immigrants arrived in the early 17th century, later followed by peoples from around the world. Early settlers cleared swaths of forest cover for timber, rural settlement, and crop and pasture lands. Connecticut River Valley farms soon emerged as important sources of food for a growing regional population. The area's oldest cities are Hartford, Connecticut, founded in 1635, and Springfield, Massachusetts, founded in 1636.

The river is navigable up to Hartford and is used for various shipping functions. The first dam was built along the Connecticut River in 1798. Beginning with the Industrial Revolution of the 19th century, forests were cleared for development, while the river and its tributaries were dammed for flood control and energy generation. The area also has a long history of outdoor and recreational uses, including walking and biking trails, fishing, boating, and rowing. By the 21st century, the ecosystem had a human population of more than 2.3 million.

Human habitation in the region has vastly altered the ecosystem. Historical settlements polluted the ecosystem's waters with raw sewage, while agricultural and industrial chemical pollutants such as DDT, insecticides, and herbicides began to damage the ecosystem in the mid-20th century. Other major regional water pollutants include sewage discharge and overflow; nuclear-plant and stormwater runoff; and toxic spills from nuclear plants, factories, ships, and barges.

Dams along the river and its tributaries have disrupted the natural seasonal water flow, blocking the passage of migratory aquatic species, notably fish heading to spawning grounds. The Connecticut River features more than a dozen dams, and its tributaries host more than 1,000, many of which are no longer operational, but which still present obstacles to local habitat and migratory species. Water-flow changes from such constructions render useless the natural instincts that tell fish when to spawn, and trigger plants to disperse their seeds.

Threats and Conservation

The resulting environmental problems include disrupted fish runs and lowered numbers of migratory fish traveling to spawning grounds;

soil erosion along riverbanks that results in sediment filtration in the water, lessened travel corridors, reduced food sources, and loss of forest cover for area wildlife; and change and loss of habitat and the endangerment of flora and fauna. Water pollution can affect the vitality of contaminated plants and animals, and can concentrate toxins as it spreads through the food chain as one species feeds on another; this is known as bioconcentration of toxins. Continued population growth and ongoing development pose deepening threats. The effects of global warming, which are projected to reduce snowpack, alter upstream snowmelt flow, disrupt growing seasons, and add to habitat fragmentation, will tend to keep this ecosystem off balance.

Threats to the Connecticut River biome began to attract widespread attention in the mid-20th century, with the river commonly described as a well-landscaped sewer. The environmental movement and subsequent federal legislation such as the Clean Water and Endangered Species Acts have heightened efforts to preserve and restore the Connecticut River ecosystem. It is today home to 10 federally protected endangered or threatened species. The watershed was further protected by its 1991 congressional designation as the Silvio O. Conte National Fish and Wildlife Refuge and its 1998 designation as a American Heritage River by President Bill Clinton, becoming the first complete watershed to receive such designation. Both the Nature Conservancy (1993) and the international Ramsar Convention (1971) have recognized the watershed's tidal wetlands as areas of international significance.

Organizations and programs dedicated to the ecosystem's rehabilitation include the Connecticut River Conservation District Coalition (CRCDC), the federal Wildlife Habitat Incentives Program (WHIP), the U.S. Army Corps of Engineers, the Nature Conservancy, the U.S. Fish and Wildlife Service, and the Connecticut River Watershed Council. Rehabilitation efforts have included riverbank-erosion control, planting or protection of riparian buffer zones, improvement of water quality through wastewater treatment and other methods, and dam removal and the return of more natural water flow. Successes have included federal designations that the Connecticut River is safe for both swimming and fishing, and that portions of the river are potable. Redevelopment efforts have also sought to rehabilitate waterfront communities such as Hartford as well as recreation areas, seeking a better balance between human use and the natural ecosystem.

MARCELLA BUSH TREVINO

Further Reading

Commonwealth of Massachusetts, Executive Office of Energy and Environmental Affairs. "Connecticut River Watershed." 2012. http://www.mass.gov/eea/air-water-climate-change/preserving-water-resources/mass-watersheds/connecticut-river-watershed.html.

Hammerson, Geoffrey A. *Connecticut Wildlife: Biodiversity, Natural History, and Conservation.* Lebanon, NH: University Press of New England, 2004.

Nedeau, Ethan Jay. *Freshwater Mussels and the Connecticut River Watershed.* Greenfield, MA: Connecticut River Watershed Council, 2008.

Tripp, Nathaniel. *Confluence: A River, the Environment, Politics, and the Fate of All Humanity.* Hanover, NH: Steerforth Press, 2005.

Coongie Lakes

Category: Inland Aquatic Biomes.
Geographic Location: South Australia.
Summary: This freshwater wetland remains unpolluted, providing an unaltered environment for the plant and animal life of Australia in the heart of its salty desert region.

Located in the Cooper Creek floodplain in the eastern region of Lake Eyre Basin, Coongie Lakes are a complex freshwater system in the heart of Australia's salty desert. Cooper Creek and the Coongie Lakes are part of one of the world's largest endorheic, or internal, drainage basins. Comprised of various forms such as swamps, lakes and channels, the ever-changing composition this region is dependent upon the amount of rainfall received each year—which changes the environment accordingly.

The area includes more than 100 lakes that vary from near-permanent to rarely flooded lands. Cooper Creek is the primary source of freshwater supply for Coongie Lakes. This aquatic habitat is dependent upon the rainfall to sustain its variety of plant and animal life; floodwaters promote the growth of supported plant life, while the dry periods temporarily suspend this growth, resulting in an ever-changing environment.

In the Lake Eyre Basin, the average annual temperature varies from 70 degrees F (21 degrees C) in the south, to 75 degrees F (24 degrees C) in the north. The average maximum temperatures are 64 degrees F (18 degrees C) and 75 degrees F (24 degrees C), respectively, in July. These average maximums climb to 97 degrees F (36 degrees C) and 102 degrees F (39 degrees C) in January, summer in the Southern Hemisphere. It is estimated that there is less than 6 inches (15 centimeters) of rainfall per year on average in the Lake Eyre/Coongie Lakes Basin. The highest rainfalls, with annual averages of about 16 inches (40 centimeters), occur along the northern and eastern margins.

Wildlife

There are about 205 bird species, both aquatic and terrestrial, found around the Coongie Lakes. In addition, this wetland is a significant breeding ground for an enormous number of birds that migrate through the area. There are many teal and pink-eared ducks, along with great numbers of maned-duck, Eurasian coot, red-necked avocet, and pelican. One can also find the black-winged stilt, hoary-headed grebe, pied cormorant, tern, silver gull, heron, ibis, and various spoonbills.

The biome is home to at least 17 raptor species. In the region of the lakes, there are at least 100 species of terrestrial birds, dependent on the freshwater supply and the riparian woodland of river red gum (*Eucalyptus camaldulensis*) and coolibah (*E. microtheca*). More than 470 plant species, 52 of which are considered very rare, are sustained in this ecosystem.

The fish population must survive many changing variables to survive here as well. Changes in salinity, temperature, oxygen and water levels all play a significant part in their daily life. Coongie Lakes are only about 6 feet (2 meters) deep, and may dry up if they receive little or no water from Cooper Creek for months at a time. This is believed to be the reason for a general lack of diversity among the fish in the Coongies. Of the approximately 26 fish species found in the Lake Eyre Basin, only 13 are found in Cooper Creek. Native animals that are present in the areas surrounding the lakes include the dingo and red kangaroo, as well as various turtles, frogs, and water-rats. The inland taipan, among the world's most venomous snakes, also makes its home here.

Human Impact

Coongie Lakes was once home to the Aboriginal people, and remains a spiritual site for them. Coongie Lakes was an important habitat for this population because of the richness of its resources. The Yandrumandha, Yauraworka, Dieri, Wangkangurru, and Wangkumara Aboriginal groups have been associated with the wetlands here. The area was once heavily populated as these peoples traveled up and down the waters using the many resources. Food in the form of fish, shellfish, birds, reptiles, and mammals was plentiful. In 1988, the Aboriginal Heritage Act was established to protect objects, sites, and remains according to traditional practices. The registry for Coongie Lakes lists 127 sites recorded during 1982–86, including archeological sites, burial sites, art sites, ritually significant sites, tool manufacturing areas, grindstone quarries, *wiltjas* (temporary shelters), early historic campsites, and stone arrangements. Several hundred more sites have yet to be added to the registry.

Cattle farming is of prime economic importance in Australia, with the first cattle farm being established in the Coongie Lakes area in 1873. The development of premium chemical-free beef is supported by the naturally irrigated pastures. However, grazing cattle also cause land degradation, destruction of habitat, changes in vegetation patterns, weed and pest infestation and pollution. The Wilderness Society is currently campaigning for the removal of cattle grazing in Coongie Lakes, with the preservation of these wetlands considered of prime importance.

Besides cattle farming, continuous threats to the Coongie Lakes today include exploration of gas and petroleum reserves, pollution, overfishing, and land development. Large reservoirs of oil and gas are thought to exist under Coongie Lakes. The value of these resources has not gone unnoticed for their potential provision of energy and economic resources to Australia. Mining these fields could drastically alter this ecoregion by degrading the environment and destroying or disrupting habitat cohesion.

In 2003, the state government announced the protection of 108 square miles (280 square kilometers) of the Coongie wetlands within National Park status, to thus be protected from petroleum exploration and production. In June 2005, the state government designated nearly all of the core of the wetlands as the Coongie Lakes National Park. Assuming day-to-day operations are properly regulated, the new park and petroleum exclusion zone effectively ends two decades of controversy over petroleum exploration in the area.

Current threats to this arid wetland include the potential consequences of climate change. Although the floodplains of the Eyre Basin have changed little since the first explorers, the public focus has been on preserving the extensive natural wetlands and the native flora and fauna that inhabit these areas. A possible threat to maintaining this biodiversity would be a reduction in the volume and frequency of beneficial floodwaters due to climate change.

Sandy Costanza

Further Reading

Department of Environment, Water and Natural Resources. "Coongie Lakes National Park." 2012. http://www.environment.sa.gov.au/parks/Find_a_Park/Browse_by_region/Flinders_Ranges_and_Outback/Coongie_Lakes_National_Park.

Hill, Robert S., ed. *History of the Australian Vegetation.* Cambridge, UK: Cambridge University Press, 1994.

Monroe, M. H. "Coongie Lakes Complex." 1993. http://austhrutime.com/coongie_lakes_complex.htm.

Cooper Creek

Category: Inland Aquatic Biomes.
Geographic Location: Australia.
Summary: This iconic watercourse through arid Australia fills with biodiversity when it floods, and provides a reminder of the devastating difficulties that Australia's early explorers faced.

The Cooper Creek is one of Australia's most iconic rivers, flowing from the western slopes of the Great Dividing Range around Longreach/Charters Towers as the Thomson and Wilson Rivers, and through the Queensland channel country from Barcaldine, as the Barcoo. Then it continues on a slow meander westward along the flat lands past Innamincka, and ultimately into Lake Eyre. It is perhaps the only waterway in the world with three rivers flowing into it to form a creek. But this flow rate is deceptive; in most years, the Cooper is an ephemeral channel that flows briefly after rainfall—with a mean precipitation of five inches (125 millimeters) per year—that evaporates long before it flows into Lake Eyre. When it does flood, however, the Cooper can be 25 miles (40 kilometers) wide at Windorah—and a cataract equal to three times the amount of water in Sydney Harbor flows over the Cooper Creek causeway at Innamincka daily in such conditions.

The Cooper, 882 miles (1,420 kilometers) long, is the second-longest inland river system in Australia (after the Murray-Darling). Cooper Creek drains the eastern side of the catchment of the Lake Eyre Basin; at 451,740 square miles (1,170,000 square kilometers), the catchment is the world's third-largest terminal lake system and is thought to be the world's largest salt pan.

The flooding rains of the summer of 2010–11 brought the Cooper into very heavy flood. The flat topography of the region meant that the water took six months to flow the 250 miles (400 kilometers) from Innamincka to Lake Eyre.

An Australian pelican with a family of black swans, which are endemic to the area. Cooper Creek draws large numbers of waterbirds like these during times of flooding, when the river can expand as wide as 25 miles. (Thinkstock)

Vegetation

During the Tertiary, the Cooper flowed through an environment of forest, but as Australia dried during the continent's slow drift northward, this environment turned into a narrow band of coolabah (*Eucalyptus coolabah*) and lignum (*Muelhenbeckia* spp.) snaking between the sand dunes and gibber plains of the Strzelecki, Sturt's Stony, and Tarari Deserts. Further from the coolabah riparian zone and lignum thickets, the vegetation becomes dominated by chenopod shrublands and hummock grasslands, with Mitchell grass on the gibber plains and *Dodonea* spp., sandhill wattles, and sandhill canegrass on the dunes.

Wildlife

During floods, the Cooper becomes filled with waterbirds that fly in from great distances away to plunder the creek's bounty. When the creek is full of water, invertebrates like brine shrimp quickly breed to supply food for Australia's only inland fishery, an industry that targets golden perch, bony bream, catfish, grunters, and yabbies (a freshwater crawfish). Bird species such as Australian pelicans, cormorants, darters, egrets, ibises, grebes, stilts, and the endemic Australian black swans move in to access these riches. Countless duck species join them; these include pink-eared, shelducks, hard-heads, and grey teals.

Following flooding times, the surrounding areas become resource-rich environments, with increases in rodent numbers and huge flocks of black kites (*Milvus migrans*) soaring overhead, looking for a meal of mouse or fish. Many of the farms in the region target dingoes (*Canis lupus dingo*) for fear of their effects on livestock. Dingoes, however, probably reduce the effect of foxes and cats on native fauna, so their retention in the environment can certainly be beneficial to wildlife.

Human Activity and Conservation

The Cooper was named after the chief justice of South Australia, Charles Cooper, by explorer harles Sturt during his travels through inland Australia in 1845. The Burke and Wills expedition to the Gulf of Carpentaria essentially ended on the Cooper, when Robert O'Hara Burke and William John Wills failed to follow the preparation methods of the local Yauraworka and Yantruwantra Aboriginal people in making nardoo sporocarp seedcakes; they died alongside a waterhole of beriberi, likely of thiamine deficiency. John King, the sole survivor of the expedition, accepted the Aboriginal people's offers of assistance and survived to tell the tale.

Despite the region's aridity, isolated permanent water sources allowed large populations of Aboriginal people to exist along these inland waterways. Within a decade of European exploration, pastoral stations were established, and by 1900, the Aboriginal population had been reduced by 90 percent, to 30 survivors.

Nowadays, the majority of the Cooper is flanked by pastoral stations until it flows into Lake Eyre. The Cooper affords little to the pastoralists unless it is flooding; then it yields a boom time for cattle, which normally receive their water from bores dug into the Great Artesian Basin. The Birdsville Track crosses the Cooper on Mulka Station, and during flood times, a two-vehicle punt is needed to cross the creek. Helicopter tours and river cruises are available from the campsite at the punt, attesting to the amount of tourism the region experiences during times of flood.

The Cooper has been proposed for listing as a Wild River under Queensland government protection. This listing would improve the conservation status of the iconic river, which currently is poorly conserved and heavily affected by pastoralism. Innamincka Regional Reserve, at 3.2 million acres (1.3 million hectares), is the primary conservation area on the Cooper, but farming and mining are both allowed in the reserve, reducing its value for conservation.

Coongie Lakes National Park is within the regional reserve; it is a wetland of international importance under the Ramsar Convention, and was identified as an Important Birding Area because it supports more than 1 percent of the global populations of 12 species of waterbirds and waders, as well as significant numbers of Bourke's parrots, Eyrean grasswrens, gibberbirds, banded whitefaces, chirruping wedgebills, and cinnamon quail-thrushes.

Climate change in coming decades is expected to tax the Cooper Creek hydrological system to the tune of about 10 percent of its water volume. Lower rainfall, higher average temperatures, and greater evaporation rates may combine to reduce the life-supporting capacity of a number of its stretches and waterholes; various grass species and many animal types will be under duress as this scenario progresses. The floodplain itself may contract, as flooding events may not reach the extremes that have been recorded in the past on a regular basis—this trend, however, could readily be disrupted by short-term extreme weather events much like the summer flood of 2010–11.

Matt W. Hayward

Further Reading

Badman, F. J. *The Birds of Middle and Lower Cooper Creek in South Australia.* Adelaide: Nature Conservation Society of South Australia, 1989.

Cobon, D. H. and N. R. Toombs. *Climate Change Impacts on the Water Resources of the Cooper Creek Catchment.* Brisbane, Australia: Department of Natural Resources and Water, Climate Change Centre of Excellence, 2006.

Ellis, R. *Boats in the Desert.* Rockhampton, Australia: Queensland University Press, 2006.

Weidenback, K. *Mailman of the Birdsville Track: The Story of Tom Kruse.* Adelaide, Australia: Hodder, 2006.

Copper Plateau Taiga

Category: Forest Biomes.
Geographic Location: North America.
Summary: The Alaskan Copper Plateau Taiga is a subarctic extensive, low-elevation forest characterized by areas of discontinuous permafrost. This biome is largely intact, but is threatened by development and logging.

The Copper Plateau taiga is located in southeastern Alaska within the United States. Taiga is the predominate biome within the Alaskan interior, and it is part of the subarctic ecoregion The taiga is also known as a boreal forest. The low-elevation terrain consists of a large plateau featuring a mixture of flatland and gently rolling plains marked by discontinuous permafrost. Coniferous forest stands predominate.

The Copper Plateau Taiga biome is approximately 6,600 square miles (17,000 square kilometers) in size, and lies at an elevation between

1,600 to 3,000 feet (400 to 900 meters) above sea level. It is part of the larger taiga biome stretching from Alaska and Canada into the northeastern and midwestern United States. The taiga was once the site of a large lake and massive glaciers during the Great Ice Age of the ancient Pleistocene Era. High mountainous regions surround the area and it borders on the tundra ecosystem to its north.

Geography and Climate

The land of the taiga is shallow and low-lying, with poor drainage. The soils are generally acidic, resulting in a dearth of organic matter or nutrients in the topsoil. The taiga ecosystem includes the Copper and Susitna Rivers, as well as central regional lakes such as Lake Louise, Tyone Lake, Paxson Lake, and Susitna Lake. There are also vast numbers of other small lakes, pools, wetlands, bogs, and marshes.

Long, dark, cold winters and short, warm summers are characteristic of this region. Sunlight hours are greatly reduced in winter while summer sun exposure can last up to 20 hours per day. Annual temperatures average a daily minimum of minus 17 degrees F (minus 27 degrees C) in the winter and a daily maximum temperature of 70 degrees F (21 degrees C) in the summer. In the region's western reaches, average temperatures are often below freezing for more than half the year. Annual precipitation consists of summer rains and winter snows, ranging from 10 to 18 inches (25 to 46 centimeters). The northern taiga receives the most precipitation; the summer is the wettest season.

A black-spruce forest rising from the mist in Alaska. Closely spaced conifers, especially black spruce like these, are characteristic of forests found within the Copper Plateau Taiga biome. (U.S. Fish and Wildlife Service)

Vegetation

The area's plant life is well adapted to survival in the cool, poorly drained soils characteristic of the taiga environment. Vegetation varies according to factors such as sun exposure, drainage, and the absence or presence of permafrost. Permafrost is not present throughout the taiga, allowing for habitat diversity. Some trees, such as aspen, cottonwood, and white spruce, only grow in the areas of permafrost absence. Hills with southern exposure tend to be permafrost-free, while those with northern exposure contain permafrost. The predominant conifer and evergreen trees generally do not lose their leaves or needles in the fall season, hold water, and have shapes that prevent large accumulations of snow within their branches.

The Copper Plateau Taiga biome is dominated by stretches of subarctic forest featuring trees of varying species and short to moderate heights. Coniferous evergreen forests and woodlands compose the majority of the treescape. Dense forest canopies and closely spaced trees are common within the conifer forests. Black spruce (*Picea mariana*) is the dominant tree within these areas. Other characteristic tree species include white spruce (*Picea glauca*), dwarf birch (*Betula glandulosa* and *Betula nana*), black cottonwood (*Populus trichocarpa*), tamarack, and quaking aspen (*Populus tremuloides*).

Below the taiga forest canopy is a ground cover composed of shrubs, mosses, lichens, and ferns.

The forest ground cover contains accumulated plant materials and acts as a storehouse for carbon. The many wetland areas support low scrub bog communities dominated by birch and ericaceous (heath) shrubs. The taiga also supports a variety of sedges, herbs, and berries. Common cottongrass (*Eriophorum angustifolium*), bogbean (*Menyanthes trifoliata*), Arctic sweet coltsfoot (*Petasites frigidus*), and marsh cinquefoil (*Potenilla palustris*) are among the ecosystem's native plant species.

The northern edges of the Copper Plateau Taiga feature another distinct habitat. The North American taiga biome lies just south of the tundra biome. The two biomes meet within a transitional zone featuring a blend of taiga and tundra habitat characteristics and species, known as a tree-line ecotone. Here, the tree canopy thins as it transitions from the taiga to its neighboring tree-less tundra.

Animal Types

Animals have adapted various survival mechanisms for life in the taiga, such as seasonal migration, seasonal cycles of fur growth and shedding, camouflage, hibernation, and burrowing under the snow. Brown bears, badgers, beavers, reindeer, mice, moles, foxes, wolverines, squirrels, moose, and caribou are all native to the taiga region. The area's large lakes and rivers support fish habitats and nesting sites. Salmon live within the Copper River; notable species are the chinook, king, and sockeye.

The area is home to both permanent and migratory bird populations. Birds nest along the taiga's wetland areas such as Lake Louise and Paxson Lake. Birds include the trumpeter swan, among the largest birds found within the North American continent. Migratory birds visit the area during the spring season, nesting among the numerous lakes and rivers. The warmer temperatures signal the hatching of insects, which provide a readily available, well stocked food source.

Invasive animal species introduced into the Copper Plateau Taiga ecosystem include the mallard), Canada goose, great horned owl, red fox, wolf , least weasel, American beaver, ermine, and snowshoe hare. The taiga does not house any significant numbers of endangered species, although the state of Alaska maintains the Nelchina Caribou Special Management Area. Wildfires are not a significant threat, as their rates of occurrence are low and those that do occur tend to be small in size. The heat generated by wildfires, however, can change soil characteristics in the affected area.

Human Impact

The Copper Plateau Taiga biome and its environs are home to several ecologically protected areas, such as the Nelchina and the Wrangell-St. Elias National Park. Conservation organizations working within the area include the Copper Country Alliance, the Copper River Watershed Forum, and the National Parks and Conservation Association. While little of native habitat here has been lost, development and the timber industry are significant potential ecological threats. Scientists estimate that total habitat loss is at approximately 10 percent.

Development and road building is most extensive in the region's Glenallen area. The northwestern portion is accessible by the Denali Highway. The timber industry operates within the Copper River Valley and on Alaskan Native Regional Corporation lands within the vicinity of Chitina. Natural areas are popular sites for tourism and recreation; they also support both sport and subsistence hunting and fishing.

Global climate change has already been affecting this ecoregion; the boreal forests have been experiencing some of the greatest temperature increases on the planet. Warmer summers may account for stunted tree growth, and summer warming trends may cause some plant habitats to metamorphose into others: for instance, a forest land may over time become more suitable as a grassland. Tree-damaging insects may also benefit from fewer days per year of colder temperatures that would normally keep their populations down. Future concerns for the Copper Plateau Taiga include climate change, likely increases in timber harvesting, and potential use of unsustainable industry practices.

Marcella Bush Trevino

Further Reading

Beresford-Kroeger, Diana and Christian H. Kroeger. *Arboretum Borealis: A Lifeline of the Planet.* Ann Arbor: University of Michigan Press, 2010.
Chapin, F. Stuart. *Alaska's Changing Boreal Forest.* New York: Oxford University Press, 2006.
Kasischke, Eric S. and Brian J. Stocks. *Fire, Climate Change, and Carbon Cycling in the Boreal Forest.* New York: Springer, 2000.
Van Cleve, K. *Forest Ecosystems in the Alaskan Taiga: A Synthesis of Structure and Function.* New York: Springer-Verlag, 1986.

Coral Triangle

Category: Marine and Oceanic Biomes.
Geographic Location: Asia.
Summary: The biologically rich and expansive Coral Triangle is threatened by climate change and human activity. Conservation programs are in place to study the reefs and educate the local population on ways to preserve and safeguard the area's natural resources.

The Coral Triangle is a large marine habitat area in southeast Asia. The roughly triangular area of 2.3 million square miles (6 million square kilometers) covered by the Coral Triangle encompasses the national waters of six countries: Philippines, Indonesia, Malaysia, Papua New Guinea, Solomon Islands, and Timor Leste.

The Coral Triangle is a true nursery of the seas, containing more than 600 coral species, or about 75 percent of all coral species in the world; six of the seven sea-turtle species found in the ocean; and more than 2,200 species of fish, or about one-third of the world's known coral-reef fish species. At least 15 of the corals in this biome have been identified as endemic, meaning they exist nowhere else on Earth. The epicenter of coral diversity here, with 553 species, is the Raja Ampat region near Bird's Head Peninsula in the West Papua province of Indonesia.

The reefs of the Coral Triangle sustain more than 120 million people, who depend directly on the reef's resources for food and income. More than 100 million people beyond the Coral Triangle also benefit.

Environmental Threats

The local threats are mainly pollution from coastal development and poor sewage treatment infrastructure; overfishing, especially of species that play critical roles in the ecosystem; destructive fishing methods such as cyanide poisoning and dynamite blast fishing, which kill corals and destroy habitat, lowering the productivity of the reef and its resources over time; and bycatch, which is the unintended catching of sea turtles, dolphins, sharks, seabirds, and the like in the context of long-line fishing for tuna. Many of the local threats persist due to high poverty levels, poor resource-management practices, and lack of political will, combined with high market demand outside the region for many of the area's reef and open-ocean fish.

The growing live-fish trade for restaurants in Asia and aquariums around the world is driving a dramatic increase in fishing efforts in the Coral Triangle, which has led to declines in many ecologically important fish species. These are the same species of fish that the local subsistence, artisanal fishers also consume for food. In addition, the Coral Triangle has immense areas for spawning and juvenile tuna, so the region harbors what is perhaps the largest tuna fishery in the world—one of the single most significant economic drivers in the area. Heavily fished tuna species in the region include southern bluefin, big-eye, albacore, yellowfin, and skipjack. While a tremendous economy has been fueled by commercial fishing and tourism, the Coral Triangle's fish stocks are declining dangerously and are in need of effective management.

Population growth and continued reef resource exploitation can significantly weaken not only the health of the marine ecosystem, but also the food security in the region. Well over 100 million people depend on the reef for food. The resources need to be cared for to preserve them for future generations in order that a food crisis may be averted.

Fish swarming around coral in the Raja Ampat Islands in the Coral Triangle. Southeast Asia's Coral Triangle contains 75 percent of all coral species in the world, and 15 of these are seen nowhere else. (Thinkstock)

Effects of Climate Change

The reefs of the Coral Triangle also are subject to global threats, just like all other reefs around the world. The rising global demand for reef-derived fish is accompanied by rising ocean temperatures and acidity levels. The unprecedented rates of the global rise of carbon dioxide in the atmosphere is leading to fast rates of ocean warming and acidification. These climate-change factors can severely undermine the ability of coral reefs to grow, reproduce, and continue to act as hubs of marine life.

When sea-surface temperatures warm the local corals above usual highs and remain at that level for several months, severe coral bleaching and die-offs can occur. During coral bleaching, the corals expel from their tissues tiny photosynthetic organisms, zooxanthellate algae, that contribute significantly to bringing food and energy to the coral animal. These algae, living inside the coral, also impart colors to the outside of the corals. The reefs in the Coral Triangle are already living close to the thermal edge of what scientists have measured. If the waters continue to warm, the corals—the engineers of the whole bountiful ecosystem—will continue to be threatened.

Temperatures are rising just as levels of ocean acidity are increasing. Unfortunately, acidification adds stress to corals and other organisms on the reefs that make their shells out of calcium carbonate. The lower the pH (higher the acidity) of the seawater, the weaker corals and other calcifying organisms are in building their skeletons. Eventually, flimsy skeletons make for a more fragile overall habitat and a less productive ecosystem, which in turn could lead to fewer and smaller fish, lower fish catches, and less food, trade, and tourism.

By the end of the 21st century, the Coral Triangle is expected to experience some of the fastest rates and greatest amounts of acidification in the world. Continued scientific studies of the area exposed to the hottest temperatures and highest acidity levels will shed light on which communities are most vulnerable to deterioration and those that could be the most resilient. Such information can be used to design conservation strategies and efforts that are more robust and likely to resist the effects of climate change.

Conservation Efforts

To inspire environmental stewardship of the local and regional reefs, many conservation organizations have developed strong, active programs that include working with local communities to sustain healthy reefs without compromising the human needs of food, tourism, and recreation.

For example, environmental education programs aimed at raising awareness for ocean stewardship have been an important focus in Raja Ampat, in West Papua, Indonesia. There, Conservation International maintains an educational sailing ship, *Kalabia,* that travels to approximately 100 coastal villages in the region every year to educate children about the importance of the reefs on which their communities depend. On another level, the World Wildlife Fund and The Nature Conservancy work with the six nations of the Coral Triangle to eliminate illegal and destructive fishing and to develop successful joint programs that focus on protection of the shared waters and reef resources. Also, many small community conservation organizations continue to work tirelessly with governmental bodies toward greater protection of these unique and valuable marine resources.

On land, there has been significant success for conservation, with wilderness areas set aside to preserve and to safeguard the resources. The Coral Triangle is now a top priority for marine conservationists because it is one of the most beautiful, biologically productive—and threatened—areas on the planet.

Lida Teneva

Further Reading

Carpenter, Kent E., et al. "One-Third of Reef-Building Corals Face Elevated Extinction Risk From Climate Change and Local Impacts." *Science* 321, no. 1 (2008).

Conservation International."Coral Triangle Initiative." http://www.conservation.org/global/marine/initiatives/oceanscapes/cti/pages/overview.aspx.

Veron, J., et al. "Delineating the Coral Triangle." *Galaxea, Journal of Coral Reef Studies* 11, no. 1 (2009).

Córdoba Montane Savanna

Category: Grassland, Tundra, and Human Biomes.
Geographic Location: South America.
Summary: The isolation of this diverse mountainous ecoregion has protected it from environmental threats; it is thriving, but only one small region has been declared protected land.

The Córdoba montane savanna ecoregion includes parts of the Córdoba and San Juan provinces of Argentina, roughly between 29 degrees and 33 degrees west longitude and 63 degrees and 66 degrees south latitude. The ecoregion consists of the isolated mountain chain of the Sierra Centrales, about 250 miles (400 kilometers) east of the main Andes chain, from which it is separated by a dry lowlands area. The Córdoba itself comprises the Sierra Grande de Córdoba, Sierra de San Luis, Sierra de Valle Fertil, Sierra de Volasco, and four smaller mountaintops.

The area includes woodlands and rocky areas, grasslands, deep gorges, and plains such as the Pampa de Achala. As a result of the mountains' isolation, the local flora are rich in endemic species (those not found elsewhere on Earth). The temperature varies but is predominantly temperate to cool; annual rainfall varies from 20 to 43 inches (500 to 1,100 millimeters), depending on the local topography.

Potential Environmental Threats

The ecoregion faces no current jeopardy, thanks largely to its isolation, but it is fragile. Because the savanna developed in isolation, any development is likely to cause significant changes to the ecosystem and its diversity. It is likely that a large percentage of the species living in delicate balance with one another would be unable to adapt swiftly to major changes such as newly introduced species (whether competitors or predators), or diversion of natural resources.

Habitat loss from overgrazing is considered one of the biggest potential threats. Studies of natural montane grasslands responses have found that both high-intensity agricultural activity over a short period (such as commercial endeavors by major business interests), and lower-intensity high-frequency activity (such as subsistence farming and more traditional smaller-scale farming) have a major impact on species richness and ecosystem balance. By contrast, simple grazing, so long as it does not lead to overgrazing, has comparatively little effect on the ecosystem and is much less likely to result in the invasion of new species.

Additional stress may come in the form of global warming. Higher average temperatures will cause greater evaporation rates, faster glacial and snowpack melt, and lower average water levels along riparian stretches of the biome. Species migration or disappearance will be a real concern, once these trends begin to accumulate. With the already-isolated nature of these montane biological niches, these factors are projected to contribute to the acceleration of habitat fragmentation.

Because the ecoregion faces no current severe threats, the only protected area is Quebrada del Condorito National Park.

Wildlife

The Córdoba montane savanna region is home to a large number of endemic species. A total of 12 bird species and subspecies are native to the savanna; species found here include the Andean tinamou (*Nothoprocta pentlandii*), olive-crowned crescentchest (*Melanopareia maximiliani*), cliff flycatcher (*Hirndinea ferruginea*), stripe-capped sparrow (*Aimophila strigiceps*), and rufous warbling finch (*Poospiza nigrorufa*).

Among other songbirds, the Córdoba Cinclodes (*Cinclodes comechingonus*), found at elevations of 5,250–9,200 feet (1,600–2,800 meters); and Olrog's Cinclodes (*C. olrogi*), found at 4,920–7,880 feet (1,500–2,400 meters) among grassy rock formations near lakes and streams, are endemic to the region and evolved similarly to the Eurasian and American bird family *Cinclidae*. Cliff walls are used as nesting sites by the peregrine falcon (*Falco peregrinus*) and black-chested buzzard eagle (*Geranoetus malanoleucus*). Also present is the Andean condor (*Vultur gryphus*).

The Hensel's short-tailed opossum (*Monodelphis henseli*) is believed to be endemic to the region. Temperate families of bats (*Vespertilionid*s and *Molossid*s) also are found in the Córdoba montane savanna. Although the latitude is too high for most tropical bats, the exception is the vampire bat (*Desmodus rotundus*), which can be found in most places in Latin America where there are cattle on which they can feed. The plain is home to the endemic Achalan frog (*Odontophrynus achelensis*), which faces the threat of habitat loss due to grazing, and the two Achalan toads *Rhinella achalensis* (also called the Córdoba toad) and *Chaunus achalensis.*

The Andean fox (*Pseudalopex culpaeus*), a false fox (fox-like canid outside the *Vulpes* genus) also called the red fox or *zorro rojo,* is one of the endangered mammals of the region. It preys on the many rodents and hares that roam the area. One of the red fox's competitors is the puma (*Felis concolor*), a vulnerable species. Other animals of the Córdoba montane savanna are Geoffroy's cats (*Oncifelis geoffroyi*); ferrets (*Mustela furax*); and colocolos (*Oncifelis colocolo*), spotted and striped wildcats that may be nocturnal and that prey on the plentiful wild guinea pigs of the region.

In 2010, a new species of *Akodon,* a genus of rodent, was discovered. The *Akodon* genus is unique to South America and found mainly along the Andes. The new species, *Akodon viridescens,* is not quite endemic to the Córdoba montane savanna, as it can be found in northwestern Argentina's Cuyo region.

Surrounded by the mountains is the plain of Pampa de Achala, which enjoys a microclimate and is home to some endemic species, most of which may still be undiscovered.

The Andean Condor

The Andean condor (*Vultur gryphus*), the only member of the *Vultur* genus, has one of the largest wingspans of any land bird: 10.5 feet (3.2 meters). Principally a scavenger feeding on large mammal carcasses, it is exceptionally long-lived; specimens in captivity have lived for more than a century.

Like the other new-world vultures (in the genera *Coragyps, Cathartes, Sarcorhampus,* and *Gymnogyps* of the *Cathartidae* family), the Andean condor is not related to the old-world vultures of Europe but simply evolved to occupy a similar ecological role, and for that reason, it is an important subject of study in parallel evolution. The fossil record, though incomplete, suggests that the Andean condor has lived in the area since the Pliocene Era and is largely unchanged since then.

Vegetation

Even apart from the endemic species, much of the diversity here comes from the meeting of the surrounding mountains' plant and animal populations in this common space. Some of the trees found on the plain are found only on mountain slopes, and the same is true of various small mammals and insects that use them for habitats or food.

The most prominent vegetation of the Córdoba Montane Savanna biome is the grass, which is coarse and of low nutritive value for grazing livestock or wild animals. Stands of tabaquillo (*Polylepis australis*) and mayten (*Maytenus boaria*) trees grow here as well. The tabaquillo is endemic to the ecoregion and has adapted to the cold winters by producing rough sheaths of loose, papery bark that surround the trunk and insulate the tree from cold temperatures. Tabaquillo forests are among the most endangered in the world; there is a strong negative correlation between seed viability and the proximity of human activity, and individual trees grow poorly, thriving only when many of their number are present. Maytens have fared better and are found throughout Argentina as well as Chile. Honey from the tree is collected commercially, and grazing cattle feed on the tender leaves. The Argentine maytens are especially hardy and known for their drought resistance. They are often found growing in conjunction with the endangered understory plant *Adiantum gertrudis,* one of the many species of ferns growing in the region, noted for the fine hairs covering its fronds.

Bill Kte'pi

Further Reading

Braun, J., M. Mares, B. Coyner, and R. Van Den Bussche. "New Species of Akodon From Central Argentina." *Journal of Mammalogy* 91, no. 2 (2010).

Diaz, Sandra, Alicia Acosta, and Marcelo Cabido. "Community Structure in Montane Grasslands of Central Argentina in Relation to Land Use." *Journal of Vegetation Science* 5, no. 4 (1994).

Stattersfield, A. J., M. J. Crosby, A. J. Long, and D. C. Wege. *Endemic Bird Areas of the World: Priorities for Biodiversity Conservation.* Cambridge, UK: Burlington Press, 1998.

Costa Rican Seasonal Moist Forests

Category: Forest Biomes.
Geographic Location: Central America.
Summary: This unique tropical forest type is characterized by intense seasonal changes, but much has been lost to clearing for agricultural crops and cattle farms, leaving just 10 percent of the land as intact forest.

Costa Rican seasonal moist forests exhibit dramatic changes between wet and dry seasons, including loss of canopy leaves. The landscape differs so completely between seasons that casual observers may not believe it is the same forest. This ecosystem hosts many rare species and serves as a migration site for others. Unfortunately, the loss of forest to farmlands threatens these diverse ecosystems.

Seasonal moist forests occur in two areas of northwestern Costa Rica over a range of elevations. The first is a strip of land bordered on the west by the seasonal dry forests of Guanacaste and on the east by the Tilarán Mountain peaks. Because this mountain range marks the Continental Divide, rain from these forests drains into the Pacific Ocean. A smaller area of seasonal moist forests lies on the Nicoya Peninsula, which curves southward and also drains into the Pacific Ocean.

Day length and temperature remain relatively constant throughout the year, but tropical Costa Rica experiences distinct wet and dry seasons. Seasonal moist forests receive about 59 inches (1,500 millimeters) of rain during the wet season (May to November) and less than 6 inches (150 millimeters) of rain from December to April. The especially severe dry season in this region results from the constant activity of the trade winds during winter months. The winds blow across the country from east to west, but the high Tilarán Mountains trap the moisture they bring from the Caribbean. Thus, dry winds sweep the northwestern Pacific slope, eliminating moisture from the land.

In the neotropics, seasonal changes in precipitation affect reproductive schedules, migration patterns, soil nutrient cycles, and other ecosystem factors. In seasonal moist forests, the annual canopy loss and accumulation of leaves on the forest floor give seasonality even greater importance. Many organisms have specialized mechanisms to cope with the changes in light penetration and nutrient cycling, giving seasonal moist forests a unique composition of species.

As in most of the neotropics, the species of this region are poorly documented. Species composition changes greatly over short distances in these forests due to significant alterations in elevation and overall rugged topography. Therefore, protecting seasonal moist forests throughout the region is critically necessary to conserve biodiversity.

Vegetation and Wildlife

Seasonal moist forest trees and shrubs consist of a mix of deciduous broadleaf and conifer species such as clusia, cedar, pochote, and the national tree of Costa Rica: the guanacaste (*Enterolobium cyclocarpum*). Orchids, bromeliads, and lianas create diverse habitats on tree trunks and branches. These epiphytes occur with less abundance in the region than in neighboring montane forests, however.

The unseen and unsung heroes of the seasonal moist forest are fungi. Though visible above ground only during reproductive phases, they occur in soil, plant roots, and on the surfaces of plants and animals. The forest requires fungi for the breakdown of leaf litter and provision of nutrients to plants.

The forest floor here abounds with diversity and activity. Agoutis, coatamundis, jagarundis, and tyras roam in search of food. There also are innumerable species of invertebrates such as beetles, centipedes, scorpions, and spiders, many of which remain unidentified. Though often invisible in their underground chambers, ants are estimated to have a higher total biomass than all other animal species combined in these forests; they play essential roles in nutrient cycling and in carving out and tending to various habitat niches.

A Costa Rican tree frog clings to a branch of bamboo. Tree frogs inhabit the canopy and understory of Costa Rican seasonal moist forests, of which only 10 percent are intact. (Thinkstock)

The understory and canopy host characteristic animals such as tree frogs, butterflies, vine snakes, bats, howler monkeys, and spider monkeys. Hundreds of birds, including white-fronted parrots, laughing falcons, and oropendulas make their home in the forest here. Seasonal moist forests also serve as a migration destination for notable bird species like resplendent quetzals and three-wattled bellbirds, which travel from the east to eat from later-fruiting trees here.

Threats and Conservation

Global warming is the source of future trouble for this biome. Disruption to wind patterns, long-term temperature rise, and higher frequency of violent weather events may combine to drive soil erosion, altered seasonality, failed reproductive cycles, and other habitat-endangering trends. Plants and animals that have evolved specifically to take advantage of the predictable dynamics of the seasonal moist forest will be jeopardized as these climate changes work to unravel the timing, temperature, and humidity regimes that they have grown to depend upon.

Seasonal moist forests are among the Costa Rican ecosystems most affected by habitat loss and degradation. In the past century, lowland seasonal moist forests have been cleared for beef cattle, while higher-elevation forests have been replaced by coffee, beans, corn, and dairy cows. Only 10 percent of the seasonal moist forest region remains intact forest.

Although conservation efforts are relatively high in Costa Rica, only 3 percent of land in seasonal moist forest regions are protected, making them the least-represented ecosystem in the national park system. Protected areas include Cabo Blanco on the Nicoya Peninsula, established in 1963 as the country's first nationally protected land; and Parque Nacionál Rincón de la Vieja, which protects a significant tract of eastern seasonal moist forests.

Many birds require continuous habitat to complete migratory paths. Thus, deforestation and fragmentation have dramatically reduced populations. The Bellbird Biological Corridor Project works to connect forest fragments and ensure the future of these species.

Recently, international ecotourism has become a reason for promoting preservation and reestablishment of these forests. The revenue brought in by bird-watchers, hikers, and other nature-lovers has fueled several conservation-related subsidy programs, such as the payment of farmers to allow their fallow land to reforest. Ideally, increased interest in this region will allow seasonal moist forests to persist for future generations to enjoy.

Kimberly M. Kellett

Further Reading

Janzen, Daniel. *Costa Rican Natural History.* Chicago: University of Chicago Press, 1983.

Kricher, John C. *A Neotropical Companion: An Introduction to the Animals, Plants, and Ecosystems of the New World Tropics.* 2nd ed. Princeton, NJ: Princeton University Press, 1997.

Olson, David M., et al. "Terrestrial Ecoregions of the World: A New Map of Life on Earth." *BioScience* 51, no. 11 (2001).

Rappole, John H. *The Ecology of Migrant Birds: A Neotropical Perspective.* Washington, DC: Smithsonian Institution Press, 1995.

Cyprus Mediterranean Forests

Category: Forest Biomes.
Geographic Location: Eastern Mediterranean.
Summary: Cyprus is an island of diverse altitudinal regions and strong habitat features, but logging, grazing and agricultural clearance has curtailed the vitality of some native plants and animals.

The Cyprus Mediterranean Forests biome comprises an island ecoregion encompassing the hills, low plains, and massif of the Troodos Mountains, on the island nation of Cyprus. Though the island, situated in the eastern reaches of the Mediterranean Sea, is relatively small, it features a wide range of altitudes, resulting in multiple forest zones. Despite the similarity, it is unclear whether the nation of Cyprus is named for the Mediterranean cypress tree so common on the island. Linguists have suggested the island's name was derived from the ancient word for copper or henna, equally prevalent on the island and similar in sound.

Vegetation

Juniper woodlands (*Juniperus foetidissima*) and forests of mountain pine (*Pinus pallasiana*) are most common at the highest elevations here. The juniper is known for the strong, even offensive odor of its crushed leaves. At medium elevations, the endemic Cyprus cedar (*Cedrus brevifolia*) can be found, primarily around the Troodos mountains. At that same elevation is the national tree of Cyprus: the golden oak (*Quercus alnifolia*), an evergreen oak. The golden oak grows only on the igneous geological substrate of the Troodos Massif, between elevations of 1,312 to 5,906 feet (400 to 1,800 meters), either in a dense maquis by itself or in association with Turkish pine (*Pinus brutia*). Cypriot law protects the golden oak, while European Union law protects the scrub and low forest vegetation of the *Quercus alnifolia* biota, encompassing its entire habitat.

The black pine and cedar forests are much smaller than they once were; black pine is intensively managed for timber, and cedar now occupies only a patchwork of stands. The poor soils of the Troodos range and the predominance of ultrabasic substrates make soil restoration difficult. Human activity remains the biggest threat to the ecoregion: tourism and urban development along the coasts, and grazing and deforestation for pasture creation inland.

At low elevations, sclerophyllus evergreen and semi-deciduous oak forests (*Quercus coccifera* and *Q. infectoria*) dominate. *Q. coccifera,* or the Kermes oak, is the food plant of the Kermes scale insect (*Kermes vermilio*), which in earlier eras was the source of the dye crimson. There also are forests of Mediterranean cypress (*Cupressus sempervirens*), an ancient evergreen that has been used as a symbol of mourning since antiquity. The cypress's association with death may originate from its failure to regenerate when overzealously pruned, or it may relate to the tree's longevity; some cypress trees are thousands of years old.

The lowlands are home to the evergreen shrub *Arbutus andrachne,* also called the Greek strawberry tree, known for its strawberry-shaped, cherrylike fruits and the honey that bees make from its nectar. In the driest lowlands, the vegetation is primarily shrubs, including jujube lotus (*Zizyphus lotus*), carob (*Ceratonia siliqua*), and wild olive (*Olea europaea*).

About 7 percent of the plant species are endemic here (found nowhere else of Earth); most, 87 of 128 species, are distributed throughout the Troodos Mountains. Included among these are *Origanum cordifolium,* wild sage (*Salvia willeana*), wild thyme (*Thymus integer,* an important food plant to butterfly larvae), *Teucrium cyprium, Nepeta troodi, Gagea juliae,* Loch's glory of the snow (*Chionodoxa lochiae*), crocus (*Crocus veneris),* Crocus cyprius *(C. cyprius),* Cyprus tulip (*Tulipa cypria*), and Cyprus cyclamen (*Cyclamen cyprium*).

Many of the plants found in Cyprus grow together in a community called garrigue, typically found in mediterranean ecosystems near the coast where annual summer droughts occur. Garrigue is a discontinuous association of bushes, herbs, and isolated trees such as lavender, thyme, rosemary, sage, Artemisia shrubs, Kermes oak, and juniper. A similar plant community in California is called chaparral.

Animals

While Cyprus's wildlife is diverse, endemism is much lower than among plant life. Most endemic species are insects or arachnids. The notable exceptions are the Cyprus moufflon and the Cyprus warbler. The moufflon (*Ovis aries ophion*) is a wild sheep, related to the ancestors of all modern domestic sheep. Its horns are curved nearly a full revolution, and it stands about 3 feet (0.9 meter) tall at the shoulder. The Cyprus moufflon is a descendant of moufflons introduced to Cyprus by Neolithic pastoralists, adapting over thousands of years to the mountainous environment. Today, there are about 3,000 individuals. The Cyprus warbler (*Sylvia melanothorax*) is a short-distance migrant that winters in Israel, Jordan, and Egypt. When in Cyprus, this bird survives on insects and berries.

Other endemic species include the Cyprus whip snake (*Coluber cypriensis*), Troodos lizard (*Lacerta troodica*), Cypriot mouse (*Mus cypriacus*), the longhorned beetles *Purpuricenus nicoles* and *Agapanthia nicosiensis,* the jumping spider *Aelurillus cypriotus,* the scorpions *Mesobuthus cyprius* and *Buthus kunti,* the land snails *Assyriella bellardi* and *Trochoidea liebertruti,* the bush cricket *Isophya mavromoustakisi,* the stag beetle *Dorcus alexisi,* and several endemic butterflies: Cyprus grayling (*Hipparcia cypriensis*), Cyprus meadow brown (*Maniola cypricola*), and Paphos blue (*Glaucopsyche paphos*).

Environmental Concerns

Cyprus has signed nine international environmental agreements and ratified the Cartagena Protocol on biosafety and the Kyoto Protocol on climate change. The island nation is committed to sustainable development, waste management, integrity of its water sources, and the treatment of agricultural and industrial waste. Forest clearing for vineyards and orchards over the last century has led to strict forest protection and management practices. Animal grazing has since been brought under control, while hunting is now prohibited in the Paphos Forest and part of Troodos in order to help reduce the decline of these forests.

Global warming is projected to deliver the equivalent of two extra weeks of summer conditions to Cyprus, while winter conditions will be tempered somewhat, although the influence of the Mediterranean waters will tend to modify extremes. There is likely to be a decrease in annual rainfall, but not as severe as in mainland areas of the eastern Mediterranean. Habitat stresses from the above trends are likely to be modest—however, with the biome already under assault from expanding human population pressures, mitigation measures need to be applied vigorously.

Bill Kte'pi

Further Reading

Blondel, Jacques. "The 'Design' of Mediterranean Landscapes: A Millennial Story of Humans and Ecological Systems During the Historic Period." *Human Ecology* 34, no. 5 (2006).

Blondel, Jacques, and James Aronson. *Biology and Wildlife of the Mediterranean Region*. New York: Oxford University Press, 1999.

Fabbio, Gianfranco, Maurizo Merlo, and Vittorio Tosi. "Silvicultural Management in Maintaining Biodiversity and Resistance of Forests in Europe—the Mediterranean Region." *Journal of Environmental Management* 67, no. 1 (2003).

Merlo, Maurizo and Leila Croituru. *Valuing Mediterranean Forests: Towards Total Economic Value*. London: CABI, 2005.

Danube Delta

Category: Inland Aquatic Biomes.
Geographic Location: Eurasia.
Summary: This freshwater-dominated estuarine deltaic complex is threatened by land use, pollution from upstream areas, and climate change; conservation efforts are in place to address further decline.

Central Europe's Danube River flows through or acts as a border for 10 countries, from the Schwarzwald (Black Forest) mountain range in Germany southward and eastward to the Black Sea. The modern Danube Delta, where the river meets the Black Sea, is a primarily freshwater-dominated estuarine ecosystem complex about 6,000 years old. The Danube Delta is home to more than 300 bird and 45 fish species, is the 22nd-largest delta in the world, and spans two countries and 1,603 square miles (4,152 square kilometers). The southern portion, in Romania, encompasses approximately 1,331 square miles (3,446 square kilometers), and the much less extensive northern section, in Ukraine, covers about 273 square miles (706 square kilometers).

Water systems in the delta include the Danube arms, the main distributaries that flow into the Black Sea, and secondary dead-end distributaries (creeks, backwaters, and dredged channels), which do not reach the Black Sea. There are three primary arms of the Danube Delta: the Chilia, the northernmost arm and the only arm that flows through Ukraine; the Sulina, the most direct arm; and the Sfântu Gheorghe arm, the river's farthest point south. Together with the nearby Razelm-Sinoie, an ecologically similar complex of lagoons physically separated from the delta by a major sandbar, the Danube Delta covers approximately 2,042 square miles (5,290 square kilometers).

Based on climate alone (without water inputs from flooding), the delta would be characterized by xeric (arid-evolved) scrub and herbaceous-dominated ecosystems. For this reason, elevation relative to both sea level and the river itself is an important factor in hydrology and ecology. Variation in elevation ultimately results in 30 distinct ecosystem types (23 natural, seven anthropogenic), with natural ecosystem types ranging from open-water lakes to upland forests. Although somewhat degraded by dams, land use, eutrophication (excess nutrient input), and overfishing, the Danube is considered the best-preserved large deltaic system in Europe.

The Danube Delta contains more than 300 bird and 45 fish species and is considered the best-preserved large deltaic system in all of Europe. The portion of the delta shown in the photograph is in Romania and is edged by marshes full of tall reeds; about 70 percent of the Danube Delta is similar marshland. (Thinkstock)

Biodiversity

Areas near the Black Sea are affected by various amounts of tidal salt inflow and typically are dominated by halophytic (salt-loving) plants, such as glasswort and alkaligrass. Estuarine wetlands provide critical habitat for many fish and bird species that rely on tidally influenced ecosystems for foraging and nesting. The coastal lagoons of Razelm-Sinoie vary in salinity and provide habitats in a broad range of conditions, leading to greater biodiversity.

Running waters encompass a small but important part of the delta. These waters serve as habitat for many plant species, including attached varieties such as mint, submerged aquatic ones like hornworts, and floating macrophytes including floating fern and duckweed. The underlying substrates are important habitat for invertebrates, such as worms and mollusks, which serve as important food sources for historically important commercial fish species such as the Danube mackerel and Danube sturgeon (fishing for the latter fish has been suspended due to overexploitation).

The ecological structures of lakes, ponds, and lagoons can be quite complex; submerged aquatic vegetation, such as Eurasian watermilfoil, may dominate shallow ponds. Vegetation in deeper lakes and lagoons generally is characterized by floating macro- and microphytic species such as the edible water chestnut, the carnivorous bladderworts, and phytoplankton. In these deeper systems, detritus and associated bacteria are particularly important components of trophic structure, making up a large part of the base of a food web that supports important fish species such as tench and perch.

A large expanse of the delta, as much as 70 percent, is covered by marshes, primarily dominated by reed, *Carex* sedge, and cattail. Although many of these reed swamps or *plaur* are attached to the underlying substrates, a large portion of this marshland is actually rooted in a thick peat mat that floats on top of the underlying water table. The plaur are generally located along lake margins. It is expected that over time, the plaur will move inland, over the lake edges, and the areas that are currently occupied by these floating mats margins will succeed to drier, more upland habitat.

These transitional marsh zones are an extremely productive and particularly important ecosystem type, serving as habitat for many native waterbirds (such as ducks, geese, cormorants, egrets, pelicans, and pheasants) and carnivorous mammals (such as otters, mink, foxes, and wolf species). Since human settlement in the region, and particularly over the past 50 years, exotic mammal species (tanuki, muskrat, nutria, feral cat, and feral hog) have established populations and could pose problems for native species.

Where the banks of the primary and secondary distributaries are shallow, riparian wetlands dominate. Areas of the floodplain closest to the river may be inundated frequently and are often dominated by marsh vegetation. These riparian marshes are similar in ecological function to the plaur at lake margins. Farther away from the river, woody shrub and small tree species begin to dominate as the marshes grade into periodically flooded swamplands, with primarily early successional species, such as willow and white poplar.

In some areas, banks are steep enough that the swamps eventually grade into forested stands that are rarely flooded. These forests may contain a network of *jaspas*—low-lying ephemeral pools that remain under standing water following river flooding. Sand dunes along river banks protect forested stands on the highest levees from flooding, allowing later-successional oak species to grow in the intervening low-lying spaces, called *hasmace*. Letea and Caraorman forests are notable for their extensive hasmace, which encompass oak and ash forests replete with a diverse array of understory species, such as blackthorn and wild hop. These areas also serve as habitat for the endangered Danubian meadow viper. The dunes are sparsely vegetated with stress-tolerant grasses such as fescue.

Effects of Human Activity

Most of the delta is not densely populated. The Romanian city of Tulcea, however, located at the western edge of the area, has a population of approximately 92,000 people, while the Ukrainian city of Izmail, on the northeastern rim, is home to some 85,000 people. There are several smaller towns, villages, and rural settlements in the region's less densely populated areas that are more integrated into the surrounding landscape and habitat matrix.

Even areas that are not currently settled may have been converted for human use. Many of these constructed ecosystems are the result of draining surrounding marshes and other wetlands. Drained wetlands are quite fertile and are used for farming and forestry. In some cases, such converted lands later become abandoned, resulting in ecosystems with novel biological communities. Elsewhere, overfishing of many lakes has caused declines in fish stocks. As a result, some wetlands are not drained but dammed, resulting in human-made ponds used for fish farming. These fish-farm ecosystems may be quite unstable, and nutrient-rich feeds may result in eutrophication.

Major alterations of the delta channels have occurred since at least the 19th century, when the Sulina arm was channelized to ease navigation. For a large part of the 20th century, the delta was controlled by the Soviet Union. As part of plans to convert the entire delta to an agro-industrial complex, the Soviet Union oversaw extensive wetland drainage.

An additional problem is nutrient loading from agricultural runoff far upriver. Nutrients in fertilizers from fields as far away as Germany may find their way to the Danube Delta, where they contribute to wetland eutrophication, a process that may be particularly devastating to floating plaur, for which buoyancy depends on slow decomposition. Similar to the problem of agricultural runoff, other forms of pollution from the many cities upriver may be deposited in the delta. Also, erosion is slowly driving loss of land, as sediments become deposited in the Black Sea. The effects of erosion will become exacerbated over the next century by rising sea levels, which will also cause longer periods of inundation with saltwater.

Conservation Efforts

In anticipation of and response to these challenges, the United Nations has established the entirety of the Danube River Delta and the Razelm-Sinoie coastal lagoon complex as a United Nations Educational, Scientific and Cultural Organization (UNESCO) World Heritage Biosphere

Preserve site. Similarly, the Romanian government established the delta as a Natura 2000 site, barring actions that could significantly affect its species and habitats. Still, the system is not safe from alteration. Despite the delta's unique ecology and natural beauty, the Ukrainian government currently has long-term plans to further expand navigational channels.

JESSE FRUCHTER

Further Reading

Panin, Nicolai and Dan Jipa. "Danube River Sediment Input and its Interaction With the North-Western Black Sea." *Estuarine, Coastal and Shelf Science* 54, no. 3 (2002).

Sarbu, A. "Inventory of Aquatic Plants in the Danube Delta: a Pilot Study in Romania." *Archiv fur Hydrobiologie Supplement* 147, no. 1–2 (2003).

Unluata, Umit. *Environmental Degradation of the Black Sea: Challenges and Remedies.* Dordrecht, Netherlands: Kluwer Academic, 1998.

Danube River

Category: Inland Aquatic Biomes.
Geographic Location: Europe.
Summary: From western Europe to the Black Sea, the Danube River and its catchment area join 19 nations in conservation efforts to ensure the region's ecology and prosperity.

Second only to the Volga in size among Europe's major rivers, the Danube River extends over 1,770 miles (2,850 kilometers) with a drainage basin area of 315,000 square miles (817,000 square kilometers). This watershed area receives effluents from hundreds of tributaries and rivers. Approximately 82 million people live within its catchment area.

The Danube River headwaters begin where the Brigach and Breg streams conjoin at the far southwest corner of Germany. Its headwaters are memorialized at a site known as Karstaufstoßquelle on the outskirts of Donaueschingen in the state of Baden-Württemberg. The Danube flows through 10 countries: Germany, Austria, Slovakia, Hungary, Croatia, Serbia, Bulgaria, Romania, Moldova, and Ukraine. Its basin includes lands in Poland, Switzerland, Italy, Czech Republic, Slovenia, Bosnia and Herzegovina, Montenegro, Republic of Macedonia, and Albania. The Danube first flows easterly, then cuts sharply southward near Budapest, capital of Hungary, before resuming its eastward flow where the Vuka River joins it at Vukovar, Croatia. The Danube flows through three other European capital cities—Vienna, Bratislava, and Belgrade—on its way to the Black Sea.

Ecoregions Along the River

The Middle European Corridor—the full stretch of the Danube—is a fertile zone with thick loess soils supporting high populations within cities, farming towns, and hamlets. Distinct ecosystems unfold across an ancient panorama of verdant alpine mountain ranges, mixed boreal and temperate forests, grasslands, open prairie, meadows, and marshlands. The Upper Danube refers to the stretch of river extending from its source in Germany's Black Forest to the edge of Bratislava at the entry to the Carpathian Mountains. For centuries, vast tracts of oak, beech, spruce *(Picea abies* and *P. excelsa),* and fir *(Abies alba* and *A. pectinata*) supported thriving lumber and woodworking industries. Mining and hunting also were important Danubian occupations.

Commercial silviculture and extensive logging radically altered the ecology of the region, replacing old forest stands with uniform plantations largely unsuited to the shrubs, herbs, plants, and wildlife that originally thrived on the soils and understory of mixed forests. In 1999, the Naturpark Südschwarzwald Association was formed to protect the beauty of the Southern Black Forest region; the reserve area includes the Southern Black Forest Nature Park. It is a cooperative project with participants from more than 100 local communities. Residents of the park strive to protect the region's unique balance of traditional lifestyles within the ancient forest.

Roughly two-thirds of all runoff waters into the Danube originate in the mountains and foothills of the Alpine Space ecoregion. During the past 100 years, alpine temperatures have risen 1.5 degrees C (2.7 degrees F), which is double the global average. Future changes in climate are predicted to alter the relative seasonal abundance and scarcity of mountain precipitation, which in turn will accelerate weather extremes such as occurrences of extreme flooding and drought in many areas dependent on alpine runoff.

The Alpine mountain region connects with the Carpathian Mountains at Bratislava; from there, the Carpathians extend across central and eastern Europe. The Middle Danube basin extends from the Gate of Devin just outside of Bratislava to the Iron Gate at the border of Serbia and Romania. It includes the Gemenc Béda-Karapancsa wetlands of Hungary and the Kopački Rit in Croatia, as well as the Gornje Podunavlje of Serbia and Montenegro, lower Drava-Mura wetlands, and the Tizsa River Basin. The area provides habitats for a varied animal population, including white-tailed eagles, terns, black storks, otters, beavers, and sturgeon. More than 250,000 migrating waterfowl fill the skies of the floodplains encompassed in this singular international flyway.

Waterways, Wetlands, and Wildlife

The Danube-Carpathian region defines a watershed area whose boundaries include Germany, Rumania, Ukraine, Poland, and Bulgaria. Approximately 80 percent of the water draining from the Carpathians flows into the Danube River basin. This region includes some of the most important wilderness, old-growth forests, and wetland areas in Europe. The Danube is a prominent landmark that has served as an important rallying point for international protests against intensive channelization of the river for navigation and hydroelectric power. In 1984, the Duna Kör (Save the Danube) movement originated in response to the announcement of the $3 billion Gabčíkovo-Nagymaros hydroelectric power project on the Czech Republic-Hungary border.

At the mouth of the Black Sea the Danube opens into three tributaries, the Chilia, Sulina, and Sfântu Gheorghe, which feed into the Danube Delta, one of Europe's largest wetlands. These wetlands are renowned as breeding and resting grounds for critical populations of pelican (including the Dalmatian pelican), wild geese (including 90 percent of the world's population of red geese), ibis, cormorant, swallow, as well as freshwater fish. More than 5,000 terrestrial species have been documented in the delta and lower Danube region, including the shy otter, mink, foxes, bears, wolves, muskrats, tortoises, and hares.

Across the Danube River realm, some of the animal species identified by the World Wildlife Fund (WWF) as endangered are: yellow-legged clubtail dragonfly (*Gomphus flavipes*), moor frog (*Rana arvalis*), Danube sturgeon (*Acipenser gueldenstaedti*), European pond turtle (*Emys orbicularis*), great white pelican (*Pelecanus onocrotalus*), grey heron (*Ardea cinerea*), kingfisher (*Alcedo atthis*), and the beaver (*Castor fiber*).

Conservation Efforts

There have been a wide array of protection schemes enacted along the Danube. The Trans-European Transport Network was established in 1990 to integrate planning for comprehensive, transnational systems of transportation, energy, and communications. Inland waterways, while vital to regional economies, also are dynamic transport systems for infectious diseases, nonnative invasive species, and non-point-source pollutants, all of which have had alarming impacts along the Danube corridor and the multi-dimensional ecological systems that depend on it for survival. Dredging and channelization have affected the river's hydrology; redistributed silts and altered rates of fluid dispersion continue to have wide-ranging impacts across the catchment area. Today, more than 700 dams and weirs regulate the Danube and its major tributaries, tending to limit or damage the biodiversity of riverine, floodplain, and wetlands areas.

In 2000, the Lower Danube Green Corridor Agreement was signed into law, binding the governments of Bulgaria, Romania, Ukraine, and Moldava to protect and restore some of the vital wetlands along the lower river and the delta. More than 5,405 square miles (1.4 million hectares) of land have been placed under protection for the

A coal barge on the Danube River in Melk, Austria. Europe's second-largest river after the Volga, the Danube travels through 10 countries on its way to the Black Sea. (Thinkstock)

sustainable maintenance of wildlife, water quality, and resources for recreation.

Working in response to mandates established by the European Territorial Cooperation program, the Alpine Space Programme 2007–13 was initiated by seven countries to promote regional development in sustainable ways. The program area covers 173,746 square miles (450,000 square kilometers) and includes about 70 million people.

In 2009, eight countries signed the Declaration of Vienna, establishing a cooperative partnership to strengthen the conservation of 12 designated sites. The Danube River Network of Protected Areas includes: the Danube Riparian Forest at Neuburg-Ingolstadt in Germany; the Donau-Auen National Park in Austria; the Zohorie Protected Landscape and the Danube Floodplain Protected Landscape in Slovakia; the Duna-Ipoly National Park and the Duna-Drava National Park in Hungary; the Nature Park Kopacki Rit in Croatia; the Special Nature Reserve Gornje Podunavlje in Serbia; the Persina Nature Park, the Kallmok-Brushlen Protected Site, and the Srebarna Nature Reserve in Bulgaria; and the Danube Delta Biosphere Reserve in Romania, designated a World Nature Heritage in 1991.

These network sites were established to protect the last free-flowing areas of the river and to conserve wildlife and migratory bird habitats. In addition, they were intended to begin restoring the Danubian floodplain, more than 80 percent of which has been damaged or destroyed by dams and navigation projects since the early 1800s.

In 2011, the WWF announced the formation of a transboundary United Nations Educational, Scientific and Cultural Organization (UNESCO) biosphere reserve to protect the lands and wildlife along the Mura, Drava, and Danube Rivers in the Carpathian region, creating Europe's largest riverine protected area. The declaration was signed by Austria, Croatia, Hungary, Serbia, and Slovenia, establishing a five-country protected zone. It covers an area of 3,090 square miles (800,000 hectares) that features extensive floodplain forests, river islands, and gravel banks.

A comprehensive listing of European documents protecting the Danube and other European inland rivers is included in the Joint Statement on Guiding Principles for the Development of Inland Navigation and Environmental Protection in the Danube River Basin. This statement was published in 2007 by the International Commission for the Protection of the Danube River, the Danube Commission, and the International Sava River Basin Commission. Other guiding directives include the Danube River Protection Convention, the Black Sea Commission, the Danube-Black Sea Region Task Force, and Black Sea Synergy.

Victoria M. Breting-Garcia

Further Reading

Bachmann, Jasmine. "Can Ecology and Water Transport Coexist?" International Commission for the Protection of the Danube River, January 2007. http://www.icpdr.org/main/publications/can-ecology-and-waterway-transport-coexist.

Baltzer, Michael. *Inland Navigation in the New EU—Looking Ahead: Corridor VII or Blue Danube?* Sofia, Bulgaria: World Wildlife Fund Danube-Carpathian Programme, 2004.

Damm, Marion, ed. *Climate Adaptation and Natural Hazard Management in the Alpine Space.* Munchen, Germany: AdaptAlp, 2011.

Eiseltova, Martina, ed. *Restoration of Lakes, Streams, Floodplains, and Bogs in Europe: Principles and Case Studies.* New York: Springer, 2010.

Mauch, Christof and Thomas Zeller. *Rivers in History: Perspectives on Waterways in Europe and North America.* Pittsburgh, PA: University of Pittsburgh Press. 2008.

Darling River

Category: Inland Aquatic Biomes.
Geographic Location: Australia.
Summary: The Darling River, key to natural and human life in southwestern Australia, is in jeopardy from development, irrigation extraction, changes in river flow, and global warming.

The Darling River system including its tributaries is the longest river system in Australia, at 1,700 miles (2,740 kilometers). As the northern half of the Murray-Darling River system, it is the major force watering the Australian state of New South Wales (NSW) and supporting both natural habitats and human settlements across the region. Not including the tributaries, the Darling is Australia's third-longest river. The Darling flows southwestward, with its tributaries rising in southern Queensland; it joins the Murray River at Wentworth, NSW. Downstream from the junction, the river is called the Murray as it flows into South Australia and into the Indian Ocean.

Annual flow in the Darling is some 80 million cubic feet (2.3 million cubic meters). Much of the basin is less than 650 feet (200 meters) above sea level, making for extremely flat gradients. Rainfall in the watershed varies from 47 inches (1,200 millimeters) per year at the heights of the Great Dividing Range to less than 8 inches (200 millimeters) per year in the western reaches downstream.

The Darling River has a high level of evaporation in the largely semi-arid environment of the basin, varying from 39 inches (1,000 millimeters) per year in the eastern mountains to 79 inches (2,000 millimeters) per year in the west. Rainfall exceeds evaporation in a very small area of the catchment. With the flat gradients and high evaporation levels, several of the westward-flowing tributary rivers in the basin's center end in deltaic wetlands systems, which have environmental management significance. Because much of the Darling's course runs through extensive saltbush pastures, receiving an average of less than 10 inches (250 millimeters) of rain annually, the river often loses more water from evaporation than it gains from its tributaries—many of which from time to time fail to reach the main stream. Several of these flow into salt flats for years at a time, only to emerge as full-fledged rivers again in wet years, and rejoin the Darling as tributary streams.

Topography and Climate

The Darling Riverine Plains bioregion occupies most of the upper catchments of the Darling and its tributary the Barwon in northern NSW and southern Queensland. The upper-catchment landscape is a series of overlapping, low-gradient alluvial fans; the lower zone comprises more narrow floodplains confined between bedrock landscapes or by extensive sand plains and dune fields; channels cut through these areas to a considerable degree. Discharge from past and present streams control patterns of sediment deposition, soils, landscapes, and vegetation.

These Darling riverine plains lie in a semi-arid climatic zone: hot and persistently dry. This semi-arid zone also extends across most of the western arm of the Darling River biome, accompanied by small patches of either more arid or less hot semi-arid climates. Some subtropical climate crops up in the eastern, mountainous zone; subhumid areas are found in the southeast.

Thus, the Darling River biome features a range of microclimates: subtropical conditions in the far north; cool, humid eastern uplands; high alpine country in the Snowy Mountains; temperate conditions in the southeast; and hot and dry in the semi-arid and arid western plains that make up much of the basin as a whole.

Biodiversity

The Darling River and its tributary region is home to a large number of plants and animals, including 35 endangered species of birds and at least one dozen endangered mammal species. The basin also includes wetlands that are listed internationally for their importance to migratory birds that fly in from within the basin from other parts of Australia, and also from overseas.

Modern river channels in the bioregion support river red gum (*Eucalyptus camaldulensis*) and river cooba (*Acacia stenophylla*) forest communities. There are some areas of river paperbark (*Melaleuca trichostachya*), especially along the tributaries of the Barwon. These species grow on the channel margin in the annual flood zone. Coolabah (*Eucalyptus microtheca*) can be found on the northern rivers here.

The bioregion is home to at least 25 amphibian species, 104 reptile species, 319 bird species, and 58 mammal species. Records of amphibian species in the Darling Riverine Plains region include seven that are either endemic (found in no other biome on Earth) or largely restricted to this bioregion: long-thumbed frog (*Limnodynastes fletcheri*), giant banjo frog (*Limnodynastes interioris*), crucifix toad (*Notaden bennettii*), *Crinia parinsignifera, C. sloanei, Neobatrachus sudelli*, and *Cyclorana verrucosa*.

Reptiles here that are either endemic or largely restricted to the bioregion include: leaden delma (*Delma plebia*), Australian coral snake (*Simoselaps australis*), grey snake (*Hemiaspis damelii*), two-clawed worm-skink (*Anomalopus leuckarti*), *Emydura macquarii, Ctenotus allotropis, C. brachyonyx, Egernia modesta, Pseudechis guttatus,* and *Anomalopus mackayi*.

Waterbirds are a significant component of the Darling River fauna; a total of 35 such species in the bioregion are classified as endangered or vulnerable. Many waterbirds are known to breed here, including the freckled duck (*Stictonetta naevosa*). The plains along the river are a stronghold of such species as the spotted bowerbird (*Chlamydera maculata*), striped honeyeater (*Plectorhyncha lanceolata*), and plum-headed finch (*Neochemia modesta*).

Among the significant wetlands in the region, the Namoi River floodplain provides important habitat for the endangered bush stone curlew (*Burhinus grallarius*), even though its condition has been described as poor and still declining. Several vulnerable species—such as the koala (*Phascolarctos cinereus*), painted honeyeater (*Grantiella picta*), and brolga (*Grus rubicundus*)—have been sighted on this floodplain.

Environmental Threats

Land degradation, river-water salinity, land salinization from irrigation, water-quality problems, and loss of biodiversity are among the major environmental problems of the Darling River basin system. The prominent land-degradation problems include soil erosion, acidic soil, native-vegetation decline, and invasive weeds and noxious plants. The basin also faces the problem of biodiversity loss because of the changes in the pattern of the river flow by human actions, and due to the reduction in the volume of water flow.

Massive extractions of water for irrigation in the flatter reaches of the Darling River basin have degraded the ecological health of the river country, transforming relationships previously sustained by the seasonal flows of freshwater. Some Aboriginal landowners are even experiencing the events caused by upstream overextraction of water as a contemporary dispossession of their country, so severe have some of the soil- and habitat-degradation problems become.

Climate change is expected to affect the Darling River biome with significantly reduced water availability in southern parts of the basin due to projected declines in both winter and spring rainfall. Stream flows in both snow-affected as well as snow-free stretches of the river are predicted to be altered toward potentially earlier and more intense spring events and likely drier summer and fall regimes.

Manoj Kumar
Hukum Singh

Further Reading

Kearle, A., C. Gosper, H. Achurch, and T. Laity. *Darling Riverine Plains Project Background Report.* Dubbo, Australia: New South Wales National Parks and Wildlife Service, 2002.

Quiggin, John. *Environmental Economics and the Murray-Darling River System.* Canberra: Australian National University, 2000.

Stern, H., G. de Hoedt, and J. Ernst. *Objective Classification of Australian Climates.* Melbourne: Australian Bureau of Meteorology, 2000.

Dead Sea

Category: Inland Aquatic Biomes.
Geographic Location: Middle East.
Summary: The Dead Sea is a deep salt lake located at one of the lowest spots on Earth's surface. A few simple organisms thrive here; climate change is predicted to increase the region's dry spells.

A hypersaline lake, or inland sea, located on the border between Israel and Jordan at an altitude of 1,329 feet (405 meters) below sea level, the Dead Sea is often referred to as the lowest place on Earth's surface. The climate of the sea is characterized by high temperatures, in the range of 90–102 degrees F (32–39 degrees C) in the summer, with low humidity; and an average annual rainfall of less than two inches (50 millimeters).

The Dead Sea is 42 miles (67 kilometers) long and 11 miles (18 kilometers) wide, spread around 84 miles (135 kilometers) of shoreline. Its maximum measured depth is 1,083 feet (330 meters), and overall salt concentration is 13 ounces per 0.3 gallon (340 grams per liter). Therefore, it is considered the world's deepest and saltiest lake. It contains 33.7 percent salts, compared with 3.5 percent in most oceans. The Dead Sea is a substantially dense water body, enabling people to float easily on its surface. The mineral composition, too, is different from oceans, as only 12 to 18 percent of Dead Sea salt is sodium chloride, compared with 97 percent in most oceans. Instead, the sea has large quantities of four diverse minerals: sodium, potassium, bromine, and magnesium.

The low altitude of the Dead Sea creates unique effects like high atmospheric pressure, which is beneficial for people with some respiratory illnesses. It also absorbs most of the sun's ultraviolet rays, which prevents most skin burns throughout the area's 330 sunny days each year. Also, pollen and allergens are of minimal concern here.

The lake was formed as a result of a 62-mile (100-kilometer) movement along the geologic rupture called the Dead Sea Rift. It runs along the boundary of two tectonic plates: the Arabian and African plates, reaching from the northern tip of the Red Sea to southeastern Turkey. Because of tectonic activity, parts of the sea's bottom are still sinking; the surface-water level of the Dead Sea also has varied throughout its history. Evidence exists that the sea's surface at one point historically decreased to at least 2,297 feet (700 meters) below ocean level and that the lake almost dissipated. Further observations of cliffs near the lake indicate the past existence of multiple sea levels several feet (meters) above the current point, representing extended rainy periods.

Hydrology

The Dead Sea is naturally fed by freshwater from the Jordan River, its main tributary. In addition, 26 minor rivers descend from the desert canyons on the Israeli and Jordanian banks and spill into the Dead Sea. Most of these streams, however, are seasonal. The water level of the lake is determined by the ratio between freshwater insertion and evaporation. Dams and canals built along the Jordan River to divert water for agriculture and drinking water have dramatically limited the volume of water that annually reaches the lake, from 52.9 billion cubic feet (1.5 billion cubic meters) previously to about 14 billion cubic feet (400 million cubic meters) currently. Natural evaporation exceeds 35 billion cubic feet (1 billion cubic meters) annually.

In the past 40 years, the lake's water level has fallen by more than 82 feet (25 meters), and today it is losing elevation at a rate of about three feet (one meter) per year. Because the bottom of the southern basin is much shallower than the rest, the decreasing water level has resulted in the separation of the southern basin from the rest of the lake. The southern basin of the Dead Sea is used by Israeli and Jordanian salt factories as major evaporation pools to increase the precipitation rate of salts, which are used for food and fertilization industries. To maintain the water volume of the southern basin—visited by tourists drawn to the lake's therapeutic qualities and beautiful landscapes—water is pumped from the northern basin.

The water-level decline has been followed by a groundwater level drop. Normally, groundwater is replaced by freshwater streaming toward the lake; this causes some underground salt layers near the shoreline of the Dead Sea to be dissolved. The salt layers were formed during past sea-level retreats and are covered in clay-containing minerals with high water adsorption, creating an elastic consistency. Under these circumstances, subsurface cavities are rapidly created; these cavities might result in gravitational collapses. This condition is believed to be the cause of the recent appearance of large sinkholes along the western shore, the biggest hole of which is 36 feet (11 meters) deep and 82 feet (25 meters) in diameter.

Plants encrusted with salt at the edge of the Dead Sea. While aquatic life in the Dead Sea is limited by the high salinity, which can reach 33.7 percent, plants and animals do inhabit areas surrounding the sea. (Thinkstock)

A canal to pump and transfer water from the Red Sea into the Dead Sea has been considered for the past few years. The 326-mile (525-kilometer) canal would transfer 14 billion cubic feet (400 million cubic meters) of water yearly at its first stage, and several hydroelectric power plants and water desalination facilities would provide electricity and potable water to Jordan, Israel, and the Palestinian Authority. This would facilitate new settlements and tourist centers along the Arava Valley, a desert area separating the two seas.

The project's estimated costs vary from $2 billion to $10 billion, depending on its structure and stages. The project is being carefully considered, as the transfer of mass volumes of water may have dramatic consequences, including a fracture in the pipeline resulting from an earthquake. This could result in a large spill of Red Sea water, adding a saline threat to the crops and aquifers along the Arava Valley. Furthermore, the different chemical composition of the salts in both seas may interact, increasing precipitation of calcium sulfate (plaster) in the Dead Sea, possibly changing evaporation rates. Massive outflow of water might affect the current distribution of the Red Sea and damage the ecology and marine life. In 2005, Jordan and Israel signed an agreement to initiate a study funded by the World Bank providing a detailed estimate of the project's influence.

Aquatic Life

The Dead Sea's high salinity results in scarce aquatic life, but several studies have found organisms. During rainy winters, reduction in salinity to 28 percent is interrelated with a bloom of algae called Dunaliella, which contains red pigments and shifts the dark-blue color of the water. At the bottom of the Dead Sea, a complex system of springs was discovered, running hundreds of miles (kilometers) and extending to a depth of 98 feet (30 meters). These springs burst through craters at intervals of up to 49 feet (15 meters) and depths of 66 feet (20 meters). In a study that took place in the 1970s, a bacterium with high tolerance for magnesium chloride (one of the properties of halobacteria, or salt-loving bacteria) was isolated from bottom sediment of the Dead Sea.

In 2010, a diving expedition to the springs revealed diverse communities of microorganisms, sometimes in surprisingly thick mats, in the bottom sediment layer covering large areas of the seafloor. These communities thrive near thin plumes of freshwater that shoot out from the springs. Some of the bacteria species are phototrophs; they use sunlight energy to oxidize sulfide, naturally occurring in the springs, for their metabolism. The organisms discovered at these depths are different from those responsible for the surface blooms.

In the area surrounding the Dead Sea, a high diversity of animals and plants are found to thrive in the warm, dry climate. Among these are camels, ibex, jackals, hares, hyraxes, and foxes. Many birds find the climate of the Jordan River and the Dead Sea suitable for breeding, including the Dead Sea sparrow. Even without the impact of humans on the area, the Dead Sea could still disappear—as it nearly did about 100,000 years ago—by virtue of the effects of global warming. Man-made factors such as irrigation and water diversion projects, coupled with climate change, could make the area more arid than it already is and apply further pressure to the region's limited freshwater resources.

Itay Cohen

Further Reading

Enzel, Yehouda, Amoz Agnon, and Mordechai Stein, eds. *New Frontiers in Dead Sea Paleoenvironmental Research.* Boulder, CO: Geological Society of America, 2006.

Niemi, Tina M., Zvi Ben-Avraham, and Joel Gat, eds. *The Dead Sea: The Lake and Its Setting.* New York: Oxford University Press, 1997.

WysInfo. "Dead Sea Flora." 2006. http://www.wysinfo.com/Dead_Sea/dead_sea_flora.htm.

Deccan Plateau Dry Deciduous Forests

Category: Forest Biomes.
Geographic Location: India.
Summary: The dry forests of the southern Deccan Plateau provide a habitat stronghold for many species, but the area faces pressures from slash-and-burn agriculture as well as looming climate change.

The Deccan Plateau is a vast geographic feature encompassing most of central and southern India. The term *Deccan* comes from the Sanskrit word *dakshina,* meaning "the south." It comprises the whole of southern peninsular India, encompassing parts of eight states—most of Andhra Pradesh, Karnataka, Kerala, and Tamil Nadu, as well as portions of Maharashtra, Madhya Pradesh, Gujarat, and Orissa. The Deccan Plateau is bounded on all sides by mountains: Western Ghats to the west and Eastern Ghats to the east, Nilgiris to the south, and the Satpura and Vindhya ranges to the north.

Undulating terrains are common here; the elevation ranges from 1,500 to 2,500 feet (457 to 762 meters). Several major rivers—including the Cauvery, Godavari, Krishna, and Penner—flow across the plateau before reaching the Bay of Bengal. The Deccan Plateau has a dry season that lasts six to nine months. The maximum temperature in the region varies from 77 to 90 degrees

F (25 to 32 degrees C), while the minimum varies from 55 to 70 degrees F (13 to 21 degrees C). Annual rainfall can vary from 47 to 197 inches (1,200 to 5,000 millimeters).

Vegetation

The forest types of the Deccan Plateau fall into six categories: tropical wet evergreen, south montane wet temperate, tropical semi-evergreen, tropical moist deciduous, tropical dry deciduous, and tropical thorn. Three ecoregions dominate the plateau: the southwestern hill ranges, with rich, dense montane rainforests; southwestern hill ranges, with moist deciduous forests; and the southern Deccan Plateau, with dry deciduous forests.

Dry deciduous forests form the largest tract of the Deccan Plateau. The tall mountain range toward the western part of the Deccan Plateau intercepts the moisture from the southwestern monsoons assuring that the eastern slopes receive very little rainfall, resulting in totally different vegetation formations here, the dominant being dry deciduous forests. The interior districts of Andhra Pradesh, for example, contain pure tropical dry deciduous forest stands with pockets of semi-evergreen and moist deciduous forests.

The dry deciduous forests of the Deccan Plateau have a three-storied structure: an upper canopy reaching from about 50 to 100 feet (15 to 30 meters), an understory of 33 to 50 feet (10 to 15 meters), and an undergrowth of 10 to 16 feet (three to five meters). The top canopy is mostly open. Stand density varies from 200 to 350 stems per 2.4 acres (one hectare). The *Tectona grandis-Pterocarpus santalinus* forests may be considered a climax forest community in the region, while the *Anogeissus latifolia-Terminalia alata* community is most common. Some of the most characteristic tree species in the dry forests include *Albizia amara, Anogeissus latifolia, Tectona grandis, Chyloroxylon swietenia, Lannea coromandelica, Lagerstroemia parvifolia, Diospyros* spp., *Hardwickia binata, Holarrhena pubescens, Pterocarpus santalinus, P. marsupium, Terminalia chebula, Terminalia tomentosa,* and *Dalbergia latifolia*. The undergrowth contains a variety of these species.

The important grass types include *Cymbopogon* and *Themeda* spp. The deciduous forests of the Deccan Plateau also host a significant number of climbers and diverse liana. The most common kinds include *Acacia sinuata, Combretum albidum, Cissus quadrangularis, Coccinia grandis, Tinospora cordifolia, Carissa spinarum, Ziziphus oenoplia, Capparis brevispina, Hugonia mystax,* and *Ipomoea staphylina*.

Several of these forest species have both timber and medicinal value. The deciduous nature of vegetation helps to reduce evapotranspiration and rehydration of stems, which is useful for subsequent flushing or flowering. Leaf flushing occurs after the first rains and the onset of flowering following the cessation of rains.

Wildlife

In addition to flora, the dry deciduous forests of the Deccan Plateau support significant faunal wealth and are of some significance from a zoogeographic point of view. Among the mammals the dry forests of the region support are: the blackbuck (*Antilope cervicapra*), blacknaped hare (*Lepus nigricollis*), blue bull (*Boselaphus tragocamelus*), common langur (*Presbytes entellus*), common leopard (*Panthera pardus*), Indian gazelle (*Gazella gazella*), sambar (*Cervus unicolor*), sloth bear (*Melursus ursinus*), spotted deer (*Cervus axis*), tiger (*Panthera tigris*), and wild pig (*Sus scrofa*).

About 350 bird species have been recorded in the region, including endangered or rare birds like Jerdon's courser (*Rhinoptilus bitorquatus*) and Blewitt's owl (*Athene blewitti*). The tree sparrow (*Passer montana*) is recorded in the Vizag region of Andhra Pradesh and is unique to the area.

Threats and Protection

Deciduous forests in the region are undergoing rapid rates of degradation. The main reasons include fire, cattle grazing, and wood extraction for fuel. Indigenous peoples in several regions on the Eastern Ghats side practice slash-and-burn agriculture, locally called *podu*. This practice significantly affects the vegetation and, most importantly, plant succession. The immediate colonizers after podu agriculture include *Aca-*

cia spp., *Cleistanthes collinus, Kydia calycina, Dichrostachys cinerea,* and *Tarenna asiatica.* Invasive species such as *Eupatorium odoratum, Lantana camara, Ageratum conyzoides,* and *Cassia occidentalis* also dominate at some of these farmed sites.

For the conservation of wildlife, as many as 27 areas in the Deccan Plateau have been demarcated as protected zones. Some of the popular national parks in the region are Nagerhole, Bandipur, Periyar, Kanha, Biligiri Rangan, and Nagarjunasagar National Parks. Although the core areas in these preserves are well protected, the buffer zones are characterized by high anthropogenic influence and resource use. Land-use conflicts are common around park areas, and more stringent policy measures are needed to conserve floral and faunal diversity in the dry deciduous forests here.

Climate change can potentially affect both the wet and dry areas of the Deccan Plateau in ways that could adversely impact each forest stand. More rain in the wet areas and longer droughts in the more arid areas each have been projected. Either scenario would impact the habitats and species distribution, leading to forced migration and new openings for invasive species.

Krishna Prasad Vadrevu

Further Reading

Champion, H. G. and S. K. Seth. *A Revised Survey of the Forest Types of India.* New Delhi: Government of India Press, 1968.

Meher-Homji, V. M. "Tropical Dry Deciduous Forests of Peninsular India." *Feddes Repertorium* 88, no. 1–2 (1977).

Singh, K. P. and C. P. Kushwaha. "Emerging Paradigms of Tree Phenology in Dry Tropics." *Current Science* 89, no. 6 (2005).

Dee Estuary

Category: Marine and Oceanic Biomes.
Geographic Location: Europe.

Summary: A major estuary on the Irish Sea, the Dee supports important bird populations; its history of human development, along with rising sea levels, call for significant mitigation and protective measures.

Emptying into the Irish Sea, the Dee Estuary divides England and Wales on the northwest coast of Britain. It occupies a large and wide basin of some 37,000 acres (15,000 hectares) in area, originally formed by glacial erosion that took place in the ice ages that shaped much of northern Europe. The current geomorphology of the estuary is a result of its more recent geological history; sediment deposition; and a large tidal cycle, which is characterized by a rapid flood and a much longer ebb. Like many of the estuaries around the coasts of Britain, the Dee is an important area for wildlife, particularly for breeding and wintering populations of wildfowl and waders. As a result, it is designated as a legally protected wildlife site at both national and international levels.

The most widespread habitats within the estuary are intertidal areas of mudflat, sand, and salt marsh. The salt marsh forms extensive stands in the Dee Estuary, especially in sheltered areas with low levels of wave action and tidal scour (an erosion process that is carried out by the tidal movement of water).

The salt-marsh vegetation often occurs in a series of zones, from the sea toward land. On recently colonized bare seaward fringes is pioneer vegetation, consisting of glassworts and annual sea blite. This vegetation develops into a more mature salt marsh at a middle elevation on the shoreline. Finally, there is a transition to grassy vegetation at the highest levels, including swamp vegetation usually dominated by common reed or sea club-rush.

In contrast to these large areas of soft substrate, any rocky intertidal habitat is relatively scarce in the estuary. The rocky shores of the sandstone Hilbre and Little Hilbre Islands (immediately offshore and often connected to the mainland at low tide) support a variety of habitats, including rock pools, bedrock ledges, gullies, and crevices. Reefs formed by honeycomb worm are a particular feature of the lower shore of Hilbre Island. Small but important areas of maritime cliff, heath, and grassland vegetation also occur above the tide line.

Wildlife

The Dee Estuary is of major international importance for birds, with more than 130,000 individuals supported in the winter months and during migration. These birds are mostly waders, such as redshank, dunlin, and knot, although wildfowl such as teal and pintail also are common. The birds make use of the intertidal flats and salt marshes for feeding and roosting, and they commute to neighboring land, shingle spits, and islands for roosting at high tide.

In summer, the site supports breeding populations of common tern and little tern at levels of European importance. An area by Gronant, near the mouth of the estuary, is Wales's only colony of breeding little terns; they winter off the shoreline of western Africa before returning to the estuary to nest.

Besides birds, a range of other notable species are supported by the estuary, including grey seal, Atlantic salmon, lamprey, twaite shad, the nationally scarce thumbnail crab, and the endemic (found nowhere else on Earth) rock sea lavender, a perennial herb.

This red sandstone formation stands on the rocky shore of England's Hilbre Island. Areas like this provide such wildlife habitats as bedrock ledges, gullies, crevices, and rock pools. (Wikimedia/Benkid77)

Environmental Threats

Potential influences on the conservation front of the Dee Estuary come from both natural and anthropogenic processes. The Dee Estuary is a naturally accreting ecosystem, with a net inflow of sediment from Liverpool Bay. As a result, a steady increase in the extent of salt-marsh habitats is expected over time, at the expense of intertidal flats. This will decrease the amount of open, invertebrate-rich habitat on which the wintering and migratory bird populations currently depend.

Flocks of birds, whether breeding, wintering, or on passage, need areas free from disturbance in which to nest, feed, and roost. To maintain such populations that currently use the estuary, human activities such as wildfowling, recreation (e.g., angling, sailing, water skiing, windsurfing), and the operation of the shellfish fishery need to be managed appropriately. All these activities have the potential to disturb birds and, in bad weather, are likely to increase the mortality of birds, especially at favored feeding and roosting sites. To minimize this risk, attempts are being made to reduce disturbance in the estuary by voluntary agreements such as no-shooting zones, education, and visitor management.

Erosion by waves and tides threatens coastal areas here; the situation at the Hilbre Islands is exacerbated by the rising sea level resulting from warmer temperatures, a local instance in the ongoing threat of global climate change. Historical sea defenses are present, but their ability to cope with changes in the estuary will reduce over time, and the scope for managed retreat needs to be developed.

However, the ability of the estuary shores to adapt to this change is constrained in many places by coastal defenses. As a result, the intertidal area may become squeezed between rising sea levels and fixed flood barriers. This issue will require innovative approaches of coastal management, with flood defenses being relocated, or land being set aside and allowed to develop naturally into new intertidal habitats such as salt marsh.

The current and historical commercial uses of the Dee Estuary have had, and will continue to have, an effect on the estuary through their effects on the surrounding land as well as on the quality of water and sediment that enters the coastal waters here. Heavy industry, coal mining, and metal smelting have left a legacy of contaminated land around the estuary. Although some of this activity has been subject to remediation, other sites remain in poor condition.

An obvious concern for such areas is their possible erosion by the rising sea and the consequent leaching of noxious pollutants into the environment. Possible new risks also have been introduced for the bird populations in the estuary, through the development of three offshore wind farms close to the mouth of the Dee at North Hoyle, Burbo Bank, and Gwynt-y-Mor. There is concern that these sites could disturb roosting birds or increase mortality through collision with turbine rotors, especially if located in critical flight lines or on migration routes. To minimize this risk, their potential effects on the ecology of the estuary have been subject to detailed environmental impact assessment and ongoing monitoring programs.

Carlos Abrahams

Further Reading

Allen, J. R. L. and K. Pye, eds. *Saltmarshes: Morphodynamics, Conservation and Engineering Significance.* Cambridge, UK: Cambridge University Press, 2009.

Archer, Mike, Mark Grantham, Peter Howlett, and Steven Stansfield. *Bird Observatories of Britain and Ireland.* London: Poyser, 2010.

Kirby, J. S. *The Ornithological Significance of the Mostyn Docks Area of the Dee Estuary to Wildfowl and Waders.* Thetford, UK: British Trust for Ornithology, 1993.

Prater, A. J. *Estuary Birds of Britain and Ireland.* London: A & C Black, 1981.

Diamantina River

Category: Inland Aquatic Biomes.
Geographic Location: Australia.
Summary: An extensive ecosystem depends on this frequently and erratically flooding river; the diverse plant, fish, and animal species here have adapted to the extreme weather regime.

The Diamantina River flows through Queensland and South Australia, draining into Lake Eyre, the lowest point in Australia. The Diamantina begins in the Swords Range northwest of Longreach, and flows southwest to form the Warburton River at its confluence with the Georgina River. The river reaches as far as Lake Eyre only during wet years. At 560 miles (901 kilometers) in length, nearly all the Diamantina River lies in a basin used as agricultural land, especially for the grazing of cattle and sheep. It is, with Cooper Creek and the Georgina River, one of the three principal rivers of the Australian Outback's Channel Country.

The river basin is quite flat, ranging from Lake Eyre's 52 feet (16 meters) below sea level to a high point in the northeast of about 1,600 feet (488 meters). Most of the rivers in this basin flow toward the cattle country of Birdsville in the southwest. The Diamantina River flows sluggishly and has a series of wide shallow channels with no main channel for much of its length. The climate is hot and arid, settling in around 100 degrees F (38 degrees C) on summer days, with summer night averages of 75 degrees F (24 degrees C) and winter temperatures of 52–77 degrees F (11–25 degrees C). Parts of the basin sometimes see frost.

Rainfall here is concentrated in the wet season, from December to March (Australian summer), with significant rainfall exceptionally rare the rest of the year. The basin always has been subject to both erratic floods and catastrophic droughts, and in wet years, rainfall totals almost four times higher (43 inches or 1,092 millimeters) than average (12 inches or 305 millimeters) have been reported in the wettest areas. During floods, the river has reached widths of nearly 20 miles (32 kilometers). In dry years, on the other hand, total rainfall tends to be less than four inches (101 millimeters) for nearly the entire river basin.

Soils in the Diamantina River region have low phosphate content, but adequate levels of other nutrients; they are mainly brown and gray vertisols (moisture-sensitive clays), with some fluvents (flood deposits). Grasses flourish in the rainy season.

Biodiversity

Floods along the Diamantina River leave behind scattered seasonal waterholes upon which numerous plants and animals depend. The bilby (*Macrotis lagotis*) and other small animals frequently depend on waterholes to tide them over until the next unpredictable major rainfall event. Another of the animals that depends on these waterholes is the endangered dusky hopping mouse (*Notomys fuscus*), a nocturnal rodent a few inches (millimeters) long that makes its home in dunes, drinking from waterholes and eating seeds, small plants, and insects. Overgrazing and predation from feral cats have jeopardized the hopping mouse, along with competition with the more recently introduced house mouse and rabbit. The plains rat (*Pseudomys australis*), about twice the size of the hopping mouse, is likewise dependent on waterholes and also endangered as a result of habitat loss and predation by feral cats. Plains rats live in complex but shallow burrow systems in the soft soil around low shrubs. They don't drink from waterholes generally, but obtain their moisture by eating many of the small animals that do drink from them.

The Diamantina River floodplain has been designated an Important Bird Area by Birdlife International. When flooded, the river supports approximately 500,000 waterbirds. It is home to, or supports for a period of migration, significant numbers of the Nankeen night-heron, royal spoonbill, little curlew, Australian bustard, grey grasswren, inland dotterel, cinnamon quail-thrust, and pied honeyeater.

The desert gobi fish is found throughout the small ponds and creeks of the Lake Eyre basin, including the banks of the Diamantina. Rabbits are a serious threat to many of the endemic species of water-dependent plants alongside these riverbanks, which are in great jeopardy as soil erosion becomes more likely when the rabbits overgraze.

In and around the water, the environments that are home to freshwater fish, small mammals, and migratory birds are particularly vulnerable to climate change. At stake is habitat loss or fragmentation; the spread of alien species is another vector generally driven by global warming. Many of Australia's freshwater fish have historically adapted to variable or unpredictable water flow conditions; they may continue to do so. However, the rate and magnitude of projected change could outpace their ability to adapt. This is especially true for species with limited ranges or unusual habitat requirements.

Among the plant communities here, the invasive gum arabic tree (*Acacia nilotica*), a spinescent woody legume, was introduced decades ago for ornamental purposes and has since become aggressively invasive. It has been spreading almost unchecked, in part because of grazing cattle that consume the plant's nutritious pods and essentially reseed the land with their waste.

Protected Areas

Halfway between Winton and Birdsville is Diamantina National Park. It was established in 1993 and received a 2007 award from the World Wildlife Fund for its work as a nature reserve over the course of the decade, particularly in protecting the bilby, a nocturnal omnivore endemic to Australia and a threatened species. It is related to the lesser bilby (*Macrotis leucura*), which did become extinct in the 1950s.

Bilbies are about the size of rabbits, 1 to 2 feet (0.3 to 0.6 meter) long, and live in extensive burrows and tunnels. Though they do not need to drink water, they do well in riverside areas where the food sources that provide their necessary moisture thrive, including spiders, fruit, fungi, small

The Night Parrot

Diamantina National Park is one of the few sites that is home to the critically endangered night parrot (Pezoporus occidentalis), a small parrot endemic to Australia that has been seen so rarely since 1979 that no accurate guess can be made about its population. The night parrot stays close to the spinifex grass and never strays far from the river. The most recent confirmed sighting of a night parrot was in 2006 in the park.

birds and mammals, and larvae. They compete with rabbits for many of their staple food sources—and they are preyed upon by the feral cats of the region.

Diamantina National Park supports large bird populations of such species as the plains-wanderer, straw-necked ibis, white-necked heron, Bourke's parrot, black honeyeater, gibberbird, Hall's babbler, chestnut-breasted quail thrush, spinifexbird, and the critically endangered night parrot.

Bill Kte'pi

Further Reading

Bull, L. J. and M. J. Kirby, eds. *Dryland Rivers—Hydrology and Geomorphology of Semi-Arid Channels.* Hoboken, NJ: Wiley, 2002.

Kingsford, Richard. *Ecology of Desert Rivers.* New York: Cambridge University Press, 2006.

Walker, K. F., F. Sheldon, and J. T. Puckridge. "A Perspective on Dryland River Ecosystems." *Regulated Rivers: Research and Management* 11, no. 1 (1995).

Dnieper River

Category: Inland Aquatic Biomes.
Geographic Location: Eastern Europe.
Summary: The Dnieper River is Europe's fourth-longest river, the main freshwater source in Ukraine, and a key feeder of the Black Sea.

Originating in the Valdai Hills, approximately 93 miles (150 kilometers) northeast of Smolensk, Russia, the Dnieper River travels generally southward for more than 1,422 miles (2,290 kilometers) before releasing its waters into the Black Sea. The Dnieper River watershed lies in three Eastern European countries: Russia, Belarus, and Ukraine. Considered the fourth-longest river in Europe, and draining an area of 196,526 square miles (509,000 square kilometers), the river system is known for the dams and reservoirs that provide energy, industrial development, and irrigation water to urban and agrarian populations of Eastern Europe.

However, the increase in large-scale industrial activities has come at a steep price for this hydrologic system, as pollution concentrations and unsustainable resource-use levels have degraded the waters of the Dnieper. This, in turn, has resulted in deleterious impacts to the entire riparian ecosystem and to the Black Sea.

Climate and Hydrology

Dominated by a moderate temperate continental climate, temperature and precipitation varies from north to south along the Dnieper riparian system. Temperatures upstream are more affected by this continental system, whereas at the delta and mouth of the river, the climate is moderated by the inland Black Sea. On average, temperatures are cold in the winter, with average December through January temperatures of 27 degrees F (minus three degrees C), and trend from warm to very warm in the summer. The average July temperature is 75 degrees F (24 degrees C).

Precipitation is variable, but predominantly arrives as rain in the summer and fall. A moderate amount of snow collects in the lower portion of the region in the winter; in the upper reaches, the winter is long and areas of the river may freeze. The mean annual precipitation in the upper Dnieper basin is approximately 28 inches (71 centimeters).

The Dnieper River hydrology has been documented for more than 250 years. Spring snowmelt in the upper basin area historically provided a pulse of freshwater for higher flows. Approximately 60 percent of the annual runoff arrives in March through May. The confluence of many large tributaries, such as the Inhulets, Sozh, Berezina, Ros, Sula, and the Pripyat Rivers, contributes to the seasonal input. Along the route, the river channel narrows and widens from 131 feet (40 meters) to over 1,968 feet (600 meters). The annual average discharge is around 2,184 cubic yards (1,670 cubic meters) per second, and carries large quantities of dissolved sediment into the Black Sea.

The river was once known for the zones of rapids that made it difficult to navigate. The rapids in the upper reach cut through limestone cliffs; the famous Dnieper rapids in the middle region were situated above Zaporizhzhya. However, the rapids and the natural flow of the Dnieper have long been altered by hydroelectric dam and reservoir construction. This affects the natural flow rate and peak discharge times, which now accommodate irrigation and energy production.

Geography and Plant Types

The Dnieper River is a meandering river system, which typically has a sinuous stream channel moving through alluvial (stream-deposited) sediments, turning and twisting over fairly low elevation gradients. Characteristic of many long river systems, the Dnieper has many tributaries and is divided into three reaches: the upper, middle, and lower Dnieper. The upper reaches of the river, from the source in central Russia to north of the Ukrainian city of Kiev, starts out at an elevation of 722 feet (220 meters) and gradually descends through swampy and forested terrain. The watershed here descends and carries itself through the gently sloping uplands of Russia and Belarus. Forests of willow and alder with lush meadows and marshes dominate the banks of the upper riparian ecosystem. This section is the most heavily forested of the three Dnieper regions, with as much as one-fourth of the watershed covered in forests; it is also the region with the greatest extent of swamp vegetation.

The middle portion, from the environs of Kiev to Zaporizhzhya, Ukraine, and the lower Dnieper, from Zaporizhzhya to the Black Sea, flow through Belarus and Ukraine. These two longer reaches have the most sinuosity and a predominance of features known as oxbow lakes and meander scars. Oxbow lakes are created when curves of the main river are cut off, resulting in crescent-shaped lakes. Meander scars are fertile, sediment-filled pockets along the floodplain of a river, created after the main channel of the river moves.

In the center of the basin here, an ecotone (boundary between two ecological regions) is established between the moist forests and the arid steppe. The lower reaches of the basin broadens out throughout the Ukraine and moves through the arid steppe, winding its way through swampy marshes, with reeds and sand-bar islands.

The lower Dnieper passes through more arid lands, in the form of the steppe country of the Black Sea Depression, a lowlands area of natural terraces cut by shallow rivers and also character-

Tourist boats docked along a bend of the Dnieper River in the port city of Kiev, Ukraine. The river is the main source of freshwater in the Ukraine. (Thinkstock)

ized by marshy seasonal floodplains. Here, too, is found the Oleshia Sands, formed by sedimentary deposits when the mouth of the Dnieper was situated here, well north of the present-day delta. Sparse native vegetation has been supplanted in many cases here by irrigated cultivation of crops such as grapevines and fruit trees. The Dnieper delta carries four main streams; broad floodplains separate these shallow brackish flows, known collectively as the Dnieper-Boh Estuary, as they enter the Black Sea.

Biota and Environmental Concerns

The lower reach of the Dnieper provides irrigation water to the well-known black soils (*chernozems*) of the Ukraine where wheat, corn, rice, and other agricultural commodities are grown. Fishing in the Dnieper is both a recreational and commercial activity, with more than 60 species of fish living in the waters. Commercially important species include pike, catfish, barbel, perch, common bream, vimba, roach, and ruff.

However, the Dnieper has lost many of the anadromous fish species—such as sturgeon and shad—that once prospered here, in the wake of its segmentation by aggressive dam construction in the 20th century. Characteristic fish today include the Black Sea goby (*Romanogobio belingi*), the beardless tadpole goby (*Benthophiloides brauneri*), and the estuarine perch (*Sander marinus*).

More than 120 species of mollusks—at least 50 in class *Bivalvia* and 70 in class *Gastropoda*—are known in the Dnieper River biome, many vital to the aquatic food chain here. This breadth of species includes the zebra mussel (*Dreissena polymorpha*), widely known around the world as an invasive freshwater species, but native to the Dnieper region. Other thriving mollusks here include the swollen river mussel (*Unio tumidus*) and the sand gaper (*Mya arenaria*).

Among avian species of the Dnieper River biome are great snipe (*Gallinago media*), curlew (*Numenius arquata*), marsh sandpiper (*Tringa stagnalis*), lapwing (*Vanellus vanellus*), and redshank (*Tringa totanus*).

One of the most polluted waterways in a fairly densely populated region of the former Soviet Union, the Dnieper had long been inundated with industrial, agricultural, and municipal runoff. The effects of such sources were trumped in 1986 with the core meltdown of the Chernobyl Nuclear Power plant, located on the Dnieper's tributary, the Pripyat, upstream of Kiev. Wind conditions during the disaster actually favored much of the Dnieper watershed, as radiocative contaminants were generally swept away to the northwest. Still, the immediate release and bioaccumulation of radioactive fallout materials into the Dnieper riparian ecosystem has had a large impact on the fishing, irrigation, and drinking water of the middle and lower reach of the river. Remediation efforts have had some positive impact, as have increased ecological advocacy within the post-Soviet era.

These efforts on behalf of the Dnieper continue to be balanced against the climactic impact of warming temperatures, which can increase rainfall that could lead to more frequent flooding overall and to flash floods during rainy periods. On the other hand, the warmer temperatures would likely expand the farming areas for winter wheat, the largest crop in Ukraine, as well as other crops like barley and potatoes—but also require more irrigation to compensate for the periodically decreased rainfall. The increased temperatures could also drive in insects not usually seen in this temperate climate.

Jennifer K. Lipton

Further Reading

Alexandrov, Boris, et al. "Trends of Aquatic Alien Species Invasions in Ukraine." *Aquatic Invasions* 2, no. 3 (2007).

Blinnikov, Mikhail S. *A Geography of Russia and Its Neighbors.* New York: Guilford Press, 2011.

Kovalssky, V. and G. M. Henebry. "Change and Persistence in Land Surface Phenologies of the Don and Dnieper River Basins." *Environmental Research Letters* 4, no. 6 (2009).

Onishi, Y., S. L. Kivva, M. J. Zheleznyak, and O. V. Voitsekhovich. "Aquatic Assessment of the Chernobyl Nuclear Accident and Its Remediation." *Journal of Environmental Engineering* 133, no. 11 (2001).

Dongsha Islands

Category: Marine and Oceanic Biomes.
Geographic Location: South China Sea.
Summary: The Dongsha Islands are a group of three islands forming an atoll in the South China Sea. They are threatened by development, climate change, and overfishing.

The Dongsha Islands (or *Tungsha,* literally meaning "east sand" in Chinese), formerly known as the Pratas, are a group of three islands located 211 miles (340 kilometers) southeast of Hong Kong in the northeastern South China Sea. Together, they comprise a circular coral atoll composed primarily of clastic coral and reef flats approximately 15 miles (24 kilometers) in diameter, enclosing a lagoon about 10 miles (16 kilometers) in diameter. The island group includes Dongsha Island, which is above sea level, about 1 mile (1.6 kilometers) long, and a little over 0.5 mile (0.8 kilometer) wide, and Northern Vereker and Southern Vereker atolls, both of which are below sea level.

Dongsha is governed by Taiwan, the Republic of China. The closest Taiwanese city is Kaohsiung, 276 miles (444 kilometers) to the northeast. Uniquely positioned along important trade routes, Dongsha has a human history that dates back to ancient times. Today, it faces numerous challenges primarily from human activity, such as overfishing, pollution, and development.

Known as the Imperial Crown of the South China Sea, Dongsha Island rises 25 to 30 feet (eight to nine meters) above sea level; it is covered by coarse coral sand, shrubs, tropical vines, and low bushes. The area has a subtropical monsoon climate, with year-round temperatures averaging 79 degrees F (26 degrees C). The region is affected by typhoons, the northeast monsoon, and the East Asian rainy season, commonly called the plum rains.

Biota

The island supports a wide variety of reptiles, birds, mammals, fish, coral, and flora, including 168 plant, 250 coral,140 bird, and more than 600 fish species. Flora include silver silk trees (genus Albizia), tung trees (Aleurites fordii), coconut trees, and marine plants. The majority of the bird species observed on Dongsha are waterfowl or migratory birds; they include little tern (*Sternula albifrons*), Chinese egret (*Egretta eulophotes*), turnstone (family *Scolopacidae*), and gull-billed tern (*Gelochelidon nilotica*).

The waters surrounding Dongsha support a variety of fish, such as grouper (subfamily *Epinephelinae*), parrotfish (subfamily *Scarinae*), and sea bream (family *Sparidae*), as well as a host of invertebrates such as mollusks, starfish, and crustaceans like mud crabs, hermit crabs, and rock lobsters. The beaches of Dongsha Island also provide an important nesting ground for marine turtles. In 2011, a newly recorded species of sea slug, *Ercolania subviridis,* was discovered at Dongsha, living in the chaetomorpha algae. The discovery of this tiny slug moved its known range of global distribution further south.

Human Interaction

Dongsha Island occupies a unique position between Taiwan, Hong Kong, and the Philippines; it has been an important location for sea travelers. Before becoming part of the Dongsha Marine National Park, the area was under military control, and today, the Dongsha Coast Guard Administration maintains a presence on the island. Once discovered during the ancient Han Dynasty, Dongsha Island became an important point along trade and fishing routes through the Taiwan Strait, which separates Taiwan from mainland China, and the

A Japanese paradise flycatcher (Terpsiphone atrocaudata) on Dongsha Island in 2011. Like most bird species on the island, it is mainly a migratory species. (Flickr/K.C. Hung)

Bashi Channel between Y'Ami Island of the Philippines and Orchid Island of Taiwan.

Lying on a direct route between Manila and Hong Kong, Dongsha is the site of many shipwrecks, particularly on its southeast side, which is covered by a thick fog during the northeast monsoon, which occurs annually around September. To increase maritime safety in the region, a lighthouse, radio station, and meteorological observatory were built on Dongsha in the early 1900s. During its wartime occupation by Japan, Dongsha was a site of significant hunting and phosphorous mining, activities that negatively affected bird populations.

Environmental Threats and Response

Ongoing threats to the Dongsha atoll include climate change, coral bleaching, and pollution. While Dongsha Island is biodiverse, the numerical population of each species is relatively low, due primarily to human pressures in the form of development, mining, hunting, overfishing, pollution, and to the introduction of invasive species.

Concerns about the health of this rich environment led to the formation in 2004 of the Dongsha International Research Station by the Kaosiung municipal government, followed three years later by the creation of Dongsha Marine National Park, the seventh and latest national park of Taiwan. Managed by the Ministry of the Interior, the park comprises 1,365 square miles (3,535 square kilometers) of mostly open water, with Dongsha Island representing the terrestrial portion. After Kenting National Park, it is the second marine ecological protection area in Taiwan, and represents the first area in the nation dedicated exclusively to the protection of a marine environment.

Global warming effects are a growing concern here. Rising sea level and increased likelihood of more severe typhoon events means that coral, intertidal, and shoreline habitats could suffer repeated and lasting damage, both from outright battering and flooding, and from salt intrusion. Seawater-temperature rise will also jeopardize the coral, as coral bleaching is directly associated with higher temperatures. Unpredictable changes in marine species balance is also projected, as several fish types have already been recorded as migrating further north across the South China Sea, presumably fleeing from higher average seawater temperatures near the equator.

Overfishing in particular has taken a heavy toll on Dongsha. Surveys conducted between 1994 and 1998 revealed that almost 8,000 fishing vessels from China, Hong Kong, Vietnam, and the Philippines were observed fishing near Dongsha, using destructive methods such as gill nets, longlines, purse seine nets, dynamite, and cyanide. Some estimates show that around 1 ton (0.9 metric ton) of dynamite, 9 tons (8 metric tons) of mercury and batteries, and 50 tons (45 metric tons) of cyanide are dumped in the waters around Dongsha every year.

Attention to these issues, and higher-visibility conservation efforts, may help to turn the tide. The endangered green sea turtle (*Chelonia mydas*), which had previously disappeared from the island, was spotted in 2011 laying eggs during the breeding season, indicating the success of Taiwan's "blue sea ad hoc" conservation program, which prevents foreign ships from operating near Dongsha.

Reynard Loki

Further Reading

Chen, Rogge, Jeffery Wu, and Jamie Wang. "Endangered Sea Turtle Reappears on Dongsha Island." *Taiwan News*, December 13, 2011.

Collingwood, Cuthbert. "The Natural History of Pratas Island in the China Sea." *Quarterly Journal of Science* 4, no. 1 (1867).

Lee, Ling-ling, ed. "Dongsha Marine National Park." *Society for Wildlife and Nature International Conservation Newsletter* 15, no. 1 (2007).

Don River

Category: Inland Aquatic Biomes.
Geographic Location: Eurasia.
Summary: The Don River, from antiquity considered the dividing line between Europe and Asia, is one of Russia's major rivers. The

navigable river connects central Russia with the world's oceans.

The Don River rises in the Central Russian Upland, at an elevation of about 623 feet (190 meters) above sea level, and flows generally southward through European Russia and Ukraine for 1,161 miles (1,870 kilometers) before emptying into the Sea of Azov, a lobe of the Black Sea in southeastern Europe. The primary tributary of the Don is the Donets; other tributaries include the Kasivaya Mecha, Sosna, Cornaya Kalitva, Chir, Voronezh, Khopyor, Medvedits, Ilovlya, Sal, and Manych.

Hydrology and Climate

The Don is the third-largest river in the European section of the former Soviet Union, with a drainage area of about 262 square miles (680 square kilometers). It has no natural lakes, and is fed primarily by melted snow. The average rate of the Don is about 2.16 cubic miles (900 cubic meters) per second at its mouth, with considerable variation over the course of the year.

In general, water levels are highest in the spring and lowest in the autumn and winter; about 75 percent of the Don's flow occurs in March and April. The Tsimlyansk Reservoir, completed in 1953, was constructed to control the Don's annual flooding in its central section; it has an average depth of almost 32 feet (10 meters), and a maximum width of nearly 24 miles (40 kilometers).

The most important city on the Don is the port of Rostov-na-Donu, which lies about 29 miles (48 kilometers) from the mouth of the river. The Don has been a major trading channel since the time of the ancient Scythian nomadic peoples; it is navigable for approximately 851 miles (1,370 kilometers), and seagoing ships can use it up to Rostov-na-Donu. The river's transport function is enhanced by the Volga-Don Canal, completed in 1952, forming part of the Unified Deep Water System of Russia. As the Sea of Azov is connected to the Black Sea, and the Black Sea to the Mediterranean Sea, the Don is linked to all the world's oceans and is navigable for most of the year.

The upper Don flows along the eastern edge of the upland through a narrow valley with its right bank rising more than 295 feet (90 meters) above the river. This high right bank is a consistent feature of the river over most of its course. The Don drops little in elevation, just over 620 feet (about 190 meters), particularly during its upper course, resulting in slow water flow. The upper Don valley is located in the eastern European forest steppe, and is characterized by trees such as the oak, field maple, and ash. The river broadens south of Dankov, and at the town of Voronezh is joined by its first major tributary, the Voronezh River.

Below the town of Liski, the Don runs through a trough between the Kalach uplands and the hills of Belogorye, with an extensive floodplain on the left bank and the high right bank persisting. The Don makes a sharp bend below Medveditsa and broadens. In this region, it is joined by the Donets, the river's chief tributary, about 93 miles (150 kilometers) from Rostov, and by another tributary, the Manych. The lower Don basin is part of the grassland plains, or steppes, called the Pontic-Caspian steppe. It is characterized by shrublands, meadows, and riparian forests. The Don connects with the Black Sea through a series of bays and seas, beginning with the Gulf of Taganrog in the northeastern corner of the Sea of Azov, which in turn is connected to the Black Sea through the Kerch Strait.

The climate of the Don basin is continental, with annual precipitation ranging from 23 inches (584 millimeters) in the north to 15 inches (380 millimeters) in the south; average temperatures are in the range of 66–72 degrees F (19–22 degrees C) in July, and 12–18 degrees F (minus 11 to minus 8 degrees C) in January. The upper region of the Don freezes for an average of 140 days annually, while near Rostov it freezes for an average of 90 days per year. Increased use of the Don for irrigation has resulted in reduced flow, which was estimated in 1975 already to be 20 percent below 1950 levels. Strong offshore winds at the mouth of the Don also can reduce the water level in the river's lower reaches. Reduced water flow due to intense irrigation has increased the salinity of the Sea of Azov from 10–12 percent to 13–14 percent in the 1970s, which diminished fish catches and biological productivity; salinity in the Sea of Azov has since experienced incremental decline.

Biota and Human Impacts

The variety of animal and plants species are diverse. Mammals include the Asiatic black bear, brown bear, Siberian tiger, Amur leopard, lynx, wolf, wild boar, roe deer, and moose. Sturgeon, salmon, sea trout, and herring are fairly common in the river proper. Steller's sea eagle is among the raptors patrolling the skies.

The Volga-Don Canal has a length of just over 60 miles (100 kilometers) and includes multiple locks as well as pumps to maintain the water level. It was completed in the mid-20th century (although attempts to create such a canal date back to at least the 16th century) and began operation in 1952. The city of Rostov-na-Donu, located on the Don near the Sea of Azov, is the capital of the Rostov region and a major industrial and trading area. A canal has been dredged from this point to take ships through the Don delta; the city of Azov (formerly Tanais) lies at the southern end of the Don delta.

The size of the Don River means the environmental impact of the waterway changes as it flows. The middle basin area is dotted with mines, leaving mine waste in its wake, including sulfides as well as metals (arsenic, iron, copper, lead, manganese, cadmium, chromium, zinc, nickel, mercury) that occur naturally in the ore body and leave toxins, which accumulate in plants and animals. The polluted waters, which affect life along the banks, are being addressed by local nongovernment organizations such as Clean Water Foundation, Green World, Youth for Environment, and volunteers. Other efforts include the establishment of the nongovernmental organization, the Centre for Ecologic Adversity Counteraction (CEAC), which promotes its agenda with media and educational campaigns.

Effects of global warming on the Don River biome are likely to include an acceleration of the already-documented trend toward drying of semiarid lands in the region. This, in turn, is projected to exacerbate the problem of dust storms and wind-driven erosion of ecological assets including riverbank structure, unforested hillsides, and agricultural lands. Additionally, the resultant dust storms could create higher turbidity and sedimentation within the river and its marshlands, to the detriment of established natural communities there. Forested areas of the biome are anticipated to undergo dominant-species migration, possibly with scattered islands of intrusive tree species linking up and overtaking the major stands of today's dominant trees. This will precipitate or accelerate the habitat pressures on the fauna that depend on such flora communities.

Sarah Boslaugh

Further Reading

Blinnikov, Mikhail S. *A Geography of Russia and Its Neighbors*. New York: Guilford Press, 2011.

Lebedeva, E. A. "Waders in Agricultural Habitats of European Russia." *International Wader Studies* 10, no. 1 (1998).

Levich, A. P., E. A. Zaburdaeva, V. N. Maksimov, N. G. Bulgakov, and S. V. Mamikhin. "The Search for Target Values of Quality Indices for Bioindicators of the Ecological State and Environmental Factors: Case Study of Water Bodies of the Don River." *Water Resources* 36, no. 6 (2009).

Dry Valleys Desert

Category: Desert Biomes.
Geographic Location: Antarctica.
Summary: Scientists use this unique, hyperarid polar desert in the middle of Antarctica to study life forms in extreme environments.

The Antarctic Dry Valleys, also called McMurdo Dry Valleys, comprise the largest ice-free region of Antarctica, at a total of 5,792 square miles (15,000 square kilometers). This hyperarid polar desert, located roughly 2,200 miles (3,541 kilometers) south of New Zealand and extending toward McMurdo Sound, is perhaps the most extreme desert in the world. Temperatures average 1.4 degrees F (minus 17 degrees C), and precipitation is 4 inches (100 millimeters) per year. Snowstorms here typically bring less than 0.4 inch (10 millimeters) of snow at a time.

Geology and Hydrology

The region has many special geologic features, including glaciers, mountains, lakes, seasonal streams, unique soil formations, permafrost, and sand dunes. The valleys are nestled between the 1-mile-high (1.6-kilometer-high) Transantarctic Mountains, and were formed when glaciers retreated, carving them out of the mountains. The main ice-free valleys are called Taylor, Write, McKelvy, Balham, Victoria, and Barwick; they all have similar characteristics. The valleys are each 3 to 6 miles (5 to 10 kilometers) wide and from 9 to 93 miles (15 to 150 kilometers) long. They cause dense, cold air coming over the mountaintops to sink. The temperature gradient from the bottom of the valleys to the top of the mountains creates high winds in the area, and any water vapor evaporates as the sinking cold air warms in a process called katabatic ("downhill") heating. Liquid water is the main control on biological, physical, and chemical processes in this polar desert, much as it is such a key factor in life processes elsewhere. Million-year-old ice has been found buried under soil layers here; the chemistry of this ice and of the air bubbles trapped in it can yield clues to past climate conditions and today's climate change.

The McMurdo Dry Valleys as seen from space. The extreme environment of the dry valleys has been used as a testing ground for equipment for Mars rovers. (Thinkstock)

Some of the valleys have lakes at their floor that remain frozen most of the year. Much of this glacial meltwater, an interesting component of the watersheds, fell as precipitation thousands of years ago. The largest river in Antarctica, the Onyx, runs through the Dry Valleys. For six to 14 weeks during the summer, small streams form from water melting off the glaciers; this carries down to recharge the lakes. The streams also carry scant nutrients and salts into the valleys. Smaller ponds here can contain so much salt that they remain liquid water, often under the frozen surface. The lakes of the Dry Valleys contain very high concentrations of nitrous oxide, higher than measured in any other lakes in the world, because they are not well mixed and the gases are trapped.

The Dry Valleys have three microclimate zones—the coastal thaw, inland mixed, and stable upland—which are differentiated by summertime measurements of air temperature, soil moisture, and relative humidity. These climate factors vary subtly but create noticeable differences in geologic features, including large-scale gullies, midsize ice wedges, and fine-scale features such as wind erosion effects.

Biodiversity

Little can survive in these harsh conditions except extremophiles, organisms that thrive in exceedingly harsh environments. Some of the extremophiles found in the Dry Valleys include lichen and mosses, yeasts, microbes, and nematodes (microscopic worms that live in the soils). Rotifers and tardigrades are other tiny invertebrate animals found in the valleys. The Dry Valleys may be the only place in the world where nematodes are found at the top of the food chain—indicating a very simple food chain indeed. Many of the invertebrates within the Dry Valleys are known to travel and spread primarily by wind.

The valleys have a special designation as a protected area because of some of the unique bacterial life forms here, including specially adapted cyanobacteria. Scientist Craig Cary has reported that the soils of the Antarctic Dry Valleys have a surprisingly wide variety of unique microbes. Much of the research in this area has been conducted using molecular DNA analysis; results have in part debunked the old hypotheses that the soils here had almost no biota.

Some bacteria live in the moist interiors of rocks, where they produce food and energy directly through metabolizing the mineral nutrients within their surrounds; this process has yet to be fully understood. These rock dwellers, or endoliths, are cyanobacteria, an algae that can survive in numerous tough environmental niches; here it lives in rocks or in the pores between minerals in the soil. One remarkable bacterium found within the Taylor Glacier apparently uses no oxygen as it metabolizes iron and sulfur deposits.

It was once thought that various of these organisms turned the water red here, whereas scientists now know that iron oxides provide the red tint to the glacial meltwaters that form the waterfalls known as Blood Falls. Fungal microflora, too, have been found in the Dry Valleys, living in small bits of soil in cracks of rocks, inside colonies of moss, in sediment deposited along shorelines, in calcium-carbonate formations, and on microbial mat formations. Lichens grow primarily in an altitude range of 3,281–6,562 feet (1,000–2,000 meters), as this zone has more moisture from frequent cloud cover.

Research

Many scientists travel to the Dry Valleys to study unique life forms. The United States maintains a Long Term Ecological Research site in the McMurdo Dry Valleys, which allows researchers to work together on studies of aquatic and terrestrial ecosystems in this unique setting. The study site joined the National Science Foundation's Long Term Ecological Research Network in 1993 and is funded through the Office of Polar Programs in six-year funding periods. Strict rules govern what types of activities can be carried out, as well as who can carry them out and where, in order to protect this fragile ecosystem.

Studying the changing climate in Antarctica is important to understanding global climate change. It is expected that global warming will be faster and more intense in polar regions, yet records show that the Dry Valleys region has cooled by 1.3 degrees F (0.7 degree C) each decade since 1986. The Dry Valleys have their own weather monitoring network: the Long Term Ecological Research Automatic Weather Network, consisting of 13 permanent weather stations that record the weather every 30 seconds. The data provides scientists with long-term weather records to use in climate studies.

Geologists and astrophysicists also use the Dry Valleys to infer what may one day be found on Mars as the valleys are considered the most similar place on Earth to the Martian surface. Both areas have generally cold, dry climates and similar landforms, such as mountains, valleys, and ancient stream beds. The arid climates create a high concentration of salts in the soils that can affect the temperature at which water changes from liquid to ice. Geologists are interested in the Dry Valley's land forms, such as ice wedges and how they change over time; this too enables researchers to infer more details about the history of Mars and its climate. Additionally, new scientific equipment is tested in the Dry Valleys before being deployed on rovers to Mars so that it is put through its routines in an extreme environment before being deployed, providing the chance to uncover shortcomings and adapt the technology to correct form before it is sent to Mars.

Gillian Galford

Further Reading

Cary, S. Craig, Ian R. McDonald, John E. Barrett, and Don A. Cowan. "On the Rocks: The Microbiology of Antarctic Dry Valley Soils." *Nature Review Microbiology* 8, no. 1 (2010).

Friedmann, E. Imre and Roseli Ocampo. "Endolithic Blue-Green Algae in the Dry Valleys: Primary

Producers in the Antarctic Desert Ecosystem." *Science* 193, no. 4259 (1976).
Long Term Ecologic Research Network. "McMurdo Dry Valleys." 2012. http://www.mcmlter.org.
Marchant, D. R., et al. "Antarctic Dry Valleys: Microclimate Zonation, Variable Geomorphic Processes, and Implications for Assessing Climate Change on Mars." *Icarus,* 192, no. 1 (2007).

East African Montane Forests

Category: Forest Biomes.
Geographic Location: Africa.
Summary: This group of mid- to high-altitude forests hosts species diversity in unique habitat niches—but growing human pressures are stressing their sustainability.

The East African Montane Forests biome is located in Sudan, Kenya, Uganda, and Tanzania. These woodlands exist only in moderate- to high-altitude regions of the volcanic mountain chain called the Eastern Arc Mountains. They thrive where altitude is roughly 3,300–11,500 feet (1,000–3,500 meters), which moderates the prevailing climate. Temperatures often fall below 50 degrees F (10 degrees C) here in July and August, when frost can occur. In the warm season, temperatures can exceed 86 degrees F (30 degrees C).

Typically, rainfall is in the range of 47–78 inches (1,200–2,000 millimeters) per year. Rainfall occurs mostly during the bimodal rainy seasons, which occur from March to June, and then October through December. Because of their elevation, these locations receive more rainfall than the surrounding lowlands, making them a good home for montane and submontane forests. Rain does not fall much above 9,840 feet (3,000 meters) on these mountains; the highest elevation areas are considered desert. The soil is typically very fertile and desirable for agriculture.

Geologic movement of the Africa tectonic plate has caused the Rift Valley to form, essentially as a large crack in the Earth's crust, which gave way to the formation of volcanoes and ultimately the current vegetation regime.

The East African Montane Forests biome today is spread over more than 25 forest patches, the northern extent of which is Mt. Kinyete of the Imatong Mountains in southern Sudan. The forest forms the headwaters of the White Nile; the southern range extends through southern Kenya into Tanzania. The biome includes Mt. Meru, Mt. Kilimanjaro, Ngorongoro Crater, and the Marang forests; parts of the ecosystem extend eastward into eastern Uganda. Many of these locations drain into Lake Victoria, and eventually flow into the Nile River.

Endemism

The East African Montane Forests biome has moderately high species richness, and some endemic species (those found nowhere else on Earth). Scientists believe that these environments have not existed long enough for speciation to create as rich a biodiversity level as seen in other zones in this part of Africa.

Still, there are an estimated 2,000 plant species growing in the East African Montane Forests biome; at least 800 of them are apparently endemic here. These forests are the centers of global endemism for various species of the African violet (*Saintpaulia*) and busy lizzy (*Impatiens*). Other notable endemic plants are Usambara violet (*Saintpaulia ionantha*), msambo tree (*Allanblackia stuhlmanni*), and a large wild nutmeg (*Cephalosphaera usambarensis*). There are many endemic bryophytes, or nonvascular plants, within the region. Many of the them are found within a single mountain range, such as, the Usambara Mountains in Tanzania, which are home to 50 endemic tree species.

There are eight endemic bird species, some of which occur only within two or three montane locations, including Aberdare cisticola (*Cisticola aberdare*), Abbott's starling (*Cinnyricinclus femoralis*), Kenrick's starling (*Poeoptera kenricki*), Aita thrush (*Turdus helleri*), Usambara akalat (*Sheppardia sharpie*), Usambara eagle-owl (*Bubo vosseleri*), and the banded sunbird (*Anthreptes rubritorques*).

There are many endemic small mammals, including mostly shrews (*Crocidura gracilipes, C. raineyi, C. ultima, Surdisorex norae,* and *S. polulus*), and rodents (*Grammomys gigas, Tachyoryctes annectens, T. audax*). There are more small mammals that are close to endemic, including Jackson's mongoose (*Bdeogale jacksoni*), Abbot's duiker (*Cephalophus spadix*), the sun squirrel (*Heliosciurus undulatus),* and eastern tree hyrax (*Dendrohyrax validus*). Other notable, nonendemic mammals include the Angolan black-and-white colobus (*Colobus angolensis*), and forest-dwelling populations of the African buffalo (*Syncerus caffer*) and Harvey's duiker (*Cephalophus harveyi*).

As many as 19 endemic reptiles are known here, mostly chameleon species (*Chamaeleo* and *Rhampholeon* families), worm snakes (*Typhlops* family), and colubrid snakes including a montane viper (*Vipera hindii*). There are just two endemic amphibian species.

The distribution of species through the biome, or its biogeography, suggests that northern Kenyan forests were once linked with forests of west and central Africa. The fact that some of the mountains have endemic species not found anywhere else in the biome suggests that species evolved in that ecoregion independent of other areas. A classic example is a group of 45 species otherwise found only on the island of Madagascar. This suggests the East African Montane Forests biome has a very different biogeographical history than other East African mountain zones.

Protected Lands

Much of the East African Montane Forests biome exists today in forest reserves and national parks. The most prominent locations are Mt. Kenya, the Aberdare Range, the Mau Complex, Kakamega Forest, Mt. Kilimanjaro, Mt. Meru, Mt. Elgon, and Ngorongoro. Mt. Kenya and Mt. Kilimanjaro are listed by the United Nations as World Heritage sites. The previous forest biome outside of protected areas has been converted almost entirely to agriculture or other human uses. At lower elevations, such lands are now tea and coffee plantations, conifer plantations, and other types of agriculture. Much of the habitat loss occurred during the colonial period in the late 19th and early 20th centuries.

Pressure on these lands has increased still more since then, due to rising local populations and their needs for more agricultural land. Another problem has been conflicts between humans and wildlife. Humans are developing agricultural land near the protected areas—and large mammals wandering outside the zone can do a lot of damage to the adjacent farm. Humans also enter the protected areas and illegally harvest bushmeat, timber, and other commodities.

Climate-change-driven temperature increases are exerting pressure on chronically underfed human populations here. Intensification of these circumstances in years ahead could cause social unrest, political upheaval, and continental hostilities—little of it good for habitat sustainability.

Women harvesting green beans for export at a vegetable cooperative in East Africa in 2005. Conversion of forests to agricultural land is threatening the East African Montane Forests biome. (USAID/ Kristina Stefanova)

Political issues have been rising, for example, over the illegal harvesting of wood for charcoal. Charcoal is created in kilns near the forests and sold to urban and rural communities that use it as a fuel source for cooking. Corrupt politicians and local townspeople can make a lot of money through charcoal production or through bribes. Poverty in villages can be a factor in the degradation or deforestation of these areas. Regions receiving little investment or development aid will tend to use the abundance of the forests to gather resources they need, such as food sources, fuel, and timber.

Some communities use ecotourism as a way to solve local poverty and protect the environment. To be successful, ecotourism must be supported by the local village; it is often villagers who lead the tours. The system invests the people in their community, keeps money out of the hands of large companies and corrupt officials, and lets money remain in local hands. Townspeople then have more incentive to preserve the forest and keep the revenue stream within the community, reducing their need for illegal gathering within the forest.

GILLIAN GALFORD

Further Reading

Hamilton, A. C. and R. A. Perrott. "A Study of Altitudinal Zonation in the Montane Forest Belt of Mt. Elgon, Kenya/Uganda." *Plant Ecology* 45, no. 2 (1981).

Lovett, J. C. and I. Friis. "Patterns of Endemism in the Woody Flora of Northeast and East Africa." In L. J. G. van der Maesen, X. M. an der Burgt, and J. M. van Medenbach de Rooy, eds. *The Biodiversity of African Plants.* Wageningen, Netherlands: Kluwer Academic Publishers, 1996.

Lovett, J. C. and S. K. Wasser. *Biogeography and Ecology of the Rainforests of Eastern Africa.* Cambridge, UK: Cambridge University Press, 2008

Sayer, J. A., C. S. Harcourt, and N. M. Collins. *The Conservation Atlas of Tropical Forests: Africa.* Cambridge, UK: IUCN and Simon & Schuster, 1992.

East China Sea

Category: Marine and Oceanic Biomes.
Geographic Location: Asia.
Summary: The East China Sea is a highly productive, temperate body in the heart of east Asian waters. Overfishing and climate change present current and future issues of concern.

The East China Sea is the major sea off the eastern coast of China; it is a marginal sea of the North Pacific Ocean. The East China Sea is bound on the west by mainland China, where the large estuary of the Yangtze River flows in at Shanghai, via Hangzhou Bay. To the south, it is bound by the Taiwan Strait (the channel to the South China Sea) and by Taiwan. To the east, the East China Sea is separated from the Philippine Sea and greater North Pacific Ocean by the arc of the Ryukyu Islands chain, and to the northeast by Kyushu, southernmost of Japan's "big four" main islands. Directly north lies the Korea Strait (channel to the Sea of Japan), the South Korean island of Jeju, and the Yellow Sea.

The East China Sea is most often considered as including the Yellow Sea, which is tantamount to a large gulf opening off its northwest, between mainland China and the Korean Peninsula. The total area of the sea is about 482,000 square miles (1.25 million square kilometers). In addition to the Yangtze, about 40 other large rivers flow into the sea, including the Qiantang, Ou, and Minjiang. Most of the sea lies over the continental shelf, which extends out as much as 300 miles (480 kilometers) to the east. The East China Sea therefore is relatively shallow, seldom more than 656 feet (200 meters) deep. The Okinawa Trough, parallel to the Ryukyu Islands, is the deepest area, with depths to about 8,858 feet (2,700 meters).

Because the East China Sea is in a subtropical zone, the average annual temperature is 68–75 degrees F (20–24 degrees C); the annual temperature differential is 45–48 degrees F (7–9 degrees C). The average salinity of the East China Sea is 31–32 percent, and rises to 34 percent in the eastern areas. This is somewhat less salty than most global oceans, but not out of the ordinary for coastal seas.

One of the vital characteristics of this biome is the constant cycling of water in the East China Sea, which benefits by river inflow and by exchange with other oceanic regions. Colder waters enter the sea from the rivers and from upwelling along coasts. The Kuroshio Current, a major warming influence, moves north from near Taiwan. A flow known as the Taiwan Warm Current enters the Taiwan Strait from the Philippine Sea, through the Bashi Channel. Another flow, the Yellow Sea Warm Current, enters the northern part of the East China Sea to create an ideal habitat for fish and shellfish breeding, feeding, and winter survival.

Biodiversity

With its temperate climate; warm, shallow but well-stratified waters; seasonal flux in water exchange with other seas; freshwater supply from its rivers; and multiple areas of strong tidal action, the East China Sea offers a very robust, differentiated, and broad base for a multitude of plant communities and for many types of invertebrates, fish, marine mammals, and birds. It has been estimated that some 13,000 species exist here, of which nearly 48 percent are endemic, that is, found nowhere else on Earth.

The food web here benefits from the widespread propagation and growth of phytoplankton and zooplankton. In turn, these creatures form the basis of life for great populations of crustaceans and mollusks, which provide sustenance to all types of fish here.

As such, the East China Sea is the most productive marine area around China for seafood; it is especially rich in both big and small yellow fish, beltfish, flatfish, croaker, chub and spotted mackerel, Pacific cod, bluefin and bigeye tuna, a full range of sharks, squid, broadclub cuttlefish, and more. Most of the fishing is done by small local boats, but larger trawlers are a used in some cases. The Zhoushan Islands fishing grounds within the East China Sea are known as the treasure trove of China seafood production.

Spotted seal, finless porpoise, minke whale, and the critically endangered gray whale are among key marine mammals in these waters, roaming seasonally from the deeps to preferred grounds along the coasts.

The Yellow Sea-East China Sea region is an absolutely vital component of the East Asian-Australasian Flyway, with a reported two million shore birds stopping over on their northward migrations. Two globally threatened species are among these: spotted greenshank and spoon-billed sandpiper. Fulmars, storm petrels, alcids, geese, mallards, ducks, and other birds make up some of the 160 distinct marine bird species either resident in or that migrate through the East China Sea.

The short-tailed albatross, a vulnerable species that favors coastal upwelling zones for feeding areas, has important colonies in the East China Sea. The eastern curlew and Asian dowitcher are two additional vulnerable migrants here.

Environmental Threats

Environmental pollution caused by fast economic growth in mainland China as well as other nearby industrialized nations has led to conditions such as eutrophication (excess nutrients), low-oxygen zones, and frequent red tides along the coast of the East China Sea. Similarly, there have been repeated events of giant jellyfish swarms. Overfishing is a perennial challenge. Most of these conditions tend to be exacerbated by increasing water and air temperatures, with the onset of global warming. Sea-level rise, too, is a growing concern as it threatens coastal habitats with accelerated erosion and saltwater intrusion.

These factors have significantly affected the marine ecological web. In addition, air pollution from southeastern China is thought to form acid rain when passing above the East China Sea, thus precipitating heavy metal and toxic chemical fallout into the sea and along the north coast of Taiwan.

The East China Sea's continental shelf bears rich petroleum and natural gas deposits that have led to disputes among the bordering countries in the region, including China, Japan, South Korea, and Taiwan. The tension is focused on the control of the areas with potentially exploitable hydrocarbon reserves, particularly regions near deep trenches, straits, rocks, and uninhabited islands. Such disputes are incremental to already-existing conflicts over fishing rights in many of the same areas.

While diplomacy is required in struggle over fossil-fuel ownership, there are single-nation efforts to protect some of the East China Sea biome. Within the areas of the sea and coastal zones that it legally controls, the government of China has established 17 nature reserves, seven special marine reserves, and a handful of fishery conservation areas. The total area of these marine protected areas (MPAs) is approximately 40,000 square miles (103,000 square kilometers). The intent of setting up and enforcing MPAs is to provide a future environment that preserves and restores biodiversity, habitats, and fishery resources.

ZHIQIANG CHENG

Further Reading

Ding, H., et al. "An Overview of Spatial Management and Marine Protected Areas in the East China Sea." *Coastal Management* 36, no. 5 (2008).

Hayes, Dennis E., ed. *The Tectonic and Geologic Evolution of Southeast Asian Seas and Islands.* Washington, DC: American Geophysical Union, 1983.

Ishimatsu, Atsushi and Heung-Jae Lie, eds. *Coastal Environmental and Ecosystem Issues of the East China Sea*. Tokyo: Terrapub and Nagasaki University, 2010.

Park, Chul. *Yellow Sea and East China Sea Reported by PICES and Korean Monitoring Program.* Daejeon, South Korea: Chungnam National University, 2012.

Qu, J., Z. Xu, Q. Long, L. Wang, X. Shen, J. Zhang, and Y. Cai. *East China Sea—Global International Waters Assessment.* Kalmar, Sweden: Linnaeus University and United Nations Environment Programme, 2005.

Zhu, Qian. *Biological Assessment of the Ecologically Important Areas for Marine Mammals in the Yellow Sea Ecoregion.* Weihai, China: Shandong University, 2008.

East Siberian Sea

Category: Marine and Oceanic Biomes.
Geographic Location: Arctic Ocean.
Summary: This shallow, foggy, ice-filled marginal sea in the Arctic Ocean is marked by little activity, either organic or tidal. Global warming is making dramatic changes in sea-ice cover here and is affecting the habitats of the planet's northernmost reaches.

A marginal sea of the Arctic Ocean, located off the northern coast of Siberia, the East Siberian Sea is bordered to the east and west by the Chukchi and Laptev Seas; it is noted for its numerous ice fields, most of which are fully melted for less than two months a year.

An obscure area in many respects, the East Siberian Sea has been less studied than much of the Arctic area, and commercial fishing has been little developed within it. The sea did not even receive a distinct name until 1935, when the government of the former Soviet Union gave the sea its present designation. Before that, the East Siberian Sea was variously called Kolymskoe, Sibirskoe, Ledovitoe, and other names, and sometimes went without specific designation on maps.

Because the shallow sea is frozen for most of the year, it is navigable only in August and September, limiting human presence. Even the coast is the site of several of Siberia's ghost towns, including the former trading post of Logashkino and the abandoned mining town of Valkumey.

Rivers flowing into the sea include the Indigirka, Alazeya, Ujandina, Chukochya, Kolyma, Rauchua, Chaun, and Pegtymel. There are two distinct hydrological zones in the sea: the west, which is influenced most strongly by the water of the Lena River (and marine- and terrestrial-derived sediments), and the east, influenced by Pacific Ocean waters. The main gulfs of the East Siberian Sea are in the south, with the north opening into the Arctic Ocean. It is home to few islands, many of which consist only of gradually eroding sand and ice.

Most of the remaining islands are grouped along the coastal limits, including the tundra island of Ayon off the coast of Chukotka, used by the Chukchi people to pasture herds of reindeer; the small adjacent islands Ryyanranot, Chengkuul, and Mosey; and the uninhabited Medvezhyi Islands in the west of the Kolyma Gulf. The Medvezhyis, the name of which means *bears* in Russian, are both covered by ice, the proximate sea generally staying frozen even in the summer. Though home to minor seasonal commercial fishing, the islands serve mainly as a breeding ground for polar bears.

The Pacific Walrus

The East Siberian Sea is home to a population of Pacific walruses, an endangered species also found in the Bering and Chukchi Seas. Melting Arctic sea ice has made the East Siberian Sea more critical to this species, as it relies on floating sea ice as passive transport to feeding areas; as shelter from predators; and as a place to sleep, give birth, and nurse calves without worrying about disruptions. The walrus can use land-based habitats when sea ice is unavailable, but this makes it more vulnerable to predators and disease; it also exposes calves, juveniles, and females to increased stress and energy costs.

Pacific walrus bulls like this one have been known to weigh as much as 3,700 pounds. (U.S. Fish and Wildlife Service)

Climate and Ice Cover

One reason the ice is so tenacious is because the sea is so shallow. Seventy percent of the area is less than 164 feet (50 meters) deep, and the deepest point is 509 feet (155 meters) deep. The summer air temperature averages just above freezing, 32 degrees F (0 degrees C), and in winter plummets to minus 20 degrees F (minus 29 degrees C). The ice cover prevents sunlight from penetrating far into the sea, and along with the seasonal change of the seawater salinity and chemical content due to the gradual freezing and thawing of the ice, this situation causes considerable complications for the ecosystem.

The entire phytoplankton life cycle takes place during the brief summer, lying dormant the rest of the year. Seasonal coastal erosion and river discharges provide key nutrients, especially in the Kolyma River, which contributes millions of tons (metric tons) to the sea every year.

Climate change appears to have increased the summer period in the East Siberian Sea. In 2007, the National Aeronautics and Space Administration (NASA) satellite imagery confirmed that as early as June 16, melting was well underway, and patches of ocean water had become visible surrounding small slabs of ice in the northeast. By the end of July, the area between the Siberian coast and the New Siberian Islands had become almost entirely ice-free, and throughout the sea, ice was sparse.

While sea-ice loss was detectable throughout the Arctic during summer 2007, the majority of it occurred in the East Siberian Sea—but even this event was just a prelude. The summer melting in 2012 exceeded anything previously seen in the record of Arctic satellite imagery, with the East Siberian Sea especially void of surface ice; scientists have issued more urgent warnings about the need for environmental protection measures, especially the curtailment of burning fossil fuels.

The water in the sea is fairly clean by 21st-century standards. While the snow and sediment can be found to contain hazardous contaminants, including heavy metals, radionuclides, and hydrocarbons, their concentrations are very low except in areas immediately surrounding mineral-resource extraction zones. The Siberian rivers contribute industrial and agricultural runoff, and coastal currents carry pollutants along the continental shelf, but the levels of these substances are low compared with other parts of the ocean.

Biodiversity

The sea is home to a limited number of gray whales in the summer months, populations of which may migrate between the East Siberian Sea and the western Chukchi Sea, wherever conditions are most favorable and ice conditions permit movement. Gray whales have been spotted in the early fall, before ice coverage is complete. Some scientists have speculated that the whales may be reoccupying habitats that had been abandoned after commercial whaling decimated the whale population.

Other fauna living on the coast and the drift ice include Pacific walruses, polar bears, Arctic foxes, seals, and seabirds. Prominent fish include salmon, pollock, halibut, and crab, mainly in the subestuary zones.

Bill Kte'pi

Further Reading

Akulichev, Viktor Anatol'evich. *Far Eastern Seas of Russia*. Moscow: Nauka Press, 2007.

Butler, William. *Northeast Arctic Passage*. New York: Sijthoff & Noordhoff, 1978.

International Hydrographic Organization. *Limits of Oceans and Seas, Special Publication, Vol. 28*. Monte Carlo: International Hydrographic Organization, 1953.

Miller, R. V., J. H. Johnson, and N. V. Doroshenko. "Gray Whales in the Western Chukchi and East Siberian Seas." *ARCTIC: Journal of the Arctic Institute of North America* 38, no. 1 (1985).

Robinson, Allan R. and Kenneth H. Brink. *The Global Coastal Ocean: Regional Studies and Syntheses.* Cambridge, MA: Harvard University Press, 2005.

Ecotourism

Category: All Biomes.
Geographic Location: Global.
Summary: Nature-based tourism has both positive and negative effects on the world's ecosystems, drawing travelers to experience natural wonders, promote environmental activism, and stimulate the economy of local communities, but exposes habitats to the risks of high traffic and disruption.

Many of the world's ecosystems still contain areas that remain mostly in their natural state. The term *ecotourism* refers to traveling to such areas to enjoy the recreational and educational experiences they provide. People have found pleasure in nature throughout history, but the expansion and intensification of human activities have degraded many of the world's ecosystems. This human footprint has reduced the possibility to benefit from visiting well-preserved ecosystems, forcing tourists to travel longer distances to see nature in its pristine, unspoiled state.

Modern transportation systems enable travelers from around the world to visit distant locations to enjoy the remaining areas of wilderness. Countries that protect their ecosystems can benefit from the inflow of ecotourists, but an increase in visitors also can put pressure on natural areas and on the human communities that depend closely on the resources that they provide.

Roots and Movements

In Asia, reverence for natural places is a major tenet of religions such as Buddhism, Hinduism, Shintoism, and Taoism. Pilgrimages to key historical sites in the development of these philosophies can be considered a form of ecotourism. Native cultures of Africa, the Americas, and Australia place natural areas in high regard; certain places considered sacred attract the faithful on pilgrimages.

In Europe and its former colonies in the Western Hemisphere, some ecosystems are seen as places of special value; many can be traced to the Romantic Movement. This philosophical movement encompassed parts of the 18th and 19th centuries; it considered nature to be a source of well-being unrivaled by any object or structure created by humans. The naturalists of the Romantic Movement insisted on protecting ecosystems (although this term was introduced much later) from human encroachment, and their ideas helped give rise to the modern conservation movement.

One key figure in the latter development of this ecologically oriented movement was John Muir (1838–1914), an author and activist who helped influence the creation of the national parks system in the United States. Muir's writings enthralled Americans, who wanted to see and experience nature's beauty for themselves. Their travels were made possible by the development of the cross-country railroad system, and later by the automobile.

By the 1980s, the term *ecotourism* had come into the vernacular as a way to refer to nature-based tourism. Sectors of the travel industry, both domestic and international, had previously begun specializing in nature travel since the middle of the 20th century. Wildlife-watching trips to African national parks, for example, have been offered in Europe by specialized travel agencies since the end of World War II, and marine ecotourism has become a popular travel experience ever since scuba-diving equipment began to be mass-produced in the early 1970s.

Further, some countries began to embrace ecotourism as a major source of external revenues. Costa Rica has become an ecotourism leader for travelers who are interested in exploring mountain areas, tropical rainforests, coral reefs, and other ecosystems in this politically stable Central American country. Such travel is concentrated mostly in the country's protected areas, which include privately owned reserves. International tourism, driven mainly by ecotourism, has become one of the country's leading sources of foreign currency. Costa Rica is easily accessible from the United States and Canada, which are among the largest sources of tourists worldwide.

Beginning in the 1990s, the Costa Rican government began offering incentives for companies investing in environmentally responsible ecotourism there. These include a certification program for hotels with a low environmental footprint, and a certification program and rating system for coastal communities based on marine water quality, access to potable drinking water, beach cleanliness, wastewater treatment, environmental education, safety, and management. Environmentally conscious tourists can select their destination based on coastal communities with high ratings.

Economic Benefits

Several African countries also have adopted ecotourism as a path to socioeconomic development. In Kenya, hunting-based tourism, which was popular in the early 20th century, largely transitioned to wildlife-viewing tours in the newly created national parks, starting in the 1950s. After the country's independence from Britain in 1961, the government began to develop programs to attract international investment to develop tourism facilities that cater mainly to wildlife-viewing and beach ecotourists. Ecotourism offered a way to supplement the country's dependence on the export of agricultural products, whose prices fluctuated greatly, as a source of foreign currency.

Some local communities in Kenya have partnered with both governmental and nongovernmental organizations (NGOs) to offer ecotourism services such as lodging and wildlife-viewing tours. South Africa, too, offers specialized services for ecotourists; wildlife viewers there can choose among a wide range of accommodations, from tents (typically available in the national parks) to luxury hotels that are available in some of the private natural reserves.

A growing number of developing countries have been able to diversify their economic bases and benefit from a generally stable inflow of foreign currency brought by ecotourists. Tropical rainforests, savannas, and coral reefs are ecosystems that are particularly sought out, and they are coincidentally present in a large number of developing countries. The activity has led to new industries and local jobs; governments and private landowners share the incentive to protect the ecosystems, sustain tourism, and protect and enhance their local economy.

In developing and developed countries, ecotourism provides financial resources that help maintain the systems of protected areas. In the United States, visitor fees in the parks and preserves managed by the federal government and individual states account for approximately one-fifth and one-third, respectively, of the park systems' budgets. Funds from visitor fees can be used to create and maintain tools for interpretation, such as information boards and multimedia visitor centers, which help educate visitors on the area's ecosystems and explain which activities should be avoided to prevent damages to the resources. Such enhancements tend to increase repeat visits.

Negative Impacts and Mitigation

Ecotourism can help educate visitors and residents on the importance of protecting ecosystems, and it can provide incentives for the creation of protected areas—but it also can have a negative effect on natural resources. An increase in the number of ecotourists is generally accompanied by an increase in the number of hotel rooms to accommodate them, which results in higher water consumption and wastewater disposal. An increase in visitors to protected areas also leads to more solid and liquid wastes in these ecologically sensitive areas. The physical impact of tourism and the interference of ecotourists with wildlife also can become problematic.

Corals located in shallow waters are frequently trampled by visitors to marine protected areas, and corals can die from diseases that affect their injured areas. In many protected areas, the feeding

An ecotourism guide looks out from the bow of his boat as he brings a tourist back from a trip to El Salvador's first marine protected area in 2009. (USAID/Angela Rucker)

and breeding behavior of birds has been affected by tourists. In national parks of the United States, the natural feeding habits of bears are affected by the presence of humans, and bears that search for food left by visitors are occasionally injured or killed by vehicles.

The impact of these activities can even intensify the impact of climate change, by hastening the deterioration and vulnerability of already-sensitive areas. As coral reefs are adversely impacted by warming waters, human activity in these areas can hasten their decline. Visitors in search of bears, other wildlife, trees, flowers, and more can be drawn to higher elevations, deeper into forests and protected areas, bringing their influence to formerly remote areas, increasing the disruption of their habitats.

Ecotourism can have negative social consequences as well. Local communities can feel that their traditional way of life is affected by the influx of visitors to protected areas in their vicinity. In Kenya, some indigenous Maasai communities have developed resentment toward ecotourism, which they see as favoring only a few people and causing social divisions between those who benefit and those who feel affected by the intrusion of tourism. In the Ecuadorian Amazon, ecotourism has caused conflicts within some of the Secoya and Siona indigenous communities. The distribution of benefits from tourism is seen as unfair by some, and the adoption of customs and technologies introduced by tourists also is viewed negatively by members of these communities.

There are a number of initiatives at the national and international levels whose goal is to make ecotourism a sustainable activity from both the ecological and social perspectives. NGOs in many developing countries have been established specifically to help design strategies for sustainable ecotourism, and universities worldwide offer courses related to ecotourism management. University research tracks on ecotourism cover areas such as ecologic, economic, socioeconomic, and cultural effects and benefits; visitor preferences; and visitor willingness to pay for visits to protected areas. Several international organizations, including the International Ecotourism Society (IES) and the World Tourism Organization (WTO), promote sustainable ecotourism and disseminate information on ecotourism research.

The WTO and the United Nations Environment Programme organized the World Ecotourism Summit in 2002 in Quebec City, Canada. The summit brought together representatives from academic institutions, governments, NGOs, tourism companies, and indigenous communities to share their experiences in ecotourism and to identify priority areas for the development of sustainable ecotourism. The IES organizes the annual Ecotourism and Sustainable Tourism Conference, the goal of which is "providing practical solutions to advance sustainability goals for the tourism industry."

Jeffrey Wielgus

Further Reading

Fennell, D. A. *Ecotourism.* 3rd ed. Abingdon, UK: Routledge, 2008.

Higham, J., ed. *Critical Issues in Ecotourism: Understanding a Complex Tourism Phenomenon.* Oxford, UK: Butterworth-Heinemann, 2007.

Honey, M. *Ecotourism and Sustainable Development: Who Owns Paradise?* 2nd ed. Washington, DC: Island Press, 2008.

Wild, R. and C. McLeod, eds. *Sacred Natural Sites: Guidelines for Protected Area Managers.* Gland, Switzerland: International Union for Conservation of Nature and Natural Resources (IUCN), 2008.

Edwards Plateau Savanna

Category: Grassland, Tundra, and Human Biomes.
Geographic Location: North America.
Summary: These brushlands are regionally unique in geology and terrain. Historically used for livestock production, human population is exerting pressure on its habitats and species.

The Edwards Plateau Savanna biome encompasses 20 million acres (8 million hectares) in central

Texas. The Cretaceous-epoch limestone forming the bedrock of the area creates a region that is distinctively different from surrounding areas in geology, terrain, and ultimately its land-use culture. The region has shallow soils and receives low annual precipitation of 16–32 inches (406–813 millimeters). Despite these limitations, the savanna has been extensively used for livestock production, generally goats, sheep, and cattle. The vegetation that existed here before livestock grazing is unclear but is believed to have consisted of mixed woodlands, oak savannas, and grasslands. Overgrazing has increased the spread of unpalatable brush species, and many areas are experiencing increased population growth and residential development.

Soil and Water Regimes

Soils in the savanna are often very shallow, with areas of exposed bedrock. The limestone bedrock also causes the soils to have high clay content. Caliche (redeposited calcium carbonate) is likely to be found in areas with deeper soils; this makes agricultural production difficult.

Several plant communities occur in the Edwards Plateau savanna, including mixed woodland, tree and shrub savanna, cedar brake, bottomland forest, and riparian forest. Except on steep, rocky slopes, the historical vegetation in the region is considered to be open savanna or grassland, and the brush species present are encroaching on the area due to changes in historic grazing and fire regimes.

The eastern Edwards Plateau savanna is commonly referred to as the Texas Hill Country. Many rivers and streams in this region have cut through the layers of limestone to produce steep hillsides and valleys. Major rivers in the region include the Colorado of Texas, Pecos, Nueces, Guadalupe, and Rio Grande. The limestone layer allows water from streams to lose water into underground caves and springs, thereby recharging the Edwards Aquifer that lies beneath the ground layers.

The Edwards Plateau watershed and Edwards Aquifer provide municipal water resources for several of the largest cities in the region, including Austin, Del Rio, San Antonio, and San Marcos. There are no natural lakes in the region, but many human-made reservoirs have been built to provide flood control, municipal water sources, and recreational opportunities.

Fauna and Flora

Underneath the limestone bedrock are many caves, sinkholes, and springs that are home to endemic (found nowhere else on Earth) vertebrate species, including the widemouth blindcat (*Satan eurystomus*), San Marcos salamander (*Eurycea nana*), comal blind salamander (*Eurycea tridentifera*), Texas blind salamander (*Typhlomolge rathbuni*), blanco blind salamander (*Typhlomolge robusta*), and Texas salamander (*Eurycea neotenes*). Invertebrate species endemic to the region include the Tooth Cave pseudoscorpion (*Microcreagis texana*) and the Kretschmarr Cave mold beetle (*Texamaurops redelli*).

Many bat species roost in the caves; the area is world renowned for the millions of Mexican free-tailed bats (*Tadarida brasiliensis*) that gather in hordes in the caves to raise their young. Among avians, the endangered golden-cheeked warbler (*Setophaga chrysoparia*) and black-capped vireo (*Vireo atricapilla*) nest in the region, which is listed in the Top 20 Most Threatened Bird Habitats in the United States by the American Bird Conservancy.

Endemic plants here include netleaf swamp-privet (*Forestiera reticulata*), plateau milkvine (*Matelea edwardsensis*), basin bellflower (*Campanula reverchonii*), Lindheimer crownbeard (*Verbesina Lindheineri*), low loosestrife (*Lythrum ovalifolium*), Buckley's fluffgrass (*Tridens buckleyanus*), Texas snowbells (*Stryas texana*), bracted twist-flower (*Streptanthus bracteatus*), and cliff bedstraw (*Galium correllii*).

Effects of Human Activity

Before settlement by people of European and African descent in the mid-1800s, Native American peoples inhabited the area. After the introduction of horses to the region, Apache and Comanche tribes developed a culture based on horses and bison. Due to the tough conditions for growing crops, settlers depended on grazing of livestock for subsistence and the trade economy. Today, the region remains one of the top sheep-producing areas in the United States.

Recreational hunting of white-tailed deer, turkey, and quail provide significant revenue to landowners in the region. Various antelope and deer species of African and Indian countries have been introduced to provide further hunting opportunities. Bird-watching also brings revenue into the region, with thousands of visitors annually arriving to see the black-capped vireos and golden-cheeked warblers. The U.S. Fish and Wildlife Service established the Balcones Canyonlands National Wildlife Refuge to protect 80,000 acres (32,375 hectares) of Edwards Plateau habitat.

Many changes have occurred in the Edwards Plateau savanna since the time of settlement, some of which threaten its rustic foundation. Rapid population growth in the region has resulted in increased urban and suburban development in formerly rural areas. As much as 90 percent of the Edwards Plateau savanna has been converted either to urban development or agricultural uses.

This rapid and pervasive growth has led to a loss of natural ecological processes, such as fire, and has led to changes in vegetation. Changes in water use and runoff from development have altered the dynamics of riparian ecosystems. Overgrazing of domestic livestock and high white-tailed deer populations have led to an increase in shrub and tree species at the cost of losses in herbaceous vegetation throughout many areas. Invasive exotic species introduced for agricultural and domestic purposes affect native plants and animals.

All of these factors, coupled with warming temperatures, are projected to alter plant growth cycles, as well as increase and expand the naturally occurring fire regime, and change the landscape over time. The opposite is true, too, as the absence of natural fires, compounded by overgrazing, oil and gas development, and tillage has already spread many species of woody plants across most of the Edwards Plateau. How this newly altered and still-reshaping ecosystem responds to long-term lower precipitation and higher temperatures remains to be seen—but the dynamics of habitat impact on the native flora and fauna here seem to rest upon the changing balance between surface conditions and the underground aquifer and cave infrastructure that caused many of theses species to evolve.

Matt Bahm

Further Reading

Fowler, Norma. "An Introduction to the Vegetation and Ecology of the Eastern Edwards Plateau (Hill Country) of Texas." University of Texas, 2005. http://www.sbs.utexas.edu/fowler/docs/FowlereasternEP.pdf.

Johnson, E. H. "Handbook of Texas Online—Edwards Plateau." Texas State Historical Association, 2012. http://www.tshaonline.org/handbook/online/articles/rxe01.

Lockwood, Mark. *Birds of the Texas Hill Country.* Austin: University of Texas Press, 2001.

Elbe River

Category: Inland Aquatic Biomes.
Geographic Location: Europe.
Summary: The Elbe River passes through the heart of northern Europe; its basin holds an extensive, contiguous floodplain forest, making for a very fertile area for vegetation and animal habitats.

The Elbe River is the third-longest river in Europe; more than 679 miles (1,093 kilometers) long, it is geographically represented in three regions. The Upper Elbe headwaters and tributaries rise in the Czech Republic, deep within the Krkonoše Mountains, flowing downstream through the Bohemian basin, where its waters bend toward the northwest. It merges with the Vlatava (Moldau) River before passing through the Bohemian uplands, then flows over volcanic bedrock at Ustí nad Labem, just below the picturesque ruins of the Schreckenstein Castle, a 14th-century icon overlooking the last weir impounding the Elbe's waters in the Czech Republic. The river here passes through a canyon down into the north German lowlands.

From Ustí nad Labem to the Geesthacht weir, a 386-mile (622-kilometer) stretch, the Middle Elbe flows unimpeded through one of the largest contiguous floodplain forests in central Europe. From Pardubice to Hamburg, the river is used extensively for navigation. The Lower Elbe flows past Hamburg and Geesthacht into an estuary leading out to the North Sea at Cuxhaven-Kugelbake. The fluvial interaction of the Elbe River with the North Sea tides has been substantially altered by deep dredging projects initiated to accommodate large ocean ships docking inland at Hamburg.

The river's catchment area of 57,400 square miles (148,650 square kilometers) includes lands in Germany and Czech Republic, as well as very small parts in Austria and Poland. Its tributaries include the Vltava, Saale, Havel, Mulde, Black Elster, and Ohre Rivers. Nearly one-quarter of Germany's land mass and more than half the area of the Czech Republic drain into the Elbe's alluvial waters. Its basin is home to more than 24 million people. The river is one of three (the Ems and Weser included) that transverse the western German landscape toward the German bight on the North Sea coastline.

Located at the northeastern periphery of Germany, the Elbe for centuries has served as the main commercial waterway for central and eastern Germany. Since ancient times, the Elbe was considered an important geographic landmark separating lands on its eastern and western flanks. For centuries, it served as the eastern border of the Holy Roman Empire. All of these roles as boundaries were important culturally—and therefore have had profound feedback ecologically to the river and its surrounding habitats.

The Elbe is one of several rivers that flow into what is known as the northern European drainage system. Other rivers that contribute to this catchment area include the Saale, Weser, Rhine, Meuse, Scheldt, Thames, Somme, and Seine. Over long geologic history, the course of the Elbe River was modified by tectonic activity; climate variability; the formation of massive glacial lakes; and complex depositions of clays, gravels, silts, and sediments. During the last ice age, a sheet of ice advanced to the central borders of Germany and Poland. Meltwaters trapped behind the sheets could move only east and west, creating the valleys, bottomlands, and marshes so familiar today along the North Sea coastline.

Habitats and Species

The basin area surrounding the Middle Elbe holds one of the largest contiguous floodplain forests in Europe. A temperate broadleaf forest, this region (locally called the Garden Kingdom of Dessau-Wörlitz) is included in the United Nations Educational, Scientific and Cultural Organization (UNESCO) World Cultural Heritage list. The species-rich aquatic vegetation results in a dense population of such key mammals as the Elbe beaver (*Castor fiber albicus*).

The Flusslandschaft Elbe Biosphere Reserve includes five German states, including Brandenburg, Mecklenburg-Vorpommern, Niedersachsen, Sachsen-Anhalt, and Schleswig-Holstein. Particular care is given here to the restoration of lakes, the protection of beaver habitats, and the reintroduction of Atlantic salmon. In April 1998, a fish pass was created at the Geesthacht weir in Germany;

The rare fern family Salvinia *is found in backwater areas of the Elbe River and includes the endangered floating fern* (Salvinia natans)*, shown here. (Thinkstock)*

another pass was built in 2002 at the Střekov lock in Ústí nad Labem in the Czech Republic. Continued restoration of some segments of the river makes it possible to build spawning grounds for other anadromous and freshwater fish to propagate.

The Elbe's expansive floodplain forests are popular breeding sites for black storks (*Ciconia nigra*) and a habitat for numerous raptor species including the lesser-spotted eagle (*Aquila pomarina*) and white-tailed eagle (*Haliaeetus albicilla*). The area is of national and international significance due to the number of bird species that breed, rest, hunt, or pass through it.

The area also is home to endangered species; some are threatened with extinction, such as the water chestnut (*Trapa natans*), floating fern (*Salvinia natans*), Siberian iris (*Iris sibirica*), and water soldier (*Stratiotes aloides*) among the flora; and the fire-bellied toad (*Bombina bombina*) and moor frog (*Rana arvalis*) among the fauna.

Wandering bats use the Elbe floodplain on their long flights between nursery roosts and wintering grounds. Large contiguous old-hardwood riverside woodlands offer breeding places to the middle spotted woodpecker, red kite, and white-tailed eagle.

More than 1,000 species of plants grow along the floodplain forests and meadows, and within the backwaters of the Elbe. Common tree species of the hardwood floodplains include English oak, elm, and field maple. The plants of the riverside meadows are adapted to the periodic changes in the river's water levels. Native species include the rare fern family *Salvinia*, the genus *Rorippa*, and the protected species water caltrop and squincywort.

Human Impact

Small-scale, locally intensive horticulture was common in central Europe until approximately the 10th century, when the scope and technology of agricultural production expanded to meet the needs of a rising urban population. Early communities transplanted wild cereal grains to moist soils on lands where groundwater was plentiful. Water also provided the power needed for a variety of mill technologies that supported the initial expansion of commercial trade networks. Plowed fields eventually dominated woodland terrains, affecting alluvial depositions and patterns of discharge into streams and rivers, particularly during winter rains and snowmelts.

Three-crop rotation systems became the norm in northern Europe. During this period, fish became a popular commodity. During the medieval warm period, stresses on fisheries were noted, with remarkable declines in anadromous and cold-water populations. Subsequent privatization and commercialization of marine fish stocks along the coasts contributed to a steep rise in medieval market values for wild freshwater fish.

Central Germany is remarkable for its fertile loess soils that lie atop extensive beds of lignite, a resource whose extraction conflicted with the region's agricultural production. Because of the proximity of Germany's prime agricultural lands to industrial centers like Magdeburg, Hanover, and Breslau, extensive scientific planning and research went into raising the sugar yield of beets. By the early 20th century, the country grew more than 25 percent of the world's sugar beet production. This success was due to the country's wide-scale production and distribution of potash from its interior and the cheap labor provided by women and children. Salt, chemicals, and fertilizers were leading downstream industries. Other products included paper, paper pulp, glass, lignite, and processed grains. Upstream traffic included porcelain, cattle food, bituminous coal, cereal grains, petroleum products, raw textile materials, and mineral ores.

During the 1970s and 1980s, the German Democratic Republic—Communist East Germany—mined more coal than any other country in the world; its burning emitted more than 5.6 million tons (5.1 million metric tons) of sulphur dioxide per year. Acid rain destroyed 9,000 lakes and nearly 40 percent of East German forests. Chemical, industrial, and agricultural effluents were discharged into area waterways, including salts, heavy metals, acids, carbon compounds, mercury, and dioxins. The nuclear plant disaster at Chernobyl in 1986 drew world attention to the unprecedented decline of the eastern European environment.

Following reunification of the German state, the International Commission for the Protection of the Elbe River was established in 1990, to restore the river's ecosystems and to mitigate the effects of its pollutants on the North Sea and adjacent estuaries and wetlands. Other protocols that govern the management and protection of the Elbe River include the Ramsar Convention on Wetlands (adopted in 1971), the International Commission for the Protection of the Elbe River Agreement, the European Union (EU) Water Framework Directive, the Natura 2000 directives, and the EU Nature Legislation (the Birds Directive as adopted in 1979 and amended in 2009, and the Habitats Directive as adopted in 1992).

The exploitation of the area's resources, coupled with man-made disasters and the effects of climate change, are contributing to erosion problems along much of the contemporary Elbe River. A lack of reliable rainfall due to global warming has already been recorded; its continuance will jeopardize the river's future and threaten the integrity of its floodplain broadleaf forests and other habitats. The mouth of the Elbe will have to be protected against the ravages of sea-level rise, which are projected to include marshland inundation, saltwater intrusion, and invasive species expansion.

Victoria M. Breting-Garcia

Further Reading

Blackbourn, David. *The Conquest of Nature: Water, Landscape, and the Making of Modern Germany.* New York: W. W. Norton & Co., 2006.

Dominick, Raymond H. III. *The Environmental Movement in Germany: Prophets and Pioneers 1871–1971.* Bloomington: Indiana University Press, 1992.

Jones, Merrill E. "Origins of the East German Environmental Movement." *German Studies Review* 16, no. 2 (1993).

Lekan, Thomas and Thomas Zeller, eds. *Germany's Nature: Cultural Landscapes and Environmental History.* Piscataway, NJ: Rutgers University Press, 2005.

Marshal, Jan. "Salmon Swim Again in Czech Elbe River." Phys.org, May 13, 2011. http://www.physorg.com/news/2011-05-salmon-czech-elbe-river.html.

Tockner, Klement, Urs Uehlinger, and Christopher T. Robinson, eds. *Rivers of Europe.* 1st ed. Waltham, MA: Academic Press, 2009.

Erie, Lake

Category: Inland Aquatic Biomes.
Geographic Location: North America.
Summary: Among the North American Great Lakes, Lake Erie is the fourth-largest, the shallowest, and has the highest biodiversity—but it has suffered the most damage from agricultural runoff, industrialization, and urbanization.

Lake Erie was named after the indigenous Erie tribe that lived along its southern shore when French explorer Louis Jolliet discovered the area in 1669. By surface area, Lake Erie is the fourth-largest of the North American Great Lakes, and the smallest in terms of volume and average depth. It is the 11th-largest lake in the world by surface area. It forms an international boundary between the United States and Canada. Lake Erie is downstream from Lake Huron, and upstream from Lake Ontario.

The Canadian province of Ontario borders Lake Erie on the north; the states of Ohio, Pennsylvania, and New York line the southern shores; and Michigan adjoins the lake to the west. Lake Erie has a mean elevation of 569 feet (173 meters) above sea level. It has a surface area of 9,923 square miles (25,700 square kilometers), with a length of 241 miles (388 kilometers) and breadth of 57 miles (92 kilometers) at its widest point. Including its offshore islands, Lake Erie's total shoreline extends 871 miles (1,402 kilometers).

Geology and Depth

Like that of the other Great Lakes, Lake Erie's basin was carved from limestone and shale bedrock by glacial ice between 1.6 million and 10,000 years ago during periodic ice ages. Its shallow depth allows it to warm significantly in summer, and freeze over during cold winters. Lake Erie is rich in nutrients that enter its waters in runoff from the 34 inches (860 millimeters) of precipitation that annually falls upon its immediate region and flows through a watershed composed of highly fertile soils, agricultural fields, and large urban areas. Consequently, Lake Erie is the most productive of all the Great Lakes and boasts the highest biodiversity—but also suffers the most from pollution. Its relatively small size makes it more vulnerable than its sister lakes to these negative effects.

The deepest point in Lake Erie is 210 feet (64 meters), while the average depth is 62 feet (19 meters). The lake has a water retention time of 2.6 years, the shortest of all the Great Lakes.

More than 12,000 years ago, the basins of the individual Great Lakes were cut from soft Paleozoic limestone. However, the glaciers were largely unable to cut through the harder dolomite at the west end of Lake Erie, making this end of the lake very shallow, with depths averaging only 30 feet (9 meters). The central and eastern parts of Lake Erie lie on beds that are tilted to the south. These basins were carved from softer Devonian shales and are deeper than the western basin: up to 60 feet (18 meters) for the central basin and 80 feet (24 meters) for the eastern basin. In the narrow eastern part of the basin, where the angle of tilt is steeper still, the Devonian shales were eroded even more, to form the deepest basin, at 210 feet (64 meters). There are 31 islands in Lake Erie (13 in Canadian waters, 18 in the United States), most of them in the western basin of the lake.

Rich Biota

High nutrient levels make Lake Erie a biologically productive and species-diverse lake. Pelee Island, for example, at the western end of the lake, is home to rare insects, snails, mammals, birds, reptiles, and amphibians, including the endangered Lake Erie water snake (*Coluber constrictor foxi*) and the related blue racer (*Coluber constrictor*). Almost one-third of the vascular plant diversity of all Ontario is represented on Pelee and its neighboring islands. Pelee Island lays claim to one of the rarest of North American ecosystems: the alvar. Alvars are defined as communities of grasses and other herbaceous plants, accompanied by sparse shrubs that grow in extremely thin soils atop limestone or dolomite. These unusual ecosystems, some dotted with trees, are found in an arc roughly following the limestone and dolomite of the Niagara Escarpment.

Pelee Island lies at the confluence of two bird migration routes, is a significant stopover site for more than 70 species of migrating birds, and is designated as a Globally Important Bird Area. Pelee also is a temporary rest stop during the fall migration of monarch butterflies. Thousands of monarchs rest and feed there before continuing their long trek across the open water of the Great Lakes.

Lake Erie has sustained an important commercial and sport fishery for a succession of species since the 1800s. Most recently, yellow perch and walleye have been the most economically important species. Among its native fish are walleye, sucker, yellow perch, channel catfish, drum, blue pike (now extinct here), northern pike, smallmouth bass, white bass, lake sturgeon, gizzard shad, and emerald shiner. Among the introduced fish species are rainbow trout (steelhead), rainbow smelt, brown trout, alewife, sunfish, coho salmon, carp, round gobies, and sea lampreys. Asian carp are species of great concern, capable of causing ecosystem damage as they invade.

Radical changes in Lake Erie during the 1990s raised concerns about the future and predictability of the fish communities. Walleye stocks declined by about 80 percent from their maximum levels in the 1980s, and the supply of smelt to the commercial fishery in eastern Lake Erie declined by about 60 percent. Habitat for the fish is shrinking, especially in the central basin of the lake where the bottom layer of water has greatly reduced oxygen levels. Populations of fish-eating birds such as ducks, loons, and mergansers have been depleted from botulism poisoning. Such changes have been linked to a variety of factors, but increasing urban and suburban development pressures have led to problems with sewage overflows and nonpoint source pollution from agriculture.

The invasion of zebra and quagga mussels changed the way nutrients were cycled and the timing of nutrient availability in Lake Erie. These mussels reject blue-green algae, but consume green

Cleveland Harbor on Lake Erie showing the entrance to the infamous Cuyahoga River in the foreground, which was so polluted that it caught fire in 1969. Lake Erie still suffers from worse pollution than any of the other Great Lakes. (U.S. Army Corps of Engineers/Ken Winters)

algae and nutrients, and then release nutrients in their fecal pellets late into the summer, when nutrient levels would normally be dropping. This process crashes populations of the green algae that normally would be available to zooplankton, which are a good source of food for juvenile sport and commercial fish and other small fishes. Concentrations of blue-green algae also can lead to taste and odor problems with drinking water from the lake.

Water Condition

Before 1950, Lake Erie was relatively clean and clear, and had low to moderate nutrient levels. In the 1950s, that changed: The water became so overloaded with phosphorus from wastewater plants, agricultural runoff, and industrial processes that massive diatom blooms developed and used up the lake's stores of silica, which these algae need to build their intricate cell walls. The diatoms were replaced with green algae, which are more dependent on phosphorus and nitrogen rather than silica.

As phosphate levels increased even further, the green algae blooms became so extensive that they used up the nitrogen and began to die. The phosphates released from these dead algae fed generations of blue-green algae (cyanobacteria) that could harvest their own nitrogen from the air and water. The cyanobacteria, in turn, became so dense that the surface layers began to shade the deeper ones. This resulted in the deaths of algae at lower depths. These cells sank to the bottom, eventually covering it in thick decaying layers. Their decomposition consumed large quantities of oxygen, causing dead zones.

The normally abundant mayfly larvae, which live and feed in the bottom sediments of clean lakes, completely died out at this stage; their demise led to an onslaught of species of worms that thrive in the anoxic conditions. Native fish that fed on the mayfly nymphs, such as chub and cisco, began to die off, allowing invasive species such as carp and alewives (which also tolerate lower oxygen levels) to thrive.

During the 1960s, Lake Erie was often called a dying lake or a dead lake. In July 1969, Ohio's most polluted river, the Cuyahoga, became so loaded with oil and debris that it caught fire in Cleveland's factory area. The Federal Water Quality Administration began a $1.5 billion municipal sewage-treatment program for the Lake Erie basin. Detergent manufacturers decreased the amounts of phosphates in their products, which had previously found their way into the lake via sewage treatment plants. The same year, Lake Erie was found to be polluted with the toxic insecticide DDT, and more than 64,000 pounds (29,030 kilograms) of coho salmon were seized from commercial fisheries on a charge that the fish were contaminated with DDT. Later, the even more poisonous chemical dioxin was found in the lake. Both these toxins had significantly harmful effects on fish and birds.

Further Threats

Lake Erie had been damaged severely, but its rapid rate of water exchange allowed it to flush the pollutants through the Niagara River and out to the Atlantic Ocean. With cleaner waters, the ecological community began to recover. However, the recovery still faces many complex obstacles.

Mostly due to human carelessness, Lake Erie has become home to an increasing number of nonnative plants, animals, and microorganisms that threaten the balance of the entire ecosystem. The zebra mussel, quagga mussel, spiny water flea, fishhook water flea, and European water milfoil, among many others, have begun once again to threaten the lake. Blue-green algae such as *Microcystis* is again forming a thick scum on the lake and producing toxins that can be fatal to animals that drink the contaminated water. Although much cleaner now than in the 1960s, Lake Erie is still classified as eutrophic, or beset with too-high nutrient levels.

Climate change also will impact the area, which could lead to a steep drop in water levels over the coming decades, a change that could cause the lake's surface area to shrink by up to 15 percent. If that occurs, the drop could undo years of shoreline abuse by allowing water to resume its natural coastal circulation, which has become blocked by man-made structures over the years. If the water level of the lake drops to that extent, there could be a significant positive impact on plant-loving species such as northern pike (*Esox lucius*), muskellunge (*Esox masquinongy*), and largemouth bass (*Micropterus salmoides*), particularly if subsequent submersed vegetation accompanies the change.

LIANE COCHRAN-STAFIRA

Further Reading

Ashworth, William. *The Late Great Lakes: An Environmental History.* Detroit, MI: Wayne State University Press, 1987.

Dempsey, Dave. *On the Brink: The Great Lakes in the 21st Century.* East Lansing: Michigan State University Press, 2004.

Dennis, Jerry. *The Living Great Lakes: Searching for the Heart of the Inland Seas.* New York: St. Martin's Press, 2003.

Government of Canada and U.S. Environmental Protection Agency. *The Great Lakes: An Environmental Atlas and Resource Book.* 3rd ed. Toronto: Government of Canada and U.S. EPA, 1995.

Grady, Wayne. *The Great Lakes: The Natural History of a Changing Region.* Vancouver, Canada: Greystone Books, 2007.

Spring, Barbara. *The Dynamic Great Lakes.* Baltimore, MD: Independence Books, 2001.

Sproule–Jones, Mark. *Restoration of the Great Lakes: Promises, Practices, Performances.* East Lansing: Michigan State University Press, 2002.

Ethiopian Montane Moorlands

Category: Grassland, Tundra, and Human Biomes.

Geographic Location: Northeastern Africa.

Summary: The Ethiopian montane moorlands are less developed than other East African moorland areas, most of which are losing habitat to expansion of high-altitude agricultural activities and livestock grazing.

The Ethiopian Montane Moorlands biome is a rugged series of mountain-based habitat located in Ethiopia, Eritrea, and northern Somalia. The highlands are divided into three ecoregions that are zoned by elevation: the Ethiopian montane forests, located at an elevation of 3,608–5,905 feet (1,100–1,800 meters); Ethiopian montane grasslands and woodlands, found at 5,905–9,842 feet (1,800–3,000 meters); and Ethiopian montane moorlands, which are located above 9,842 feet (3,000 meters) and compose more than three-fourths of the land above this altitude.

This is the largest Afro-Alpine region in Africa, extending across an area of 9,700 square miles (25,120 square kilometers). The climate is considered complex, with annual rainfall at its greatest in the southwest, at 98 inches (2,500 millimeters), and as little as 38 inches (1,000 millimeters) in the north. The annual mean temperature range on the higher peaks is 43–54 degrees F (6 to 12 degrees C). Frosts are common throughout the year, especially in the months of November through March.

Vegetation

These semiarid montane moorlands lie above the tree line, and vegetation, known as *wurch* to Ethiopians, consists of grasslands and moorlands filled with herbs. Trees are all but absent at such high elevations. The main vegetation types are scrub, grassland, and herb meadow. Much of the montane vegetation is a heathland scrub around 19–39 inches (0.5–1.0 meter) high, dominated by *Philippia,* the tree-heather *Erica arborea,* and other shrub species; along with the small plants *Helichrysum, Alchemilla,* and *Cerastium;* and the grasses *Koeleria* spp.

There is very little vegetation on the steep rocky slopes and cliffs in the high-elevation regions. Flat swamp areas are dominated by the sedge *Carex monostachya.* The giant *Lobelia rynchopetalum,* which grows to nearly 20 feet tall when fully flowered, is a distinctive feature of plant life at this elevation. All vegetation found in the Afro-Alpine zone show xeromorphic characteristics, that is, leaves and other features adapted to retain water and reduce transpiration.

Characteristic Animals

The high-altitude environment means that endemic species (those found nowhere else on Earth) often display a number of unique behavioral, morphological, and physiological adaptations. Notable species at this elevation include highly threatened mammals such as the Ethiopian wolf (*Canis simensis*), one of the rarest canids in the world. Also known as the Simien fox, the Ethiopian wolf lives in open moorlands higher than 9,842 feet (3,000 meters). Bale Mountains National Park, which has one of the highest proportions of endemic species of terrestrial habitats in the world, contains the largest population of these wolves.

The Ethiopian wolf may have evolved from a shared ancestor of the grey wolf (*Lupus lupus*) that made its way to northern Africa from Europe in the late Pleistocene Period. The northern population of Ethiopian wolf is considered a separate subspecies and is unique among canids as it relies primarily on subterranean rodents for food. Because of the animal's adapted diet, the Ethiopian wolf has been able to thrive in these high-altitude areas where rodent biomass is extremely high.

The giant mole-rat (*Tachyoryctes macrocephalus*) and Starck's hare (*Lepus starcki*), also endemic to the park, compose the largest portion of the Ethiopian wolf's diet. Other mammals in the ecoregion include giant Nikolaus mouse, Ethiopian narrow-headed rat, gray-tailed narrow-headed rat, black-clawed brush-furred rat, the critically endangered Walia ibex (*Capra walie*), mountain nyala, and gelada baboon. Klipspringer and rock hyrax (*Procavia capensis*) are also found in the rocky habitats of this ecoregion.

This avian-friendly biome provides ample water in the form of alpine lakes and streams. Significant numbers of Palaearctic birds winter here, with several thousand wigeon (*Anas penelope*) and shovelers (*A. clypeata*) observed in the Bale Mountains, along with waders such as ruffs (*Philomachus pugnax*) and greenshanks (*Tringa nebularia*).

Globally threatened species found in the Bale Mountains include the spotted eagle (*Aquila clanga*), imperial eagle (*Aquila heliaca*), lesser kestrel (*Falco naumanni*), and wattled crane (*Grus carunculatus*), as well as the near-threatened Rouget's rail.

The Bale Mountains are among the few known breeding sites outside the north temperate zone for such boreal species as the golden eagle (*Aquila chrysaetos*), red-billed chough (*Pyrrhocorax pyrrhocorax*), and ruddy shelduck (*Tadorna ferruginea*). The near-threatened pallid harrier (*Circus macrourus*) and Abyssinian longclaw (*Macronyx flavicollis*) also are found in this region.

Plant and animal colonists traveled several routes to get to these remote highlands, mostly from the dry lowlands surrounding the region. A number of tropical species arrived from the moist areas in the south and southwest, passing through the formidable barriers of the Kenyan deserts in the south and the White Nile floodplains in the west. The low trench of the East African Rift Valley may have further isolated species, by dividing the northern and southern massifs. The ecoregion is naturally very fragmented because it occurs only in the highest portions of the Ethiopian Highlands.

Conservation

Ethiopia has established two major protected areas: Bale Mountains National Park in the southern highlands; and Simien Mountains National Park in the north, which includes Ras Dashen, the highest point of Ethiopia at 15,578 feet (4,620 meters). The Bale Mountains are the largest continuous area above 9,852 feet (3,000 meters) on the African continent.

Numerous alpine lakes are found in this region, some of which persist year-round, providing valuable habitat for migratory Palaearctic waterbirds. These higher elevations are sparsely populated, although agricultural activities continue at considerable elevations wherever it is still possible to grow sufficient food to survive. There are approximately 950 vascular plant species with at least 15 site endemics in the Bale Mountains.

The Simien Mountains National Park is currently designated a World Heritage in Danger site by the United Nations Educational, Scientific and Cultural Organization (UNESCO) in part to protect it from extreme overgrazing. Outside the protected areas, other habitat blocks are relatively intact when they are too high to be used by people for cultivation or grazing. Agriculture decreases above 10,500 feet (3,200 meters), where barley is the only crop that can be cultivated.

The lower reaches of the range are more suitable for growing barley than the upper, less vegetated areas, but mostly the area is being used for cultivation and pastoralism.

Overgrazing and other human-driven activities have disrupted plant and wildlife habitats throughout the region. Rising temperatures due to climate change have further impacted the area by shrinking the size of the ecoregion, which could even lead to its eventual disappearance as average temperature increases push habitat zones further upslope. Habitat loss due to climate conditions, and threats such as hybridization of native ungulates with free-ranging domestic goats, are negatively impacting the populations of endemic animal species, such as the Walia ibex, and taking its toll on other animal species as well.

Muraree Lal Meena

Further Reading

Henze, Paul B. *Layers of Time*. New York: Palgrave Macmillan, 2000.

Kingdon, Jonathan. *Island Africa: The Evolution of Africa's Rare Animals and Plants*. Princeton, NJ: Princeton University Press, 1989.

Nievergelt, B., T. Goos, and R. Güttinger. *A Survey of the Flora and Fauna of the Simien Mountains National Park, Ethiopia*. Zurich, Switzerland: Pano-Verlag, 1998.

Vuilleumier, F. and M. Monasterio. *High Altitude Tropical Biogeography*. New York: Oxford University Press, 1986.

Yalden, D. W. and M. J. Largen. "The Endemic Mammals of Ethiopia." *Mammal Review* 22, no. 3–4 (1992).

Euphrates River

Category: Inland Aquatic Biomes.
Geographic Location: Middle East.
Summary: The Euphrates, part of the Mesopotamian "cradle of civilization," includes a unique wetland ecosystem; the river is threatened by the construction of dams and irrigation canals.

The Euphrates River is the longest river in southwest Asia, exceeding 1,678 miles (2,700 kilometers) in length. It flows through Turkey, Syria, and Iraq, and is one of the area's two major rivers. Together with the Tigris, they constitute the largest and the most important river system in the Middle East. The construction of dams and irrigation canals across the Euphrates has reduced the river's outflow and drained its lower reaches, especially the marshes, and threatened its biota.

The Euphrates River emerges from the confluence of two major headstreams, Murat Nehri and Karasu, near the city of Keban in Turkey. It flows through steep canyons and narrow gorges in the river's upper reaches, then enters Syria near Jerablus, where it receives water from two tributaries: Nahr al-Khabur and Nahr Balikh. The Euphrates continues its flow in a southeasternly direction until it enters Iraq. Northwest of Basra in southern Iraq, the river unites with the Tigris to form the Shatt-al-Arab, running about 124 miles (200 kilometers) in a southeastward direction before it drains into the Persian Gulf.

Before merging with the Tigris, the Euphrates flows in an alluvial low basin. It forms a unique wetland area called the Hammar Marsh, the largest water body on the lower Euphrates, which encompasses 1,081 square miles (2,800 square kilometers). The river floods to 1,737 square miles (4,500 square kilometers); the marsh is about 75 miles (120 kilometers) long and 16 miles (25 kilometers) wide. Hammar Marsh is a eutrophic swamp with a major area of open water, forming Lake Hammar. It is one of three marshes that constitute the Mesopotamian Marshes.

The other two marshes are the Central (Qurnah) and the Hawizah, both fed by the Tigris and/or its distributaries. Encompassing an area of about 6,564 square miles (17,000 square kilometers), the Mesopotamian Marshes are among the top 10 marshlands in the world. Located in the lower portion of an area between the Tigris and Euphrates, this area epitomizes Mesopotamia, "the land between the two rivers," which has long been known as a moist and fertile place, attracting and sustaining a series of early civilizations.

Flora and Fauna

The Euphrates River is a unique watercourse with different aquatic biomes: rivers, streams, lakes, and wetlands. It provides various habitats for aquatic floral and faunal elements. Aquatic vegetation includes emergent plants, the most dominant of which are the common reed (*Phragmites australis*) and reed mace (*Typha domingensis*). In addition, there are floating and submerged plants such as hornwort, eelgrass, pondweed, duckweed, stonewort, and water lilies. Papyrus (*Cyperus* spp.) has both grown naturally and been cultivated here since the earliest historic records.

Mammals of the marshes include the signature subspecies smooth-coated otter (*Lutra perspicillata maxwelli*). A variety of amphibians, such as toads and frogs; reptiles such as turtles, including the threatened softshell turtle (*Rafetus euphraticus*); and crustaceans, such as *Potamon* crabs are found in the rivers and wetland areas here. The shrimp *Metapenaeus affinis* is both commercially and ecologically important in the river and marshlands.

Approximately 66 fish species exist in the Tigris-Euphrates river system, most of which are types of freshwater carp, barbel, and minnow. The Iraq blind barb (*Typhlogarra widdowsoni*) is endemic, that is, evolved specifically in this habitat and found nowhere else. Another endemic

A member of the Al Shakamra tribe, which lives by the marsh waters in the village of Al Kuthra outside Basra, Iraq, sits by his boat in April 2010. The destruction of most of the Hammar Marsh, which was 94 percent drained in 2001, was an ecological disaster but restoration efforts are ongoing. (U.S. Army/Adelita Mead)

species is a local form of the catfish *Glyptothorax steindachneri*.

A few Euphrates fish are of marine origin, such as the yellowfin seabream (*Acanthopagrus latus*); several are anadromous species, including Hilsa shad (*Tenualosa ilisha*).

Many kinds of birds, such as pelicans, cormorants, herons, waders, storks, ducks, and waterfowl, reside along or migrate via the river and the marshes, feeding on fish, amphibians, and reptiles. Characteristic avians here include the Basra reed warbler (*Acrocephalus griseldis*), an endangered endemic species with a range now thought to be limited to the marshlands. Two other endangered endemic waterfowl species here are the Iraq babbler (*Turdoides altirostris*) and grey hypocolius (*Hypocolius ampelinus*). Among those birds making migratory stopovers in the southern reaches of the Euphrates River biome are marbled teal, white-headed duck, and Dalmatian pelican.

Effects of Engineering Projects

In an attempt to divert water from the Euphrates River for irrigation, drinking, hydroelectric power generation, and other purposes, Turkey, Syria, and Iraq have over the years initiated water development and drainage projects. Construction of the first modern water diversion structures started in Iraq with the disposition of the Hindiyah Barrage, which diverted the water of the Euphrates into irrigation canals. Another dam, Ramadi Barrage, was erected in 1951, diverting water from the Euphrates into Lake Habbaniyah. Following that, the Haditha Dam and others were built on the Euphrates in Iraq.

Later, Turkey and Syria began constructing their own dams, such as the Keban and Atatürk in Turkey, and Tabqa in Syria, on the Euphrates and its tributaries. With the launch of the Southeastern Anatolia Project in the early 1980s, Turkey planned to incorporate 21 dams, including the major Keban and Atatürk, and 19 hydroelectric facilities into a project largely designed to use the waters from Euphrates and Tigris headwaters to revitalize economic, social, and cultural life.

Aside from barrages and dams, the construction of the Main Outfall Drain (MOD), later called the Third River/Saddam River project, began in 1953 to drain the area between the Euphrates and Tigris south of Baghdad. Two other canals were constructed near the MOD, the Mother of Battles Canal and Fidelity to the Leader Canal. The former was used to divert the flow of the Euphrates southeast to the Hammar Marsh, and the latter transported water from the MOD to the south of Basrah.

To reclaim land for agriculture and suppress Saddam Hussein's opponents, who used the marshes as a refuge, drainage of the marshes began with the erection of a dike to divert Central Marsh overflow away from Hammar Marsh. This was aggravated by dam construction in Turkey and Syria. The Iran–Iraq War also played a role in damaging the wetlands, and by 2001, 94 percent of Hammar Marsh had been converted to bare ground.

The construction of dams and canals decreased the water flow of the Euphrates, especially in its lower reaches, by 60–90 percent, reducing the amount of water available for human and habitat use, and increasing its salinity. This led to tension between Turkey on one hand and Syria and Iraq on the other hand. The building of numerous dams and irrigation projects also led to the displacement of a large number of people, changing their way of living.

In addition, it posed serious threats to the Mesopotamian marshes and their biota. The destruction of Hammar Marsh was a major ecological disaster, putting several bird populations—such as the marbled teal, Basra reed-warbler, sacred ibis, and African darter—at risk. Also, 28 percent of the historic number of fish species was lost. After the 2003 invasion of Iraq, efforts were undertaken by the new Iraqi government, the United Nations, and various U.S. agencies to restore the marshland and its ecosystem. Partial recovery of the wetland is underway.

Climate Change

In addition to the dams and other water diversion projects, climate change makes the water supply sensitive to variations in rainfall and availability. Rising sea levels, due to warming waters, could have an immediate and devastating impact on coastal areas, including saltwater incursion on habitats at the mouth of the Shatt-al-Arab, and in the marshlands further inland. Extended drought could affect agricultural output and further stress water availability and food supply in the Middle East, which could further destabilize the social and political fabric of the area.

Nisreen H. Alwan

Further Reading

Coad, Brian W. *Freshwater Fishes of Iraq*. Sofia, Bulgaria: Pensoft Publishers, 2010.

Ettinger, Powell. "Iraq's Marshes Show Progress Toward Recovery." *Wildlife Extra*, March 11, 2009. http://www.wildlifeextra.com/go/news/good-iraqimarshes.html#cr.

Rzóska, Julian. *Euphrates and Tigris, Mesopotamian Ecology and Destiny*. The Hague, Netherlands: Dr. W. Junk Publishers, 1980.

Scott, Derek A., ed. *A Directory of Wetlands in the Middle East*. Gland, Switzerland: International Waterfowl and Wetlands Research Bureau, 1995.

European Broadleaf Forests, Western

Category: Forest Biomes.
Geographic Location: Europe.
Summary: These characteristic temperate deciduous forests are a vital, living piece of the European habitat puzzle.

Western European broadleaf forests represent a group of temperate forest communities that cover a large area of central-western Europe, mainly in the midmountain ranges of (from west to east) France (Central Massif), Switzerland (Jura Mountains), Germany (Central German Uplands), Austria (Bohemian Massif), and the Czech Republic. Their eastern limits reach approximately to the Caucasian and Baltic mountains and Asia Minor territories.

In these areas, the general climate is Euro-Siberian, characterized by moderate temperatures (average annual temperature of 55 degrees F or 13 degrees C), with cool summers and relatively mild winters, except in certain eastern areas, where very cold winters occur. Most areas present a uniform annual rainfall regime without water-scarcity problems for forest communities. This regime averages 39–71 inches (1,000–1,800 millimeters) annually. In areas influenced by closer mediterranean conditions, a slight summer drought can be present.

Forest Composition

The Western European Broadleaf Forests biome is represented by three main communities: deciduous oak, including sessile oak and pedunculate oak; hornbeam, including common hornbeam and hop-hornbeam; and beech.

Stands of sessile oak (*Quercus petraea*) and pedunculate oak (*Quercus robur*) are spread across the basal areas of central-western Europe's midmountain ranges, mainly from 2,300–3,300 feet (700–1,000 meters) above sea level, and influenced by Euro-Siberian bioclimatic conditions. Although their distribution ranges are similar, these two deciduous oaks present slight differences. Sessile oaks are not lobed and have a longer stalk, up to 3 to 4 inches (80–100 millimeters). The oaks also produce different fruits. *Quercus robur* acorns have a long peduncle; *Quercus petraea* acorns don't have it or have a very short one. As to habitat, pedunculate oak manifests more resistance to continental conditions and shows a major expansion over central eastern Europe to the Urals.

Community composition comprises a dense understory, composed of a diverse group of taxa characteristic of wet-temperate conditions, such as wood sage (*Teucrium scorodonia*), purple betony (*Stachys officinalis*), spurge-laurel (*Daphne laureola*), and sanicle (*Sanicula europaea*). There are also some shrubs that provide edible fruits, such as blackthorn (*Prunus spinosa*) and wild strawberries (*Fragaria vesca*); and ferns like *Pteridium aquilinum* and *Blechnum spicant*.

Otherwise, these deciduous oak lands are enriched with other companion trees such as common howthorn (*Crataegus monogyna*)—sometimes in shrub form—as well as holly (*Ilex aquifolium*), whitebeam (*Sorbus aria*), and shepherd whitebeam (*Sorbus torminalis*). When deciduous oak lands become altered, the forest community tends to be replaced by heath lands that are composed mainly of secondary succession species like heathers and brooms.

Hornbeams

Common hornbeam (*Carpinus betulus*) and hop-hornbeam (*Ostrya carpinifolia*) stands usually occur in intermediate zones between deciduous oak lands and beech forests, located at 2,300–4,000 feet (700–1,200 meters), and mainly in shady and humid soils of the central Europe mid-mountain areas. A few relict populations of these species are present on the islands of Corsica, Sardinia, and Sicily.

Morphological differences between these two species of the *Betulacea* family are manifested mainly in bark, fruits, and leaves. Whereas common hornbeam has a greenish-gray bark, even in older trees, hop-hornbeam has a grayish color that becomes darker with time. Also, common hornbeam forms seed brunches composed of involucres divided into three segments that partially surround one small nut, whereas hop-hornbeam forms seed bunches in which each seed contains two to four small nuts. Hop-hornbeam leaves have lateral nerves ramified to the edge.

Community composition includes an usually rich and dense understory, with wet-temperate shrubs like common dogwood (*Cornus sanguinea*), European cornel (*Cornus mas*), and butcher's broom (*Ruscus aculeatus*); and perennial herbs like black hellebore (*Helleborus niger*) or different cyclamen species. Also, the formation of mixed stands is very usual, as common hornbeam and hop-hornbeam are accompanied by other tree species, mainly deciduous oaks and beech, but also by common hazel (*Corylus avellana*) and South European flowering ash (*Fraxinus ornus*).

Fauna

The region is home to a variety of animal and bird species, but most large mammal populations are in decline, with 20–25 percent of mammals and 15–40 percent of forest birds listed as threatened in central Europe. The population of European bison (*Bison bonasus*) was down to 12 animals when it was rescued from extinction; its population is still too small to sustain itself. The lynx (*Lynx lynx*) is endangered because it needs a large home range in a remote habitat. Other threatened mammals include the wolf (*Canis lupus*), steppe polecat (*Mustela eversmannii*), and spotted souslik (*Spermophilus suslicus*).

Two eagles of the region, the white-tailed eagle (*Haliaeetus albicilla*) and the greater spotted eagle (*Aquila clanga*), both need large tracts of undisturbed forests, lakes, or rivers, and often wetlands to thrive. They are threatened by the loss of unfragmented habitats, and by poaching along their migration routes in the southeastern part of Europe. Other threatened birds include the corncrake (*Crex crex*), lesser kestrel (*Falco naumanni*), and aquatic warbler (*Acrocephalus paludicola*).

Human Use of Forests

Historically, due to their hardness, deciduous oaks have been used in construction of houses, traditional shipbuilding, and in railway industries. This is especially true of sessile oak, which has a straight growth compared to pedunculate oak. Sessile oak wood also is greatly appreciated in the wine industry, due to the organoleptic qualities that the barrels provide to wine. Also, oak wood is used for firewood and charcoal.

At the same time, oak is an important symbolic tree for various European cultures, representing values such as strength and commitment. Greeks associated it with Zeus; Romans did the same with the god Jupiter. Germanic tribes worshiped the great god oak, while Gauls chose oak forests as cult places (in fact, *druid* means *oak man*).

Hop and common hornbeam wood also have been used in construction, as well as in traditional medicine. Hornbeam leaves are boiled to obtain eye drops, and the boiled bark is a well-known cough remedy.

Climate change further threatens the biodiversity of these forests. Warming temperatures, which are projected for the region, will tend to impose

Beeches

Beech (*Fagus sylvatica*) is the only endemic species of the *Fagus* genus in Europe, differentiated from Oriental beech (*Fagus orientalis*), whose distribution range extends from the eastern boundary of the *Fagus sylvatica* influence area to the Caucasian region, Iran, the Balkans, and Asia Minor. Beech stands are perhaps the most significant examples of central and western European broadleaf forest communities, due to their present wide distribution range.

This deciduous forest community tends to occur in midmountain ranges, usually at 2,625–5,250 feet (800–1,600 meters), in places where rainfall is more than 31 inches (800 millimeters) per year. Morphologically, beech bark is similar to common hornbeam bark, but its color is whiter, and it has no ribs. Beech leaves also differ from common hornbeam and hop-hornbeam leaves in their edges, due to the characteristic absence of serrate margins.

Community composition is highly influenced by the very dense tree canopy that remains during the spring and summer months, as it limits light. As a result, the understory usually is poor, composed first by a dense, thick carpet of leaves and then by an abundance of low, shade-tolerant plants such as wavy hairgrass (*Deschampsia flexuosa*), woodruf (*Galium odoratum*), spurge-laurel (*Daphne laureola*), green hellebore (*Helleborus viridis*), perennial herbs, ferns such as *Pteridium aquilinum*, and mosses. Sometimes, beech is accompanied by other tree species, forming mixed forest stands, including evergreen trees such as holly (*Ilex aquifolium*) and yew (*Taxus baccata*), and deciduous trees like silver birch (*Betula pendula*) and chestnut (*Castanea sativa*).

Beech wood has been widely used in construction, and in making items such as carts, ships, oars, furniture, and utensils. Before the wide use of fossil fuels, beech firewood and charcoal were popular. Nevertheless, current use is very low. Beech fruit (beechnuts) were traditionally eaten by people, and their importance as a food for Indo-European cultures was described in the 6th century by St. Isidore of Seville in his *Memories*.

habitat shifts, generally moving some species to higher, cooler elevations—as well as foment some expansion of habitat for invasive species and pests. On the other hand, the warmer climate would tend to boost production of some of the most desirable commodity trees here. Trending against this is the increased likelihood of more frequent and more severe drought events.

Francisco Javier Gómez Vargas

Further Reading

Dudley, Nigel, Daniel Vallauri, Helma Brandlmaier, Gerald Steindlegger, and Duncan Pollard. *Deadwood—Living Forests: The Importance of Veteran Trees and Deadwood to Biodiversity.* Gland, Switzerland: World Wide Fund for Nature, 1996.

Heath, M. F. and M. I. Evans, eds. *Important Bird Areas in Europe: Priority Sites for Conservation.* Cambridge, UK: BirdLife International, 2000.

Kirby, Keith and Charles Watkins, eds. *The Ecological History of European Forests.* Oxfordshire, UK: CABI Publishing, 1998.

Linder, Marcus, et al. "Climate Change Impacts, Adaptive Capacity, and Vulnerability of European Forest Ecosystems." *Forest Ecology and Management* 259, no. 1 (2010).

Röhrig, E. and B. Ulrich, eds. *Temperate Deciduous Forests. Ecosystems of the World, Vol. 7.* Amsterdam: Elsevier Science, 1991.

Everglades Wetlands

Category: Inland Aquatic Biomes.
Geographic Location: North America.
Summary: The Florida Everglades is a flowing "river of grass" and an international treasure, threatened by agricultural and residential development and by climate change.

The Everglades was once an 11,000-square-mile (28,500-square-kilometer) wetlands and body of water flowing through much of southern Florida and into the Gulf of Mexico and Atlantic Ocean.

While somewhat diminished from its original state, this still-vast ecosystem supports an abundance of wildlife, including such hallmark species as roseate spoonbills, wood storks, alligators, crocodiles, and manatees. The habitats here are rich and diverse—but the Everglades was once viewed as a large swath of underdeveloped real estate.

This established the goal of drastically altering the area's hydrology to create dry land that could be used for agriculture and municipal water supply endeavors. While this process has supported a large sugarcane industry and great population growth, the ecosystem has suffered lost habitat and declines in water quality. A slow but continuous national effort has helped to restore water flow through the Everglades, and to ensure the viability of this ecosystem is sustained.

Ecologic and Cultural History

As early as 128,000 years ago, during the Pleistocene Epoch, most of the Everglades were completely submerged as a lagoon that rested upon an expansive limestone bed. As the sea levels then receded, the Everglades were formed approximately 5,000 years ago. From near the center of the vast and extremely flat peninsula of Florida, a vast expanse of freshwater seeped and flowed through much of the southern half, traveling 100 miles (161 kilometers) from Lake Okeechobee to Florida Bay. Known by different cultures as The Lake of the Holy Spirit, Grassy Water, River Glades, and Everglades, this unusual wetlands ecosystem has withstood a number of different land uses.

The first to discover the value of the Everglades, about 1,000 B.C.E., were the Calusa Indians, a highly organized people who crafted shell tools and extensive canoe trails, consuming aquatic animals such as fish, turtles, alligators, and shellfish. After being challenged by other tribes, the Calusa, as well as the Miccosukee, sought refuge in the Everglades and safety in the peninsula's vast prairies and mangrove swamps.

When European colonizers discovered the marshy expanse, the development of farmland and water draining began. By the mid 1800s the consensus was that the Everglades was a useless swamp and should be drained; in 1850, the U.S. Congress authorized the Swamp and Overflowed Lands Act, transferring 20 million acres (8.1 million hectares) of such land from federal to state control for purposes of drainage.

The Big Drain

In 1904, recently elected Florida governor Napoleon Bonaparte Broward pledged to create a dry Empire of the Everglades. The state established drainage districts and began installing canals to remove water from the land. Farmers began converting the River of Grass to a sea of vegetables, sugarcane, and sod. Sugarcane was one of the first large commercial industries in the region, established by 1920. Towns boomed, but in the late 1920s two hurricanes devastated built-up areas. As a result, the Hoover dike was built in the mid-1930s, to reduce the impact of storm flooding.

Natural disasters were common in this region. In the 1950s, residents of southern Florida experienced considerable flooding and hurricanes, large wildfires, and droughts. These weather episodes led Congress to authorize the Central and Southern Florida Project, a massive flood control undertaking that would ensure sustained supplies of water to the growing population and to the agricultural industry. Development meant that less water would be available for the ecosystem and the habitats within it, however. While this was a large project, designed to serve a population of two million, the rapid growth in this region ultimately mandated providing for the water needs of some six million people.

Drastic Impact

The Everglades are half the size they were in 1900. Many species that once called the Everglades home are now extinct or extirpated in the area. The Everglades Wetlands biome receives less than one-third of its historic water flow, and much of this water is contaminated by fertilizer and other runoff.

Wading bird populations have declined by as much as 95 percent, as water and habitat availability shrank away. Nearly 68 plant and animal species today are threatened and endangered. Approximately 1.7 billion gallons (6.4 billion liters) of water per day that used to be stored and utilized by the ecosystem is now channeled directly into the ocean. Because this runoff is often polluted with nutrients and toxic products used in agriculture, lawn care, and municipal functions, the fish populations and seagrasses are declining in the Florida Bay estuaries. Lower flow of water in some areas, due to its diversion for human use, has set the scene for saltwater intrusion into the supply system, wreaking havoc upon some habitats.

In the mid-1980s, phosphorous originating from the sugarcane fields in the Everglades Agricultural Area and in Lake Ochechobee led to large algal blooms. For four years, 1988–92, the cane industry resisted accepting responsibility for this pollution, and fought attempts by both state and federal governments to regulate this nutrient material. Separately, mercury contamination was accumulating from power plant emissions.

Fauna

In addition to development and agriculture, the pet trade has disrupted this ecosystem. In particular, nonnative snake species such as the Burmese python and boa constrictor are commonly released by owners when they become too large. A report by the National Academy of Sciences found that these escapees had, within a decade, wiped out 99 percent of raccoons, opossums, and other small and medium-sized mammals across great swaths of the Everglades, as well as 87 percent of bobcats. In early 2012, the federal government approved a bill to ban the sale of imported snakes, including Burmese python, northern African python, southern African python, reticulated python, green anaconda, yellow anaconda, Beni or Bolivian python, DeSchauensee's anaconda, and boa constrictor.

An International Treasure

Even through the decades of heavy damage, not everyone viewed the Everglades as a place to drain and develop. In 1934, Congress officially declared portions of the Everglades a national park. Later, President Harry S. Truman dedicated 1.3 million acres (526,000 hectares) of land to the park. Internationally, the Everglades is designated a Biosphere Reserve, a World Heritage site, and a Wetland of International Importance. The Everglades National Park is one of the rare places in the world where alligators and crocodiles coexist.

The Everglades is interspersed with a number of unique habitats. Many of its marshes are sawgrass marshes, places where few animals or other plants thrive, though alligators often choose these locations for nesting. Freshwater sloughs are deeper than marshes and are areas of free-flowing water in between the marshes. The borders between these systems are called "ridge-and-slough" landscapes. Aquatic animals such as turtles, young alligators, snakes, and fish live in sloughs; they usually feed on aquatic invertebrates, such as the Florida apple snail. On the ridge zones are wet prairies, which being elevated tend to support a higher diversity of creatures than sawgrass marshes.

Drier lands surround these habitats. Tropical hardwood hummocks are elevated areas that support a wide range of trees, including strangler fig and mahogany, as well as other terrestrial life. The prairies and sloughs are bordered by pine forest and cypress swamps. Pine forests are highly susceptible to fire, but require this process to maintain their health. The Big Cypress ecosystem measures about 1,200 square miles (3,100 square kilometers) and supports an abundance of associated vegetation and animals.

Water from Lake Okeechobee and The Big Cypress eventually flows to the ocean. At this point, the Everglades transitions to mangroves, which thrive throughout a range of freshwater, brackish water, and saltwater. The Everglades have some of the most extensive contiguous systems of mangroves in the world. Besides supporting many faunal communities, these groves buffer inland systems from hurricanes and sea-level rise.

Restoring the Flow

In 1983, newly elected Governor Bob Graham announced the formation of the Save Our Everglades campaign with the goal of restoring the Everglades to its condition as of 1900. A part of the plan was the restoration of the Kissimmee River, in which Graham lifted the first shovel of backfill in 1985 to transform the canal into a

An alligator rests after a meal in Everglades National Park, which is one of the few places in the world where alligators and crocodiles both live. In addition to past damage caused by attempts to drain and develop the Everglades, sea-level rise could cause saltwater inundation of 10 to 50 percent of the park's freshwater marshes. (Thinkstock)

meandering river again. Later, in 1994, Governor Lawton Chiles introduced the Everglades Forever Act, aimed specifically at water-quality improvement within the Everglades Agricultural Area, which was still heavily burdened by pollutants and excess nutrients released into the lower Everglades.

In 1995, a Commission for a Sustainable South Florida report detailed the degradation of the Everglades and noted that continued deterioration would cause a significant decrease in both tourism (loss of 12,000 jobs and $200 million annually) and commercial fishing (loss of 3,300 jobs and $52 million annually). The report stated that the ecosystem and surrounding areas had reached a tipping point, where the health and economy of millions of people were at risk.

In 2000, the federal government approved the Comprehensive Everglades Restoration Plan (CERP) and included more than 60 components, which would take 30 years and cost $9.5 billion to implement. The goal of the CERP is to preserve, protect, and restore the Everglades, along with other parts of central and southern Florida, covering more than 18,000 square miles (47,000 square kilometers). The CERP plan was designed to capture freshwater that currently flows into the Gulf of Mexico or Atlantic Ocean and store or redirect it to areas of need, particularly to fragile ecosystems. The rest of the water will go toward human supplies.

In 2012, Department of the Interior Secretary Ken Salazar established a preserve named Everglades Headwaters National Wildlife Refuge and Conservation Area. It was created to filter pollutants, protect wildlife, reduce development, and provide water for Lake Okeechobee. Property owners, primarily in the agriculture industry, will have the option of making easements available for restoration, while allowing for room to raise livestock or plant crops. If the property owners take full advantage of this opportunity, the total area could encompass as much as 150,000 acres (61,000 hectares).

Environmental Future

The Everglades was once completely submerged by freshwater. Rising emissions, especially carbon dioxide from fossil fuel consumption over the last few decades, is leading to an overall warming trend that will increase ocean water expansion and contribute to the melting of ice sheets. As this occurs, sea levels will rise globally—and certainly will contribute to the loss of some of the Everglades.

For example, the entire Everglades National Park lies at or close to sea level. Sixty percent of the park is less than 3 feet (1 meter) above mean sea level, and the highest portion is 11 feet (3.4 meters) above sea level. The 2007 report of the Intergovernmental Panel on Climate Change projects that sea level could rise from seven to 23 inches (18–58 centimeters) by the end of this century, or sooner.

If this occurs, 10–50 percent of the park's freshwater marsh would be transformed by saltwater pushed landward by rising seas. If the transition occurs slowly, mangroves may be able to hold off some of the impact, but this is unlikely if change is more rapid. The future of the Everglades is unknown, but restoration efforts will contribute to its health and its resilience to the effects of the changing climate.

Marcus W. Griswold

Further Reading

Batzer, Darold P. and Andrew H. Baldwin, eds. *Wetland Habitats of North America: Ecology and Conservation Concerns*. Berkeley: University of California Press, 2012.

Grunwald, Michael. *The Swamp: The Everglades, Florida, and the Politics of Paradise*. New York: Simon & Schuster, 2006.

Lodge, Thomas E. *The Everglades Handbook: Understanding the Ecosystem*. 3rd ed. Boca Raton, FL: CRC Press, 2010.

McVoy, Christopher W., Winifred Park Sid, Jayantha Obeysekera, Joel Van Arman, and Thomas Dreschel. *Landscapes and Hydrology of the Predrainage Everglades*. Gainesville: University Press of Florida, 2011.

Eyre, Lake

Category: Inland Aquatic Biomes.
Geographic Location: Australia.
Summary: A desert lake that is most often dry, Lake Eyre is home to hardened species with special adaptations.

Located in the deserts of Australia, Lake Eyre provides an extremely unique habitat for its inhabitants, as the lake does not always contain water. On average, the lake fills up with water every two to three years, and on a very irregular basis, making it difficult for most plants and animals to survive. Mostly, the lake is a string of water holes, leaving very little water to dependably support its inhabitants. Only the hardiest species of plants and animals are able to survive in this environment. The small populations of flora and fauna that do call Lake Eyre home have learned to adapt and thrive in this unusual environment, and they've created a complex ecosystem in the process.

Geography and Climate

Lake Eyre is located in the northern part of South Australia. The lake is generally flat, with the lowest portion about 50 feet (15 meters) below sea level. The lake is surrounded by sandy and stony desert, and the climate is arid to semi-arid—when it rains. During the year, the temperature fluctuates from 70 degrees F (21 degrees C) in July, the coolest time of year, to 99 degrees F (37 degrees C) in January, the area's summer.

Lake Eyre receives about 5 inches (127 millimeters) of rainfall per year, which is about three times less than the world average rainfall. The lake's evaporation rate is about three times higher than the world average, with the evaporation rate exceeding 140 inches (3,556 millimeters) per year. When filled, Lake Eyre is Australia's largest saline lake, containing a 15-inch- (38-centimeter-) thick salt crust on the lake's floor.

Lake Eyre is divided into two parts, north and south, connected by the Goyder Channel. The northern part of the lake is 90 miles (145 kilometers) long and 47 miles (76 kilometers) wide, while the Lake Eyre South is 40 miles (64 kilometers) long and 14 miles (23 kilometers) wide. The combined volume of both parts is 19 cubic miles (79 cubic kilometers).

The lake's basin overlies the Great Artesian Basin, one of the largest underground aquifer basins in the world; the Great Artisan Basin's mound springs are a crucial water source for the biome. During excessive rainfall, the Lake Eyre tributaries—Cooper Creek, Diamantina River, and Georgina River—will feed the lake, but only in very small amounts. Most of the moisture is lost to evaporation prior to reaching Lake Eyre.

Biota

Since the lake's discovery in 1840, it has rarely filled. The first recorded filling was in 1949—and before that, it was thought to be permanently dry. As it fills irregularly, the lake cannot be relied upon to sustain most plants and animals. As a result, the lake is home to few aquatic plants, mainly including a few species of diatoms, green algae, and blue-green algae.

Surprisingly, there are about 20 species of fish that live in the lake. During dry periods, these fish survive in water holes, some of which are extremely small—but the largest are up to 12 miles (19 kilometers) long. These water holes have an average temperature of 95 degrees F (35 degrees C) or higher for days at a time, and are high in salt, making it possible for only the hardiest species to survive.

These fish, which include six species of goby as well as the bony bream, spangled perch, rainbowfish, and silver tandan, have adapted. Many have evolved into smaller versions of their species, enabling them to thrive in small spaces for long periods of time. They can thrive in the harsh conditions and are able to live by consuming the lake's few algae species for food. Their reproductive cycles are adjusted to match the lake's flood years.

There are some invasive species in the lake area, including crayfish, lobsters, shrimp, and crabs. All of these species have had to adapt, as well; the crabs that now live in the lake can survive long periods with little or no water, and dwell in the desert sand. During flood years, the environment can drastically change, and several species of nomadic birds will make the lake home during flood years; these species include the black swan, Australian pelican, cormorants, silver gull, and pink-eared duck. These birds rely on the fish as well as the invasive species for food sources, and use the lake as breeding grounds.

Environmental changes such as rising temperatures from global warming could alter the composition of Lake Eyre—with desert plants filling in the dry riverbeds instead of water. Extended dry spells could endanger the species that have adapted to the river's current cycle, as hardened as they may be. If, however, rainfall increases enough to counter evaporation, then aquatic activity would spring to life in the altered life cycles of the river region's inhabitants. Associated shifts in habitats beyond the lake could alter species distribution in ways that would impact all of the Lake Eyre life forms.

ELIZABETH STELLRECHT

Further Reading

Kerezsy, Adam. *Desert Fishing Lessons.* Crawley, WA: UWA Publishing, 2011.

Kotwicki, Vincent. *Floods of Lake Eyre.* Adelaide, Australia: Engineering and Water Supply Department, 1986.

McMahon, Thomas A, et al. *Hydrology of Lake Eyre Basin.* Malvern, Australia: Sinclair Knight Merz, 2005.

Measham, Thomas G. and Lynn Brake, eds. *People, Communities, and Economics of the Lake Eyre Basin.* Alice Springs, Australia: Desert Knowledge Cooperative Research Centre, 2009.

Timms, Brian V. "The Biology of the Saline Lakes of Central and Eastern Inland of Australia: A Review With Special Reference to Their Biogeographical Affinities." *Hydrobiologia* 576, no. 1 (2007).

Fernando de Noronha Archipelago

Category: Marine and Oceanic Biomes.
Geographic Location: Atlantic Ocean.
Summary: The islands of Fernando de Noronha and nearby Atol das Rocas form part of the mid-Atlantic volcanic chain of islands and seamounts; they are both important nesting areas for seabirds and sea turtles.

Fernando de Noronha Archipelago is a volcanic mountain chain that developed along the east-west ocean fracture zone, forming 21 islands and islets in the equatorial Atlantic, approximately 224 miles (360 kilometers) off the eastern coast of Brazil. The main island, Fernando de Noronha, measures 7 square miles (18 square kilometers), or about 90 percent of the total area of the chain. The islands of Cabeluda, Rata, São Jose, and Sela Gineta together make up most of the remaining land area. The highest point on the archipelago is Morro de Pico at 1,053 feet (321 meters).

Currently, exotic shrublands and secondary forest, consisting of trees introduced by humans, dominate the vegetation. During the 19th century, when the island was used as a penal colony, the original native forests were cleared, apparently to prevent prisoners from building rafts and escaping. Only 5 percent of the original forest remains, near the isolated Cape Sapata.

Due to this clear-cutting, although more than 400 plants have been recorded, today only three are endemic, or found only in this biome: the gameleira (*Ficus noronhae*), mulungo (*Erythina velutina*), and burra leiteira (*Apium escleratium*). The current vegetation includes vine and shrub species that were introduced to the area, along with a few trees, principally of the families *Nyctaginaceae, Bignoniaceae, Anacardiaceae, Rubiaceae,* and *Euphorbiaceae,* as well as some small planted secondary forest fragments.

The coastline is complex, with numerous high cliffs and sandy beaches used as nesting areas for seabirds and sea turtles, respectively. In fact, Fernando de Noronha is the home of one of the largest bird breeding colonies among the islands of the tropical South Atlantic, and Atol de Rocas is Brazil's second-largest nesting area for green sea turtles (*Chelonia mydas*).

Because of the island's volcanic origins, it has never been connected to the mainland; thus, all species arrived on wing or over water, which has

The photo shows Lion Beach on Fernando de Noronha Island, which has a population of about 3,000 people. Birds and sea turtles nest and lay eggs on smaller offshore islands like the one seen here, while the waters surrounding the islands serve as breeding grounds for tuna, billfish, sharks, and marine mammals. (Thinkstock)

resulted in fairly high species endemism among animals. Several bird species are believed to have colonized the archipelago naturally: endemic Noronha vireo (*Vireo gracilirostris*), endemic Noronha elaenia (*Elaenia ridleyana*), eared dove (*Zenaida auriculata noronha*), and cattle egret (*Bubulcus ibis*).

Two species of lizards are found on the islands: the endemic Noronha skink (*Mabuya masculata*) and the introduced tegu (*Tupinambis teguxim*). There are two terrestrial mammals found on the islands: the introduced rock cavy (*Kerodon rupestris*) and domestic sheep. The rich waters around the islands are breeding and feeding grounds for tuna, billfish, and shark, as well as the various cetaceans and turtles. There are numerous marine mammals, too, in nearby waters, including such whales as pilot (*Globicephala melas*), minke (*Balaenoptera acutirostrata*), and humpback (*Megaptera novaeangliae*); and such dolphin varieties as common (*Delphinus delphis*), spotted (*Stenella attenuate*), and melon-headed (*Peponocephala electra*).

Conservation

Fernando de Noronha Marine National Park, covering just over 42 square miles (110 square kilometers) of the archipelago, was established in 1988, and a management plan was implemented in 1990. Despite the serious loss of terrestrial habitat that occurred during the prison era, many protective measures have been established, allowing a variety of introduced and native vegetation to recover and persist. The offshore islets are still relatively undisturbed, making them excellent seabird habitats. The archipelago has a stable human population of about 3,000 inhabitants, concentrated on Fernando de Noronha Island, and a small transient population of tourists and researchers.

Invasive species are common on these islands, so much so that all future conservation efforts must take them into account. Among such non-native biota are linseed, intended for use as cattle feed; the tegu lizard (*Tupinambis merianae*, known as *teju*), imported to control rats but now considered a plague since it feeds mostly on bird eggs; and the rock cavy, (*Kerodon rupestris*, known as *mocó*), a rodent that looks a lot like a squirrel without the bushy tail.

Despite all this, the area has remained relatively unaffected by other destructive forces, including climate change. Global warming is expected to punish some of the low-lying coastal areas of the archipelago. However, the nature and scope of changes in average temperature, moisture, and precipitation remain unknown.

JAN SCHIPPER
JOSÉ-F. GONZÁLEZ-MAYA

Further Reading

Carleton, Michael D. and Storrs L. Olson. "Amerigo Vespucci and the Rat of Fernando de Noronha: A New Genus and Species of Rodentia (Muridae, Sigmodontinae) From a Volcanic Island off Brazil's Continental Shelf." *American Museum Novitates* 3256, no. 1 (1999).

Claudio, B., M. A. Marcovaldi, T. M. Sanches, A. Grossman, and G. Sales. "Atol das Rocas Biological Reserve: Second Largest Chelonia Rookery in Brazil." *Marine Turtle Newsletter* 72, no. 1–2 (1995).

Sanches, T. M. and C. Bellini. "Juvenile Eretmochelys Imbricata and Chelonia Mydas in the Archipelago of Fernando de Noronha, Brazil." *Chelonian Conservation and Biology* 3, no. 2 (1999).

Schulz Neto, A. "Aspects of Seabird Biology at Atol das Rocas Biological Reserve, Rio Grande do Norte, Brazil." *Hornero* 15, no. 17–28 (1998).

Stattersfield, Alison J., Michael J. Crosby, Adrian J. Long, and David C. Wege. *Endemic Bird Areas of the World: Priorities for Conservation.* Cambridge, UK: BirdLife International, 1998.

Fisheries

Category: Inland Aquatic Biomes; Marine and Oceanic Biomes.
Geographic Location: Global.
Summary: The world's fisheries face threats from overfishing, destructive methods, and climate change that could impact global food sources.

Fisheries is the term used to define industries involved in the harvesting, processing, and marketing of aquatic products. In a broad sense, *fisheries* is a term used to describe areas of water bodies, species of organisms being harvested, methods of harvesting, and the people involved in any of these processes. Fisheries products are not limited to fish; rather they include the entire span of aquatic organisms, both marine and freshwater. There is a great variety of such resources—and great challenges in terms of the sustainability of major fisheries in the face of regional and global ecosystem changes.

Fisheries Products

Aquatic organisms such as fish are among the most readily available protein sources in many countries. Fish and other aquatic organisms provide approximately 15 percent of global protein intake for humans, and fisheries provide the general population with access to these protein sources. Aquatic organisms included in the fisheries designation include seaweeds, cnidarians (jellyfish), echinoderms (sea cucumbers), crustaceans (crabs and shrimp), mollusks (snails, limpets, bivalves, squid, cuttlefish, and octopuses), marine reptiles (sea turtles), fish (cartilaginous and bony), marine mammals (manatees, dugongs, porpoises, sea lions, and whales), and even the marine worms (peanut worms) that are harvested in parts of Asia.

Aquatic organisms that are sold alive or chilled command the highest prices. Live seafood is extremely popular in countries in east Asia, such as China (especially in Hong Kong), Singapore, and Taiwan, where they fetch prices multiple times higher than those of frozen fish. Fresh fisheries products are those that have been chilled over short periods; these also cater to the people of Japan and Japanese restaurants all over the world as sashimi. To extend shelf life, aquatic organisms are processed in various ways, often via freezing, canning, fermenting, or drying, which dramatically reduces their cost as compared to fresh goods.

Fisheries products also refer to parts, by-products, or processed aquatic organisms. Caviar, caviar substitutes (roe from nonsturgeon fish), and roe from other marine organisms such as shrimp are included. Mollusks are harvested for nacre, a lustrous substance from which pearls and mother-of-pearl are made. Shark and ray skins are by-products that can be made into leather. Other by-products of fisheries include trash fish or unwanted fish parts that are ground up to make fish meal, which is used in feed for pets, cultured fish, and livestock.

Capture Fisheries

Capture fisheries is a term used for exploiting resources from the wild. Capture fisheries in marine environs occur at near-shore areas such as estuaries, mangroves, and coral reefs; or at offshore areas such as open oceans and seamounts. Capture fisheries also occur in lotic (moving water) habitats like freshwater streams and rivers, or in lotic freshwater lakes. There are several scales at which capture fisheries occur: commercial, artisanal, and recreational. Commercial or industrial fisheries are typically large-scale, and meet the demands of a broader national or international market. Artisanal or subsistence fisheries cater to local demands or are self-sufficient. Recreational fisheries are exclusively for pleasure and not for commercial purposes.

Fishing Down the Food Web

The post–World War II period saw an increase in commercial fishing that has resulted in greater (by tonnage) yields. The ability to capture more fish through better-equipped commercial vessels and more effective means of capture allowed for increased exploitation of aquatic resources. The sharp increase in the number, size, and engine power of vessels allowed for extended periods at sea. The spatial and temporal intensity of exploitation, along with evolved fishing gear, enhanced skills of fishers, and high catch efforts contributed to increased landings. Computers installed in fishing vessels to locate positions of shoals allowed for precision in harvesting. Intensive commercial fishing from the 1960s through the 1990s thereby saw the rapid depletion of global fisheries resources.

The compositions of captured fisheries transited from organisms high in the trophic level, such as carnivorous fishes, to those low in the trophic level, such as invertebrates and planktivorous fishes. The term *fishing down the food web* has been coined to describe this phenomenon.

Initially, global landings of organisms lower in the trophic levels were high, but soon stagnated or declined as overharvesting persisted. Countries where fisheries biomass decreased moved their efforts to deeper waters. Scientists have arrived at a consensus that the propensity for slow growth rates and high longevity of deepwater organisms render this resource type unsustainable.

Some researchers argue that collapses in various fisheries occurred before the 20th century. This conclusion was reached via paleoecological records from marine sediments; archaeological records from ancient human settlements; historical written records, journals, and documents; and ecological records from scientific literature for more than a century. A group of scientists led by Jeremy Jackson (Scripps Institution of Oceanography at the University of San Diego), for example, studied patterns of population abundances and crashes of harvested marine organisms.

Such scientists postulated that larger marine organisms were already trending toward population crashes before the 20th century. Citing, among others, the example of the Steller's sea cows that became extinct in 1768, the scientists stated their belief that aboriginal hunting was responsible for decimating populations throughout the northern Pacific Rim.

Similar patterns were observed for organisms such as whales, dugongs, fur seals, and turtles, the difference being that small isolated populations of those species still exist. Global colonial infiltrations and the advent of vessels capable of long-haul cruises sealed the fate of these organisms. The dramatic reduction in populations of these organisms altered coastal ecosystems significantly and eventually led to the crashes presently observed. These authors did not discount the fact that the intensity and persistence of current exploitation of fisheries resources are unsustainable and, if continued unchecked, will lead to systemic ecosystem collapses in the future.

Current trends show that global fishing efforts are estimated to exceed the optimum factor by three- to fourfold. Excess fishing efforts contribute to losses estimated at $50 billion per year through four main avenues: increased fishing, declining fishing stocks, stagnating or declining resource prices, and fishing subsidies that support overexploitation. The Food and Agriculture Organization of the United Nations (FAO) report based on data from 2007 showed that global fisheries were facing inevitable crashes. A total 52 percent of all fish stocks were already overexploited, 28 percent were overexploited or depleted, 20 percent were moderately exploited, and only 1 percent showed signs of recovery.

In 2009, capture fisheries saw a global landing totaling more than 98 million tons (88.9 million metric tons). If the current levels of capture fisheries continues, some scientists estimate that global fisheries will collapse by the mid-21st century, with only smaller fish surviving.

Fisheries Collapse

At present, there are two major persistent problems related to fisheries: bycatch and bottom trawling.

Bycatch is the incidental capture of nontarget fisheries that are often of no economic concern due to their low commercial value. At times, these are nontargeted species or species that are discarded due to their size. Often, organisms such as echinoderms, crustaceans, turtles, marine mammals, and birds are entangled in the nets set out for commercially targeted fish. Marine cetaceans often get entangled and consequently drown in tuna fisheries. Marine birds, and large marine organisms such as turtle and sharks, often are entangled in longlines when trying to obtain baited hooks targeting tuna, swordfish, and halibut.

Bycatches have been identified as the cause of the dramatic decline in many marine species. The unintentional yet systematic removal of nontarget species from the environment has strong ecological implications; it is considered one of the greatest threats to the marine environment. The highest incidence of bycatches is associated with shrimp capture fisheries, contributing approximately 2 percent of global fisheries landing (by tonnage). Shrimp capture is responsible for 33 percent of total global bycatch.

Trawling is a fishing method involving the use of large nets and mechanized arms on fishing vessels. The nets are set into the water at specified depths and pulled through the water column over a period of time. Bottom trawling occurs when the net is dragged over the seabed, targeting marine organisms living on the seabed or just above it, such as bottom fish, shrimp, and squid. This method of fishing has been associated with heavy, negative effects on seabed communities. Nets dragged over the seabed not only cause seabed sediments to be suspended in the water column, but also physically damage reefs and deepwater corals. In addition, catches made by this method are nondiscriminatory in nature, resulting in large volumes of bycatches.

Global Change and Fisheries

Scientists believe that in the chronological chain of events, factors such as pollution, physical destruction of habitats, and anthropogenic climate change are recent factors affecting fisheries. However, human-induced climate change has been a particularly confounding factor. Coupled with overexploitation, the effects of climate change on fisheries are only now beginning to be understood.

The oceans, being open systems, are governed by many interacting factors, which are often difficult to elucidate in isolation. It is generally agreed, however, that changes in the climate will cause distinct changes in the patterns of distribution of important fisheries resources. Marine organisms are likely to respond to global warming through shifts in latitudinal ranges, for instance. Consequently, local extirpations, invasions by other organisms, and changes in the population structure of remaining or newly invaded species are likely to occur. In short, resultant community interactions for such ecosystems also are likely to change.

Mitigation Measures

Farmed fisheries for both freshwater and marine organisms have been touted as the solution to meet increased demand for a waning natural supply. Economically important aquatic organisms are farmed either ex-situ in facility tanks; in-situ at enclosed inlets or bays; or in sea cages and pens. In 2009 alone, farmed fisheries contributed 80 million tons (73 million metric tons) to global fisheries production. Farmed fisheries relieve fishing pressures from declining wild stocks when correctly implemented.

However, several concerns about this industry have arisen. Large tracts of mangrove forests, particularly in southeast Asia, have been cleared to make way for unsustainable coastal shrimp and fish farms. With increasing bioaccumulation and toxin buildup over time, these areas are abandoned, and new mangrove areas are then deforested. Many aquafarms also are fundamentally nonsustainable, requiring that fingerlings and hatchlings of aquatic organisms are still harvested from the wild and are only grown and fattened in farmed areas.

Cleaning fish from a small boat close to shore in Canada's Maritime provinces. In 2007, 52 percent of all fish stocks were found to be overexploited, while only 1 percent appeared to be recovering. (Thinkstock)

The introduction of antibiotics to farmed organisms, especially if these areas are connected to natural water bodies or other natural ecosystems, is a concern among many environmentalists who fear that resistance to parasites and disease is undermined by the overuse of antibiotics. The biological discharge from aquatic farms to the natural environment raises similar concerns. In recent years, yet another issue surfaced when it was found that fish farms consumed more wild fish than their final output. This situation is being monitored, and scientists are still refining ways to produce farmed aquatic resources with minimal environmental effect.

A major challenge faced by managers of aquatic ecosystems is illegal fishing. For example, in some reefs around southeast Asia, dynamite fishing was rife in the past few decades. Increased efforts by managers of marine ecosystems to curtail bombing of coral reefs have caused dynamite fishing to subside somewhat. At present, both the live-seafood and marine-aquarium trades are booming, with much of the resource base centering on southeast Asian seas. The use of low doses of cyanide to stun and capture live marine organisms is common in parts of southeast Asia.

The effects of this method of marine-organism collection are poorly understood. In addition to the obvious removal of selected organisms, cyanide remains in the ecosystem and affects the remaining sea life. Catch efforts in intertidal pools are especially deleterious for all creatures left in the pools postcollection. Corals are known to be affected by cyanide, with reports of decreased productivity by symbiotic *zooxanthallae.* Managers of aquatic ecosystems are aware that these activities persist but face constraints in the implementation of maritime laws, due largely to enforcement agencies' insufficient capacities to patrol the areas under their purview.

The Way Forward

Stakeholders in the industry are making some progress in dealing with bycatches. Trawl nets have been redesigned to aid in excluding several species of organisms, notably turtles, for which turtle excluder devices have been created. These nets are equipped with a device above or below the nets so that large organisms can escape unharmed. Efforts such as those by World Wildlife Fund (WWF) in 2005–09 to reward innovative ideas to reduce bycatches are seen as a positive step toward solving global problems as a global community. With WWF and interested stakeholders such as the U.S. National Oceanic and Atmospheric Administration (NOAA) and the U.S. Department of Fisheries, some of these novel ideas are being refined and tested to assess their suitability for broad use by the industry.

According to the Database of World Protected Areas, there are at present more than 6,000 marine protected areas, and there are different scientific recommendations for their use. In some areas, partial exploitation is allowed. In others, there are strict no-take zones where no fisheries activities are allowed. There also exists direct evidence of positive spillover effects, where larvae borne from organisms in protected areas are able to seed neighboring reefs.

The constant efforts by aquatic ecosystem managers and scientists to understand the processes behind the ecosystems under their care serve to improve the management of these imperiled ecosystems and, by extension, enhance the management of fisheries resources. These efforts are especially pertinent in open systems such as the oceans, where connectivity and common resources call for the participation of all stakeholders. Global initiatives led by organizations such as FAO are beneficial in elucidating ways in which fisheries-dependent countries can work together to achieve sustainable resources for the future.

Zeehan Jaafar

Further Reading

Barange, M., J. G. Field, R. P. Harries, E. E. Hofmann, R. I. Perry, and F. E. Werner, eds. *Marine Ecosystems and Global Change.* New York: Oxford University Press, 2010.

Food and Agriculture Organization of the United Nations. "Fisheries and Aquaculture Department Website." 2012. http://www.fao.org/fishery/en.

Jackson, B. C., K. E. Alexander, and E. Sala, eds. *Shifting Baselines: The Past and the Future of Ocean Fisheries.* Washington, DC: Island Press, 2011.

Pauly, D. "Global Fisheries: A Brief Review." *Journal of Biological Research—Thessaloniki* 9, no. 3–9 (2008).

Worm, B., E. B. Barbier, N. Beaumont, J. E. Duffy, C. Folke, S. Halpern, et al. "Impacts of Biodiversity Loss on Ocean Ecosystem Services." *Science* 314, no. 1 (2006).

Florida Keys Coral Reef

Category: Marine and Oceanic Biomes.
Geographic Location: North America.
Summary: North America's only coral barrier reef features a rich biota; its attractiveness to human activities must be balanced against the limits and requirements for the health of the habitat.

The Florida Keys is a 158-mile (254-kilometer) chain of keys or islands that begins south of Miami on the Florida mainland, and ends at the Dry Tortugas group at the westernmost tip of the chain. About 6 miles (10 kilometers) offshore, and paralleling the keys, is North America's only coral barrier reef: a natural wonder that supports tourism and commercial fishing industries even as it has declined in terms of ecosystem health. There are few if any other coral reefs in the world so close to such a heavily populated area. The barrier reef here extends 220 miles (354 kilometers) from North Key Largo to the Dry Tortugas.

Biologic Diversity

Coral reefs are among the most biodiverse ecosystems, rivaling even the tropical rainforests. The Florida Keys Coral Reef biome is home to 107 species of corals—80 percent of all coral species in the tropical western Atlantic Ocean—and 500 species of fish. This interdependent ecosystem boasts stony and soft corals, fish, sponges, anemones, jellyfish, snails, crabs, lobsters, rays, eels, conch, sea turtles, dolphins, and seabirds among its richness. Together with mangroves and seagrass beds, one-third of Florida's threatened and endangered species call the reef home.

The Florida Keys coral barrier reef is characterized by spur and groove formations, or linear coral formations flanked by sand channels, in waters 3 feet to more than 100 feet (1 meter to more than 30 meters) deep. Patch reefs also grow in shallow waters on both sides of the keys. Typical stony coral types include brain, star, staghorn, elkhorn, and pillar. These reef-building creatures create rigid limestone exoskeletons, or corallites, that protect the living polyp within.

Soft corals, or gorgonians—such as sea fans, sea whips, and sea rods—are also abundant in this biome. These filter feeders sway with the ocean currents; they lack an exoskeleton, although they contain tough and spiky skelites that give them rigidity and help them anchor to and build up reefs.

Common reef fish here include edible species such as grouper, yellowtail snapper, grunt, hogfish, and barracuda. Tropical fish are numerous, among them angelfish, beau gregory, blue tang, butterfly fish, cowfish, damselfish, drum, parrotfish, porcupine fish, sergeant major, trunkfish, and trumpetfish.

Living Animals

Corals are comprised of millions of tiny animals, coral polyps, that form massive colonies. They grow by producing buds that create new layers above the skeletons of older ones. Corals also reproduce during an annual spawning event that occurs in August after the full moon in the Florida Keys. Polyps eject sperm, and others eject eggs into the water that join and then eventually recruit on hard substrate to form new colonies. Corals grow slowly—typically, a half-inch (13 millimeters) per year. One exception is elkhorn, a branching coral that grows up to 6 inches (152 millimeters) every year. Once abundant in the Florida Keys, elkhorn is now an endangered species.

Corals are scientifically classified in the phylum *Cnidaria.* They are characterized by their tentacles, with nematocysts or stinging cells to feed and defend themselves; a central digestive cavity for food; and radial symmetry. Coral polyps extend their tentacles at night to feed, especially on zooplankton, while during the day, the polyps retract.

Zooxanthellae is an algae that lives within the coral in a symbiotic relationship, giving it color and using photosynthesis to produce 90 percent of the food for the coral while assimilating waste products. Corals require clear, clean, tropical waters with temperatures of 75–85 degrees F (24–29 degrees C). Growth stops when temperatures exceed 90 degrees F (32 degrees C). They also have trouble when temperatures drop below 64 degrees F (18 degrees C) for any extended length of time.

Ancient Reef Events

The Florida Keys were formed over an extended geologic time. The Lower Florida Keys, from Big Pine Key to Key West, is a zone comprised of ancient coral reefs, shifting sand shoals, and tidal sandbars. During the period 125,000–200,000 years ago, these sandbars were shaped by tidal currents flowing between Florida Bay and the Atlantic Ocean. (The Middle and Upper Keys, from Big Pine Key to Key Largo, were formed by corals and calcareous algae.)

Sea level was 20 to 50 feet (6 to 15 meters) higher than now, and most of present-day Florida was under water. The water started dropping 100,000 years ago, down to 300 feet (91 meters) below present levels, causing the corals to die, and creating coral rock. Gradual erosion from air, sun, wind, rain, and wave action solidified the sand shoals and flattened the reefs to slope down toward receding sea levels.

During a glacial period 15,000 years ago, sea level rose again. It stabilized 5,000 years ago, and corals started growing on former reefs. Today, those ancient reefs and sand shoals provide the foundation for land and reefs alike in the Florida Keys.

Interdependent Ecosystem

The ecosystems of the Florida Keys Coral Reef biome include biota in habitats within and around the reefs, mangroves, and seagrasses—each of which depends on the others. Mangroves are salt-tolerant trees that stabilize the shoreline and provide habitat and a nesting area for seabirds. Their roots are nurseries for sea life that migrates to the reef. Mangroves trap and produce nutrients, filter land-based pollutants, and provide an important sink for carbon dioxide. This region has three species of mangroves: white, black, and red. The seaward side is dominated by red mangroves.

Red mangroves have two types of branching aerial roots: prop roots that develop from the trunk, and drop roots that develop from branches. Breathing pores called lenticel deliver oxygen from above the water surface to submerged roots. Red mangroves are salt excluders; they inhibit salt absorption through their roots. To reproduce, a fertilized flower develops into an embryo, then an elongated seed or propagule falls into the water and floats vertically until it roots in sediment. Black and white mangroves follow a similar pattern, but with differently shaped propagules.

Black mangroves occur shoreward of red mangroves. Their leaf undersides are covered with excreted salt crystals. These mangroves produce pneumatophores—small, pencil-like vertical roots—that extend through the ground, enabling the trees to obtain oxygen from air.

White mangroves, the smallest of the three types, grow above the high-tide mark on dry ground. The leaves are thick, rounded at both ends, and the same color on both sides. Their root system resembles that of terrestrial trees, and they extrude salt crystals at the leaf base. These mangroves are protected by state law.

Protected Status

Because of its accessibility from the continental United States, the coral reefs of the Florida Keys have become the world's most heavily visited reef, attracting millions of divers, snorkelers, photographers, ecotourists, and fishermen annually. It is the world's biggest dive destination, with the most charter boats, combined with Florida's largest commercial spiny-lobster and stone-crab fishery. Key West, the most populated key, hosts more cruise-ship guests than any other site in North America. Many of those guests visit the reef.

In recognition of the reef's value, Looe Key and Key Largo National Marine Sanctuaries were created and then expanded in 1990 to form the Florida Keys National Marine Sanctuary, encompassing 2,800 square nautical miles (9,616 square kilometers). Management of state and federal waters is shared by the National Oceanic and Atmospheric Administration (NOAA) and the State of Florida. The sanctuary implemented a zoned management system to reduce user conflicts. Some reefs are off-limits, and fishing is restricted. Special rules apply in Sanctuary Preservation Areas, and a Sanctuary Advisory Council provides citizen input. These sanctuaries also have a water-quality-protection program headed by the U.S. Environmental Protection Agency (EPA), which launched a coral-reef monitoring program. Keys environmental centers encourage reef-friendly actions, but physical effects still occur from heavy use.

Reef Mooring Buoys

One seemingly simple—but highly effective—preservation method is the reef mooring buoy, which relieves boaters from having to drop anchor on the fragile coral. Sanctuary biologist John Halas and then-U.S. Geologic Survey geologist Harold Hudson together designed the reef mooring buoy. The idea began with the holes left when geologists extract small cores of fossilized coral to study growth circles; these holes provide perfect foundations for a mooring. To create a reef-mooring buoy, a stainless-steel eyebolt is inserted into one of these holes and then packed with hydraulic cement or epoxy. After the cement or epoxy cures, a downline with a buoy and pickup line are attached to the eyebolt. Boaters retrieve the floating pickup line to secure their positions on the reef. Today, there are more than 500 reef-mooring buoys here, and the design has been exported to reefs worldwide.

Endangered Ecosystem

The Florida Keys reef is a fragile ecosystem that requires clean, clear waters. In the 1990s, a spate of new coral diseases and algae blooms appeared, resulting in extensive coral and seagrass loss. As ocean waters clouded, photosynthesis was difficult, and nuisance algae outcompeted corals and seagrasses. The source of the water-quality decline was traced to high nutrients—nitrogen and phosphates—found in wastewater, storm water, and

Keys Seagrasses

The Florida Keys is home to the world's largest documented contiguous seagrass bed. These flowering marine plants stabilize the ocean bottom, trap sediments, and release oxygen into the water, reducing sediment on the reef and improving water quality. Keys seagrasses provide habitat for foraging sea urchins and sea cucumbers, and are a breeding ground for pink shrimp, spiny lobster, snapper, and other sea life.

The Florida Keys hosts three common species of seagrass. Turtle grass has thin, flattened, lance-shaped leaves; it forms extensive meadows. Manatee grass has rounded leaves and occurs mixed with other grasses. Shoal grass has thin, flat leaves and is the only species in the keys to colonize high-sediment areas. Great swaths of seagrasses have been lost to propeller damage, accidental boat groundings, and poor water quality.

A seagrass meadow growing in the Florida Keys National Marine Sanctuary in 2010. The sanctuary protects 2,800 square nautical miles (9,616 square kilometers) through a system of management zones. (NOAA/Heather Dine)

agricultural runoff from Florida Bay. Reef Relief, a nonprofit organization dedicated to improving and protecting coral reef ecosystems, began monitoring Key West reefs and alerting scientists, who since have pioneered much research into coral diseases.

The city of Key West led efforts to improve water quality by upgrading sewage treatment to advanced nutrient-stripping levels and then upgraded storm water. A keys-wide No Discharge Zone was adopted, and 31 pump-out facilities plus mobile pump-out boats are available to boaters. The Everglades Restoration plan recognizes the need for improved water quality for the downstream reefs. Nonetheless, conditions are significantly reduced from the gin-clear waters and vibrant coral reefs of the past.

Additional threats have emerged. Climate change raises water temperatures, increasing coral bleaching, disease vulnerability, and fish die-offs at shallow keys reefs. A major challenge is ocean acidification, which can prevent coral growth. Oceans become acidic as carbon dioxide dissolves into them. The waters are 30 percent more acidic than they were before today's fossil-fuel-burning era. This effect is unprecedented over 20 million years—and will result in widespread loss of corals and other marine life.

Local, national, and international efforts are needed if coral reefs in the Florida Keys and elsewhere are to survive. Millions of people visit and enjoy them each year. At risk are the local tourism and fishing economies that provide healthy revenue, high property values, and quality of life for keys residents. But most at risk is the biologic diversity of life itself.

DeeVon Quirolo

Further Reading

Humann, Paul and Ned DeLoach. *Reef Coral Identification—Florida Caribbean Bahamas.* Eugene, OR: New World Publications, 2002.

Lawhorn, Larry. "Ten Things We Can Do for the Oceans." Eco Del Mar, 2011. http://ecodelmar.org/shift/Ten_Things.php.

Porter, James and Karen Porter, eds. *The Everglades, Florida Bay and Coral Reefs of the Florida Keys,*

An Ecosystem Sourcebook. Boca Raton, FL: CRC Press, 2001.

The Royal Society of London. *Ocean Acidification Due to Increasing Atmospheric Carbon Dioxide.* Cardiff, UK: The Clyvedon Press, 2005.

Fraser River

Category: Inland Aquatic Biomes.
Geographic Location: Canada.
Summary: The largest river in British Columbia, the Fraser is a profoundly rich habitat for salmon, migratory birds, and a host of other animal and plant life; it is vulnerable to overuse by humans.

The Fraser River biome in western Canada is the largest river system in British Columbia; the river flows some 850 miles (1,375 kilometers) from its sources in the Rocky Mountains to its mouth at the Strait of Georgia, a channel between the mainland and Vancouver Island in the Pacific Ocean. The catchment basin is approximately 90,000 square miles (230,000 square kilometers). Beginning near Mt. Robson Provincial Park, the Fraser streams through the Rockies and cuts through the Cariboo Mountains and the Coastal Range on its way to the sea. Major tributaries include the Nechako-Stuart River system, the Quesnel, the Chilcotin, the Thompson-Clearwater system, and the Lillooet.

On its way, the Fraser winds through alpine habitat, rugged boreal forests, several major plains, tundra, arid sagebrush scrublands, grasslands, long-tamed farmlands, canyons and white water, marshland, and coastal rainforest zones.

Flora

The central valley area of the Fraser River is a fertile floodplain that features such plants as Indian hemp (*Apocynum cannabinum*), western white clematis (*Clematis ligusticifolia*), blanketflower (*Gaillardia aristata*), and sand dropseed (*Sporobolus cryptandrus*), as well as the occasional ponderosa pine (*Pinus ponderosa*), and Rocky Mountain juniper (*Juniperus scopulorum*) that sprout from cones that get washed down the river. Further downstream, the waters slow and finer materials accumulate to form sandy bars and shores where more stable plants are located, including red alder (*Alnus rubra*), paper birch (*Betula papyrifera*), tatarian dogwood (*Cornus stolonifera*), and black cottonwood (*Populus balsamifera*).

The receded shorelines have rounded stones stabilized with such perennials as Columbia River wormwood (*Artemisia lindleyana*), Canada wildreye (*Elymus canadensis*), brown-eyed Susan (*Gaillardia aristata*), Douglas aster (*Symphyotrichum subspicatum*), and reed canarygrass (*Phalaris arundinacea*). In lower runs, where the waters are calm and salt-influenced species thrive, are found seacoast bulrush (*Bolboschoenus maritimus*), sedge (*Carex lyngbyei*), and common three-square (*Schoenoplectus pungens*).

Fauna

The estuary of the Fraser River is home to more than 400 species of vertebrates and 20 species of birds of prey, including the bald eagle. Waterfowl and shorebirds from Siberia, Alaska, Yukon, and other arctic and prairie areas all stop along the Fraser to refuel. A vast proportion of the continental population of western sandpipers stops here to rest along their migration route. The Fraser River delta is located midway along the Pacific coast, providing an international crossroad for bird migration.

One of the largest salmon rivers, the Fraser provides spawning habitat for Pacific salmon including chinook, chum, coho, pink, and sockeye. Chinook salmon, while the largest, is the least abundant of the Pacific varieties. Pink salmon is the smallest species in the river. Most salmon are anadromous—they mature and live in the saltwater of the ocean, but return to the freshwater of their native river to spawn.

Millions of hatched young salmon spend their early life cycle in the Fraser River before entering the ocean to grow to full adult size. Estuarine marshes, mudflats, floodplains, and river channels of the Fraser are critical transition habitat for the young salmon. Some species spend several years in the Fraser before heading out to sea. The white sturgeon, considered the largest freshwater fish in North America, and another anadromous type, also spawns in the Fraser.

Mammals making their home in the Fraser River biome include beavers, coyote, mink, deer, cougar, black bear, grizzly bear, moose, and caribou. Seals are frequent marine mammal visitors to the delta area. Among the waterbirds breeding within the Fraser River basin are western grebes, mergansers, great blue herons, and mallards. Hawks, owls, wrens, and blackbirds are active at the shore and well inland.

Environmental Issues

The ecosystem of the Fraser and the surrounding lands have been compromised by municipal development, farming, mining, logging, hydroelectric power production, paper manufacturing, fishing, and tourism. Overfishing, damming for hydroelectric power, pollution, and runoff from deforestation also have impacted the Fraser River biome. As a result of water diversion for hydroelectric production and crop irrigation, the river flow and depth have changed in various segments of the river. Global warming seems to have already increased the overall temperature of the river, threatening some salmon spawning grounds, and stressing numerous types of habitat in the biome.

About two-thirds of the people of British Columbia live and work in the Fraser River biome; approximately two-thirds of the province's income is generated here. Half of British Columbia's timber yield is from this area, as well as 60 percent of the region's metal mine production. Ecological conservation and preservation measures are in constant tension with these human land-use facts.

The Fraser delta has been proposed as a Western Hemispheric Shorebird Reserve Network (WSHRN) site. This is an international initiative designed to identify and protect habitats in stopover points used by shorebirds during migrations across North and South America. One of the key sites here for bird migration is the Boundary Bay-Roberts Bank-Sturgeon Bank wetlands; it has been recognized and is protected by provincial, federal, and international bodies. The Alaskan National Wildlife Area on the Fraser River delta has been identified as one of two Ramsar wetlands sites in British Columbia. The Fraser delta wetlands is regulated by several different agencies within the Canadian government; salmon fishing is monitored by Fisheries and Oceans Canada.

Sandy Costanza

Further Reading

Benke, Arthur C. and Colbert E. Cushing, eds. *Rivers of North America*. Maryland Heights, MO: Academic Press, 2005.

Bocking, Richard, C. *Mighty River: A Portrait of the Fraser*. Seattle: University of Washington Press, 1997.

Klinkenberg, Brian, ed. "E-Flora BC: Electronic Atlas of the Plants of British Columbia." Lab for Advanced Spatial Analysis, Department of Geography, University of British Columbia, 2012. http://www.geog.ubc.ca/biodiversity/eflora.

Fundy, Bay of

Category: Marine and Oceanic Biomes.
Geographic Location: North America.
Summary: With the greatest tidal range of any place on earth, the Bay of Fundy is filled with rich marine and terrestrial life, as well as a robust variety of coastal plants.

Aside from the striking beauty of its imposing rocky shores, one of the first things a visitor notices about the Bay of Fundy, on Canada's Atlantic Coast between New Brunswick and Nova Scotia, is the incredible changes brought on by its tides. At high tide, a boat may be even with the dock so that a passerby can easily step onto the deck. At low tide, the boat may be resting on the mud below and that same passerby may be able to attach a flag to the tip of its mast.

With a tidal range of up to 50 feet (15 meters), the Bay of Fundy, which extends northeast for nearly 200 miles (320 kilometers) from its mouth in the Gulf of Maine, has the greatest tidal range of any place on Earth. The tidal bore, or incoming tide, can sound something like an oncoming train. In fact, the volume of water entering the bay at flood time is similar to that flowing through the Gulf Stream.

The extreme differences in height between high and low tides and the violence with which the waters ebbs and flows here create numerous challenges for organisms living in the Bay of Fundy's intertidal zone. The bay's intertidal zone—the area alternatively flooded and exposed by the semidiurnal (twice-daily) tides—is a mosaic of sandy or rocky beach, mudflat, and salt marsh. On steeply sloping shorelines, low tide reveals a damp zone as high as a two- or three-story building. On gently sloping shorelines, one may (at one's peril) wander thousands of yards out over exposed mud flats before reaching the waves.

The Bay of Fundy's climate supports organisms characteristic of cooler seas. Classified as a cold mid-latitude climate with warm summers and no dry season, the area has a mean monthly temperature of 17.4 degrees F (minus 8.1 degrees C) in January to 62.8 degrees F (17.1 degrees C) in July at Saint John, New Brunswick, a port city about midway up the bay's western coast. Monthly precipitation ranges from 3.5 inches (89.6 millimeters) in August to 5.5 inches (139.4 millimeters) in January. The mean water temperature is 32 degrees F (0 degrees C) in winter and 54 degrees F (12 degrees C) in summer.

Terrestrial and Sea Life

The shores of the bay, in the Kejimkujik National Park in southwestern Nova Scotia, are home to hundreds of species of vascular plants, including 23 fern, 15 orchid, and 37 aquatic plant species. On Brier Island, so named for the Brier rose that grows freely across its undulating surface, such rare plants as bluebead lily, pink ladyslipper, fireweed, yellow iris, shrubby cinquefoil, pitcher plant, and the endangered eastern mountain avens are found. A large portion of the island is a nature preserve administered by the Nature Conservancy of Canada.

In terms of the proportion of surface cover, bare rock and tar lichen (*Verrucaria marra*) dominate the region of the high tide line around the bay. Other species reach their highest-percent cover at the high tide line; these include a green alga (*Ulothrix flacca*), a red alga (*Hildenbrandia rubra*), and an acorn barnacle (*Semibalanus balanoides*). The occurrence of brown alga (*Fucus spiralis*) increases with depth until the middle of the intertidal zone, where it dominates.

Knotted wrack (*Ascophyllum nodosum*) is the leading species in terms of surface cover from about 30 percent of the tidal range above mean low water to about 80 percent of the range. Bladder wrack (*Fucus vesiculosus*) has the next-highest cover in the middle of the zone. A red alga, wrack siphon weed (*Polysiphonia lanosa*), reaches its own highest percentage cover in the middle portion of the zone. A hydrozoan, *Dynamena pumila*; a bryozoan, *Flustrellidra hispida*; and a polychaete, *Fabricia sabella*, are often associated with the knotted wrack.

Several species increase in abundance with increasing depth. These include brown alga, such as bladderlocks (*Alaria esculenta*) and *Fucus evanescens*; a golden brown alga, *Ectocarpus siliculosus*; a green alga, *Acrosiphonia arcta*; and red algae including carrageen (*Mastocarpus stellatus* and *Chrondrus crispus*), dulse (*Palmaria palmata*), laverbread (*Porphyra umbilicalis*), *Devaleraea ramentacea*, *Rhodochorton purpureum*, and several crustose species such as *Petrocelis middendorffi*. Tortoise-shell limpets (*Testudinalia testudinalis*) are frequently associated with red algae.

Marine animal life in the Bay of Fundy biome is rich with species, including a range of sharks, including spiny dogfish *(Squalus acanthias)*, the most common shark in the bay; basking shark *(Cetorhinus maximus)*, the second-largest shark; great white shark *(Carcharodon carcharias)*, usually found between April and November; thresher *(Alopias vulpinus)*; and porbeagle *(Lamna nasus)*. The sharks feast on squid, herring, mackerel, small cod, hake, and cusk. They all share the waters with porpoises, sea turtles, and such whales as the right and humpback.

Smaller marine species include the rough periwinkle (*Littorina saxatilis*), the common flat periwinkle (*L. obtusata*), and common periwinkle (*L. littorea*), all grazers. Green sea urchins (*Strongylocentrotus droebachiensis*) are abundant at lower levels of the intertidal zone. The dog whelk (*Nucella lapillus*), a predatory snail, prowls the center of the zone.

Salt marsh is most extensive at two locations in the head of the Bay of Fundy: New Brunswick's Cumberland Basin and Nova Scotia's Minas Basin. The marshes can be divided into two basic types: high marsh, which is infrequently inundated; and low marsh, which is flooded frequently. Cordgrasses (*Spartina* spp.) dominate both marsh types.

Other high marsh species include quackgrass (*Agropyron repens*), saltmeadow rush (*Juncus gerardi*), goose tongue (*Plantago maritima*), American alkaligrass (*Puccinellia americana*), and seaside alkaligrass (*P. maritima*), and both smooth and prairie (*Spartina pectinata*) cordgrass. Other low marsh species include American alkaligrass, slender grasswort (*Salicornia maritima*), and herbaceous seepweed (*Suaeda maritima*).

Along the shores, sandpipers and plovers, white-tailed deer, and moose can be found. Patrolling the skies are bald eagles, ospreys, and peregrine falcons.

Environmental Challenges

As the earth warms, the impact on terrestrial and aquatic life in the Bay of Fundy becomes vulnerable to mutations. Long-term changes in average water temperature could affect the entire life cycle of the fish swimming in and out of the area. Their distribution range could shrink or expand, depending on a species's temperature tolerance and preference. In the 1960s, when water temperatures south of Newfoundland fell slightly, capelin, a cold water fish, moved south toward the Bay of Fundy—and then moved back again in the 1970s when the ocean warmed once more.

Cod, halibut, and plaice altered their distributions in a similar way. It also appears that changing ocean temperatures might cause some fish to alter their migration routes, departure and arrival times, or even their final destinations. Temperature changes can influence the reproduction cycle of fish, the number of eggs laid, incubation time, survival of their young, growth and feeding rates, and the time it takes for them to reach maturity. For example, cod grow faster in warmer water.

The impact of climate change on salmon and shellfish in the Bay of Fundy biome is likely to be both beneficial and detrimental to the species. The salmon, like the cod, grow faster and bigger in warm waters and could reduce the costly fish kills that occasionally happen when the water temperature falls too low in the winter. The benefits, however, might be more than offset by a greater incidence of toxic blooms, disease, or parasites. Significant temperature change in coastal waters might alter the types of fish and shellfish that can be successfully farmed in the region.

Weather changes due to climate change impact both the number and severity of storms along the coastal regions of the Bay of Fundy. As a result, storm surges could be more detrimental to the land, biota, and people of the bay, with the potential to devastate fishing, farming, and forestry industries in its environs.

David M. Lawrence

Further Reading

Gordon, Donald C. Jr., Peter J. Cranford, and Con Desplanque. "Observations on the Ecological Importance of Salt Marshes in the Cumberland Basin, a Macrotidal Estuary in the Bay of Fundy." *Estuarine, and Shelf Science* 20, no. 2 (1985).

Leslie, Scott. *Bay of Fundy: A Natural Portrait.* Toronto: Key Porter Books, 2007.

Mathieson, A. C., C. A. Penniman, and L. G. Harris. "Northwest Atlantic Rocky Shore Ecology." In A. C. Mathieson and P. H. Nienhuis, eds. *Intertidal and Littoral Ecosystems.* Amsterdam: Elsevier, 1991.

Thomas, M. L. H. "Littoral Communities and Zonation on Rocky Shores in the Bay of Fundy, Canada: An Area of High Tidal Range." *Biological Journal of the Linnean Society* 5, no. 1–2 (1994).

Galápagos Islands Xeric Scrub

Category: Marine and Oceanic Biomes.
Geographic Location: Pacific Ocean.
Summary: The Galápagos archipelago is a veritable laboratory for species diversity and evolution, with habitat niches that support a complex web of distinct life forms.

Though it straddles the equator, the Galápagos archipelago is not a typical tropical island paradise. Life there is molded by the availability of freshwater and dictated by two very different seasonal weather patterns and one climatic anomaly. Though the archipelago has remained relatively pristine due to its recent age and isolation, threats include introduced species, loss of native biodiversity, effects from humans, and the effects of climate change.

The Galápagos Islands are famous for their endemic species (found nowhere else on Earth) of mammals and birds. Because many of the mammals and birds here are adapted to the harsh and specific conditions of the Galápagos, any changes in climate affecting the terrestrial and marine life impact strongly upon the six mammalian species on the island and the many bird species. Changes in weather patterns, for example, the El Niño/Southern Oscillation (ENSO) climatic anomaly, benefit land creatures thanks to abundant rainfall, but make life difficult for seabirds and marine life due to warmer waters. Climactic changes favoring warmer El Niño years could be disastrous for the reproductive cycles of such animals.

Climate

The volcanic islands of the Galápagos archipelago lie at the intersection of several major oceanic currents, the seasonal dominance of which influence weather patterns at any one time. From January to June, weather patterns from the north bathe the area in warm winds and currents, creating clear skies with occasional heavy rain showers. These showers provide the only major source of precipitation to the islands' lowland regions.

From July through December, cold currents from the south hold sway, creating a cool, moist season (the *garúa*) characterized by strong upwelling of nutrient-rich waters at the coast and the formation of thick fog at upper elevations. While this fog provides continuous moisture to the highlands, the lowlands receive virtually no precipitation during this season, perpetuating their harsh, desertlike conditions. Transition periods between the two seasons are temporally and spatially variable and highly unpredictable.

Coastal or Littoral Zone

The coastal or littoral zone supports a variety of salt-tolerant plants that are mainly sea-dispersed, and thus often found elsewhere in the Pacific. Mangroves are common in sheltered coves, while creeping vines, grasses, and succulent shrubs grow in sandy areas.

In the coastal or littoral zone, seabirds are prevalent. Approximately 90 percent of the entire population of Galápagos penguins (*Spheniscus mendiculus*) lives on the islands of Isabella and Fernadina. Other endemic bird species include the waved albatross (*Diomedea irrorata*), the flightless cormorant (*Nannopterum harrisi*), the swallow-tail gull (*Creagrus furcatus*), and the rare lava gull (*Larus fuliginosus*). Some birds, such as the waved albatross, may spend most of their time out at sea, while others are well adapted to blending in with the lava-formed landscape.

The Galápagos sea lion (*Zalophus wollebaeki*) lives exclusively along the coastal areas of the islands, as does the Galápagos fur seal (*Arctocephalus galapagoensis*). The Galápagos sea lion and Galápagos fur seal are two of only six mammalian species found on the islands. The others include two species of rice rats and two species of bats, the hoary bat and the Galápagos red bat.

Arid Zone

The arid zone is the most extensive, and the most unique, covering the majority of the low islands and a good portion of the high islands as well. Species here are drought-tolerant and adapted to extreme conditions; many are endemic. As one might expect, the Galápagos's arid zone supports several species of cactus. One of these, the giant prickly pear cactus (*Opuntia* spp.), is an excellent example of adaptive radiation. In this process, one ancestral prickly pear-like colonist is believed to have evolved over time and across the different islands into six species and 14 varieties, radiating out from the original.

Prickly pear cactus pads, fruits, flowers, and seeds are important resources for many Galápagos animals, including a variety of insects, the Galápagos dove (*Zenaida galapagoensis*), the four species of Galápagos mockingbirds (*Nesomimus* spp.), two species of Darwin's finches (*Geospiza scandens* and *G. conirostris*), land iguanas (*Conolophus subcristatus*), and the many island-specific varieties of the Galápagos giant tortoise (*Geochelone elephantophus*). Indeed, *Opuntia* cacti on islands with land iguanas and tortoises are often taller than varieties found on other islands, presumably an adaptation to reduce predation by these herbivorous reptiles. Other species, such as the Palo Santo tree (*Bursera graveolens*), cope with drought by remaining dormant most of the year, putting out a new flush of leaves when the rains come. These trees are often regularly spaced across the landscape, an arrangement that minimizes competition for water.

Transition Zone

Above the arid zone lies the transition zone, characterized by an ever-decreasing abundance of arid-zone plants amid a dense and diverse community of tangled shrubs, perennial herbs, and epiphytes. The mainly deciduous, lichen-festooned forest here is dominated by two endemic trees: the broad Pega Pega tree (*Pisonia floribunda*), which produces sticky fruit that is widely distributed by birds, and two varieties of *guayabillo* or Galápagos guava (*Psidium galapageium*) that produce small, apple-like fruit. The *matazarno* (*Piscidia carthagenensis*) is the tallest tree in the transition zone, possessing an extremely durable and valuable wood. Lichens typify this zone, as they are drought-tolerant but able to take advantage of occasional *garúa* mists that penetrate down into this region.

Moist Uplands Zone

Above the arid and transition zones lies the moist uplands zone, characterized by stands of the softwood genus *Scalesia*; the prickly-ash (*Zanthoxylum* spp.); the flowering fruit source *Miconia* spp.; and coverings of pampas grass, ferns, and sedge. These stands are supported by the *garúa* fogs,

A broad tree spreads its mostly bare branches over an arid area of the Galápagos Islands. Because the only major source of precipitation to the islands' lowland regions are the occasional rain showers during the first six months of the year, much of the vegetation in arid regions of the Galápagos is adapted to drought conditions. (Thinkstock)

which supply a relatively consistent source of moisture, often in the form of droplets that condense on the leaves.

Endemic to the Galápagos, *Scalesia,* like the *Opuntia* cactus, is a fine example of adaptive radiation. The 15 species and six subspecies of *Scalesia* shrubs and trees, members of the sunflower family, have adapted widely to different zones and different islands. The majority of species are found in the transition zone, but the most dominant and easily identified species, *Scalesia pedunculata,* a tree that can reach 49 feet (15 meters) or more, is found in the moist uplands, which indeed are sometimes called the *Scalesia* zone.

The prickly-ash stands are often seen with a dry-season drapery of mosses, liverworts, and ferns festooning an open forest of spiny evergreen cat's claw (*Zanthoxylum fagara*) and other xerophytic shrubs.

The Miconia version of the moist uplands zone is found only on the islands of Santa Cruz and San Cristobal and is named for its dense, almost impenetrable stands of *Miconia robinsoniana,* a shiny-leaved shrub up to 16 feet (5 meters) in height. In this zone, ground burrows or cavities provide a nesting habitat for the endangered dark-rumped petrel (*Pterodroma phaeopygia*).

At the highest elevations of the tallest islands, often on the very rims of the volcanoes, lies the zone of pampas or mixed fern-sedge. During the *garúa* season, this zone can receive up to 98 inches (2,500 millimeters) of rain in certain years, supporting a lush profusion of grasses, sedges, and ferns, its tallest being the Galápagos tree fern (*Cyathea weatherbyana*), which can reach heights of up to 10 feet (three meters.)

Among the animals living in the humid zone are the famous Darwin's finches, also known as Geospizinae or Galápagos finches. They live on many of the islands at the higher elevations. Thirteen species of Darwin's finches are found among the Galápagos Islands. Two live in the arid zone, while the 11 other species prefer the higher elevations of the moist uplands zone.

Preservation

Because of their isolation and relatively recent discovery, the Galápagos have escaped the destruction so often wreaked on oceanic islands. Currently, the biggest threats to the archipelago include introduced species, loss of native biodiversity, and human visitors. For this reason, the Ecuadorean government has declared all the islands, except areas already colonized by humans, to be a national park, managed in conjunction with the non-profit Charles Darwin Foundation and Research Station.

MELANIE L. TRUAN

Further Reading

Fitter, Julian, Daniel Fitter, and David Hosking. *Wildlife of the Galápagos.* Princeton, NJ: Princeton University Press, 2000.

Jackson, Michael H. *Galápagos: A Natural History.* Calgary, Canada: University of Calgary Press, 1993.

Kricher, John. *Galápagos: A Natural History.* Princeton, NJ: Princeton University Press, 2006.

McMullen, Conley K. *Flowering Plants of the Galápagos.* Ithaca, NY: Comstock Publishing Associates, 1999.

Gambia River

Category: Inland Aquatic Biomes.
Geographic Location: West Africa.
Summary: The last remaining undammed river in west Africa is an important reserve of biological diversity and support for human well-being, but is threatened by human activities.

The Gambia River winds over 700 miles (1,130 kilometers) from the Fouta Djallon highlands of northern Guinea through southeastern Senegal and The Gambia into the Atlantic Ocean. It is one of the largest rivers in west Africa and is composed of two distinct habitat types: estuary at the mouth and freshwater further upstream. Each of these supports a collection of flooded areas, swamps, mudflats, and riverine forests.

As part of the broader Senegambia Catchment area, the Gambia River is an important ecosystem for biological diversity, though it has few endemic species (those found nowhere else on Earth) in its own waters. The river supports a range of economic activities from agriculture to fishing, hunting, and transportation. It also shows promise for potential hydroelectric power. Like many of the world's aquatic and coastal systems, it is under direct threat from these anthropogenic factors.

Mangroves, Salt Marsh, Wetlands

The estuarine and freshwater zones, which are largely a function of the reach of the tide and saltwater from the sea, create different communities of flora and fauna. At its mouth, the Gambia is an 8-mile-wide (14-kilometer-wide) permanently flooded ria, or funnel-shaped, estuary formed from the submergence of the lower portion of the river valley. This estuary is a matrix of mangrove swamps and creeks with riverside mudflats on elevated ground formed from silt deposition.

The river narrows as it travels further inland, though it remains about half a mile (1 kilometer) wide, even 124 miles (200 kilometers) from the mouth. As the influence of the tide wanes, so does the level of flooding. The middle section of the river is only seasonally flooded, and the early stages of the river in the Fouta Djallon and southeastern Senegal—before the Gambia meets with any of its important tributaries—do not experience any flooding.

The basin of the Gambia River contains about 1,500 species of plants, 80 species of mammals, 330 species of birds, 26 species of reptiles, and 150 species of freshwater fish. This diversity results from the broad range of habitats including mangroves, freshwater swamps, salt mudflats, and Sudanian-Guinean riverine forest. Though the Gambia is important for biodiversity, it is not unique from other river ecosystems in the area. It possesses only a few endemic species. Most species are held in common with the Senegal, Bafing, and Faleme Rivers and are known together as the Senegal-Gambia Catchment, or the Senegambia Catchment. Thus the Gambia River ecosystem is part of the broader Sudan-Guinean

Savanna biome, and shares several species from the Guinea-Congo Forest biome and the Sahelian biome as well.

The mangroves and wetlands of the Gambia River reach 62 miles (100 kilometers) inland from the Atlantic and comprise one of the 200 most biologically valuable ecoregions of the Earth, as listed by the World Wildlife Fund. The river supports nearly 111,000 acres (45,000 hectares) of mangrove swamp dominated by trees such as *Avicennia africana*, *Sesuvium portulacastrum*, and *Rhizophora* spp. Much of this area is covered with salt marsh herbs and halophytic (salt-loving) grasses. The mangroves support a diverse community of avifauna, both as breeding and wintering grounds.

More than 560 bird species are found here, including many migratory species arriving from Europe each winter who find refuge among the mangroves. The yellow-billed stork (*Mycteria ibis*) and African sacred ibis (*Threskomis aethiopicus*) can be spotted along the Gambia's banks, along with the long-crested eagle (*Lophaetus occipitalis*), pelicans, and other birds that find food and shelter in the area. Owls such as the African scops owl (*Otus senegalensis*) may also be found in this area.

Small-bodied mammals, such as the spotted-necked otter and greater cane rat, and larger-bodied mammals such as duikers, crocodiles, hippopotamuses, and the vulnerable African manatee call this area home.

As the influence of the Atlantic tide and saltwater wanes, the *Rhizophora* mangroves thin and transition into a collection of freshwater swamps, salt flats, and seasonally flooded grasses, trees, and agriculture. The freshwater swamps are dominated by *Phragmites karka* grasses and are habitat for many species of wintering birds. The swamps create a breeding ground for mosquitoes and tsetse flies, which are vectors for malaria and sleeping sickness, respectively. The most efficient malarial vector in sub-Saharan Africa and host to the most deadly malarial parasite is *Anopheles gambiae*, named for the region in which it is common. Native fish diversity and abundance is an important control on mosquito populations and provides a key ecosystem service by regulating the malaria vector.

Human Interaction

The importance of the Gambia River to human communities is apparent from the names of the five administrative divisions of The Gambia: Western, North Bank, Lower River, Central River, and Upper River. Many Gambians outside of the capital depend directly on agriculture and fishing for their well-being. Wood is the most common fuel source and comes from the riverbanks and floodplain. These economic activities both depend on and severely impact the health of the river.

Nearly one-third of The Gambia has been converted to arable land dedicated to the production of rice, maize, groundnuts, millet, and sorghum. Similar land transformations have occurred in southeastern Senegal and are beginning in the mountainous Fouta Djallon of northern Guinea. These transformations began centuries ago. Flood rice cultivation has been the dominant land use on the banks of the Gambia River for recorded history. The system of flood rice cultivation used in the American south was a direct application of the system used along the banks of the Gambia; evidence suggests that American slave owners and traders targeted slaves of particular ethnic groups because of their experience in rice cultivation along the Gambia's banks.

In addition to its role in cultivation, the river and its tributaries are fished from source to mouth. Prawn is caught in the estuary, while fish are consumed over the river's entire course. Fish populations in the river remain relatively healthy. Hunting for wildfowl and mammals of all sizes is common in the riverine wetlands. Crocodiles, African manatee, and hippopotamuses have been hunted nearly to the point of local extinction in The Gambia, though hippopotamuses can be found in southeastern Senegal, where human populations are less dense.

Though the Gambia River remains the last major undammed river in west Africa, there have been recent efforts to build a hydroelectric dam on the border between Senegal and Guinea, in the region of Kedougou, Senegal. So far the governments of The Gambia, Guinea, Senegal, and Guinea-Bissau have raised $700 million through the African Development Bank for construction. The dam has potential to produce 400 gigawatt hours and would connect to power networks in all four countries, though most of the power would go to Guinea and Senegal.

On the negative side, the dam is expected to impact the sedimentary balance in the river and estuary as well as modify the salt front. It will also influence water quality, deplete mangroves, and generally lead to habitat loss and changes in the populations of some species. These concerns have led several conservation groups to begin planning for monitoring ecosystem impacts of dam construction. As a potential source of electricity as well as an important contributor of biodiversity and other ecosystem services, the Gambia River is a frontier for the management of ecosystems for both environmental and human well-being.

One effect of adding this dam to the Gambia River—the modification of the salt front—will likely be exacerbated by global warming. With higher average temperatures and projected lower annual rainfall will come faster average evaporation rates along the river, which in turn will drive greater rates of salt deposition. Global warming effects will also lead to more acute demand for irrigation water, another factor that will lead to drawing water out of the mainstem of the river and its tributaries. The spread of drier and saltier areas will cause a shift in habitat toward more halophytic plants, and thereby alter the range patterns of animals that rely on vegetation that is less salt-tolerant.

STEPHEN WOOD

Further Reading

Jones, Michael. *Flowering Plants of The Gambia*. Boca Raton, FL: CRC Press, 1994.

Louca, Vasilis, Steve W. Lindsay, Silas Majambere, and Martyn C. Lucas. "Fish Community Characteristics of the Lower Gambia River Floodplains: A Study in the Last Major Undisturbed West African River." *Freshwater Biology* 54, no. 2 (2009).

Verkerk, M. P. and C. P. M. van Rens. *Saline Intrusion in Gambia River After Dam Construction.* Enschede, Netherlands: University of Twente, 2005.

Webb, James L.A. Jr. "Ecological and Economic Change Along the Middle Reaches of the Gambia River: 1945–1985." *African Affairs* 91, no. 365 (1992).

Ganges River

Category: Inland Aquatic Biomes.
Geographic Location: Asia.
Summary: The lifeline of India, the Ganges River supports massive and diverse habitats but is threatened by climate change and pollution.

The national river of India, the Ganges, originates at Gaumukh as a stream called Bhagirathi, in the Gangotri glacier in the Himalayan mountains at an altitude of 13,451 feet (4,100 meters) above mean sea level. The first town in the course of the Ganges, Gangotri, is 14 miles (23 kilometers) from its source of origin. The main stream of the river flows through the Himalayas until another two streams—the Mandakini and the Alaknanda—join it at Devprayag, the point of confluence. The river is also known as Ganga Ma, Mother Ganges, and may be called Ganges or Ganga.

The river takes its course through the Himalayan valleys until it reaches the plain at the town of Haridwar. From there, the Ganges flows southeast through the Indian states of Uttar Pradesh, Bihar, Jharkhand, and West Bengal. At Allahabad, the Yamuna River joins the Ganges. The main tributaries of the Ganges are the Yamuna, Ram Ganga, Gomati, Ghaghara, Son, Damodar, and Sapt Kosi Rivers. The largest tributary, the Ghaghara, meets the Ganges in Bihar near Patna. Another important Himalayan tributary is Gandak, which flows from Nepal. The Upper Ganges supplies water to extensive irrigation works.

The Ganges flows through some of the most populous cities in India, such as Kanpur, Allahabad, Varanasi, Patna, and Kolkata. At Bhagalpur, it meanders past the Rajmahal Hills and changes its course southward. At Pakaur, the Ganges has its first distributary, the Bhagirathi River, followed by the river Hoogly. In central Bangladesh, the Brahmaputra and Meghna Rivers join the Ganges. These three rivers combined are called the Padma River, which forms a delta 220 miles (354 kilometers) wide when it empties into the Bay of Bengal. The delta is also called Sundarbans; its plains are among the most fertile and densely populated regions of the world.

The Ganges River basin covers 332,590 square miles (861,400 square kilometers) and encompasses 26 percent of the land area of India, as well as parts of Nepal, China, and Bangladesh. The annual flow of the river is subject to local variations. The predominant water-flow pattern, however, is a low-flow dry season from January to May and a wet season from July to November, with peak flows usually occurring in August. The waters of the Ganges River carry one of the world's highest sediment loads, at a mean annual total of 1.8 billion tons (1.6 billion metric tons), compared with 0.5 billion tons (0.4 billion metric tons) for the Amazon River.

Small boys by the Ganges River. While the river provides water for over a third of India's population, as much as 264 million gallons (1 billion liters) of often untreated sewage still enters the river every day. (Thinkstock)

Biodiversity

The Ganges supports a rich variety of flora and fauna, including the endangered Ganges River dolphin or Susu (*Platanista gangetica*). The Ganges River dolphin is one of only four freshwater dolphins found throughout the world, and is considered threatened; pollution and changes to its habitat from dams and human activity have caused the Ganges River population to drop to a quarter of what it was a mere 15 years ago. Nine other species of aquatic mammals, and three species of crocodiles including the mugger crocodile and the gharial, also call the Ganges home. There are 11 species of freshwater turtles and one minor lizard as well.

The riparian zone supports many plant species of ecological and economic importance. Many crops are grown along the banks of the Ganges River, including staple crops such as rice, sugarcane, potatoes, cotton, wheat, oil seeds, and various legumes. Among the flora of the region are approximately 450 medicinal plants used in Ayurvedic medicine, India's system of traditional healing.

The river has the richest diversity of freshwater fish anywhere in India. Fish are an important part of life here, providing sustenance for people, animals, and birds living near the sacred river. Among the fish in this biome are featherbacks (*Notopteridae* family), common to the Bengal region; barbs (*Cyprinidae*); the walking catfish (*Clarias batrachus*); and others. The Ganges shark (*Glyphis gangeticus*) is critically endangered.

Parrots, crows, myna birds, kites, patridges, and various types of waterfowl such as ducks and snipes find refuge along the Ganges River, but there are no endemic (found nowhere else on Earth) birds in the upper Gangetic Plain. Many birds cross the Himalayas to reach the Ganges and migrate to wetland areas. Among the bird species found along the Ganges River, there are two threatened species: the great Indian bustard (*Ardeotis nigriceps*) and lesser florican (*Sypheotides indicus*).

Common mammals found along the Ganges River include deer, boars, wildcats, foxes, wolves, and jackals. At the Ganges Delta in Bangladesh, the clouded leopard (*Neofelis nebulosa*), the Indian elephant (*Elephas maximus indicus*), and the endangered Bengal tiger (*Panthera tigris tigris*) may be found.

Mythology and Human Activity

According to Hindu mythology, the Ganges River is believed to have come down to Earth from the heavens. The river symbolizes purification to millions of Hindus, who believe that drinking its water before death or bathing in the river at least once will lead to salvation. The river has been the final resting place for many Hindus, whose cremated ashes or partially burned corpses are placed in the river for spiritual rebirth.

It may not be an exaggeration to say that ever since humans started settling on its banks, the mighty river has steadily moved toward degeneration. The Aryans settled on the Gangetic plains around 1200 B.C.E. In the subsequent 3,200 years of human occupancy, the landscape of the Gangetic Plain has been exploited and transformed by agricultural and industrial activities. The Ganges River accounts for around 32 percent of India's annual usable water resources for agriculture, aquaculture, hydroelectric power generation, and industry, and it supplies water for an area composing more than one-third of the country's population.

The exploitation of the river by the human settlements along the river has eroded the quality of the river's waters. The populous and industrial towns, through their overextraction and water-diversion activities, have intensified the problem of river pollution at various points. About 30 cities and more than 60 towns thrive along the banks of the Ganges.

Besides sewage that is often untreated—as much as 264 million gallons (1 billion liters) per day—industrial discharges from pharmaceuticals factories, electronics plants, textile and paper industries, tanneries, fertilizer-manufacturing units, and oil refineries produce an additional load. The hazardous components of the effluents include hydrochloric acid, mercury and other heavy metals, and polychlorinated biphenyls (PCBs). There is much room for modernization of water treatment and filtration systems.

Pollution is not the only threat. A report by the Intergovernmental Panel on Climate Change warned that if the current trends of climate change continue, the size of the Himalayan glaciers could be reduced by as much as 80 percent by 2030. These glaciers are the ultimate source of the Ganges River, along with rainfall, and any changes in the rate at which these glaciers melt will impact the waters of the Ganges. Higher seasonal water levels followed by a drop in the water levels of the river may be disastrous for all living near these waters, reducing the supply of freshwater at some times, and generating uncontrollable erosion at others.

Conservation Efforts

The Ganga Action Plan (GAP) was initiated in 1984 by the government of India, supported by the Netherlands, the United Kingdom, and some voluntary organizations. The principal aims of the GAP were to reduce pollution load on the river immediately and to establish self-sustaining treatment plants.

The Indian government has declared the 84-mile (135-kilometer) stretch of the Ganges between Gaumukh and Uttarkashi to be an eco-sensitive zone, seeking specific activities to save the rich biodiversity of the region. As a result, three hydroelectric projects that were proposed to be initiated along the river—Bhaironghati, Pala Maneri, and Loharinag Pala—were discontinued. In April 2011, the Cabinet Committee on Economic Affairs approved a river-cleaning project to be implemented by the National Ganga River Basin Authority (NGRBA). The central and state governments of Uttarakhand, Uttar Pradesh, Bihar, Jharkhand, and West Bengal agreed to share the expenses.

The pilot study conducted at Chandernagore Municipality by Birol and Das in West Bengal reveals that it would be economically viable to invest in infrastructure that would treat higher quantities of wastewater to create higher quality. The pilot experiment stated that all households, regardless of their income levels, showed willingness to pay higher taxes to ensure the full capacity of the sewage treatment plant for primary treatment and to upgrade the technology needed to enable secondary treatment. The results of the study can be adapted to similar municipalities along the Ganges with the use of the benefits-transfer method. The results support enhanced investments to improve the water quality of the Ganges River in order to reduce the environmental hazards and health risks currently threatening the sustainability of ecosystems across the biome.

The Union Ministry of Environment and Forests has asked the Indian Institute of Technology to prepare a work plan for the NGRBA. A carefully designed environmental policy may help prevent a great deal of environmental damage and associated cleanup costs in the future, as well as save the Ganges from shrinking in the face of global warming and glacier contraction.

Rhama Parthasarathy

Further Reading

Birol, Ekin and Sukanya Das. "Estimating the Value of Improved Waste Water Treatment: The Case of River Ganga, India." *Journal of Environmental Management* 91, no. 1 (2010).

International Institute for Environment and Development. "Up in Smoke? Asia and the Pacific." 2007. http://pubs.iied.org/10020IIED.html.

Markandya, Anil and Maddipati Narasimha Murty. "Cost-Benefit Analysis of Cleaning the Ganges: Some Emerging Environment and Development Issues." *Environment and Development Economics* 9, no. 1 (2004).

Ministry of Environment and Forests, Government of India. "Annual Report 2011–12." 2012. http://moef.nic.in/report/report.html.

Ghats Montane Rainforests, Northwestern and Southwestern

Category: Forest Biomes.
Geographic Location: Asia.
Summary: These montane rainforests in western India support very high biodiversity that is under threat from habitat fragmentation.

India's main extent of tropical montane rainforests are found on a continuous chain of mountains along the west coast of the country, a region known as the Western Ghats. The mosaic of vegetation here arises from interactions between gradients of rainfall, latitude, and elevation. This biome is part of the Western Ghats and Sri Lanka biodiversity hot spot, and is home to many endemic (found nowhere else on Earth) species.

The Southern Ghats have higher peaks, above 6,560 feet (2,000 meters); these high mountains have a distinct forest structure referred to as the Shola forests. These Shola are patches of relatively low forested areas interspersed with grasslands. The Northern Ghats do not have high mountains, but have flora and fauna that are also distinct to this region. Given the pressure on the resources of this geographic region due mainly to a growing human population, this ecosystem needs to be protected in order to ensure that its unique flora and fauna are conserved.

Climate change is of particular concern to this region, with the Western Ghats having been identified as one of several areas in Asia most likely to be impacted by global warming. Some studies suggest significant forest dieback will occur, as a result of higher average temperatures, disrupted timing of rainfall, and altered seasonal storm patterns, which combined would be devastating to many of the habitat niches here.

The Western Ghats

The Western Ghats are a continuous chain of mountains found along the southwestern coast of India. The 994-mile (1,600-kilometer) mountain chain runs north-south, approximately 20 to 30 miles (30 to 50 kilometers) parallel to the west coast. The Western Ghats chain is interrupted by a 22-mile (35-kilometer) stretch of flat terrain, the Palghat Gap. The northwestern Ghats include the states of Gujarat, Maharashtra, and Goa; those in the southwest include the states of Karnataka and Kerala.

The Western Ghats have their origin in volcanic eruptions that occurred during the prehistoric movement of the formation of continental Gondwanaland and its journey toward Asia 150

The Western Ghats in fog in the state of Karnataka, India. The region is home to 51 critically endangered species, and the Western Ghats were designated a World Heritage Site in 2012. (Wikimedia/ManOnMission)

to 100 million years ago. The Western Ghats receive heavy rainfall primarily in two seasons: the southwest monsoon that spans from June to September, and the northeast monsoon that spans from October to November. Almost 80 percent of the rainfall occurs during the southwest monsoon.

Rainfall is heaviest on the western slopes, and the intensity of rainfall decreases toward the north. A similar pattern exists from west to east. The leeward, or the east-facing, slopes are rainshadow areas and receive less than 10 percent of the rainfall; a corresponding area on the windward or the west-facing slope is inundated with the lion's share of precipitation. The southernmost Ghats also have a shorter dry season, typically two months less than the northern Ghats.

Southwestern Montane Rainforests

The southern Western Ghats have numerous peaks taller than 6,560 feet (2,000 meters). These areas feature distinct vegetation types along their elevational gradients. They receive most of their rainfall during the monsoons, but due to the cloud-scraping function of the ridge lines, also receive substantial rainfall during other months, and hence have a shorter dry period. The tallest mountain peak in this region is the Anaimudi at 8,840 feet (2,695 meters). The mountains at this elevation have the Shola formation of wooded forest and grasslands. These forested stands here feature a dense understory with stunted trees. Shola forests are found along the hill slopes and valleys. Since this unique vegetation occurs only in isolated patches at the higher elevations, they are also referred to as sky islands, and serve as hot spots of endemic flora and fauna.

The transition between the forest patches and the grasslands is abrupt. The trees generally have a spread-out canopy and are covered with mosses, ferns, and epiphytes; they are particularly rich in orchids. These forests have high levels of humidity even during the dry seasons. The grasslands, however, get extremely dry during the summer, becoming highly flammable, and resulting in seasonal man-made fires and clearings. In some areas these grasslands and the montane forests have been cleared to grow plantation crops such as tea.

At lower elevations, the southwestern Ghats have evergreen forests which are structurally complex and provide niches for a diverse community of fauna. These forests are generally old growth and the canopy height ranges between 131 to 262 feet (40 to 80 meters). Many of the tree species in these forests are endemic to the southern Ghats.

At elevations in the range of 1,640–2,950 feet (500–900 meters), moist deciduous forests with bamboo thickets are seen on the leeward side of the mountains. Many of these stands continue to thrive. Low-elevation evergreen forests and marsh communities, however, have largely been decimated. Still, some small, scattered patches of water-logged *Myristica* swamps persist. Since the midelevation forests are almost contiguous in the southern Ghats, they serve as corridors for dispersal of large mammals, Asian elephants, Indian gaur, lion-tailed macaques, sloth bears (*Ursus ursinus*), Bengal tigers, Nilgiri tahrs (*Hemitragus hylocrius*), an ibex, and a type of monkey called Nilgiri langurs (*Trachypithecus johnii).*

Northwestern Montane Rainforests

Drier than the southern areas, the northern montane rainforests are also more affected and adapted to sharper seasonal contrast in precipitation. Average annual rainfall here tends to be about 98 inches (2,500 millimeters); most comes during the monsoon, while the dry season tends to extend for a full four months. This climate favors wet-evergreen and deciduous forest types.

Higher elevations in these northern Ghats areas harbor subtropical broadleaf forests; a few stands are old growth. Canopy height ranges from 130 to as high as 260 feet (40 to 80 meters).

The levels of endemism across taxonomic groups of both trees and animals decline toward the north generally in the Ghats. The northern Ghats also have a longer history of more intensive human use and exploitation, although there are some remnant forests where large-girth trees can still be seen. Portions of these forests have in the past been cleared for raising extensive plantations of teak (*Tectona grandis*).

The Khas Plateau in the northern Ghats has a unique geomorphology and vegetation. These flat hill tops are found at about 3,280 feet (1,000 meters) in elevation. A characteristically thin layer of topsoil here supports scant woody vegetation. The rocky outcrops on the hilltops support herbaceous vegetation, dominated by annuals that flower only in a narrow window of time during the monsoon season. These mountains are of volcanic origin and have rich deposits of iron and manganese, helping provide nutrients.

Biodiversity and Conservation

The Western Ghats of India are considered a global, mega-biodiversity hot spot. An astonishing three-fourths of all amphibians identified in India may be found in this area, as well as approximately half of India's reptiles. This region also accounts for more than 27 percent of all the flowering plants in India. The Western Ghats were designated a World Heritage Site in 2012.

There are 325 globally threatened species found in the Western Ghats, of which 51 are critically endangered. Apart from the high species richness and endemism, the region is also known to harbor evolutionarily distinct species surviving as relic populations in its evergreen forests. Examples of such species are found among trees, mollusks, fish, amphibians, reptiles, and mammals. Only 9 percent of the total area of the Western Ghats is protected, however, and new species of flora and fauna of great evolutionary significance are periodically still being discovered. This further enhances the conservation importance of this biome.

Habitat fragmentation is one of the most pressing issues that need to be addressed to ensure the safety of this unique ecosystem. Fragmentation usually occurs either by a change in the human land-use practice, which often entails commoditization of one or more local species of plant or animal, or from outright industrialization or urbanization of an area. Separately, during the dry season, when the grasslands and the moist deciduous forests are flammable, human-induced fires cause extensive damage to natural vegetation, sometimes undermining the ability of a particular habitat to recover. The Western Ghats essentially are undergoing predictable degradation and habitat fragmentation in order to support a large and growing population. The region is also vulnerable to climate change, as its natural infrastructure is tuned to a moisture and temperature regime of long standing.

LILLY MARGARET ELUVATHINGAL
KARTHIKEYAN VASUDEVAN

Further Reading

Champion, G. H and S. K. Seth. *A Revised Survey of the Forest Types of India.* New Delhi, India: Government of India Press, 1968.

Chandran, M. D. S.. "On the Ecological History of the Western Ghats." *Current Science* 73, no. 1 (1997).

Daniels, R. J. R.. *National Biodiversity Strategy and Action Plan: Western Ghats Eco-Region.* New Delhi, India: Ministry of Environment and Forests, 2001.
Myers, N. "Threatened Biotas: 'Hot-Spots' in Tropical Forests." *Environmentalist* 8, no. 1 (1988).
Nair, S. C. *The Southern Western Ghats: A Biodiversity Conservation Plan.* New Delhi, India: INTACH, 1991.

Gibson Desert

Category: Desert Biomes.
Geographic Location: Australia.
Summary: This largely untouched desert nevertheless faces the decline of many of its species and its biodiversity.

One of the largest deserts in Australia, the Gibson Desert covers some 60,000 square miles (155,400 square kilometers) of western Australia between the salt lakes of Disappointment and Macdonald, south of the Great Sandy Desert, east of the Little Sandy Desert, and north of the Great Victoria Desert. It was named for the explorer Alfred Gibson, who died there on an expedition in 1874. Broadly, the desert can be divided into the lateritic (aluminum- and iron-oxide-rich silica soil) plain, which takes up the bulk of the desert, and the dune field in the northeastern corner. The desert is characterized by red sand plains and dune fields, low rocky ridges, several small saline lakes, and large expanses of thin grass growing out of gravelly terrain. The upland is laterized on Jurassic and Cretaceous sandstone.

The desert's ecosystem is considered to be relatively stable, although a shift to more frequent and abundant rainfall has some climate change experts concerned. The flora and fauna of the Gibson Desert, like many in desert environments, are adapted to take maximum advantage of rainfall. If the rainfall levels increase, the range of many species may also increase, encroaching upon additional territories and destabilizing the biome.

There are few established paths through the desert, and the two main tracks—east-west to Alice Springs via Papunya, and north-south along the Gary Highway—are not always accessible. The land's principal use is for mineral exploration; most of it remains untouched. Little of the known mineral resources have been explored; there are no major mines in the region. A total of 10 percent of the desert is in conservation reserves, including the Gibson Desert Nature Reserve.

The Gibson's rainfall ranges from 8 to 10 inches (20 to 25 centimeters) a year, most of it in the summer, and the climate is generally hot. Summer maximums are around 104 degrees F (40 degrees C), while winter temperatures vary from 45 to 65 degrees F (7 to 18 degrees C). Summer rains help prevent summer wildfires from spreading, but are not always enough. Groundwater from the Officer Basin and Canning Basin feeds several springs. The quaternary alluvial is associated with paleodrainage features, supporting coolibah woodlands flora over bunch grasses.

The bulk of the human population are indigenous Australians, including a Pintupi group believed to have been the last uncontacted tribe in Australia until its 1984 departure from the central-eastern Gibson Desert when the prevailing drought depleted its springs and food sources. The few sedentary population centers are scattered around the eastern margin of the desert, such as the towns of Mantamaru, Warakurna, and Warburton—the closest weather station for the region and the source of most climatic data.

Vegetation

Vegetation varies throughout the desert, according to the soil type and landscape features. The lateritic *gibber* plains of the desert are characterized by mulga parkland (spreads of the short mulga tree or *Acacia aneura*) over hard spinifex (*Triodia basedowii*) grasslands. Mulga trees resemble the eucalyptus tree and can live for over 200 years.

The red sand plains and dunes are home to soft spinifex (*Triodia pungens,*) also called porcupine grass, on which may be found a mixed shrub steppe, including species of hakea, acacia, and grevillea. Spinifex roots burrow deeply into the soil and may be found as deep as 20 feet (6 meters) in the soil. The lateritic upland areas are home to mulga scrub in the south and scrub steppe in the north. The grasses have faced a moderate threat from the weed invasion of buffle grass.

Wildlife

No comprehensive biodiversity study has been done of the Gibson Desert. The desert has about 45 animal species, six of which are exotic mammals introduced recently. Common species in the area include rabbits, foxes, wildcats, and camels. Foxes and cats are persistent threats to smaller mammal and bird species, and have contributed to the decline and near-extinction of several. Though there has been no widespread land degradation, what little occurs is the result of camels and rabbits feeding on vegetation. Several species have disappeared from the region, including the western quoll (*Dasyurus geoffroii*), numbat (*Myrmecobius fasciatus*), red-tailed phascogale (*Phascogale calura*), golden bandicoot (*Isoodon auratus*), and lesser bilby (*Macrotis leucura*).

The desert is also home to the red kangaroo, one of the world's largest marsupials, found in Australia's hottest and driest ecoregions. The kangaroo's mode of traveling by hopping evolved to protect its feet from the hot sand and is an energy-efficient use of its powerful muscles, allowing it to travel great distances between food sources—a necessity in the desert.

The bilby, a burrowing nocturnal mammal, survives by sleeping in dens (which are at least 10 degrees F or 6 degrees C cooler than the outside air in the summer) during the day to avoid both the hot sun and active predators. It drinks no water, relying on a highly efficient digestive system that provides water from a diet of fungus, insects, and vegetation.

A lizard (*Moloch horridus*) known as the moloch, thorny devil, or thorny dragon thrives in the Gibson Desert and is distantly related to the horned lizard of American deserts.

Rare species, some of which may be found in other ecoregions, include the marsupial mole, the rock wallaby, the malleefowl, the princess parrot, and the woma python. Birds are particularly rare in the Gibson Desert, and the princess parrot is the only limited-range bird that has been seen regularly, thanks to a population thereof at Tobin Lake. Emus, hooded robins, jacky winters, and Australian bustards are also found in small numbers. The grasslands are home to a large population of striated grass wrens (*Amytornis striatus*), while the few wetland areas along Cooper Creek provide waterbird habitats.

Conservation Efforts

The faunal attrition index for the desert (a measure of loss of species richness) is high, at 0.45. The faunal contraction index, measuring the decline of range covered by a given species, is 0.44. The processes putting most species at risk are increasing fragmentation in the dune fields, changed fire regimes, the predation of or competition with feral animals, and grazing pressure.

The Gibson Desert's Gnamma Holes and Lake Gruszka are both considered to be wetlands of national significance. Lake Gruszka is considered to be in pristine condition, with no foreseeable degradation; Gnamma Holes requires minimum intervention for full recovery, as the site is no longer maintained by the indigenous people. In both cases, animals are the biggest risk to the ecosystem through predation, habitat building, or land degradation from feeding on plants.

The desert also has five seasonal freshwater/floodplain lakes in near-pristine condition, as well as a seasonal river. All are at mild risk from changes in the Gibson Desert's fire regime.

Bill Kte'pi

Further Reading

Beard, J. S. *Plant Life of Western Australia.* Kenthurst, Australia: Kangaroo Press, 1990.

Department of Sustainability, Environment, Water, Population and Communities. "Biodiversity Assessment—The Gibson Desert." Australian Government, 2009. http://www.anra.gov.au/topics/vegetation/assessment/wa/ibra-gibson-desert.html.

McInerney, Graham and Mathieson, Alec. *Across the Gibson.* Adelaide, Australia: Rigby, 1978.

Gironde Estuary

Category: Marine and Oceanic Biomes.
Geographic Location: Atlantic Coast of France.
Summary: The Gironde Estuary is the largest estuary in western Europe, with its wealth of sediments and strong tidal regime providing habitat for diverse and abundant fish and invertebrates.

Sometimes called the Gateway to Bordeaux, the Gironde Estuary is an important biological, economic, and cultural area. Estuaries are the transition zones between the river and the sea; the combined flows of sea-spawned saltwater and river-borne freshwater provides varying balances of nutrients in the water and sediment, making estuaries among the more productive of natural habitats. The Gironde exemplifies all these estuarine characteristics.

Morphology and Hydrology

The Gironde is formed from the meeting of the Rivers Dordogne and Garonne and the Atlantic Ocean on the western coast of France. It is about 47 miles (75 kilometers) long and up to 7 miles (11 kilometers) wide. At some places, the estuary is so wide that it looks and acts like an internal sea. At other places, it is narrow enough to seem to be a simple, if navigable, river.

The sea itself occupies about 75 percent of the Gironde Estuary biome, while 10 percent of its area is covered by rivers, mudflats, sandbanks, and lagoons. Forests and farmlands are 5 percent of its area, with meadows, dunes, and beaches composing the remaining 10 percent. This range of natural ecozones makes the estuary relatively unique; it is one of few such extensive—and relatively unspoiled for its size—natural estuaries in Europe.

Covering about 245 square miles (635 square kilometers), the Gironde is the largest estuary in western Europe. It transports about 264,000 gallons (1 million liters) of water per second into the Atlantic Ocean. The estuary and its two rivers are so full of sediment that it generally keeps the rivers and the estuary tan in color. The estuary has a tide of between 6.5 and 16.5 feet (2 to 5 meters). This tide increases from the mouth of the river to upstream areas on the rivers, because of the decreasing width and depth of the river. There are also strong tidal currents in the estuary. Because of the amount of sediment and the tides, sand banks and islands in the estuary are always changing.

The estuary has been altered by harbor construction, and by dredging to maintain navigation. There are about 40 harbors for industrial purposes, fishing, and recreation, most of which are located inland in the former swamps of the Gironde, offering protection from tidal currents. The navigation channels are deep, with strong currents and turbidity, and are maintained by dredging and submersible dikes. There are over 84 miles (135 kilometers) of maritime dikes throughout the Gironde area; this is more than half of all the maritime dikes in France.

Natural Communities

Much of the flora of the Gironde Estuary is found underwater. The yellow boring sponge, for example, grows to unusual sizes here, often growing to more than 3 feet (1 meter) in diameter. Elephant ear sponges may be found growing among the rocks under the waters of the estuary, as well as colonies of anemones. Along the banks, plant life includes various reeds, wild orchids, and alder forests.

The Gironde Estuary is an important migration and nesting site for birds. Birds of prey, seabirds, songbirds, and waterfowl migrate along the East Atlantic Flyway, traveling along the Atlantic coast through the estuary in the spring or autumn as they fly to and from winter destinations in Africa, Spain, and Portugal. Birds also nest year-round in the estuary, drawn by the coastal and marshland riches of invertebrates, fish, and amphibians, as well as by their preference among the fresh, brackish, or marine waters. Notable bird types found in the Gironde Estuary include skylarks, nightingales, and various stork and heron species. Minks and green turtles are some of the animals found along its banks.

The Gironde has relatively narrow diversity of fish and marine species, but those species represented are found in abundance. At least 75 fish species are known here, including important

The Mascaret

The Gironde Estuary is noted for its tidal bore, a wall-like wave at the leading edge of the incoming tide. Known locally as the Mascaret, and occurring especially with spring tides, the tidal bore when it is running twice daily applies considerable force to the banks and bottom of the waterway, extending its effects from the estuary into the narrower rivers. On the Garonne River, the Mascaret forms a barreling wave that can reach a height of 5 feet (1.5 meters), and can also break and reform. Water turbidity is a factor that native wildlife here has adapted to.

The photo shows the wave of a small Mascaret at the leading edge of a rising tide on a narrow river in New Brunswick, Canada. (Wikimedia/Fralambert)

commercial fish stocks such as flounder, mullet, salmon, sea trout, shad, smelt, and sturgeon. Other aquatic animals include various bivalves, white shrimp, eels, sea and river lampreys. The estuary is a valued food supply for both Bordeaux and other regions of France. Wild sturgeon was once plentiful in the estuary, providing local caviar and fish—but numbers have greatly decreased, mainly through overfishing.

Fish and benthic (bottom-dwelling) species in the Gironde have adapted to tolerate regular physical and chemical changes to the estuary's salinity, depth, silt, and oxygen levels. As a result, it can be difficult to measure the impacts of human stresses like dredging, fishing, and navigation, as species here typically can tolerate the large estuarine stresses already in place. Despite human encroachment, animal life continues to find the estuary a suitable habitat.

Threats and Conservation

The estuary was contaminated by heavy metals and other contaminants due to upstream mining operations, and this contamination has affected the ecosystems and human uses. For example, oysters are not produced in the estuary. The ecosystems in the Gironde Estuary may now be shifting from a longtime mix of freshwater and marine ecosystems to a more marine environment. If so, Gironde Estuary could become more suitable as a base of nursery habitats for marine species, than as a key migratory corridor and nursery habitat for species seeking lower-salinity plants and faunal communities.

In addition to concerns about contaminates, a 2012 study used various models to extrapolate the potential effects of rising sea levels on six specific sites in the estuary. The long-term impacts upon the greater Bordeaux area could be profound, with some floodplains currently not generally covered by water becoming submerged, for example. Climate change is projected to threaten the Gironde Estuary in several ways. Saltwater intrusion is seen as a constant challenge, no matter what presumed rate of sea-level rise is taken into account. Storm surge damage and coastal erosion are similar outputs, regardless of the accepted or predicted rates of sea-level rise or increase in storm severity. The size of inland floodplains is also likely to be reduced, as the maritime limit moves further upstream in every model researched; the extent is a matter of degree.

The ultimate effects of global warming on species health and diversity in the Gironde Estuary biome will be linked to the human efforts applied to slow or reverse emissions of greenhouse gases, the expense taken to build up dikes and other water control systems, and any mitigation campaign undertaken to counterplant affected areas with native plants that can best protect the soils and support those habitat areas designated as crucial.

There is a discussion of how to protect the estuary from future storms, which may be intensified by sea-level rise and increasing storm intensity and tides. A range of options are now being considered such as coastal defenses, managed realignment of the estuary, controlled flooding of areas behind the dikes, and wave damping measures in the mouth of the estuary.

The Mortagne-sur-Gironde polder and Île Nouvelle were purchased by the Coastal Protection Agency after a major 1999 storm, when they were partially flooded by the sea. The agency decided to not repair the breaches, and these two areas are now flooded with each tide cycle. These two areas have shifted their makeup to tidal salt marshes, which assist local fish in reproducing and feeding.

MAGDALENA A. K. MUIR

Further Reading

Dauvin, Jean-Claude, et al. "Benthic Indicators and Index Approaches in the Three Main Estuaries Along the French Atlantic Coast (Seine, Loire and Gironde)." *Marine Ecology* 30, no. 2 (2009).

Laborie, V., F. Hissel, and P. Sergent. "Future Extreme Water Levels and Floodplains in Gironde Estuary Considering Climate Change." *Geophysical Research Abstracts* 14, no. 1 (2012).

Lobry, Jeremy, et al. "Structure of the Gironde Estuarine Fish Assemblages." *Aquatic Living Resources* 16, no. 2 (2003).

Pasquaud, S., et al. "Impact of the Sampling Protocol in Assessing Ecological Trends in an Estuarine Ecosystem: The Empirical Example of the Gironde Estuary." *Ecological Indicators* 15, no. 1 (2012).

Gobi Desert

Category: Desert Biomes.
Geographic Location: Asia.

Summary: The Gobi is a cold desert high in the rain shadow of the Himalayas. Its growth is alarming and is greatly affecting agricultural land.

The Gobi is the fifth-largest desert in the world, stretching across northern China and southern Mongolia for more than 500,000 square miles (1.3 million square kilometers). It is a rain shadow desert created by the Himalayan mountains, which prevent rain-carrying clouds from reaching the Gobi on the prevailing southwesterly winds. Despite its temperature extremes and arid terrain, the Gobi is notable in history for making up a significant portion of the great Mongol Empire and for containing several important cities along the Silk Road from China. It is also the source of critical scientific discoveries, including the first fossilized dinosaur eggs found on Earth.

The Gobi is known for its extreme climate; sweltering midday heat can drop to freezing temperatures at night. A shift of 60 degrees F (33 degrees C) in the space of 24 hours is not unusual. In the winter, icy Siberian winds sweep across the barren landscape with few mountains significant enough to stop their progress. Summer is the rainy season in the Gobi, but temperatures that frequently surpass 100 degrees F (38 degrees C) mean that the water does not stay for long.

The Gobi Desert is comprised of five distinct ecoregions, determined by variations in climate and topography.

Eastern Gobi Desert Steppe

The Eastern Gobi Desert Steppe ecoregion covers more than 100,000 square miles (260,000 square kilometers) at an elevation of 3,300 to 5,000 feet (1,006 to 1,524 meters). Summers in this region can be very hot, and winters are extremely cold due to winds from the north sweeping across the plains. This region receives 4 to 6 inches (102 to 152 millimeters) of precipitation per year, primarily in the summer, just enough to support drought-resistant shrubs and low grasses.

Alashan Plateau

The largest Gobi ecoregion lies to the south and southwest of the steppe. The Alashan Plateau is a

260,000-square-mile (675,000-square-kilometer) region of 6,500- to 8,500-foot (1,981- to 2,591-meter) mountains that enclose intermountain basins. The basin and range topography creates an internal arid climate, but a few rivers along its eastern edge and numerous oases are fueled by mountain snowmelt in the basins. Most of this region has the typical rocky, barren Gobi landscape, but the basins along the southern edge contain shifting sand dunes that reach up to 1,000 feet (305 meters) high. This region has an unusually high diversity of plant and animal life for the Gobi.

Gobi Lakes Valley Desert Steppe

North of the Alashan Plateau lies the Gobi Lakes Valley Desert Steppe ecoregion. The long, narrow, nearly flat Gobi Lakes Valley winds among several mountain ranges. The valley has a few sand dunes interspersed with salt marshes, and several large, often dry, lake beds are fed by intermittent rivers from the nearby mountains. The land here is mostly arid, receiving 2 to 8 inches (51 to 203 millimeters) of rain per year, but temperatures are more moderate in both summer and winter. Plants and animals here have adapted either to the wetlands or to arid conditions.

Junggar Basin

To the west of the Alashan Plateau lies the fourth distinct zone, the Junggar Basin. This region also lies within several mountain ranges, but has wide openings to the northwest. The arid center of the basin still receives enough precipitation during the year to support vegetation that stabilizes the sand dunes, and the margins at the base of the mountains receive up to 10 inches (254 millimeters) of rain annually, in addition to runoff from the mountains. This total is enough to sustain several lakes. The Junggar has moderate summers but severe winters, with icy winds sweeping in from Siberia through the open passage to the northwest.

Tian Shan

Finally, the Tian Shan, or Celestial Mountains, is a large, isolated range surrounded by the desert basins of northern China. Running from east to west across central Asia, the Tian Shan stretches across more than 1,550 miles (2,500 kilometers). While the lower slopes are nearly as arid as the surrounding deserts, the mountains here—averaging about 14,000 feet (4,267 meters) high—catch arctic moisture flowing from the northwest, resulting in 16 to 32 inches (406 to 813 millimeters) of rain per year falling on the upper slopes, which support conifer forests and meadows. Heavy snowpack on the highest reaches supports rivers and streams. The wide range of elevations and environments in the Tian Shan results in more than 2,500 wildlife species making this ecoregion their home.

Flora and Fauna

Vegetation is sparse and rare across most of the Gobi. The plateaus and plains support small, salt-resistant bushes and low grasses. The semi-arid regions can sustain somewhat less hardy vegetation. One of the most important plants in the Gobi is the saxaul tree (*Haloxylon ammodendron*). Water reserves that accumulate behind this short, tough tree's bark can be squeezed out by humans and animals. Wild onions are other sources of food and water, occurring across several zones. Saltwort (*Batis, Salsola,* and *Salicornia* families) is unique to a specific region of the Gobi: the Salt Desert. The salt content of the soil here is so high that this weed is about the only plant that grows in the Salt Desert.

There is ample variety, if not great numbers, in the fauna that inhabit the Gobi. Some species are common to the deserts of central Asia, while others are unique to a specific ecoregion. The best-known large mammal is the bactrian, or two-humped, camel (*Camelus bactrianus*). Other large mammals include the Gobi bear, golden eagle, snow leopard, wolf, and ibex. A variety of rodents, lizards, and birds call the desert home.

Many species in the Gobi are endangered due to human activity. Hunting, overgrazing livestock, water diversion for irrigation, mining, and other factors affect the biodiversity in each ecoregion of the Gobi.

Desertification and Conservation

The biggest threat to the Gobi Desert biome is increasing desertification—and the root cause is human activity. The Gobi continues to grow rapidly today, causing concern among its closest neighbors. China is especially affected, losing valuable grasslands to fast-moving desertification.

Despite a very low population density of fewer than three people per square mile (2.6 square kilometers), the main occupation of nomadic animal herding stresses the fragile desert fringes where most of the inhabitants live. Cattle, sheep, and cashmere-producing goats have degraded these very limited grasslands. Overgrazing has decreased vegetation cover and increased erosion of the underlying soil, resulting in advancing desertification, which causes semi-arid lands to become even drier and eventually convert to full deserts. Overuse of the land leaves the region with little or no ability to recover naturally. Poor irrigation management and deforestation also contribute to desertification. Irrigating incorrectly results in the salinization or alkalinization of the soil, which also makes the land unproductive.

A bactrian camel in Mongolia, where 70 percent of the land is being affected by desertification. Bactrian camels are among a small variety of large mammals that can survive the harsh conditions of the Gobi Desert. (Thinkstock)

The effects of climate change are of great concern to Mongolia, as 70 percent of its land is affected by desertification. In 2010, an estimated 20 percent of the country's livestock was killed by extreme weather. Climate change and the effects of humans upon the land are the two greatest threats to the biome as a whole.

China has made plans to try to slow the expansion of the desert. A variety of programs have been developed in recent decades to reclaim lands lost to the Gobi. It is too soon to know whether those programs will work. Various labs and research centers are experimenting with techniques to stop erosion, are developing crops suitable to the desert, are seeking ways to enrich the soil, and are studying the movement of the sands and dust storms.

The most recent plan is dubbed the Green Wall of China. The Chinese government is funding a zigzagging wall of pines, oleasters, junipers, hawthorns, aspens, and other trees that it hopes will someday stretch across 2,000 miles (3,200 kilometers) and stop the desert in its tracks. The entire project is expected to take 30 years to complete and cost more than $150 million.

Despite all these efforts, the Gobi continues to grow, and to grow more dry. Major dust storms have increased in number over the past decade, reaching as far as Korea and the Pacific Ocean. Even though its expansion threatens human habitation, the Gobi remains a unique place with a rich history buried under its surface.

Jill M. Church

Further Reading

Allan, Tony and Andrew Warren, eds. *Deserts: The Encroaching Wilderness: A World Conservation Atlas.* New York: Oxford University Press, 1993.

Barta, Patrick. "To Stop Dust Bowl, Mongolia Builds 'Great Wall' of Trees." *Wall Street Journal Eastern Ed.*, October 24, 2006.

Friedman, Josh. "Mongolian Cabinet Meets in Gobi Desert to Make Stand Against Global Warming." In Asia, September 1, 2010. http://asiafoundation.org/in-asia/2010/09/01/mongolia-cabinet-meets

-in-gobi-desert-to-make-stand-against-global -warming.
Hairong, Wang. "Building a Great Green Wall." *Beijing Review* 53, no. 14 (2010).
Jianxiong, Zhou. "Holding Back the Desert." *Beijing Review* 48, no. 24 (2005).
Man, John. *Gobi: Tracking the Desert.* New Haven, CT: Yale University Press, 1997.
Oldfield, Sara. *Deserts: The Living Drylands.* Cambridge, MA: MIT Press, 2004.

Godavari Estuary

Category: Marine and Oceanic Biomes.
Geographic Location: Asia.
Summary: The globally significant, biodiversity-rich Godavari estuary and its species-rich mangrove forests are threatened because of anthropogenic stressors.

The Godavari, a perennial river in southern India, is considered to be one of the seven sacred rivers in India and is popularly known as the Ganga of the South. At 910 miles (1,465 kilometers) long, it is the second-longest river after the Ganga (as the Ganges is sometimes called), extending its catchment over an area of 120,777 square miles (312,812 square kilometers), which is nearly 9.5 percent of the total area of the country. It originates in the Trayambak, or Western Ghats, near Nasik, northeast of Mumbai in the state of Maharashtra, at an elevation of 3,501 feet (1,067 meters). The Godavari meets the Bay of Bengal on the east coast of India in the state of Andhra Pradesh.

The river bifurcates at Dowlaiswaram into two principal distributaries, the Vasishta and the Gautami, which give rise to a sprawling estuarine deltaic system fringed with tidal creeks and dense mangrove forests, before reaching the Bay of Bengal. The Vasishta-Godavari further divides at Gannavaram into two branches known as Vasishta Proper and Vainatheyam, both of which open separately into the Bay of Bengal. The Gautami-Godavari is also connected to Kakinada Bay by two channels: Coringa, arising at Yanam, and Godern, arising at Bhairavapallam.

The Godavari River has an average depth of 39 feet (12 meters), and the maximum width of the estuary is 0.6 mile (1 kilometer). The tidal influence extends up to 28 miles (45 kilometers) from the mouth. Average salinity is 34 parts per thousand. The temperature ranges from 84 to 95 degrees F (29 to 35 degrees C).

The southwest monsoon in early July brings a heavy influx of freshwater to the estuary, resulting in freshwater flowing in the upper layer toward the sea, with the seawater moving at the bottom in the opposite direction. This circulation prevents vertical mixing and disrupts the prevailing estuarine salinity, resulting in a complete scouring of the entire system. This condition generally exists from mid-July to September. The recovery phase begins in October, when the estuarine condition stabilizes. Less freshwater flow in the system results in the establishment of a vertical salinity gradient at sea, progressively shifting toward the estuary. The tidal range is about 3 feet (1 meter).

Flora and Fauna

The Godavari estuarine delta is characterized by dense mangrove forests and associated floral and faunal species. It is the second-largest mangrove forest in India. Part of this area has been preserved as Coringa Wildlife Sanctuary. The mangrove community in the Godavari estuarine delta is represented by 35 species of mangroves, of which 16 are true mangroves and the rest are associated species. The dominant species in other mangrove forests, *Rhizophora* spp., is poorly represented here. This area is the only place in India where three species of *Avicennia* are found together: *Avicennia officinalis, A. marina,* and *A. alba.*

The Godavari estuarine system supports prawn aquaculture throughout the year. A total 23 species of *Penaeid* prawn contribute to the aquaculture industry. Dominant species are *Metapenaeus monoceros, Penaeus indicus,* and *P. Monodon.*

One dominant sessile faunal element, a wood-boring mollusk, is common to this biome. The estuary also provides an important habitat for fish, reptiles, amphibians, birds, and mammals that need a coastal ecosystem to survive. Animals such as otters, fisher cats, jackals, and sea turtles are found in creeks here. Birds such as snipes, ducks, seagulls, and flamingos are common. There is a recorded population of 119 bird species, of which 50 are migrants from eastern Europe and central and northern Asia. The relatively rare winter migrant species include the golden plover (*Pluvialis apricaria*), woodcock (*Scolopax rusticola*), common snipe (*Gallinago gallinago*), and long-billed ringed plover (*Charadrius placidus*).

Many sea turtle species, like the leatherback turtle (*Dermochelys coriacea*) and the green turtle (*Chelonia mydas*), are found in the estuary during feeding and breeding seasons. The Krishna-Godavari basin is the main nesting site of the endangered olive Ridley turtle (*Lepidochelys olivacea*).

Also documented in the estuary are a full range of microbethos, including 137 species of phytoplankton and 81 species of zooplankton; 37 groups of meiobenthos, such as nematodes and copepods; and 114 species of macrobenthos, featuring worms, sponges, and crustaceans. Dinoflagellates, blue-green algae, red algal species, and many foraminiferal species (dominated by 14 species) are found throughout the year in the Godavari estuary.

Environmental Threats

Mangroves not only provide shelter to many species, but also protect the nearby villages and the coastal area by acting as a barrier against tropical storms, tidal waves, cyclones, and other natural calamities. Over the years, much of the dense mangrove forest has been destroyed for fuel wood and feral cattle grazing.

Additional threats come from anthropogenic stresses from the adjacent rural and urban areas, including nearby Kakinada City. These stresses include habitat conversion for real estate development, maritime traffic, the discharge of heavy effluents in the estuary, unsustainable fisheries and other aquacultural activities, overexploitation of biological resources, nonadherence to seasonal fishing bans, pollution, eutrophication, and siltation. Such events are contributing to the degradation and loss of mangrove forest stands here, directly affecting the health of the Godavari estuarine ecosystem.

The Godavari mangroves also play important roles as carbon sinks. Therefore, maintaining the extent and ecosystem functionality of the mangrove forests and preventing any further retrogression is an important strategy for mitigating the effects of climate change.

Sabuj Kumar Chaudhuri

Further Reading

Jhingran, V. G. *Fish and Fisheries of India.* 3rd ed. New Delhi, India: Hindustan Publishing, 1991.

Sen Gupta, R. and E. Desa, eds. *The Indian Ocean—A Perspective, Vol. 1.* Boca Raton, FL: CRC Press, 2001.

Sharma, N. S. *Godavari Estuarine Processes.* Goa, India: National Institute of Oceanography, 2007.

Gondwana Rainforest

Category: Forest Biomes.
Geographic Location: Australia.
Summary: The Gondwana Rainforest offers a unique glimpse into the prehistoric past through its geological features, flora, and fauna that remain nearly unchanged throughout the ages.

One of the largest subtropical rainforests in the world is the Australian Gondwana Rainforest. Formerly known as the Central Eastern Rainforest Reserves, the Gondwana Rainforest includes 50 separate reserves totalling 1,415 square miles (3,665 square kilometers) across southeast Queensland and northeast New South Wales. It offers an extremely high conservation value, with 270 rare or threatened plant and animal species as well as many species quite similar to those found in the fossil records of the area, such as several types of ferns.

The Gondwana Rainforest provides an intriguing living link with the evolution of Australia. Few places on Earth contain so many plants and animals that remain relatively unchanged from their ancestors in the fossil record. Some of the oldest elements of the world's ferns and conifers are found here; there is also a concentration of primitive plant families that are direct links with the birth and worldwide spread of flowering plants more than 100 million years ago.

Geologic evolution is another field of study in the Gondwana Rainforest biome. Calderas—formed by prehistoric volcanic eruptions—form a veritable laboratory of study for the sequence of volcanic eruption and erosion over time. The Tweed Shield erosion caldera here is one of the most famous for both its age and size, with visible examples of each phase of volcano building and erosion, which continues today via the waterfalls and coastal rivers of the area.

Biota

The evolution of new species is encouraged by the natural separation and isolation of rainforest stands. Many plants and animals found in the Gondwana are locally restricted to a few sites or occur in widely separated populations. Perhaps the largest stand of the Antarctic beech (*Nothofagus moorei*) left on the planet may be found in this rainforest. Other trees include ash, elm, and palm. Ferns, too, abound in the rainforest, including climbing fern (*Arthropteris tenella*) and bird's nest fern (*Asplenium australasicum*), among many others.

The distributional limits of several species and many centers of species diversity occur in this biome. The Border Group is a particularly rich area, with the highest concentration of frog, snake, bird, and marsupial species in Australia. Notable bird species include songbirds such as lyrebirds (*Menuridae* spp.), treecreepers (*Climacteridae* spp.), scrub-birds (*Atrichornithidae* spp.), catbirds (*Ptilnorhynchidae* spp.), and others.

Mammals once thought to be extinct—but that have recently been found in this rainforest—include the Parma wallaby (*Macropus parma*) and the Hastings River mouse (*Pseudmys oralis*). Notable amphibian species include many types of frogs such as the tusked frog (*Adelotus brevis*) and the orange-eyed treefrog (*Litoria chloris*).

Conservation Efforts

The greater Gondwana Rainforest preserve system has been built up out of eight total reserve areas, the first of which was designated a United Nations Educational, Scientific and Cultural Organization (UNESCO) World Heritage Site in 1986. Considered separately, these areas also have been identified as having outstanding heritage significance to Australia; they are therefore included on the Australian National Heritage List.

The Gondwana Rainforest as a whole is managed principally by the New South Wales National Parks and Wildlife Service (part of the New South Wales Department of Environment and Climate Change) and the Queensland Environmental Protection Agency. The Queensland areas include Lamington, Mount Chinghee, Springbrook, Mount Barney, and Main Range National Parks. The New South Wales areas include Barrington Tops, Dorrigo, Mount Warning, New England, Mebbin, Nightcap, Border Ranges, Oxley Wild Rivers, Washpool, Willi Willi, and Werrikimbe National Parks.

Given the unique nature of this site, many are concerned about the impact of climate change upon the region. Any decreases in rainfall, for example, have the potential to alter the habitat distribution here by depriving the rainforest plants of much-needed moisture. Other projected global warming threats to the Gondwana Rainforest include higher average temperatures, changes in natural forest fire frequency, and an upswing in invasive species that thrive on rainforest species.

Ghazala Nasim

Further Reading

Australia Department of the Environment, Water, Heritage and the Arts. "Gondwana Rainforests of Australia." 2008. http://www.environment.gov.au/heritage/places/world/gondwana/index.html.

Australian National University. "Implications of Climate Change for Australia's World Heritage Properties: A Preliminary Assessment." Commonwealth of Australia, 2009. http://www.environment.gov.au/heritage/publications/climatechange.

Reid, Greg. *Australia's National and Marine Parks: Queensland.* South Yarra, Australia: Macmillan Education Australia, 2004.

Great Barrier Reef

Category: Marine and Oceanic Biomes.
Geographic Location: Australia.
Summary: The Great Barrier Reef is the largest coral reef system in the world, extending more than 1,200 miles (2,000 kilometers) along the northeastern coast of Australia.

Australia, an island continent with an extensive tropical coastline and 17 percent of the world's total area of reefs, has the world's largest total area of coral reefs after Indonesia. Conditions for reef development vary considerably along the island's coastline. The Great Barrier Reef runs along the northeastern coastline of Australia, extending out to the margins of the continental shelf, from the Warrior Reefs in the northern Torres Strait for well over 1,200 miles (2,000 kilometers) to the Capricorn-Bunker Group of reefs and islands in the south.

This reef system, extending to Papua New Guinea, actually is comprised of more than 3,000 individual reefs, coral shoals, and high island fringing reefs, which can range in size from scarcely visible to over 39 square miles (101 square kilometers). The shape and structure of the individual reefs are also of great variety. Two main categories have been defined: lying platform or patch reefs, resulting from radial growth; and wall reefs, resulting from elongated growth, often in areas of strong water currents. There are approximately 300 coral cays, including those with vegetation or not, and over 40 low-wooded islands. There are also 618 continental islands which were once part of the mainland.

Continental drift brought the northern coastline of Australia into tropical latitudes about two million years ago, when the Great Barrier Reef originated with some minor development, but widespread reef development is thought to have begun about 500,000 years ago.

Climate and Currents

Complex patterns of currents in Australian waters are key driving forces and contribute to the different ecological areas within the region. The Great Barrier Reef, bounded by the Coral Sea to the east and the Torres Strait to the north, is characterized by the counterclockwise gyre of the South Equatorial Current. Cyclones are seasonally common here in this tropical marine climate. Air temperatures vary between an average maximum of approximately 86 degrees F (30 degrees C) in January and 73 degrees F (23 degrees C) in July, and an average minimum of approximately 74 degrees F (24 degrees C) in January and 64 degrees F (18 degrees C) in July. Mean water surface temperature is at a maximum during February and at a minimum during July.

Current patterns across the continental shelf induce localized upwelling and are predominantly

A diver swims by a giant sea fan growing on a steep wall of coral on the Great Barrier Reef. A marine park protects 133,205 square miles (345,000 square kilometers) of the reef, leaving 80 percent of it open for general use. (Thinkstock)

driven by prevailing southeast trade winds, northwest monsoons, and patterns of tidal flow.

The continental shelf of Australia's northern shores connects across the relatively shallow waters of the Torres Strait to Papua New Guinea, which loads considerable freshwater and sediments into the strait. The westernmost areas have the shallowest and most turbid waters, and soft mud dominates the shallow surface of the reefs in the area. The easternmost areas of the strait are characterized by fringing corals. There are very extensive platform reefs around Darnley Island, stretching out toward the edge of the continental shelf and a nearly continuous line of reefs. This particular section of reefs marks the northern edge of the outer Great Barrier Reef. The northern section of the Great Barrier Reef is unique, with well-developed ribbon-type barrier reefs on the outer edge, and wide areas of shoals, banks, and fringing reefs.

The southernmost reefs of the Great Barrier Reef are known as the Capricorn-Bunker Group, and are a little more than 30 miles (50 kilometers) offshore south of the Swain Reefs Complex. At this relatively small reef complex, the continental shelf narrows rapidly, with steeply sloping reef edges, deepwaters, shallow lagoons, and well-developed coral cays. The best-known reefs of the entire Great Barrier Reef system are at One Tree Island at the Tropic of Capricorn, whose cooler waters are largely responsible for the lower coral diversities in these islands.

Biodiversity

The Great Barrier Reef has very high levels of diversity, with some 1,500 to 2,000 species of fish, more than 4,000 species of mollusks, and 350 hard coral species. The high number of coral species decreases toward higher latitudes, with the number decreasing to about 240 species in the southern sections. The differences in the higher levels of nutrient loads, sediments, and amounts of freshwater present near the mainland, and the low inputs on the outer reefs, have led to considerable variation in species assemblages in different locations. Fish populations tend to be relatively small, in that the waters of the Great Barrier Reef are generally nutrient-poor. Some nutrients are supplied by Coral Sea surface water and upwellings, and also by terrestrial runoff. Tidal mixing is a major contributor to the nutrient recycling dynamics of this ecoregion.

The Great Barrier Reef serves as a nesting ground for green, hawksbill, loggerhead, and flatback sea turtles. Raine Island and Pandora Cay are the largest breeding grounds for green turtles, whereas islands in the northern section of the area are nesting sites for hawksbill turtles. About 26 species of cetaceans are found within the region, with frequent reports of humpback, minke, and killer whales, and of bottlenose, spinner, and Irrawaddy river dolphins. Humpbacks breed along the east coast of Australia in winter, between July and October.

There are some 23 species of breeding birds on more than 55 major nesting islands in the northern and southern sections of the Great Barrier Reef, with most bird communities reported in the Capricorn-Bunker Group. Raine Island is a major seabird rookery. Saltwater crocodiles are found in mangrove swamps and river estuaries along the region's coastal fringe, but individuals are seen around offshore coral reefs. Giant clams are also common in the Great Barrier Reef.

Seagrass and mangrove communities are notably important in the Great Barrier Reef, as they are breeding and nursing grounds for many reef and marine species. Seagrass beds are important for marine turtles and provide breeding grounds for significant numbers of dugongs. Seagrass beds both shallow and deep—as much as 49 feet (15 meters) below the surface—are widespread, with an estimate of at least 1,930 square miles (5,000 square kilometers) of cover. The largest known populations of dugongs occur in the northern and southern sections of the Great Barrier Reef region, but the southern groups are declining due to an increase in the number of boat collisions and entanglement in gill nets and shark nets placed near swimming beaches. Mangroves, for the most part, are located away from coral reef communities; some mangroves have been recorded on a few fringing reef systems, with the highest levels of diversity north of Cairns.

The Great Barrier Reef is the site of a mass spawning event that takes place once a year for a few nights after a particular full moon in the late austral spring, when the majority of stony corals, sponges, sea cucumbers, marine worms, and giant clams reproduce, releasing eggs and sperm in synchronicity. Occurring globally, mass spawning on coral reefs allows cross-fertilization between colonies and increases the chances of survival of individual larvae.

Human Interaction

The types of fishing activities that are popular in the Great Barrier Reef are recreational fishing of groupers, emperors, lobsters, fish for the aquarium trade, sea cucumbers, and trochus; and trawling of fish, prawns, scallops, and crustaceans. Repetitive trawling and the size of bycatch have become big concerns associated with trawling activities, as 50 percent to 90 percent of hauls typically include unusable species of benthic (deepwater) organisms as well as fish, sea snakes, and turtles. Illegal trawling often occurs in some of the protected reef areas and seagrass communities.

Reef health varies, depending on the type and location of reefs, local conditions, and the effect of anthropogenic activities. Generally, the Great Barrier Reef has not been heavily affected by human activities, as the majority of reefs are far offshore, but some effects can be seen on reefs closest to the mainland. Increases in dominant human activities like deforestation, poor agricultural practices, high concentrations of agricultural chemicals, nutrients in terrestrial runoff, land-based sediments, overgrazing, oil spills, intensive recreational use, and commercial fisheries are of future concern.

Generally, cyclones, bleaching events, and coral predators such as the crown-of-thorns starfish are related to effects on reef structure and development. Bleaching—which can be caused by environmental stress and seawater temperature rise—occurs when corals expel zooxanthellae, a unicellular flagellate (that is photosynthetic), with which the coral have a symbiotic relationship. This expulsion causes a bleached or whitened appearance, and may eventually lead to the decline in the health of the coral. Factors such as changes in salinity, decreases in zooplankton (caused by fishing), changes in water temperatures, bacterial infections, and climate change due to global warming can all be associated with human activity. Both bleaching events and outbreaks of the crown-of-thorns starfish may be human-induced. The starfish predation of coral is of greatest concern in Australia because of the potential economic effect. The first outbreak of the crown-of-thorns starfish was observed on Green Island, off Cairns, in 1962, but most outbreaks have been recorded in the central sections of the Great Barrier Reef. There is some evidence to suggest that increased soil runoff may aggravate outbreaks and in some instances may cause them.

The great majority of Australia's reefs fall within protected areas. Most of the lagoons, and all offshore reefs from the Capricorn-Bunker Group to the northern section of Cape York Peninsula receive legal protection as the Great Barrier Reef Marine Park, managed since 1975 by the federal Great Barrier Reef Marine Park Authority (GBRMPA) and the Queensland Department of Environment and Heritage. The remaining coastal waters and offshore islands fall within other protected areas.

The marine park has a detailed zoning plan of its 133,205 square miles (345,000 square kilometers), providing areas of strict protection alongside much larger areas of multiple use. About 80 percent of its total area is open for general use, including permitted commercial fishing and trawling, and a further 16 percent is open for general use but no trawling. Only about 5 percent is closed to fishing activities, but this area includes more than 120 reefs, and there are 13 fisheries habitat reserves. The reefs of the Torres Strait are outside the jurisdiction of the Great Barrier Reef and thus do not fall under its legal protection, but a fisheries management agreement with Papua New Guinea offers these reefs some kind of loose protection. The Great Barrier Reef has been a World Heritage Site since 1981.

Lucia M. Gutierrez

Further Reading

Australian Government. "Great Barrier Reef Marine Park Authority." 2011. http://www.gbrmpa.gov.au.

Hutchings, P., Michael J. Kingsford, and O. Hoegh-Guldberg, eds. *The Great Barrier Reef: Biology, Environment and Management.* New York: Springer, 2008.

Spalding, Mark. D., Corinna Ravilious, and Edmund P. Green. *World Atlas of Coral Reefs.* Berkeley: University of California Press, 2001.

Great Basin Desert

Category: Desert Biomes.
Geographic Location: Western North America.
Summary: The Great Basin, the largest desert biome in North America, is home to numerous unique species and subspecies of plants and animals.

The semi-arid and desert region referred to as the Great Basin Desert has been defined in at least five different ways, each affecting its reported size and location. The hydrographic definition will be used here; this refers to the low elevation areas west of the Wasatch Range, east of the Sierra Nevada mountains, south of the Columbia Plateau, and north of the Mojave Desert and Colorado Plateau. This great stretch of land covers roughly 200,000 square miles (518,000 square kilometers), with elevations ranging between 5,000 feet (1,500 meters) and up to 12,000 feet (3,700 meters) above sea level. There are also many cave systems in the basin.

Geography and Climate

Since this area has no outlet to the sea, the term *basin* is used in its most literal sense: a broad area of the earth, beneath which the strata dip usually from the sides toward the center. Land area in the Great Basin includes about 50 percent of the state of Utah, almost all of Nevada, and smaller areas of southeastern Oregon, southern Idaho, and eastern California. This makes it the largest of the four desert regions of North America and, generally, composes the northern portion of the basin and Range Province.

Water is scarce and consists mainly of shallow groundwater, large regional springs, and small creeks. Each of these sources depends on healthy groundwater flows supported by precipitation in mountain ranges, which receive more moisture than the valleys. Climate here is typical of mid-latitude semi-arid lands, where the evaporation potential exceeds precipitation throughout the year.

Since the basin suffers from lack of precipitation due to the rain shadow effect of the Sierra Nevada mountains, most annual moisture reaches the basin during the seasonal runoff of snowmelt. Average annual precipitation ranges from about 6 inches (15 centimeters) at the lowest elevations to more than 30 inches (75 centimeters) at the upper elevations. At lower elevations, about half of the precipitation falls as snow during the November 1 to May 1 winter season, increasing to more than three-quarters in high elevations. June and September are the driest months; summer thunderstorms are common.

Temperatures range from minus 32 degrees F (minus 36 degrees C) to 99 degrees F (37 degrees C). Mean January temperature is 18 degrees F (minus 8 degrees C); mean July temperature is 55 degrees F (13 degrees C). Maximum and minimum daily temperature differences can range from 37 to 50 degrees F (3 to 10 degrees C) on any day, depending on elevation and site.

Flora and Fauna

Often referred to as the Cold Desert because of its high latitude and elevation, low seasonal temperatures, and the accumulation of surface snow in the winter, the basin is biologically diverse. The composition of the flora and fauna of the basin varies widely, based on available water, elevation, soil chemistry, and structure. The simplest plant systems are found in barren salt flats, starting usually below 5,500 feet (1,670 meters) in elevation. These plant communities make up what is called salt-desert shrub: greasewood (*Sarcobates vermiculatus*), shadscale (*Atriplex confertifolia*), and green molly (*Kochia americana*) appear in conjunction with creosote brush (*Larrea tridentata*), four-winged saltbush (*Atriplex canescens*), and rabbitbrush (*Chrysothamnus viscidiflorus*).

As elevation is increased, vegetation transitions into sagebrush, and these "oceans" of sage are often referred to as sagebrush steppe. The plant communities that thrive at even higher elevations are pinyon-juniper woodlands. The very highest elevations include the oldest living organisms on Earth, the Great Basin bristlecone pines, which have been recorded as living more than 4,000 years. Other plant species that can be found at various elevations throughout the Great Basin Desert include prickly pear cactus, alpine wildflowers, and aspen.

Animal wildlife is abundant in the Great Basin Desert, despite diminishing sources of water. Some of the mammals that can be found here are: elk, mule deer, bighorn sheep, mountain lion, marmot, beaver, porcupine, jackrabbit, ring-tail cat, weasel, skunk, sagebrush vole, and water shrew. There are many reptiles and some amphibians; the Great Basin rattlesnake (*Crotalus oreganus lutosus*) is the only venomous snake found here, and the chuckwalla lizard is the largest lizard in the United States.

Birds commonly found in the basin include the sage grouse, quail, several species of hawks and falcons, and three owl species. Great Basin National Park is home to four native fish species: Bonneville cutthroat trout, mottled sculpin, redside shiner, and speckled dace. It also provides habitat for four nonnative species: Lahonton cutthroat, rainbow, brook, and brown trout. These fish may all be found in creeks or lakes in the basin.

The twisted trunk of a Great Basin bristlecone pine, which can live for thousands of years and is thought to be the oldest living organism in the world. (Thinkstock)

Human Impact

Although the basin was home to many tribes of aboriginal Native Americans including, but not limited to, Paiute, Shoshone, Ute, and Panamint, they left no written records of their respective environments, cultures, or lifestyles—although a few oral histories have been recorded. These groups with their diverse cultures and languages lived off the land primarily as hunter-gathers, and had little effect on the overall ecology of the basin. The pristine nature of this land and the existing flora and fauna that the first white explorers and settlers found in the late 1700s has been completely changed due, primarily, to explosive human population growth, irrigation-based agriculture, and high levels of grazing by domestic livestock, principally cattle and sheep.

The distribution of plants and animals within the basin was initially constrained by the amount of available water and geographic latitude. However, approximately 78 percent of the basin is now administered as public land, primarily by the Bureau of Land Management, the U.S. Department of Agriculture's Forest Service, the Fish and Wildlife Service, the Bureau of Reclamation, the National Park Service, the Department of Defense, and various tribal and state agencies. Almost all of the Great Basin Desert lands are subject to governmentally subsidized livestock grazing, including areas protected as national parks, such as the Great Basin National Park. Consequently, livestock grazing in the arid west is one major cause of species endangerment.

Another major concern in the desert is water distribution. Much of the valuable and scarce groundwater is already being pumped into urban areas (such as Las Vegas), and if the amounts increase, the loss in the basin will likely impact sustainability of the wildlife here. Climate change due to global warming is expected to have a significant effect on the basin over the next several decades. Warming trends, increases in precipitation, decline in snowpack, earlier springs, and extended fire seasons may all have grave implications for water resources, and therefore for native ecosystems and biodiversity. It is thought there will be a continued shift in species communities, as the number of shrinking habitats and local extirpations increase.

Robert C. Whitmore

Further Reading

Chambers, Jeanne C. *Climate Change and the Great Basin.* Reno, NV: U.S. Department of Agriculture Forest Service Rocky Mount Research Station, 2008.

Grayson, Donald K. *The Great Basin: A Natural Prehistory.* Berkeley: University of California Press, 2011.

Natchlinger, J., K. Sochi, P. Comer, G. Kittel, and D. Dorfman. *Great Basin: An Ecoregion-Based Conservation Blueprint.* Reno, NV: The Nature Conservancy, 2001.

Trimble, S. *The Sagebrush Ocean: A Natural History of the Great Basin.* Tenth Anniversary Edition. Reno: University of Nevada Press,1999.

Wilcove, David S., David Rothstein, Jason Dubow, Ali Phillips, and Elizabeth Losos. "Quantifying Threats to Imperiled Species in the United States." *BioScience* 48, no. 8 (1998).

Great Basin Montane Forests

Category: Forest Biomes.
Geographic Location: North America.

Summary: These forests, dominated by coniferous trees on isolated mountain ranges, rise up from desert shrublands within the Great Basin Desert. Altered fire regimes and climate change are potential threats to this ecosystem.

The Great Basin montane forests occupy isolated mountain ranges within the Great Basin Desert, rising from surrounding desert shrubland at lower elevations. This desert covers roughly 200,000 square miles (518,000 square kilometers), encompassing most of Nevada, parts of eastern California, western Utah, southern Idaho, and southeastern Oregon—and on some maps, it also extends into northwestern Arizona and southwestern Wyoming. Basin and range topography typifies this desert, consisting of valley basins flanked by north-south trending mountain ranges. Hundreds of these mountain ranges that support montane forests occupy the Great Basin Desert, with more than 300 discrete ranges mapped in Nevada alone.

A total 33 Great Basin mountain ranges have peaks exceeding 10,000 feet (3,050 meters) above sea level. The higher-elevation mountains have cooler temperatures and greater precipitation, including snow, than the basins: The basins often receive only 6 inches (15 centimeters) of annual precipitation, compared with the mountains, which receive more than 20 inches (50 centimeters) per year. As a result, vegetation generally changes from desert shrubland in the basins to pinyon-juniper woodlands on lower mountain slopes, forests of various conifer species on higher mountain slopes, and alpine meadows on the highest peaks above where trees grow.

The current Great Basin montane forests partly originated from the contraction of conifer forests that during the last glacial period (more than 10,000 years ago) covered larger areas than forests do today. Petrified wood of these glacial-era trees can still be seen in areas that today support only desert shrubland.

Flora and Fauna

The plant and animal life of montane forests is not uniform from one mountain range to another. Forest composition varies with latitude, elevation, soil type, proximity to other mountain ranges (influencing species dispersal), geography (such as susceptibility to lightning that ignites wildfires), and past and present land use by humans.

The generalized vegetation pattern—which is not present on all mountain ranges—of Great Basin montane forests from low to high elevation includes: woodlands composed of the short trees juniper (such as *Juniperus osteosperma*) and pinyon pine (such as *Pinus monophylla*); ponderosa pine forests (*Pinus ponderosa*); mixed conifer forests comprised of species such as white fir (*Abies concolor*), limber pine (*Pinus flexilis*), spruce (*Picea engelmannii*), and subalpine fir (*Abies lasiocarpa*); and bristlecone pine (*Pinus longaeva*) or other conifers at the highest elevations. Bristlecone pines are among the oldest living organisms on Earth, often living more than 4,000 years on the high peaks. There also are alpine meadows near some of the highest peaks, above the tree line. Some mountain ranges contain deciduous forests of quaking aspen (*Populus tremuloides*); these trees colonize disturbed areas left by wildfires or avalanches.

The plants growing on the forest floor and in the understory also vary among mountain ranges and with elevation and soil type. Understories can be dominated primarily by shrubs, forbs, grasses, or mixtures. These plants reduce soil erosion, provide forage and habitat for animals, play an important role in fire ecology (by providing fuel), and enhance species diversity. Among the thousands of understory species are big sagebrush (*Artemisia tridentata*), currant (*Ribes cereum*), fleabane (*Erigeron divergens*), muttongrass (*Poa fendleriana*), and squirreltail (*Elymus elymoides*).

As with plants, a long-term perspective illustrates how animal species of the Great Basin have changed. During the last ice age, now-extinct camels, mastodons, sabertooth tigers, American cheetahs, and other species inhabited the area. Today, a different but still diverse array of species inhabits the region, including mammals, birds, amphibians, reptiles, fish, and invertebrates.

The 76,603-acre (31,000-hectare) Great Basin National Park in eastern Nevada, for example, lists 71 mammal, 241 bird, 8 fish, 18 reptile, and six amphibian species as living in the park and vicinity. Carnivorous mammals include mountain lions (*Pumua concolor*), gray foxes (*Urocyon cinereoargenteus*), and coyotes (*Canis latrans*). Herbivorous mammals include mule deer (*Ococoileus hemionus*) and elk (*Cervus elaphus*). Other characteristic animal species that may be found in the Great Basin montane forests include bighorn sheep, jackrabbit, porcupine, marmot, weasel, and the ring-tailed cat.

Human Threats and Conservation

Beginning in earnest in the late 1800s and early 1900s, Euro-American settlement brought livestock grazing (cattle and sheep), timber harvesting, introduction of nonnative plant species (plants introduced as livestock forage or inadvertently introduced), and other human influences to the Great Basin area. Additionally, some of the montane forests are believed to have been partly shaped since the last ice age by recurring fires (ignited by lightning and in some places by Native Americans), which could affect tree density and species composition. Humans began suppressing these fires in some areas during the 1900s, which is believed to have resulted in a buildup of fuel and thus susceptibility to severe, deforesting fires.

Nonindigenous plants such as cheatgrass (*Bromus tectorum*) also have increased fuel loads in some areas. Climate changes may further influence montane forests, especially because the mountain ranges are isolated (limiting potential for species migrations), and species currently occupying the highest peaks cannot migrate upward. The most recent climate change has already had effects on this ecoregion: warmer temperatures, decline in snowpack, earlier spring runoff, and extended fire seasons may all have grave implications for water resources, and therefore for native ecosystems and biodiversity.

Public lands containing Great Basin montane forests are managed by the U.S. Forest Service, Bureau of Land Management, National Park Service, and Fish and Wildlife Service. Examples of these areas include the Humboldt-Toiyabe National Forest and Great Basin National Park in Nevada, Death Valley National Park in Califor-

Native American Influence

Humans have influenced montane forests since the last ice age. Some of the Native American tribes that inhabited the Great Basin Desert at the time of Euro-American settlement include the Southern Paiute, Panamint, Ute, Shoshone, Bannock, and Northern Paiute. Some crops were grown, but because the climatic conditions of the Great Basin limited agriculture, hunting and gathering were critical to human survival. Native Americans hunted animals such as mountain sheep (*Ovis canadensis*), gathered edible plants, and harvested berries and nuts.

Pinyon nuts were food staples, and it is believed that a family of four could gather enough nuts in a month of harvesting during a good year to last for four months, which was critical during winter. Native Americans also may have influenced forests by lighting fires, which could have been used to drive game animals for hunting, change habitat to favor certain food plants, or communicate across distances.

nia, Cedar Breaks National Monument in Utah, Sawtooth National Forest in Idaho, and Malheur National Forest in Oregon.

Key management considerations common to all areas include understanding and accommodating fire regimes, ensuring that forests have desirable densities of trees (forests that are too dense are more susceptible to insect outbreaks and severe fires), limiting movements of undesired nonindigenous species, controlling livestock grazing areas, incorporating climate change into planning, and balancing human uses of forests with conservation strategies. These forests supply many services directly to humans, such as timber, grazing lands, water resources, carbon storage, and recreational opportunities. Montane forests within the larger desert landscape are irreplaceable resources to humans and numerous other species.

Scott R. Abella

Further Reading

Bender, Gordon L. *Reference Handbook on the Deserts of North America.* Westport, CT: Greenwood Press, 1982.

Cronquist, Arthur, et al. *Intermountain Flora: Vascular Plants of the Intermountain West, U.S.A.* New York: Hafner Publishing, 1972.

Grayson, D. K. *The Great Basin: A Natural Prehistory.* Berkeley: University of California Press, 2011.

Rickart, Eric A., et al. "Mammals of Great Basin National Park, Nevada: Comparative Field Surveys and Assessment of Faunal Change." *Monographs of the Western North American Naturalist* 4, no. 1 (2008).

U.S. Department of Agriculture (USDA), Forest Service. *National Roadmap for Responding to Climate Change.* Washington, DC: USDA, 2010.

Great Bay Estuary

Category: Marine and Oceanic Biomes.
Geographic Location: North America.
Summary: Advocates of this strongly tidal estuary, one of the most important such water bodies in the northeastern United States, have rallied around its Wildlife Refuges and Research Reserve to restore the imperiled habitats here.

Often considered New Hampshire's hidden coastline, the Great Bay Estuary is a complex embayment and one of the nation's most recessed estuaries. This waterway, situated just south of Maine's border, offers a variety of habitats that sustain a large biodiversity of wildlife. Current eutrophication, pollution, and habitat fragmentation due to human development are causing significant changes to the estuary.

Geography and Climate

This estuary is comprised of two large inland bays—Great Bay and Little Bay—and the Piscataqua River. These three bodies of water lie at the confluence of freshwater from seven major river systems, and the tidal seawater from the Gulf of

Maine. Before reaching Great Bay, this seawater travels 15 miles (24 kilometers) inland through the Piscataqua River and Little Bay. The Piscataqua forms the northern end of the Great Bay estuarine system; it is directly controlled on its east side by oceanic tides, and along the northwestern edge by its inland tributaries, the Salmon Falls and Cocheco Rivers.

The more centralized Little Bay extends from the Piscataqua River southward toward Big Bay. Characterized by its many mudflats, the deep-channeled Little Bay is fed by the Oyster and Bellamy Rivers.

Horseshoe crabs are among the marine species that inhabit the Great Bay Estuary's many saltwater-dominated habitats. (Thinkstock)

The final component, Great Bay, begins at Furber Strait, at which point seawater has already traveled inland through the Piscataqua River and Little Bay. In addition to tidal waters from the ocean, Great Bay is directly influenced by the continual input of freshwater from the Squamscott, Lamprey, and Winnicut Rivers.

The saltwater tidal exchange in the Great Bay Estuary has a significant effect on the bay's ecosystem. Despite freshwater continuously flowing from its tributaries, the estuary remains an ocean-dominated ecosystem. It takes close to 18 days during peak river flow for water to flush from one end of the estuary to the other. The tidal dynamics of this estuary, especially within the Piscataqua River, create some of the strongest currents in North America. The change in tides also produces a prominent change in the landscape, with the ebbing tide leaving more than half of Great Bay exposed as mudflats.

Temperature ranges in the summer (July) have average highs in the mid-70 to low-80 degrees F (24 to 28 degrees C), and average lows in the mid-50 to low-60 degrees F (13 to 15 degrees C). Temperatures in the winter (January) have an average high of 34 degrees F (1 degree C), and an average low of 0 degrees F (minus 18 degrees C). Precipitation in the area is roughly 40 inches (102 centimeters) annually.

Biodiversity

The entire Great Bay estuarine system provides a variety of habitats including eelgrass beds, mudflats, salt marsh, rocky intertidal, channel bottom, tidal creek, and upland forest and fields.

Because the bay is part of the North Atlantic Flyway, numerous species of waterfowl, shorebirds, and wading birds use it for migration and overwintering, including a large winter population of federally protected bald eagles. The bay area serves as New Hampshire's primary wintering area for black ducks. Many state-protected species use the refuge as well, including the common loon, pied-billed grebe, osprey, common tern, northern harrier, and upland sandpiper.

Commonly found fish species here include both saltwater and freshwater varieties. Striped bass (*Morone saxatilis*) have become one of the greatest success stories for marine fisheries in New Hampshire and Maine coastal waters. Also found along the coast are bluefish, flounder, mackerel, pollock, haddock, and Atlantic cod. The Piscataqua River is a migratory highway for herring and shad; further into the estuary, in more brackish waters, can be found more herring, striped bass, and occasionally an Atlantic salmon or lamprey. Horseshoe crabs, oysters, mussels, and clams also reside in the estuary, most notably oysters.

Human Impact

Historically, the Great Bay Estuary has been an important location for natural resources. Native American tribes including the Abenaki Nation and others inhabited the region for more than 11,000 years before the first European settlements began in the 1600s. They survived on the abundant fish, shellfish, waterfowl, and mammals that lived in and around the estuary. As the population of new, white settlers around the bay increased, new industries developed, including textiles, timbering, and agriculture. The water was used to support new industries and to move cargo further inland.

Over time, ecological changes have occurred in the bay due to increased human activity, including increased sedimentation, agricultural runoff, industrial waste, raw sewage, and other pollutants. In 1989, the Great Bay Estuary was formalized as a Great Bay National Estuarine Research Reserve by the National Oceanic and Atmospheric Administration (NOAA) and the National Ocean Service. In addition, the New Hampshire Fish and Game Department has designated Adams Point, a small parcel of land bordering the western shoreline of Great Bay, a Wildlife Management Area.

The University of New Hampshire's Jackson Estuarine Laboratory is located directly on the shoreline of Adams Point, allowing the university to specialize in Great Bay ecology through numerous ongoing research projects, including short- and long-term monitoring of the bay. Active areas of research at the laboratory include the ecology of estuarine seaweeds, nutrient cycling, fish populations, and shellfish aquaculture.

In 2009, a federally funded conservation unit, the Piscataqua Region Estuaries Partnership, generated a State of the Estuaries report on the health of Great Bay. The report warned that sprawling growth was having a negative effect on the bay, and that indicators such as increased nitrogen loading and poor dissolved oxygen concentrations were affecting water quality and threatening important submerged aquatic vegetation such as eelgrass (*Zostera marina*), a seagrass species vital for fish survival and sediment binding. As with many estuarine systems affected by human development in the last century, nonpoint pollution sources such as septic systems and fertilizers have been linked to the increased nitrogen loading in Great Bay.

Also of major concern are the impacts of climate change, which are already being studied in the estuary. These changes have substantial implications for the Great Bay ecosystem. New Hampshire and Maine are likely to experience warmer temperatures and increased flooding from storm events and a rising sea level. Species and habitats may shift in response to changes in these anticipated temperature and water levels.

Brian W. Teasdale

Further Reading

Jones, Stephan H., ed. *A Technical Characterization of Estuarine and Coastal New Hampshire.* Durham: University of New Hampshire, 2000.

Mills, Kathy, K. Loughlin, S. Miller, R. Stevens, P. Wellenberger, E. Heckman, R. Roseen, L. Poinier, and V. Young. *Ecological Trends in the Great Bay Estuary: Twenty-Year Anniversary Report.* Durham, NH: Great Bay Nation Estuarine Research Reserve, 2009.

Piscataqua Region Estuaries Partnership. "The Great Bay Initiative." 2012. http://www.prep.unh.edu/GreatBayDialogue.htm.

Great Bear Lake

Category: Inland Aquatic Biomes.
Geographic Location: Canada.
Summary: This natural water body in the far north has remained nearly unaffected and unaltered by humans and is believed to be the largest freshwater lake in the world with this degree of pristine character.

Great Bear Lake is located in Canada's Northwest Territories, with its northern waters within the Arctic Circle. This vast boreal pool is considered among the most pristine freshwater bodies in the world; its remote position has helped protect both its water and its habitats from degrading by human interaction. Yet its biological character, formed in extreme cold conditions, may prove somewhat fragile under the rigors of global warming.

Geography and Climate

Smaller than the Great Lakes Superior or Huron, but the largest freshwater lake completely within Canada, Great Bear Lake was carved by glaciers during the most recent ice age. The lake has a surface area of approximately 12,000 square miles (31,000 square kilometers). It is one of the deepest lakes in the world, with a mean depth of 235 feet (72 meters), with its deepest point at 1,460 feet (446 meters). Scattered with numerous small islands, Great Bear Lake is shaped like a jigsaw puzzle piece (similar in shape, some say, to the Starship *Enterprise*) with five main arms extending from the center. The arms of the lake are named McVicar, McTavish, Keith, Smith, and Dease. The longest stretch across is roughly 190 miles (300 kilometers) from the far reaches of one arm to another.

Great Bear Lake drains into the Mackenzie River, which flows northwest and is joined by the Arctic Red and Peel Rivers, before finally emptying into the Beaufort Sea, a part of the Arctic Ocean. The drainage basin for the lake is approximately 44,400 square miles (115,000 square kilometers). Great Bear Lake lies across two chief physiographic regions: the Canadian Shield and the northern Interior Plains. Boreal forests of the taiga surround the lake, and its northernmost arm reaches into the tundra.

Great Bear Lake temperatures are very cold except during the very brief subarctic summer. During this short summer—the lake is typically covered with ice from late November through July—the lake's surface temperature remains at or below 39 degrees F (4 degrees C). Great Bear is a polar, or cold, monomictic lake, meaning that its water mixes thoroughly from top to bottom only once a year, when it lacks ice cover and then has relatively uniform temperature and density at all depths.

Wildlife

Commercial fishing is prohibited, and sport fishing is closely regulated on Great Bear Lake, because of the slow regeneration of the fish in the icy water. As a result of the temperature and depth of the waters, fish here grow slowly; because of this slow growth rate, it is estimated that some fish found in Great Bear Lake can be over 100 years old. The trout of Great Bear Lake can take anywhere between 15 and 26 years to reach sexual maturity, and only spawn once every two to three years.

Because stocks can quickly become endangered, sport fishing is heavily regulated. Lake trout over 28 inches (71 centimeters) or 10 pounds (4.5 kilograms) are not allowed to be harvested from the lake. Lake trout, lake whitefish, Arctic grayling, Arctic char, walleye, and northern pike are some of the noted species found in the lake. Great Bear Lake has a significant reputation as the greatest trout and Arctic grayling fishery in the world, with world records (in size and weight) held for both.

Surrounding the southern and western shores of the lake are boreal forests, largely of black and white spruce, interspersed with muskeg in the lower-lying, poorly drained regions. To the north the forest declines, as it gives way to tundra. The shores of Great Bear Lake are rich in wildlife, and since there is little interaction with or interference from humans, this ecosystem remains virtually unaltered. Animals that inhabit the lake's surrounds include: musk ox, moose, Dall sheep, mountain goat, woodland caribou, barren-ground caribou, black bear, grizzly bear, bald eagle, golden eagle, gyrfalcon, elk, martens, and numerous types of water fowl.

Human Impact

The Great Bear Lake area's first known inhabitants were indigenous tribes of the Athapascan language group: the Hares, Mountain (known also as Slavey or Slave), Dogribs, and Copper Indians together making up a group recognized as the Satudene, derived from words in Chipewyan meaning *bear water people*. These distinct groups inhabited and hunted in various areas around the lake and to the south toward Great Slave Lake.

In 1799, the first European settlers arrived in the form of explorer Alexander Mackenzie, his partners, and their fur trading company, the North-West Fur Company. This company, a rival of the Hudson's Bay Company, merged with it in 1821. Except for fur trading posts, settlements were scarce around the Great Bear Lake until the discovery of pitchblende, silver, and cobalt in 1930. A mine was established, extracting radium from the pitchblende deposits; the associated uranium ore was discarded as of little commercial value. The mine eventually closed, but was reopened in 1942 to supply uranium for the Manhattan Project. Quantities of uranium, having previously been dumped in the lake, were recovered by dredging. This mine was closed in 1964, but certain workings were maintained for the extraction of silver.

The only permanent settlement in the region is in Deline, with a population of just over 500 people, mainly of Satudene heritage. There are fishing lodges in various locations around the lake, accessed by airplane. Most of the year, there are very few viable roadways. There is no current evidence of siltation, toxic contamination, eutrophication, or acidification in Great Bear Lake.

However, northern Canada has already experienced some of the most extreme climate change effects on the planet. There is concern that the pristine nature of the lake and its aquatic ecosystems could be compromised as temperature, humidity, and precipitation regimes deflect from their ages-old patterns. Habitats here, and species with their extenuated growing cycles, would be hard-pressed to adapt and adjust to the dramatic changes that seem to loom from global warming.

Sandy Costanza

Further Reading

Déline Renewable Resources Council. *Great Bear Lake Climate Change Analysis: An Assessment of Climate Change Variables in the Great Bear Lake Region, Canada.* Fairbanks, AK: Scenarios Network for Arctic Planning, 2011.

Johnson, Lionel. "The Great Bear Lake: Its Place in History." *Arctic* 28, no. 4 (1975).

Likens, Gene, E. *Lake Ecosystem Ecology: A Global Perspective.* New York: Academic Press, 2010.

Vincent, Warwick F. *Polar Lakes and Rivers: Limnology of Arctic and Antarctic Aquatic Ecosystems.* New York: Oxford University Press, 2008.

Great Black Swamp

Category: Inland Aquatic Biomes.
Geographic Location: North America.
Summary: An ancient wetlands ecosystem in the heart of North America, this biome underwent near-total undoing by human effort; today, attempts are underway to return some of these lands to their primeval condition.

At one time, the Great Black Swamp was one of the largest wetlands in the northeastern state of Ohio. That ecosystem sustained old-growth forests and habitats for many mammals, reptiles, amphibians, fish, and migratory and nonmigratory birds. It was drained by settlers in the mid-1800s to create roadways and farmlands. There are currently conservation efforts being made to reclaim some of this lost wetland acreage and to restore—or recreate—some of the preexisting habitat.

Geography and History

The Great Black Swamp was a wooded wetland formed during the Pleistocene Epoch, when the glaciers of the Wisconsin Glacial Episode began to melt and retreated, during a period some 12,000–20,000 years ago. Before the glaciers melted, their advancing movement and the enormous weight from over a mile of vertically packed ice scraped and scoured the land here into a relatively flat plain. The melting and retreating of the glaciers deposited vast amounts of water, leaving behind many lakes in the basins of the scraped plain.

These lakes eventually subsided, and the flat, flooded plain evolved into a massive, woody swamp dotted with intermittent wet prairies and savannas. The subsiding lakes also left behind sandy ridges and low moraines (accumulations of soil and rock left by glaciers) that acted as natural wetland borders. Clay sediment at the bottom of the lakes made a natural water holding area, as clay is fairly impervious to water.

The Great Black Swamp was one of the largest wetlands in the region that became Ohio, stretching approximately 30 to 40 miles (48 to 64 kilometers) in width and 120 miles (193 kilometers) in length. Its borders were within the watersheds of the Maumee, Auglaize, and Portage Rivers in northwestern Ohio. The area it covered arced from Sandusky, Ohio, bordering Lake Erie in the east; to Findlay, Ohio, in the south; and Fort Wayne, Indiana, in the west. It was home to Native Americans and farmers who worked the flood plains of the Maumee River and its tributaries.

To most of the first European settlers, the area was a dark, forbidding quagmire with huge trees, bottomless mud holes, putrid marshes, and heavy infestations of mosquitoes. Native American tribes, specifically the Ottawa, lived at the edges of the swamp until they were forced out by white settlement. The tribesmen would not enter the forbidding swamp themselves, but instead hunted along its perimeter. The Ottawa conquered the mosquito problem in several ways. Their lodges were built with holes in the top and a fire could burn inside; the smoke kept the mosquitoes at bay. The Ottawa also fabricated an insect-repelling rub. They typically wore long-sleeved garments, and hunted mainly in winter when insects were mostly dormant. Before the arrival of Europeans, indigenous peoples did not suffer from malaria.

Draining the Swamp

Settlement incursion into the area began in the 19th century, with the building of three main roads through the swamp. In 1808, the Western Reserve Road, now U.S. Route 20, stretched from Fort Meigs (now Perrysburg) to Findlay. General William Hull, on his march to Detroit during the War of 1812, blazed a trail through the swamp known as Hull's Trace, now U.S. Route 25. The third road stretched from Fort Meigs to McCutchenville; the "worst road in the nation," as it became known, was this Western Reserve road. It was so atrocious that travelers considered themselves fortunate to travel a single mile (1.6 kilometer) in a day.

The decision to drain the swamp began with the building, in the 1850s, of a road through the very heart that ran from Fremont, Ohio, to Detroit, Michigan. Initially, road construction was difficult due to erosion, until the engineers used rock beneath the road to level it. To allow for drainage, a ditch was placed on each side of the roadway. Culverts were placed beneath the roads to allow the water to be channeled to its natural drainage direction. Once the surface water was eliminated, the subsurface water that created seemingly bottomless mud holes was next. Draining the groundwater was performed by installing special clay tile drains buried 2 to 3 feet (up to 1 meter) below the surface, which carried the water to natural and man-made streams and on to Lake Erie.

It is ironic that the Great Black Swamp held the source of its own demise. The layer of clay beneath the topsoil was 60 to 200 feet (18 to 61 meters) thick; this rich resource provided the clay mills with the necessary raw materials to fabricate the clay drain pipes used in the draining process. In 1892, machines were invented that could create the ditches faster than manual labor; the draining process accelerated. Within a few decades, the Great Black Swamp that had stood unscathed for millennia was all but gone.

Impact and Conservation

The disappearance of the swamp occurred at a faster rate and left fewer traces than the demise of any other ecosystem in Ohio; it happened so fast that a systematic inventory of flora and fauna species had not been undertaken. Not only was the water removed, but also the forest. Before the draining and tree removal, the swamp was home to a wide diversity of deciduous trees such as elm, black ash, sour gum, silver and red maple, pin oak, swamp white oak, sycamore, white ash, buckeye, shell bark hickory, honey locust, black cherry, and yellow and red oak. These trees, along with their stumps, were all removed to pave the way for agricultural fields. Gone were the many varieties of trees and ground plants, along with the wildlife habitats they supported. Gone, too, were countless unknown mammal, amphibian, reptile, insect, and avian species.

The draining of the Great Black Swamp and its subsequent transformation into farmland also had unintended consequences. Wetlands act as an environmental buffer in an ecosystem. They slow soil erosion and filter harmful agricultural runoff such as herbicides, pesticides, and fertilizers before they reach streams, rivers, and lakes.

The water from the Great Black Swamp was now free to travel downstream, where it accumulated in Lake Erie. These nitrate-rich deposits from fertilizers accelerate the growth in Lake Erie of algae, creating vast blooms that deplete oxygen in the water, resulting in massive fish kills. According to the Environmental Protection Agency (EPA), 90 percent of Ohio's wetlands were lost with the destruction of the Great Black Swamp, to the detriment of towns downstream, and ultimately to the Lake Erie ecosystem. Flood damage, increased drought, invasive weed and insect populations, and a declining avian population were among the negative direct impacts associated with the disappearance of the Great Black Swamp.

State and national government agencies, private landowners, and nongovernmental organizations such as Ducks Unlimited and the Black Swamp Conservancy (BSC) have realized the importance of wetlands. These organizations and individuals are working to transform some of the land back into much-needed wetlands. The BSC has transformed 10,000 acres (4,046 hectares) back into marsh; it is on pace to add 1,000 acres (405 hectares) each year. The eventual and ongoing goals are to prevent agricultural runoff, to free some wetland water flow, and to re-create lost habitats for mammals, insects, reptiles, amphibians, fish, and birds.

David A. Douglass
William Forbes

Further Reading

Hallett, Kaycee. "History of the Great Black Swamp." *The Black Swamp Journal*, April 14, 2011. http://blogs.bgsu.edu/blackswampjournal/2011/04/14/history-of-the-great-black-swamp.

Light, Christopher. "Draining the Great Black Swamp." The Historical Marker Database, December 7, 2007. http://www.hmdb.org/Marker.asp?Marker=4025.

Platt, Carolyn V. "The Great Black Swamp." *Timeline: Ohio Historical Society Magazine* 4, no. 1 (1987).

Pollick, Steve. "Land Group Hits 10,000 Milestone." *The Toledo Blade*, May 8, 2011. http://www.toledoblade.com/StevePollick/2011/05/08/Land-group-hits-10-000-milestone.html.

Great Cypress Swamp

Category: Inland Aquatic Biomes.
Geographic Location: North America.
Summary: The Great Cypress Swamp was a rich base of diverse wetland habitats until the land was drained and clear-cut; more recently, efforts to restore and conserve the ecosystem are slowly producing positive results.

Locally known as Burnt Swamp, the Great Cypress Swamp is the northernmost cypress swamp in the United States. This wetland, located in what is now part of Delaware and Maryland, was once home to bears, wolves, and cougars, as well as many smaller mammals, and a great diversity of birds, fish, amphibians, and reptiles. With European settlement of this coastal region—and the eventual practices of logging and the associated fires, iron ore mining, and drainage of some areas of the swamp for agricultural purposes—most of the wetlands were lost. Currently, work is being done to reclaim and restore some of the vegetation and habitat that were once sustained here.

Original Ecosystem

The Great Cypress Swamp originates on the Delaware-Maryland border and flows through the state of Maryland about 75 miles (120 kilometers) before emptying into the Pocomoke Sound of Chesapeake Bay. The swamp is the largest tract of forest remaining on the Delmarva Peninsula, covering roughly 50 square miles (150 square kilometers) of this mid-Atlantic coastal region.

The Great Cypress Swamp and the surrounding forests were historically dominated by bald cypress, swamp tupelo, sweetgum, and Atlantic white cedar. As a core species, bald cypress trees are deciduous trees that have needles similar to those of pine trees—but do not remain green year-round. Bald cypress trees can grow up to 120 feet (37 meters) tall and live up to 600 years; they are easily recognized by the buttressed trunks and tall knees that aid in stability and possibly oxygen retrieval. Cypress swamps are characterized by tea-colored water, which is created from the tannic acid produced as leaves and roots from the forest decompose. Indeed, the name of the local indigenous tribe, the Pocomoke, means black water.

Human Onslaught

The swamp's first human inhabitants lived here more than 10,000 years ago, around the end of the last ice age, when the swamp covered as much as 93 square miles (240 square kilometers). Native American tribes such as the Pocomoke, Nanticoke, and Nassawattox, among others of the Algonquin

Bald cypress trees, which can grow as tall as 120 feet, in a swamp in eastern Maryland. The Great Cypress Swamp covers 50 square miles (150 square kilometers) of its original 93 square miles (240 square kilometers). (USDA Natural Resources Conservation Service)

Nation, set up villages along the borders of the swampland to take advantage of the abundance of wildlife and fertile soil that the swamp produced. By the late 17th century, these communities were displaced by European settlement.

As the region was rapidly colonized, demand for raw materials increased. Cypress is strong and rot-resistant, and settlers found this wood to be excellent for building ships, house shingles, and water tanks. From the 17th century through the early 1900s, lumber industries began to clear-cut the trees around the edges and into the heart of the swamp. By the late 18th century, cypress was in high demand. As was the case in much of the eastern United States, tree growth could not keep up with demand for wood, and many of the slow-growing cypress tree stands were lost. The last straw for the cypress swamp was an extensive peat fire in the 1930s that destroyed much of the remaining forest. This fire burned for eight months, resulting in the moniker Burnt Swamp provided by local residents.

The swamp produced other valuable resources: iron ore, which was used for making industrial, agricultural, and household products beginning in the early 19th century; and clay, used for manufacturing bricks. The extraction of these materials added to the swamp's degradation.

The ultimate clearing of the swamp left the land open to other uses, primarily agriculture. A burgeoning population in the eastern United States meant that more food was needed; land without any forest cover was more easily converted to cropland. However, the land of the Great Cypress Swamp was wet through much of the year, and needed to be drained in order to be cultivated. A large drainage canal was built in 1936 to connect the swamp to the estuary at Pocomoke Sound, rapidly draining water from the swamp, leaving behind dry, fertile soil. Though this process resulted in a large increase in farmland, the loss of the wetland had an immeasurable impact on the local ecosystem and its previously existing flora and fauna.

In more recent times, the effects of global warming are working against some of the remnant habitat of the Great Cypress Swamp. Chesapeake Bay waters are rising faster than the global average sea-level rise, meaning the estuary of the swamp is transporting more saline waters further inland. Coastal erosion, too, has been eating away at the downstream reaches of this biome. The projected rise in average temperatures will have unpredictable effects on the plant and animal species that might otherwise return to their former habitats around the swampland. Unless they adapt, they will be more likely to yield the area to invasive species.

Restoration Campaigns

Today, the Great Cypress Swamp has at last been recognized by the prevailing settlers for its ecological value. In the 1970s, two nonprofit groups—Delaware Wetlands and The Conservation Fund—took up the initiative to begin the process of reclaiming parts of the swamp. In 1980,

U.S. Senator Joe Biden of Delaware proposed that the swamp be named a national park, although many advocates feared that the mass of visitors to such a park would precipitate further harm to this extremely fragile ecosystem.

Later, however, the state of Maryland designated the Pocomoke River—which flows through the swamp—a State Scenic and Wild River, mandating the protection of the natural infrastructure associated with the river, as well as the practice of conservation measures, at both state and local levels. Because this river and surrounding wetlands provide wildlife habitat, recreational opportunities, and protection for Chesapeake Bay shellfish populations, the recognition it has attained is leading to a concerted effort to restore the swamp.

Delaware Wild Lands is working with Ducks Unlimited, the U.S. Fish and Wildlife Service, and the Center for Inland Bays to restore the swamp. Since 2009, these groups have restored 769 acres (311 hectares) of swamp by installing water-control structures and plugging ditches, all in an effort to retain more water in the swamp. When the hydrology is restored, the cypress trees, Atlantic white cedars, and other species that thrive in wet soils and standing water will be able to return. Today, one may find within the slowly returning Great Cypress Swamp increasing numbers of largemouth bass, trout, perch, shad, and bluegills; a variety of toads, frogs and salamanders; warblers, great blue herons, pileated woodpeckers, bald eagles, osprey, and other birds of prey.

MARCUS W. GRISWOLD

Further Reading

Heckscher, C. "Forest-Dependent Birds of the Great Cypress Swamp: Species Composition and Implications for Conservation." *Northeastern Naturalist* 7, no. 2 (2000).

Holden, Christina. "Southern Exposure: The Great Cypress Swamp at Pocomoke Park." Maryland Department of Natural Resources, 2005. http://www.dnr.state.md.us/naturalresource/spring2005/parkticulars.asp.

Mansueti, Romeo. *A Brief Natural History of the Pocomoke River, Maryland.* Solomons, MD: Maryland Department of Research and Education, Chesapeake Biological Laboratory, 1953.

Great Dismal Swamp

Category: Inland Aquatic Biomes.
Geographic Location: North America.
Summary: This large swamp in the southeastern United States is both an unusual and heavily damaged ecosystem and one where species diversity is on the rebound, thanks to becoming a National Wildlife Refuge in 1974.

The Great Dismal Swamp, ensconced in more than 100,000 acres (40,500 hectares) of forested wetlands on the border between Virginia and North Carolina, also holds the largest natural lake in Virginia. Centuries of logging and drainage operations devastated the swamp's ecosystems—but the Great Dismal Swamp National Wildlife Refuge was established in 1974 to restore and protect its original biodiversity.

Geography and Climate

The Great Dismal Swamp is both beautiful and hazardous, and unique in many ways. While most swamps exist in the lowest elevations of a region, where water accumulates faster than it drains, the Great Dismal Swamp is on a hillside 20 feet (6 meters) above sea level. Lake Drummond, a 3,100-acre (1,250-hectare) natural lake, is situated in the center, and seven rivers flow outward from it.

In most swamps, water, heat, and darkness combine to cause the rapid decay of vegetation. In the Great Dismal, the water is so acidic from leaching of juniper and cypress trees that there is little bacterial growth to cause decay. Debris continues to accumulate every year, making this one of the few places in the southern United States where peat forms. Dense vegetation and an abundance of food make the Great Dismal Swamp an ideal home for a wide variety of wildlife.

The area generally has mild winters and warm humid summers. Winter temperatures average above 36 degrees F (two degrees C); summer temperatures typically average above 78 degrees F (26 degrees C), with high humidity.

History and Human Impact

Human occupation of the area dates back about 13,000 years, but by 1650, only a few Native American and European settlers could be found inhabiting the edges of the swamp. In 1665, William Drummond, the first governor of North Carolina, discovered the lake that was eventually named for him. William Byrd II led a surveying party through the swamp while calculating the dividing line between Virginia and North Carolina in 1728, and is credited with naming it the Dismal Swamp.

George Washington visited the area in 1763 and had a very different opinion. He saw great opportunity, and organized the Dismal Swamp Land Company to drain the swamp and clear it for settlement. Later, the group turned to the more profitable venture of logging enormous stands of cypress, juniper, and cedar trees. A 5-mile (8-kilometer) canal was dug through the western side of the swamp to Lake Drummond for shipping shingles and other wood products. By 1795, more than 1.5 million cedar shingles were being cut per year. Additional canals followed, and a railroad was laid through part of the swamp in 1830.

Trees in the Great Dismal Swamp National Wildlife Refuge, shown here in 2008, still include Atlantic white cedar and tupelo bald cypress, but they now make up less than 20 percent of the forest. (U.S. Fish and Wildlife Service)

Washington used slave labor and hired very poor whites to work in the swamp. It was hard and dangerous work, done in muddy ooze while coping with mosquitoes, yellow flies, and poisonous snakes. Some slaves were able to gain their freedom by working as bondsmen for the loggers. The mysterious, almost mythic swamp has inspired many poems and books; its prominent role in the history of slavery and the Underground Railroad was featured in Harriet Beecher Stowe's novel about fugitive slaves hiding in the swamp, *Dred: A Tale of the Great Dismal Swamp.*

While early efforts to drain the swamp were often in vain, logging operations were far more successful. Timber was harvested continually from the 1700s through 1973. More than 200 years of exploitation has drastically altered the wetland. Agricultural and commercial development encroached on the swamp's boundaries, reducing it to half of its original size. The entire swamp has been logged at least once, and many areas have been burned regularly by wildfires. Roads and canals also disrupted the natural hydrology of the area.

In the mid-20th century, conservation groups demanded that something be done to preserve the remainder of the Great Dismal Swamp. In 1973, Union Camp Corporation, a large landowner since the early 1900s, donated more than 49,000 acres (19,800 hectares) of swamp to the Nature Conservancy. The land was transferred to the U.S. Fish and Wildlife Service the following year, and the Great Dismal Swamp National Wildlife Refuge was officially established.

The primary purpose of the refuge resource management program is to restore and maintain the natural biological diversity that existed before commercial exploitation. Water, native vegetation, and diverse wildlife are essential to the swamp ecosystem. Water levels and water movement are being returned to their natural states. Plant diversity is being enhanced by careful forest management activities, including selective cutting and prescribed burns to reestablish native species. Wildlife management includes monitoring required habitats, with selective hunting used as a tool to balance some animal populations with available food supplies.

Wildlife

Today, the Great Dismal Swamp lies entirely within the Middle Atlantic coastal forests ecoregion. Eight major plant communities compose the swamp vegetation; the forested types include Atlantic white cedar, tupelo bald cypress, sweetgum-oak poplar, maple blackgum, and pine. The nonforested types include remnant marsh, sphagnum bog, and evergreen shrub communities. Tupelo bald cypress and Atlantic white cedar, formerly predominant forest types in the swamp, today account for less than 20 percent of the total cover. Three rare plants that thrive in the swamp are the dwarf trillium, the log fern, and silky camellia. The swamp is still home to many mammal species, including whitetail deer, black bear, bobcat, otter, and weasel. More than 200 bird species and 70 reptiles and amphibians can also be found here. Although the Great Dismal Swamp can never be returned to its condition prior to European exploration and settlement, it is on the rebound and has the potential to become an outstanding natural laboratory. Currently, the wildlife refuge supports growing populations of flora and fauna, while also being maintained for recreational purposes such as camping, boating, fishing, and hunting.

Jill M. Church

Further Reading

Badger, Curtis J. *A Natural History of Quiet Waters: Swamps and Wetlands of the Mid-Atlantic Coast.* Charlottesville: University of Virginia Press, 2007.

Blackburn, Marion. "Letter from Virginia: American Refugees." *Archaeology* 64, no. 5 (2011).

Davis, Hubert J. *The Great Dismal Swamp: Its History, Folklore, and Science.* Murfreesboro, NC: Johnson Publishing, 1971.

Levy, Gerald F. "The Vegetation of the Great Dismal Swamp: A Review and an Overview." *Virginia Journal of Science* 42, no. 4 (1991).

Popular Archeology. "Digging Up the Secrets of the Great Dismal Swamp." April 2011. http://popular-archaeology.com/issue/april-2011/article/digging-up-the-secrets-of-the-great-dismal-swamp.

Simpson, Bland. *The Great Dismal: A Carolinian's Swamp Memoir.* Chapel Hill: University of North Carolina Press, 1990.

U.S. Fish and Wildlife Service, Region 5. *Great Dismal Swamp National Wildlife Refuge and Nansemond National Wildlife Refuge.* Suffolk, VA: U.S. Fish & Wildlife Service, 2006.

Great Lakes Forests

Category: Forest Biomes.
Geographic Location: North America.
Summary: Three distinct types of forest ecoregion are spread across a vast section of North America and are anchored by the Great Lakes and the St. Lawrence River. Fragmented now, the habitats here are making a slow recovery.

The Great Lakes drainage basin covers more than 200,000 square miles (518,000 square kilometers) and contains more than 11,000 miles (17,700 kilometers) of shoreline. Here, once-vast stands of boreal forest around the north of the lakes, transitioning to deciduous forests along the southern shores, are highly fragmented today because of the effects of industrialization and agriculture. Reforestation efforts in recent decades, however, have resulted in the slow increase of forest area, extending positive effects on local habitats and biodiversity.

Climate and Ecoregions

The climate of the Great Lakes Basin is affected by three factors: air masses from other regions, the location of the basin within a large continental landmass, and the moderating influence of the lakes themselves. In the winter, the Great Lakes region is affected by two major air masses, which contribute to the characteristically changeable weather here. Arctic air from the northwest is very cold and dry when it enters the basin, but is warmed and picks up moisture traveling over the comparatively warmer lakes. When it reaches the far shore, the moisture condenses as snow, creating heavy snowfalls on the lee side of the lakes in areas frequently referred to as snowbelts. The temperature of the lakes continues to drop during winter; ice frequently covers Lake Erie, but seldom fully covers the other lakes. Less frequently, air masses that are warm and humid enter the basin from the Gulf of Mexico.

The basin encompasses a variety of distinct ecoregions, including prairies, savannas, woodlands, and wetlands. The Great Lakes forests have been further subdivided into more specific forest ecoregions. The Northern Forests ecoregion is wide and crescent-shaped, extending from northern Saskatchewan, Canada, east to Newfoundland and south to the northern shores of the Great Lakes. The region is hilly and lies on the Canadian Shield-Precambrian granite bedrock that is among the oldest on Earth. The hills are morainal deposits left by the most recent glacial retreat, more than 10,000 years ago.

Boreal forests consisting of large stands of black and white spruce, balsam fir, jack pine, tamarack, and other conifers cover more than 80 percent of this region. Many small lakes and areas of exposed bedrock also dot the landscape. Along the southern reaches of the Northern Forests region, there is a transition to a mix of white birch; aspen; and varieties of poplar, sugar maple, beech, spruce, and oak.

Immediately south of the Northern Forests ecoregion lie the Eastern Great Lakes lowland forests. This ecoregion is highly fragmented, with more than 95 percent of the habitat lost to development and pollution in the watershed of the St. Lawrence River. Numerous rare ecological phenomena can be found in this transitional zone, separating the boreal forests to the north from the broadleaf deciduous zones to the south. A mosaic of freshwater marshes, dunes, bogs, fens, and both hardwood and conifer swamps exist. Very rare and unique alvar (limestone plain) communities—also called pavement barrens—support a suite of prairie communities that are globally endangered. Rare stands of ancient white cedars, some determined to be almost 800 years old, grow on the exposed limestone cliffs of the Niagara Escarpment. The wide variety of forest types and habitats makes this region one of the richest in animal species in continental North America.

The Southern Great Lakes forest ecoregion covers much of the industrial heart of the Midwest, including southern Michigan, much of Ohio and Indiana, and extreme southwestern Ontario. The area is so heavily populated and developed that essentially no large blocks of the original habitat remain. The topography ranges from very flat to low rolling hills. Once dominated by deciduous forests of sugar maple and beech, the onrush of agricultural, industrial, and urban development have altered or eliminated nearly 100 percent of the original landscape here. Wetland losses in this region have been severe. Ohio has lost 90 percent of its wetlands, and 80 percent of the tamarack swamps of southern Michigan have been eliminated.

The fourth Great Lakes forest ecoregion is the Western Great Lakes forest. Reaching across the northern portions of Michigan, Wisconsin, and Minnesota into southeastern Manitoba and northwestern Ontario, this entire ecoregion was once covered entirely by glaciers. Characteristic vegetation is a mixed forest influenced by the topography and drainage of the area. The Western Great Lakes forest includes northern coniferous forest, northern hardwood forest, boreal hardwood-conifer forest, swamp forest, and peatland. Approximately 20 percent of this region remains as intact habitat. Most of the original mature pine forests have been logged and replaced by birch and aspen, with only scattered pine; this alteration has made a deep impact on the biodiversity of the region.

Human Interaction

Native people first settled the Great Lakes Basin about 10,000 years ago, establishing hunting and fishing communities around the lakes and their tributaries. The population of the entire region is estimated to have been perhaps 100,000 in the 1500s when the first Europeans arrived looking for a passage to the Orient. By the early 1600s, Europeans had explored the St. Lawrence Valley as far as Georgian Bay, and developed a thriving fur trade. The British maintained control through the American Revolution, resulting in the Great Lakes marking the boundary between the new United States republic and the remainder of British North America; the population grew on both sides of the border as each side attempted to settle and develop the area first.

While the initial settlers had a modest effect on the ecosystem, limited to the exploitation of some fur-bearing animals, the following waves of immigrants logged, farmed, and fished commercially, turning the lakes and their waterways into major trade highways and bringing about profound ecological changes.

The southern and eastern Great Lakes forests have been the most heavily affected by human activity. Pockets of original habitat are rare and remain highly isolated, limiting the species they can support. About 20 percent of the Western Great Lakes forests, in contrast, remain as intact habitat. Forest coverage is actually greater here, but the intense harvesting of the original pine forests has resulted in the conversion to forests of aspen and birch. While adequate for supporting wildlife, these younger trees are dramatically changing the biodiversity of the area. The boreal Northern Forests ecoregion remains the most unchanged and is 80-percent forested today.

Logging Effects

Logging operations in the Great Lakes forests involved clearing the land for agriculture and building houses and barns in settlements. Much of the wood was simply burned to remove it. By the 1830s, however, commercial logging operations began in upper Canada, followed by Michigan, Minnesota, and Wisconsin a few years later. The earliest loggers harvested white pine, mainly for ship construction. Trees reached 200 feet (60 meters) tall in virgin stands, and a single tree could yield 6,000 feet (1,800 meters) in board lumber. When easily harvested pine was exhausted, other species were cut. Hardwoods such as maple, walnut, and oak were logged for the creation of furniture, barrels, and household products.

Trees were clear-cut, meaning that every tree was removed from a stand, leaving bare, easily eroded soil behind. Poor management of litter from logging operations caused fires, resulting in the additional loss of large areas of forest. Debris from logging and sawdust from mills clogged and altered the flow of streams, disrupting fish and amphibian habitat. Settlers introduced nonnative plants and trees to their yards and fields, resulting in native species being choked out by vegetation with few if any enemies to control its spread, and limiting resources for animal species that had evolved to rely on the native plants for food and shelter. Pulp and paper mills and other industry arose quickly, releasing pollutants into the environment.

Conservation Efforts

The Great Lakes Basin is a mix of public and private lands, containing portions of two sovereign nations, two Canadian provinces, and eight American states. As a result, hundreds of local, regional, and private special-purpose groups have jurisdiction over the management or protection of some aspect of the basin or the lakes themselves. The U.S. Forest Service, the U.S. Environmental Protection Agency, and a multitude of private industry programs promote stewardship of public and private lands. Forested lands can also be enrolled in sustainable forestry certification programs such as the Sustainable Forestry Initiative, the Forest Stewardship Council, and the Canadian Standards Association.

These programs require participants to manage the quality and distribution of wildlife habitats and to contribute to the conservation of biological diversity in their regions. Cooperation between and among agencies of the United States and Canada is also necessary to confront the issues of climate change. Climate in the Great Lakes region is already changing: Shorter winters, warmer annual average temperatures, heavier rain and snowstorms, and extreme heat events are occurring more frequently. Air and water temperatures are increasing; lake ice cover is decreasing. These changes are expected to alter lake snowpack density, evaporation rates, and water quality. As a result, jurisdictions in Canada and the United States are studying how to adapt to the anticipated impacts on both lake and river water and across the numerous habitat types here.

Wildlife

Some of the wildlife that still thrives in parts of the Great Lake forests include such mammals as the gray wolf, Canada lynx, coyote, marten, little brown bat, beaver, moose, and river otter. Among the birds that breed, nest, or feed here are such predators as the bald eagle and northern harrier, as well as the double-crested cormorant, loon, tern, least bittern, bobolink, common merganser, and Kirtland's warbler, an endangered species. Characteristic reptiles and amphibians here included salamanders, frogs, toads, and snakes. The fish population runs from brook trout and walleye to lake sturgeon and lake trout.

JILL M. CHURCH

Further Reading

Armson, K. A. *Ontario Forests: A Historical Perspective.* Toronto: Fitzhenry & Whiteside Limited, 2001.

Davis, Mary Byrd, ed. *Eastern Old-Growth Forests: Prospects for Rediscovery and Recovery.* Washington, DC: Island Press, 1996.

Flader, Susan L., ed. *The Great Lakes Forest: An Environmental and Social History.* Minneapolis: University of Minnesota Press, 1983.

Grady, Wayne. *The Great Lakes: The Natural History of a Changing Region.* Vancouver: Greystone Books, 2007.

Stearns, Forest W. "History of the Lake States Forests: Natural and Human Impacts." In J. Michael Vasievich and Henry H. Webster, eds. *Lake States Regional Forest Resources Assessment: Technical Papers.* St. Paul, MN: U.S. Department of Agriculture, 1997.

U.S. Environmental Protection Agency. "State of the Great Lakes 2009: Highlights." 2010. http://www.epa.gov/glnpo/solec.

Great Salt Lake

Category: Inland Aquatic Biomes.
Geographic Location: North America.
Summary: This hypersaline lake in northwestern Utah is bordered by marshlands that provide habitat for shorebirds and a resting point for migratory birds.

The Great Salt Lake is a remnant of the vast, ancient Lake Bonneville, which, at the end of the Pleistocene, covered parts of Nevada and Idaho as well as most of northern Utah west of the Wasatch Range. Lake Bonneville stretched across more than 19,000 square miles (49,000 square kilometers) of surface area. Its dimensions are estimated to have been approximately 325 miles (523 kilometers) long by 135 miles (217 kilometers) wide.

Comparatively, in its present state, this variable-sized lake covers between roughly 2,000 to 3,000 square miles (5,180 to 7,770 square kilometers); it averages about 70 miles (112 kilometers) in length and 30 miles (48 kilometers) in width. The Great Salt Lake is located on the eastern edge of the Great Basin, bordered by the Wasatch Mountains to the east and the Oquirrh and Stansbury ranges to the west and south. It is the largest saltwater lake in the Western Hemisphere; however, its size and salinity levels fluctuate considerably as they are dependent on yearly runoff, evaporation, and diversion of tributary waters for irrigation and other human uses.

Because the Great Salt Lake is located on a flat plain, its shape is that of a wide, flat basin, and even a slight rise in water level expands the surface area considerably. The first series of scientific measurements were taken in 1849, and since then the lake water level has varied by 20 feet (6 meters), shifting the shoreline in some places by as much as 15 miles (24 kilometers). Salinity levels vary greatly also, typically between three to five times greater than seawater, but can reach up to 27 percent, which is roughly eight times saltier than seawater. The lake is situated at approximately 4,000 feet (1,300 meters) above sea level; winters are cold. Because the Great Salt Lake is located in an arid region, precipitation is typically low. However, this area receives a boost from lake-effect rain and snow.

Three major river systems enter the lake: Bear River from the northeast; Weber River from near Ogden, Utah; and the Jordan River from the south. The Jordan drains Utah Lake, a freshwater lake fed by melted snow and rainwater from the Uintah and Wasatch ranges.

Geological History

As a consequence of climatic cooling and warming cycles at the end of the last major glacial advance, water from melting ice and rain began accumulating in the east side of the Great Basin. Approximately 28,000 years ago, Lake Bonneville began to rise rapidly, reaching its maximum surface area and greatest depth—of more than 1,000 feet (300 meters)—about the current size of Lake Michigan, but much deeper. A body of water of that size greatly controlled the local climate and likely resulted in a highly diverse biological community.

Lake Bonneville eventually shrank, mainly because of a warming, drying climate. However, there is geologic evidence of an enormous flood that took place about 15,000 years ago, as the lake's waters overspilled into what is now Red Rock Canyon, Idaho. This is known as the Bonneville Flood. As the climate continued warming, Bonneville Lake continued to evaporate. The Great Salt Lake and the Bonneville Salt Flats in the desert are relics of Lake Bonneville, their salts derived from millenniums of mineral accumulations and high evaporation rates in these closed basins.

Human Impact

As a result of thousands of years of evaporation, additional minerals being transported by its feeder rivers, and the fact that Great Salt Lake is a terminal lake (no outlets), the salinity and mineral content of the water have continued to rise. There are at least eight major companies located in the region around the lake that own extraction industries for the minerals and salts found here. This activity extracts about 2.5 million tons (2.3 million metric tons) of salts and ores from the lake annually. These minerals include significant amounts of sodium chloride (used primarily as common table salt), potash of sodium (low-chlorine fertilizer), magnesium chloride (used in the production of magnesium metal, premium deicing products, and for dust and erosion control), and potassium chloride (used as an ingredient in deicing and water-conditioning products).

Production of another high-profit compound, lithium chloride, is currently underway. This compound is used in the production of lithium metal for use as the anode in lithium batteries, as a brazing flux for aluminum in automobile parts, and as a desiccant for drying air streams. These products are created by pumping water from the lake through a series of near-shore drying ponds; the salts precipitate as the water evaporates.

Wildlife

The Great Salt Lake is biotically depleted because of its high salinity. The ecosystem of the lake is highly simplified, consisting of several species of salt-tolerant bacteria, protozoa, algae, brine flies, and brine shrimp. Yet another economic draw, the brine shrimp and their eggs are used as food for pet fish. While no vertebrates inhabit this lake per se, at three locations where freshwater enters the lake a series of dikes and borrow pits have been constructed to keep the two water types from mixing.

The Bear River Migratory Bird Refuge, part of the National Wildlife Refuge System, was created by capturing the water from the Bear River on the northeast edge of the lake. The Ogden Bay Waterfowl Management Area is another tract created by diverting freshwater from the Weber River. Additionally, the Farmington Bay Wildlife

Hundreds of migrating waterfowl and shorebirds of various species resting at Utah's Great Salt Lake in the spring of 1997. The lake has varying salinity levels of between three to five times greater than seawater, but it can sometimes reach a concentration of as high as 27 percent. (NOAA/Dwayne Meadows)

Management Area was created by diverting water from the Jordan River.

The latter two refuges are administered by the Utah Division of Wildlife Resources. These refuges support more than 200 species of birds and numerous other animal species.

There are between eight and 11 islands (depending on water levels) in the Great Salt Lake. Some of these provide habitat for various mammal species. Antelope Island is the largest; animals that can be found here include: bison, mule deer, pronghorn antelope, coyotes, bobcats, badgers, porcupines, jackrabbits, and several species of rodents. The refuges and marshes around the lake support between 2 and 5 million shorebirds, augmented by hundreds of thousands of waterfowl during spring and fall migration.

The Great Salt Lake region, which accounts for about 75 percent of Utah's wetlands, provides a resting and staging area for these birds, as well as an abundance of brine shrimp and brine flies that serve as their food. Some birds that may readily be found here include: Wilson's phalarope and red-necked phalarope, snowy plover and western sandpiper, American white pelican, white-faced ibis, and eared grebe. Raptors include the bald eagle and peregrine falcon. Large populations of ducks and geese also frequent the biome.

Because the lake's marshes provide a stopover point for millions of birds, there is increasing concern about how climate change in the region will affect these habitats. Winters have already been warmer. There is the possibility that an increase in precipitation would cause the salty lake water to flood into marshlands, killing off freshwater species and disrupting the entire ecosystem. An event of this type occurred in the 1980s.

Robert C. Whitmore

Further Reading

Bedford, D. "Utah's Great Salt Lake: A Complex Environmental-Societal System." *Geographical Review* 95, no. 1 (2005).

Jarrett, R. D. and H. E. Malde. "Paleodischarge of the Pleistocene Bonneville Flood, Snake River, Idaho,

Computed From New Evidence." *Geological Society of America Bulletin* 99, no. 1 (1987).
Oren, Aharon, D. Naftz, P. Palacios, and W. A. Wurtsbaugh, eds. *Saline Lakes Around the World: Unique Systems With Unique Values.* Logan, UT: S. J. and Jessie E. Quinney Natural Resources Research Library, 2009.
Wagner, Frederic H., ed. *Preparing for a Changing Climate: The Potential Consequences of Climate Variability and Change.* Logan: Utah State University, 2003.
Williams, T. T. *Refuge: An Unnatural History of Family and Place.* New York: Pantheon Press, 1991.

Great Sandy-Tanami Desert

Category: Desert Biomes.
Geographic Location: Australia.
Summary: This vast desert ecoregion consists of three smaller deserts. It is known worldwide for the landform known as Uluru or Ayers Rock, a feature sacred to the Anangu Aborigines of this region.

The Great Sandy-Tanami Desert biome extends from Western Australia into the Northern Territory. It consists of three smaller desert regions: the Little Sandy Desert, located west of the Gibson Desert and east of the Great Northern Highway; the Great Sandy Desert, north of the Little Sandy in the northwest section of western Australia; and the Tanami Desert, a rocky, hilly desert in the Northern Territory. Together, the deserts stretch across roughly 490,000 square miles (1.26 million square kilometers).

All three deserts are ecologically significant. Rainfall throughout the area is low, though high by desert standards; even the driest regions have annual precipitation of about 10 inches (25 centimeters). The desert-defining aridity is preserved not so much by a dearth of precipitation, as by the extremely high evaporation rate due to the prevailing heat. Even winter low temperatures are in the high-70-degrees F (20-degrees C) range.

The Sandy deserts contain several large ergs, also called dune seas or sand seas—long, broad, flat areas of desert in which aeolian (windblown) sand covers a large part of the surface. These sands, important features of the desert ecosystem, are formed from aeolian processes that break down, erode, shape, and deposit rock and the resulting particles. These same processes help create desert varnish—a paper-thin coating of minerals (especially clays, manganese, and iron oxides) that forms on exposed rocks, sometimes combined with a microbial film of lichens and possibly microfauna, whose acidic excretions contribute to the ongoing breakdown of rocks into the salty soils of desert sand. The sands of both the Little and Great Sandy Deserts have a notable red hue.

Flora and Fauna

The dominant vegetation of the region is spinifex, an Australian desert grass of the *Triodia* genus (unrelated to grasses of the genus *Spinifex*). The grass has awl-shaped, pointed leaves that store water during dry periods. The leaf tips are sharp and high in silica, because of the soil in which the grasses grow, and have been known to injure people. Aborigines use spinifex to build shelters and fish traps, cook seedcakes, and prepare an adhesive for use in making spears. Other plant species found here that have adapted well to the desert climate include: the Livistonia palm, native walnut (*Owenia reticulata*), and desert oak (*Casuarina decaisneana*). One endemic (found nowhere else on Earth) plant species is Wickham's grevillea (*Grevillea wickhamii*), a shrub.

Wildlife includes numerous species of lizards, such as the thorny devil (*Moloch horridus*), which is covered in uncalcified spines; the spines and ridges allow it to collect water from any part of its body, as when water condenses as dew in the night, as well as acting as a defense against predators. It moves oddly, rocking back and forth, conserving energy while searching for food, usually ants, in the spinifex sandplain. The devil can also suck water into its body from the surface of its skin, drinking in rainwater before it has a chance to run off.

Predators of the thorny devil include the goanna (genus *Varanus*), an Australian monitor lizard that grows to as much as 6 feet (2 meters) long, with sharp claws and teeth and strong jaws. The goanna principally preys on animals small enough to eat whole; when seen feeding on larger animals, it is nearly always scavenging, consuming carrion that it discovered already dead. The largest goanna species in Australia, the perentie (*Varanus giganteus*), is found in the Sandy deserts. The *Pogona* genus of lizards known as bearded dragons is also found here.

The deserts are also home to the bilby (*Macrotis lagotis*), marsupial carnivore mulgara (genus *Dasycercus*), marsupial mole (*Notoryctes caurinus*), rufous hare-wallaby (*Lagorchestes hirsutus*), red kangaroo (*Macropus rufus*), desert tree frog (*Litoria rubella*), sandy burrowing frog (*Limnodynastes spenceri*), desert spadefoot toad (*Notaden nichollsi*), feral camel (*Camelus dromedarius* and *C. bactrianus,* both introduced to Australia in the 19th century), and the dingo (*Canis lupus dingo*).

Many desert species are specially adapted to the region. The red kangaroo's gait, for example, is an energy-efficient way of traveling from food source to food source, and minimizes the amount of time the kangaroo's footpads are in contact with the hot daytime sand.

Significant endemic mammals in the Great Sandy-Tanami Desert include the long-tailed planigale (*Planigale ingrami*), the smallest of all marsupials and one of the smallest mammals at less than 2.5 inches (6 centimeters) long; the western chestnut mouse (*Pseudomys nanus*); and the delicate mouse (*Pseudomys delicatulus*). The Tanami, along with the rest of the region, is also home to the rare grey falcon (*Falco hypoleucos*), typically found in the *Triodia* grasslands and Acacia shrublands. The grey falcon is one of Australia's mystery birds, rarely seen.

Human Impact

While the Great Sandy-Tanami Desert biome remains largely intact, concerns over negative environmental impact stem from overgrazing in some areas, mining activities, tourist activities, and changes in native and endemic species due to intended or unintended introduction of foreign plant and animal species. Also of concern is how much impact global warming will have on climate patterns in this region.

BILL KTE'PI

Further Reading

Beard, John Stanley, and M. J. Webb. *Vegetation Survey of Western Australia: Great Sandy Desert.* Canberra: University of Western Australia Publishing, 1974.

Thackway, R., and I. D. Cresswell, eds. *An Interim Biogeographic Regionalization for Australia.* Canberra: Australian Nature Conservation Agency, 1995.

Wild Australia Program: Pew Environment Group. "Conservation of Australia's Outback Wilderness." 2008. http://www.pewenvironment.org/uploadedFiles/PEG/Publications/Report/Conservation%20of%20Australias%20Outback%20Wilderness.pdf.

Great Slave Lake

Category: Inland Aquatic Biomes.
Geographic Location: North America.
Summary: Great Slave Lake, among the deepest and northernmost North American lakes, supports a largely unspoiled habitat range that is, however, under pressure from global warming.

Great Slave Lake, ensconced in the boreal forest and undulating tundra of Canada's Northwest Territories, is the deepest lake in the continent, plumbed to a depth of 2,015 feet (614 meters). Except for the remnant (and some ongoing) contamination from mining on its northern shore and disruption from logging and road construction, Great Slave Lake anchors an almost pristine series of habitats, with intact forests, wetlands, shrublands, and tundra surrounding its clean, often ice-covered waters.

Contrary to popular belief, the lake's name has nothing to do with slavery; it was named for the Slavey North American Indians, a group that inhabited the region when it was discovered by Europeans. Located in the sub-Arctic, Great Slave Lake is at least partly frozen as much as eight months each year. In winter, the ice is thick enough for ice roads crossing the lake to support tractor-trailers transporting goods and fuel to the communities and mines on the north shore.

Lake Formation

During the last ice age, two huge freshwater lakes formed from the ice melting off the leading edge of the Laurentide Ice Sheet in North America. Both formed about 12,000 years ago, and lasted for a few thousand years before draining into the lakes we know today. The larger, Lake Agassiz, changed continually in shape and size until the ice sheet retreated fully, leaving the modern lakes Winnipeg, Manitoba, and Winnipegosis behind. The other, Lake McConnell, reached its greatest extent 10,000 years ago. As the ice front retreated to the northeast, three daughter lakes remained behind. Great Bear Lake, Great Slave Lake, and Lake Athabasca have continued relatively unchanged to this day.

Great Slave Lake has a very large drainage basin, about 375,000 square miles (975,000 square kilometers); in turn, it is a direct tributary source that is integral to the vast Mackenzie River catchment. Water exits at the shallow, marshy westernmost point of Great Slave Lake via the Mackenzie River, traveling north past Great Bear Lake to empty into the Arctic Ocean.

The Slave and Hay Rivers are the main tributaries feeding the Great Slave Lake, with the Slave River providing at least three-fourths of the incoming water. A tremendous sediment load is delivered through the Slave River Delta, in the southern area of the lake, greatly undercutting the water clarity at its southwestern end in spring and summer. The Hay, too, flows in from the south. The eastern arm of the lake is much deeper and clearer, and contains many islands.

Located about 320 miles (512 kilometers) south of the Arctic Circle, the lake straddles two distinct ecoregions: boreal forest and tundra. The summers are short and cool, with long hours of daylight. The long, cold winters have limited sunlight hours.

Flora and Fauna

The western shore of Great Slave Lake is covered with boreal forests, or taiga, containing aspen, spruce, and balsam fir. This is an area of alternating plateau and plains that overlaps with the Muskwa-Slave Lake Forests ecoregion. Wetlands and bogs here spread over one-fourth of the area.

A diverse and nearly intact wildlife system inhabits the boreal forest. Woodland caribou, moose, black and grizzly bear, wolf, lynx, deer, and elk all migrate naturally throughout the area. Given the remoteness, climate, and low human population of the basin, more than 75 percent of the biome remains intact. This is despite some logging operations and the construction of the Mackenzie Valley highway and pipeline corridors.

The northern and eastern shores of the Great Slave Lake transition into tundra. The Canadian Shield, among the oldest bedrock on the planet at more than four billion years old, underlies terrain here that is either flat or low rolling hills, with many bedrock outcrops. The land appears to be barren, but is rich in minerals. Vegetation consists of sparse, stunted stands of black spruce and tamarack, with ground cover of dwarf birch, shrubs, cottongrass, lichens, and moss. The combination of forests and tundra in this ecoregion results in significant overlap of animal species, notably woodland and barren-ground caribou, along with the wolves that hunt them. Most of this ecoregion is intact, with the greatest fragmentation occurring in the western portion, where both permanent and seasonal roads have been created to support mining operations.

Effects of Human Activity

The First Nations Dene population was the principal aboriginal group in the Great Slave region; the Dene have inhabited Great Slave Lake for thousands of years. Approximately 28,000 people live in the Great Slave subbasin today, and First Nations people comprise about 38 percent of this population. Most people live in small towns and hamlets; the largest community is Yellowknife, situated on the north shore of Great Slave Lake, with a population of 18,500.

The fur trade brought the first European settlers and dominated the economy until gold deposits were discovered on the northern shore in 1896. By the 1930s, bush aircraft made travel far easier, and in 1934, visible gold was found along the shores of Yellowknife Bay. That find led to the establishment of Yellowknife near the outlet of the Yellowknife River; the settlement quickly became a boom town and eventually grew to become the capital of the territory. The name refers to a local Dene tribe once known as the Copper Indians or the Yellowknife Indians, who traded tools made from copper deposits found near the Arctic Coast.

Over the years, numerous mines were developed within the basin for a variety of minerals, including zinc and lead. In 2003, there were still two operating gold mines in Yellowknife, but about 20 mines have been shut down in the region. As capital of the territory, the town's purpose shifted to government services. The discovery of diamonds north of Yellowknife in the 1990s is returning the focus to mining. An environmental assessment is currently underway for a proposed diamond mine approximately 155 miles (250 kilometers) northeast of the city.

Fishing boats and small houses on the shore of Great Slave Lake near the town of Yellowknife. While over 75 percent of the biome is intact, past mining operations near Yellowknife left behind highly toxic waste. (Thinkstock)

Environmental Threats

Despite the region's reputation as one of the most pristine lake environments in the world, the residents of Yellowknife cannot drink the water. Drinking water must be drawn from well up the Yellowknife River. Predictably, the mining legacy is a factor. Beginning in the 1940s, gold was extracted from ore by a process called roasting. This released arsenic trioxide and sulfur dioxide into the air. Pollution-control processes were belatedly added to the procedure, but contamination still proceeded. Roasting was discontinued in 1999, but 238,000 tons (215,910 metric tons) of highly toxic, water-soluble arsenic dust is stored in underground chambers close to Great Slave Lake. The Canadian government and the mine owners are currently working to evaluate strategies for managing the waste and protecting the environment.

Global climate change, with average temperature increases in Arctic areas occurring faster than worldwide averages, is already affecting the biological structure of the Great Slave Lake biome. Permafrost terrain across the northern arc of the region is experiencing longer thaws, to a greater depth, and the southern boundary of true permafrost is advancing northward. As this occurs, there is land subsidence and pooling of water to form new ponds and lakes. Marsh vegetation, shrublands, and grasslands are seen to be establishing their flora communities in previous stands of boreal forest and tundra. Associated fauna in some cases is moving poleward, tracking the migration of their favored habitat ranges.

Jill M. Church

Further Reading

Cohen, S. J. "Impacts of Global Warming in an Arctic Watershed." *Canadian Water Resources Journal* 17, no. 1 (1992).

Evans, Marlene S. "The Large Lake Ecosystems of Northern Canada." *Aquatic Ecosystem Health and Management* 3, no. 1 (2000).

Mackenzie River Basin Board (MRBB). "Great Slave Sub-Basin." In MRBB, *Mackenzie River Basin State of the Aquatic Ecosystem Report 2003.* Fort Smith, Canada: Saskatchewan Watershed Authority, 2004.

Pielou, E. C. *After the Ice Age: The Return of Life to Glaciated North America.* Chicago: University of Chicago Press, 1991.

Great Victoria Desert

Category: Desert Biomes.
Geographic Location: Australia.
Summary: Covering a vast stretch of the Outback, the Great Victoria Desert is a harsh arid environment but one that supports great diversity of plant and animal species.

The Great Victoria Desert is one of the 10 largest deserts in the world and the largest Australian desert. Stretching some 430 miles (700 kilometers) from east to west in the states of South Australia and Western Australia, the desert covers an area of at least 164,000 square miles (424,400 square kilometers). It is bound by the Musgrave Range to the north, and by other deserts and shrublands, and is separated from the Indian Ocean by the limestone Nullarbor Plain to its south.

Dunes in the Great Victoria Desert generally run east-west and can vary between 16 to 66 feet (5 to 20 meters) in height, and up to 62 miles (100 kilometers) in length. Groundwater here is largely recharged by precipitation, which is unpredictable. Yearly rainfall average varies between 8 and 10 inches (20 and 25 centimeters). During summer, days are extremely hot, with average temperature ranging between 86 and 104 degrees F (30 and 40 degrees C), while in winter the range is 68–77 degrees F (20–25 degrees C). At night, temperatures can drop to the freezing point.

In spite of its harsh arid nature, the Great Victoria Desert does encompass some wetlands, including two—Yeo Lake and Lake Throssell—certified as of national significance by the Australian government. Both are considered in ecologically good condition; ongoing restoration of their habitats consists mainly of removal of remnant herds of domesticated grazing animals and of feral fauna such as rabbits, goats, foxes, and cats.

Two additional wetlands within the desert biome are Lake Minigwal and Lake Rason, both intermittent saline lakes. Here, too, the aftereffects of pastoral land use are being mitigated. Additionally, both of these wetlands, and Lake Minigwal in particular, have seen some hypersaline discharges into their waters due to mining activity; some structural hydrological mitigation is needed at Lake Minigwal to restore its habitat viability. All four of these areas have been subjected to invasive plant expansion, another ongoing threat.

Biota

The Great Victoria Desert is full of life. After a heavy rain, flowering plants break out in dizzying displays. A characteristic tree is the marble gum (*Eucalyptus gonglocarpa*), which can grow up to 66 feet (20 meters) tall. There are several species of eucalypts, also known as malles: The Ooldea mallee (*Eucalyptus youngiana*) has red or yellow flowers from June to October, while the kingsmill mallee (*Eucalyptus kingsmillii*) has white-cream or red flowers.

Among the acacia species is the western myall (*Acacia papyrocarpa*), an endemic (found only in this biome) tree which grows on limestone plains and attains heights of up to 23 feet (7 meters). The true mulga (*Acacia aneura*) is a tall tree with a life span of about 200 to 300 years.

Shrub species endemic to Australia are the bluebush (*Maireana sedifolia*) and bladder saltbush (*Atriplex vesicaria*). The latter grows around salt lakes, coastal dunes, and limestone ridges. The desert quandong (*Santalum acuminatum*) is a hemiparasitic tall bush; its shiny red fruit ripens in late spring and is used as an exotic flavoring. The bitter bush (*Pittosporum phylliraeoides*) is a native shrub to western Australia with white-cream flowers. It grows in sand, clay, and limestone; its fruit has medicinal properties to indigenous people. Spinifex and porcupine grass are among the most widespread ground vegetation across the biome.

Animals have adapted extraordinarily to the extreme environment. However, over the past 200 years, Australia has suffered as much as half of the world's mammal extinctions. Threatened species here include the sandhill dunnart (*Sminthopsis psammophila*), crest-tailed mulgara (*Dasycercus cristicauda*), and the southern marsupial mole (*Notoryctes typhlops*).

Birds of this biome include the endangered night parrot (*Geopsittacus occidentalis*) and the chestnut-breasted whiteface (*Aphelocephala pectoralis*). The vulnerable malleefowl (*Leipoa ocellata*) is a ground-dwelling bird that rarely flies. Among the reptiles, the desert hosts at least 100 species. One of them is the vulnerable great desert skink (*Egernia kintorei*), which was recently rediscovered after being considered extinct.

Human Interaction

The former dwellers of the Great Victoria Desert are traced back to around 24,000 years ago. Aboriginal groups of the western portion of the desert shared similar social structure and language. They developed survival skills to thrive with the limited water resources in the desert, from digging many wells and rock holes to store water, to finding water in the roots of trees, to damming water on the natural clay pans. The Ooldea Soak, a permanent freshwater oasis in the southern Great Victoria Desert, was one of the few well-known reliable sources of water. It became a trading and ceremonial gathering center.

European explorers of the 19th century made the stop at Ooldea Soak, as well as the intermittent Queen Victoria Spring and Empress Spring, both in the western reaches of the desert. Development in the Great Victoria Desert has tended to be limited to agricultural grazing and some mining activity. Additionally, nuclear weapons testing began here in the 1950s. One result of that episode was the granting of land titles to displaced indigenous peoples, an outcome that consequently helps preserve some parts of the biome.

Mamungari Conservation Park, formerly known as Unnamed Conservation Park, is one of the many efforts to preserve pristine wilderness and culturally significant areas. Straddling both the Great Victoria Desert and the Nullarbor Plain, it encompasses an area of about 8,260 square miles (21,400 square kilometers). Established in 1970, Mamungari was designated a World Biosphere Reserve in 1977.

Another conservation area is Queen Victoria Spring Nature Reserve, a vital watering and nesting area for many bird species. The reserve covers 667,000 acres (270,000 hectares) of mostly yellow sandplain habitat in the rugged hills of the western desert. Largest of the protected sites is the Great Victoria Desert Nature Reserve, comprising 6.2 million acres (2.5 million hectares) in the central-southern realm of the desert.

Global warming is projected to potentially lead to harsher temperature and moisture gradients, which could unbalance the desert's fire regime. The Great Victoria Desert, except for some swaths of the eastern desert, has in general been spared the worst effects of habitat fragmentation; this could change with increased fire severity. Acacia stands across much of the desert are a key species that would be a visible early indicator of habitat stress from fire regime alteration, drastic precipitation change, or incremental temperature rise.

Rocio R. Duchesne

Further Reading

Bryson, Bill. *In a Sunburned Country.* New York: Broadway, 2001.

Cogger, Harold G. *Reptiles and Amphibians of Australia.* Fort Myers, FL: Ralph Curtis Books, 2000.

Greenslade, Penelope, Leo Joseph, and Rachel Barley, eds. *The Great Victoria Desert.* Adelaide: Nature Conservation Society of South Australia, 1986.

Guyanan Savanna

Category: Grassland, Tundra, and Human Biomes.
Geographic Location: South America.
Summary: This large region of mixed tropical savanna and forest is situated between the Amazon and Orinoco River basins.

The Guyanan savanna is a heterogeneous system of mixed tropical grasslands and forest covering approximately 21,236 square miles (55,000 square kilometers) of northeastern South America. Researcher J. G. Myers formally described the ecosystem in 1936 as the Rio Branco–Rupununi savannas, named for the two major river systems in the region. The Pakaraima Mountains (*Sierra de Pacaraima* in Spanish) bound the savannas to the north in Guyana and to the west in Venezuela. The Amazon rainforest marks the southern boundary of the Guyanan savanna.

To the east, the Kanuku Mountains bisect the northern and southern portions of the Rupununi savanna, which eventually transition to the rainforests of interior Guyana. The region receives 59 to 79 inches (1,500 to 2,000 millimeters) of rainfall per year, 70 percent to 80 percent of which falls during the wet season (May to August). Although many biologists consider these savannas to be part of the Amazon region, they are geographically isolated from the southern Brazilian *cerrados* (savannas) and therefore constitute a distinct ecosystem.

Vegetation

The vegetation of the Guyanan savanna is a mosaic of low-lying tropical grassland interspersed with low-statured solitary trees, small bush islands, and gallery forests bordering the rivers. In the Pakaraima and Kanuku Mountains, savanna vegetation transitions into deciduous and semi-evergreen forest, while cloud forests cover some of the higher peaks. Vast tracts of the savanna become inundated during the wet season, with floods often high enough that the waters from the Orinoco and Amazon basins actually intermingle. Wet-season inundation also favors the establishment of grasses and restricts forests to areas with more persistently dry soils.

During the dry season each year, frequent savanna fires—fueled by high grass abundance and steady trade winds—burn large areas of the region. These fires also maintain the boundaries between forest and savannas, by restricting the establishment of woody plants and favoring fire-tolerant plant species.

Wildlife

The Guyanan savanna is renowned for its abundant wildlife; the biome provides critical habitat for many remnant elements of neotropical megafauna. South America's largest mammals are relatively common here, including tapirs (*Tapirus terrestris*), giant anteaters (*Myrmecophaga tridactyla*), giant armadillos (*Priodontes maximus*), and capybara (*Hydrochoerus hydrochaeris*). Big cats such as jaguars (*Panthera onca*), pumas (*Puma concolor*), and ocelots (*Leopardus pardalis*) are also fairly abundant.

The habitat along the Rupununi River is especially notable for high densities of giant river otters (*Pteronura brasiliensis*) and arapaima (*Arapaima gigas*), one of the world's largest freshwater fish. Also common along the river are some of South America's largest reptile species, such as black caiman (*Melanosuchus niger*), green anacondas (*Eunectes murinus*), and giant river turtles (*Podocnemis expansa*).

The Kanuku Mountains provide important habitat for harpy eagles (*Harpia harpyja*), which are among the world's largest raptors; while scarlet macaws (*Ara macao*) and blue and yellow macaws (*A. ararauna*) occur in high numbers in the savannas. The seasonally flooded wetlands are globally important for supporting high densities of waterbirds such as stork, ibis, duck, heron, and many other avian types.

Human Activity

Several indigenous Amerindian groups exercise stewardship over lands in the Guyana savanna. The Mukushi (Macuxi) reside among scattered villages throughout the northern Rupununi savanna of Guyana and in Roraima state in Brazil. In the forested Pakaraima Mountains to the north, communities are mixed, featuring both Makushi and Patamona Amerindians. The Waipishana make up the majority of the population in the south Rupununi savannas, while the Waiwai people live in the forest to the south.

Amerindians have long played an important role in the ecology of the Guyanan savanna, primarily through the intentional use of fire. Burning was traditionally used to drive game and provide fresh grass for game species, reduce insect pests, prevent destructive fires, and clear land for farming.

At present, limited economic opportunities in Guyana lead many Amerindians to seek work elsewhere, primarily on cattle ranches in Brazil. However, among the scattered communities, many Amerindians still practice swidden cassava farming and rely on wildlife and fish resources for subsistence use. A developing ecotourism trade also contributes to the local economy.

One of the reasons for the Guyanan savanna's rich wildlife is the relative lack of development, especially in the Rupununi region. There are effects, however, from cattle ranching and soybean farms, the region's primary industries, with the latter posing a greater threat to the Guyanan savanna, in part due to the biofuels-driven boom in soy cultivation. The majority of development has occurred in the Brazilian state of Roraima, in the western part of the savannas, which is experiencing rapid economic growth.

By comparison, in Guyana, population density is much lower, and economic activities consist largely of a few scattered ranching operations. Natural rubber production was prevalent in Guyana from the late 1800s to the mid-1900s, through the harvesting of balata, the resin of bulletwood (*Manilkara bidentata*).

Although the region is rich in timber, minerals, and oil, Guyana's poor infrastructure has limited access and extraction. The route through the Guyanan savanna provides the only direct land access between the Amazon forests and the sea at Guyana's capital city of Georgetown. Thus, to expedite the export of timber and other resources, current development is quickly improving bridges and roads, which will fragment some habitat and open the entire region to exploitation. While much of the Guyanan savanna remains undeveloped at present, very little land area has been set aside for conservation.

The long-term effects of global warming are difficult to project for this region of South America. The complex interplay of temperature gradients, wind patterns, humidity, and precipitation seasonality are not well understood. However, it is believed that a less stable precipitation cycle, coupled with even moderate increase in average temperatures, will lead to greater threat from both natural and man-made fire in the Guyanan Savanna biome. If increased forest-understory and grassland damage from more severe fire events becomes a regular feature in the dry seasons in the Guyanan savanna, then habitat fragmentation will accelerate and many animal species will be pressed to migrate toward wetter niches, putting additional stress on those habitats.

Clay Trauernicht

Further Reading

Myers, J. G. "Savannah and Forest Vegetation of the Interior Guiana Plateau." *Journal of Ecology* 24, no. 1 (1936).

Ter Steege, H. *Plant Diversity in Guyana With Recommendations for a National Protected Area Strategy.* Wageningen, Netherlands: Tropenbos Foundation, 2000.

Watkins, G. *Rupununi: Rediscovering a Lost World.* Arlington, VA: Earth in Focus Editions, 2010.